ABNORMAL PRESSURES

IN

HYDROCARBON ENVIRONMENTS

An outgrowth of the AAPG Hedberg Research Conference
Golden, Colorado, June 8–10, 1994

Edited by

Ben E. Law

Gregory F. Ulmishek

Vyacheslav I. Slavin

AAPG MEMOIR 70

Published by:
The American Association of Petroleum Geologists
Tulsa, Oklahoma, U.S.A.
Printed in the U.S.A.

Associate Editor: Neil F. Hurley
Science Director: Jack Gallagher
Publications Manager: Kenneth M. Wolgemuth
Special Projects Editor: Anne H. Thomas
Production: Richard Leishman, R. Leishman Design, Carrboro, North Carolina

This and other AAPG publications are available from:

AAPG Bookstore
P.O. Box 979
Tulsa, OK 74101-0979
U.S.A.
Tel (918) 584-2555
or (800) 364-AAPG *(U.S.A.–book orders only)*
Fax (918) 584-0469
or (800) 898-2274 *(U.S.A.–book orders only)*

Geological Society Publishing House
Unit 7, Brassmill
Enterprise Centre
Brassmill Lane
Bath BA1 3JN
United Kingdom
Tel 0225-445046
Fax 0225-442836

Australian Mineral Foundation
AMF Bookshop
63 Conyngham Street
Glenside, South Australia 5056
Tel (08) 379-0444
Fax (08) 379-4634

Canadian Society of Petroleum Geologists
#160 - 540 5th Avenue S.W.
Calgary, Alberta T2P 0M2
Canada
Tel (403) 264-5610
Fax (403) 264-5898

AAPG WISHES TO THANK THE FOLLOWING FOR THEIR GENEROUS SUPPORT TO:

Abnormal Pressures in Hydrocarbon Environments

HYDROCARBON MANAGEMENT INTERNATIONAL, LTD.

Contributions are applied against the production costs of publication, thus directly reducing the book's purchase price and making the volume available to a greater audience.

CONTENTS

ABOUT THE EDITORS

Ben E. Law was born in Drain, Oregon. He earned B.Sc. and M.Sc. degrees in geology from San Diego State University. Ben has had an interest in abnormal pressures since his employment with Texaco, Inc. in 1969–1971, as an exploration geologist in the Rocky Mountain region. He has been employed with the U.S. Geological Survey since 1971. His interest in abnormal pressures became more focused in 1977 when he became involved in research on low-permeability gas reservoirs. Ben has written several papers on abnormal pressures, particularly in unconventional gas reservoirs, and is currently involved in abnormal pressure studies in the Appalachian Basin, the Timan-Pechora Basin of Russia, and the Dnieper-Donets Basin of Ukraine. He is the Regional Coordinator for South Asia in the USGS World Energy Program. Ben was co-editor of AAPG Studies in Geology No. 38, *Hydrocarbons from Coal*, and co-convenor of the AAPG 1994 Hedberg Research Conference on Abnormal Pressures in Hydrocarbon Environments.

Gregory F. Ulmishek was born in Moscow, Russia. He earned a M.Sc. degree in geology from the Moscow Petroleum Institute and Candidate of Science degree from the Institute of Geology and Exploration for Fossil Fuels in Moscow. Through most of his career he has been involved in several global scale, petroleum geology research projects and has written approximately 70 papers, mainly concerning regional petroleum geology. Since emigrating to the U.S. in 1980, Gregory has been involved in the World Energy Program of the USGS, with special interests in the Commonwealth of Independent States (CIS), China, and the Arctic. Since 1987, he has been the project chief of two cooperative research programs with Russia and Ukraine. He is also the Regional Coordinator for the CIS in the USGS World Energy Program. Gregory was co-convenor of the 1994 AAPG Hedberg Research Conference on Abnormal Pressures in Hydrocarbon Environments.

Vyacheslav I. Slavin was born in Baku, Azerbaijan. He earned M.Sc. and Candidate of Science degrees from the Mining Institute, St. Petersburg, Russia and a Doctor of Sciences degree from the Russian Petroleum Exploration Research Institute (VNIGRI) in St. Petersburg. He has been employed by VNIGRI since 1962 and is presently a Department Head in VNIGRI. His research interests include subsurface hydrodynamics, production of oil and gas from complex reservoirs, and various geological and technological problems related to exploration for and production of oil and gas in abnormally high pressured reservoirs. Dr. Slavin was co-convenor of the AAPG 1994 Hedberg Research Conference on Abnormal Pressures in Hydrocarbon Environments.

PREFACE

This book had its beginnings during an AAPG Hedberg Research Conference entitled "Abnormal Pressures in Hydrocarbon Environments" convened June 8–10, 1994, in Golden, Colorado. During the conference geoscientists from 18 countries presented papers on different aspects of abnormal pressure. It became apparent that there were very diverse opinions concerning the nature and integrity of pressure seals, causal mechanisms, transient nature of pressure systems, the global distribution of abnormal pressures, pressure modeling, detection of abnormal pressuring, drilling and completion techniques in abnormal pressure environments, and the association of abnormal pressures with hydrocarbon accumulations.

The subject of abnormal pressures encompasses a wide spectrum of topics, and an attempt to cover all aspects of abnormal pressure in a single volume would be woefully inadequate. Therefore, in an attempt to narrow the scope and provide new information, this book focuses on abnormal pressures in hydrocarbon environments. More specifically, the objectives are to provide information concerning the global distribution of abnormal pressures in petroleum provinces, relate occurrences of abnormal pressures to hydrocarbon accumulations, provide an overview of the causal mechanisms of abnormal pressures, bring into question some long-held pressure paradigms, and provide examples from a few select regions in the world of the utilization of abnormal pressures in the exploration for hydrocarbons. We are aware that we will have offended some individuals by omitting some aspects of abnormal pressure that they consider to be of extreme importance, such as abnormal pressures caused by hydraulic head.

Abnormal pressures, pressures above or below hydrostatic pressures, have been the subject of investigation for many decades. The first description of abnormal pressure known to us was by Gulishambarov in 1878. He described oil fountains (geysers) and discussed abnormally high pressures in the Balakhany oil field near Baku, Azerbaijan. We are uncertain when the first study on abnormal pressures in the western world was published. However, the work of Dickinson (1953) in the U.S. Gulf Coast initiated an era of investigation of abnormal pressure that has continued to this day. In the 1950s in the former Soviet Union there was also a surge of investigations of abnormal pressure coinciding with an increase in the number of deeply drilled wells. In China, studies of abnormal pressure were largely introduced from western countries in the late 1970s.

As a consequence of the wide interest in abnormal pressures, there has been a large number of comprehensive works on abnormal pressures. Some of the more notable overview publications in western countries include those by Fertl (1976), Fertl et al. (1994), Hall (1993), Mouchet and Mitchell (1989), and Sahay et al (1988). In the former Soviet Union, the more significant publications on abnormal pressure include those by Aleksandrov (1987), Anikeev (1964, 1971), Buryakovsky, Dzhevanshir, and Alnyarov (1986), Dobrynin and Serebryakov (1978, 1989), Kucheruk and Lyustikh (1986), Slavin, Sheverdyayev, and Khimich (1987), and Zkhus and Bakhtin (1979).

The early investigations of abnormal pressure in North America were driven by a concern for drilling and completion problems in abnormally overpressured rocks. During that time some of the earlier proposals for the cause of abnormal pressure included tectonics, mineral transformations, and hydraulic head. While these mechanisms are still important, compaction disequilibrium and hydrocarbon generation have gained prominence within the last 20 years. It is now quite apparent that many of the processes involved in the generation, expulsion, migration, entrapment, and preservation of hydrocarbons are, in some cases, the same processes involved in the development of abnormal pressure. Therefore, knowledge of the processes involved in the development of abnormal pressure may also be used as a hydrocarbon exploration tool.

Although our knowledge of pressure systems has considerably advanced during the last 20 years, there remains much work to be done. In our view, problems that warrant additional study include causal mechanisms of abnormally underpressured systems, the nature and integrity of pressure seals, dynamic vs. static pressure systems, the transient nature of pressure systems, the prediction of abnormal pressures, and the application of abnormal pressures to the exploration for hydrocarbons.

Ben E. Law

Gregory F. Ulmishek

Vyacheslav I. Slavin

REFERENCES CITED

Aleksandrov, B.L., 1987, Abnormally high formation pressures in petroleum basins: Moscow, Nedra, 216 p.

Anikeev, K.A., 1964, Abnormally high reservoir pressures in oil and gas fields: Leningrad, Nedra, 167 p.

Anikeev, K.A. 1971, Prediction of abnormally high formation pressures and improvement of deep drilling for oil and gas: Leningrad, Nedra, 166 p.

Buryakovsky, L.A., Dzhevanshir, R.D., and Alnyarov, R.Yu, 1986, Geophysical methods of study of geofluid pressures: Baku, Azerbaijan, Elm, 146 p.

Dickinson, G., 1953, Geological aspects of abnormal reservoir pressure in Gulf Coast Louisiana: American Association of Petroleum Geologists Bulletin, v. 37, p. 410–432.

Dobrynin, V.M., and V.A. Serebryakov, 1978, Methods for prediction of abnormally high formation pressures: Moscow, Nedra, 232 p.

Dobrynin, V.M., and V.A. Serebryakov, 1989, Geologic and geophysical methods for prediction of abnormal reservoir pressures: Moscow, Nedra, 287 p.

Fertl, W.H., 1976, Abnormal formation pressures: Developments in Petroleum Science 2, Amsterdam, Elsevier, 382 p.

Fertl, W.H., R.E. Chapman, and R.F. Hotz, 1994, Studies in abnormal pressure: New York, Elsevier, 454 p.

Hall, P.L., 1993, Mechanisms of overpressuring: an overview, in D.A.C. Manning, P.L. Hall, and C.R. Hughes, eds., Geochemistry of clay-pore fluid interactions, Chapman and Hall, London, p. 265–315.

Kucheruk, E.V., T.E. Lyustikh, 1986, Prediction and evaluation of abnormal reservoir pressures from data of geophysical surveys: Moscow, VINITI, 128 p.

Mouchet, J.P., and A. Mitchell, 1989, Abnormal pressures while drilling: Manuels Techniques Elf Aquitaine, v. 2, 264 p.

Sahay, B., and W.H. Fertl, 1988, Origin and evaluation of formation pressures: Boston, Kluwer, 292 p.

Slavin, V.I., V.V. Sheverdyayev, and V.F. Khimich, 1987, Methodological directions for prediction and evaluation of abnormally high formation pressures: Leningrad, VNIGRI, 135 p.

Zkhus, I.D., and V.V. Bakhtin, 1979, Lithogenetic modifications of shales in zones of abnormally high formation pressure: Moscow, Nauka, 139 p.

Law, B.E., and C.W. Spencer, 1998, Abnormal pressures in hydrocarbon
environments, *in* Law, B.E., G.F. Ulmishek, and V.I. Slavin eds., Abnormal
pressures in hydrocarbon environments: AAPG Memoir 70, p.1–11.

Chapter 1

ABNORMAL PRESSURE IN HYDROCARBON ENVIRONMENTS

Ben E. Law[1]
C. W. Spencer
U.S. Geological Survey
Denver, Colorado, U.S.A.

Abstract

Abnormal pressures, pressures above or below hydrostatic pressures, occur on all continents in a wide range of geological conditions. According to a survey of published literature on abnormal pressures, compaction disequilibrium and hydrocarbon generation are the two most commonly cited causes of abnormally high pressure in petroleum provinces. In young (Tertiary) deltaic sequences, compaction disequilibrium is the dominant cause of abnormal pressure. In older (pre-Tertiary) lithified rocks, hydrocarbon generation, aquathermal expansion, and tectonics are most often cited as the causes of abnormal pressure.

The association of abnormal pressures with hydrocarbon accumulations is statistically significant. Within abnormally pressured reservoirs, empirical evidence indicates that the bulk of economically recoverable oil and gas occurs in reservoirs with pressure gradients less than 0.75 psi/ft (17.4 kPa/m) and there is very little production potential from reservoirs that exceed 0.85 psi/ft (19.6 kPa/m). Abnormally pressured rocks are also commonly associated with unconventional gas accumulations where the pressuring phase is gas of either a thermal or microbial origin. In underpressured, thermally mature rocks, the affected reservoirs have most often experienced a significant cooling history and probably evolved from an originally overpressured system.

INTRODUCTION

Through the years there has been an evolution of ideas or concepts concerning the cause(s) of abnormal pressure, as well as reasons for studying abnormal pressures. Most studies of abnormal pressures prior to the mid-1980s were driven by the concern for drilling and completion practices, as well as safety considerations during drilling. While those concerns are still important, abnormal pressures are now important components of hydrocarbon exploration, field development, and resource assessment.

Some of the earlier proposed causal mechanisms of abnormal pressure, such as mineral transformations, osmosis, and tectonics have given way to additional causes such as compaction disequilibrium, hydrocarbon generation, and aquathermal expansion. Within the last 15 years there has been a realization that, in some cases, the processes involved in the generation, expulsion, migration, and entrapment of hydrocarbons are the same processes responsible for the development of abnormal fluid pressures. In addition, the concept of pressure compartments with vertical and lateral seals now play a major role in the exploration for hydrocar-bons. Therefore, the study of abnormal pressures is not only important for purposes of hydrocarbon exploitation, but is also now recognized as an important component of hydrocarbon exploration.

As a consequence of these ongoing developments in the evolution of abnormal pressure studies, this investigation was initiated to provide information on the global distribution of abnormal pressures, examine the relationships among various attributes of abnormal pressure, and evaluate relationships that may occur between the occurrence of abnormal pressures and the occurrence of hydrocarbon accumulations. The conclusions of this study are largely based on the evaluation of previously published literature and the authors' collective experience.

ATTRIBUTES OF ABNORMAL PRESSURES

Global Distribution

Abnormal pressures occur in a wide range of geographic and geologic conditions. Figure 1 shows the

[1] *Present Affiliation: Consulting Petroleum Geologist, Lakewood, Colorado, U.S.A.*

global distribution of abnormal pressures. This distribution reflects information available in the literature and the experience of the authors. There are undoubtedly many additional areas of abnormal pressure either not identified or not reported in the literature. In this compilation, we have attempted to show only those regions associated with petroleum provinces. In many cases, the areal distribution of abnormal pressures is not known or was not defined in our sources of information, so we have shown the entire basin or region.

Based on our compilation of the occurrence of abnormal pressures, there are approximately 150 geographic locations around the world known to be abnormally pressured (Figure 1). Hunt (1990) has indicated that abnormal pressures have been identified in about 180 basins. In many of these areas, however, there are more than one abnormally pressured stratigraphic unit or zone. For example, in the U.S. Gulf Coast region there are at least seven stratigraphic units ranging in age from Jurassic to Recent that are abnormally pressured. Nearly all of the abnormally pressured regions shown are overpressured. Only about 12 of the areas in Figure 1 are underpressured. Underpressure is much more difficult to identify during drilling than overpressure, consequently more overpressured systems have been identified than underpressured systems.

The distribution of abnormal pressures (Figure 1) appears to favor the northern hemisphere, even though there are no readily apparent reasons why there should be a preferential occurrence of abnormal pressures there. We suspect that this unequal distribution merely reflects the relatively larger number of investigations conducted in the northern hemisphere. For example, the large number of abnormally pressured areas shown on Figure 1 in the Rocky Mountain region of the United States is a consequence of several, detailed investigations of abnormally pressured, unconventional gas reservoirs. In this region and elsewhere in North America, investigators have noted the close association of hydrocarbon accumulations, particularly unconventional gas accumulations, and abnormal pressures. Conversely, the relatively few number of abnormally pressured areas in the Andean region of South America, probably reflects differences in exploration objectives and perhaps an unawareness of the association of abnormal pressures and hydrocarbons.

Causal Mechanisms of Abnormal Pressure

While it is not our intention to review all aspects of abnormal pressures, we have tabulated some of the more important attributes of abnormally pressured rocks (Table 1) in an attempt to identify those attributes that may have a bearing on the cause(s) of abnormal pressure. From an examination of this compilation, attributes such as depth to the top of abnormal pressure and structural province do not appear to render any useful information regarding the cause of abnor-

mal pressure, other than documenting the variability of depth and structural settings within which abnormally pressured rocks occur. However, attributes such as the geologic age of abnormally pressured rocks, their depositional setting, maximum pressure, nature of the seal, temperature, and thermal maturity do reveal useful information concerning the cause(s) of abnormal pressuring.

Because of this wide range of variability, the cause(s) of abnormal pressure are often difficult to determine and may involve more than one process. Swarbrick and Osborne (1998-this volume) provide a comprehensive list and discussion of the mechanisms of abnormal pressure. Some of the more notable published overviews of the different mechanisms of abnormal pressures include those by Fertl (1976), Mouchet and Mitchell (1989), and Fertl et al. (1994).

Of all the causes of abnormal pressures referred to in the literature: compaction disequilibrium, aquathermal expansion, hydrocarbon generation, mineral transformations, tectonics, and osmosis; the most commonly cited cause of abnormally high pressure is compaction disequilibrium. And in nearly all cases where compaction disequilibrium has been determined to be the primary cause of overpressuring, the age of the rocks is geologically young. Examples of areas where compaction disequilibrium is cited as the primary cause of abnormal pressure include the U.S. Gulf Coast, Niger Delta, Mahakam Delta, MacKenzie River Delta, North Sea, Adriatic Sea, the Nile Delta, and the Potwar Plateau of Pakistan (Figure 1, Table 1). In these areas, the age of the abnormally pressured rocks is Tertiary, the depositional setting is dominantly deltaic, and the lithology is dominantly shale. A notable exception is the highly overpressured Neogene rock sequence in the Potwar Plateau of Pakistan (Figure 1), where the dominant lithology is sandstone (Law et al., 1998-this volume). The most commonly cited depositional environment for abnormally pressured rocks is deltaic.

In pre-Tertiary rocks, the main causes of abnormal pressure include hydrocarbon generation, aquathermal expansion, mineral transformations, and tectonic deformation–with hydrocarbon generation cited as the most common cause. In our judgment, hydrocarbon generation as a cause of abnormal pressure has been under-evaluated.

The relationship between the cause of abnormal pressuring in young versus old rocks suggests that there may be a continuum of processes responsible for the development of abnormal pressure. We are of the opinion that pressures are time transient and that pressure causing mechanisms are also transient. For example, Law and Dickinson (1985) presented a conceptual model for the origin of abnormal pressures in low-permeability rocks that involved hydrocarbon generation. In their model, abnormal high pressures were initially caused by hydrocarbon generation. With subsequent changes of structural uplift, erosion, and temperature reduction during the burial and thermal history, the overpressured rocks evolved into an underpressured phase. And finally, at an even later burial history, Law and Dickinson theorized that the underpressured rocks would evolve into a normally pressured system. Investigations by Dickey and Cox (1977) and Doré and Jensen (1996) have also called on uplift, erosion, and cooling as a cause of underpressuring, but have not proposed an earlier pressure history of overpressuring.

We speculate that in some cases, such as in deltaic systems with high rates of deposition, abnormal pressures may be initiated by compaction disequilibrium. As these deltaic sediments are buried deeper and experience higher temperatures, hydrocarbon generation may supplant compaction disequilbrium as the main cause of abnormally high pressure. In deltaic rock sequences where the hydrocarbon source rock occurs stratigraphically below the compaction disequilibrium-affected sediments, the generation of hydrocarbons from these source rocks may result in the development of overpressure which could be physically transferred upward, via the development of a pressure gradient, into the region of compaction disequilibrium.

Hunt et al. (1994; 1998-this volume) have proposed an abnormal pressure mechanism of hydrocarbon generation for the U.S. Gulf Coast. In our opinion, the observations by Leach (1993a, b, c) of the close association of productive oil and gas fields and the top of overpressure in southern Louisiana are also suggestive of the role of hydrocarbon generation in the development of overpressure. Alternatively, basin modeling by Burrus (1998-this volume) attributes the origin of overpressuring in the U.S. Gulf Coast almost exclusively to compaction disequilibrium.

HYDROCARBON ACCUMULATIONS AND ABNORMAL PRESSURES

Hydrocarbon accumulations are frequently found in close association with abnormal pressures. In abnormally pressured, conventionally trapped oil and gas accumulations, pressures above hydrostatic are common. However, as Chapman (1994) points out, some of these "abnormal pressures" are normal for their fluids and are a function of the densities of the fluid and the height of the oil and gas column above the oil-gas/water contact. Therefore, such "abnormal pressures" are not due to processes such as compaction disequilibrium, aquathermal expansion, or hydrocarbon generation and are not considered here to be abnormally pressured. Discounting these "abnormally pressured" hydrocarbon accumulations, the association of truly abnormal pressures and hydrocarbon accumulations have been noted in several studies of conventionally and unconventionally trapped hydrocarbons.

In the U.S. Gulf Coast, Burst (1969) noted that hydrocarbon production was evenly distributed about a depth 1,500 ft (460 m) above the depths of his 2nd dehydration stage of clays (top of overpressure). Sub-

sequent studies by Fertl and Leach (1990) and Leach (1993a, b, c) in southern Louisiana have also shown spatial relationships between the top of overpressuring and the accumulation of oil and gas fields. A statistical evaluation by Leach (1993a) of oil and gas production from Tertiary reservoirs in southern Louisiana showed that almost half (46.1%) of the oil production was from an interval 2,000 ft (610 m) above the top of overpressuring and that nearly half of the gas production came from a 2,000 ft (610 m) interval centered around the top of overpressuring.

Other studies in the U.S. Gulf Coast by Timko and Fertl (1971) and Leach (1993a, b, c) noted that oil and gas production decreased with increasing pressure, and at pressure gradients approaching 0.85 psi/ft (19.6 kPa/m) there was a marked decrease in production. Leach (1993b) concluded that gradients of 0.85 psi/ft (19.6 kPa/m) or higher exceed the fracture gradients of most sandstone reservoirs. Consequently, hydrocarbons that may have originally been trapped in these high-pressure reservoirs may have been lost through pressure-induced fractures. Similar observations of the relationship between the distribution of hydrocarbons and abnormal pressures have been proposed by Dow (1984) in the U.S. Gulf Coast and by Schaar (1976) in the Baram Delta of Sarawak. In the Nile Delta and North Sinai basins of Egypt, Nashaat (1998-this volume) concluded that hydrocarbon production is precluded in reservoirs that exceed 0.85 psi/ft (19.6 kPa/m). Heppard et al. (1998-this volume) noted that oil and gas production in the Trinidad, West Indies area was restricted to reservoirs with pore pressures gradients less than 0.73 psi/ft (16.9 kPa/m). In the former Soviet Union, Belonin and Slavin (1998-this volume) observed that most oil and gas production from abnormally pressured reservoirs occurred at abnormality coefficients (measured pore pressure divided by hydrostatic pressure) less than 1.8 (assuming a hydrostatic gradient of 0.45 psi/ft [10.2 kPa/m], an abnormality coefficient of 1.8 is equal to about 0.81 psi/ft [18.7 kPa/m]). In the Sichuan Basin of China, Da-jun and Yun-ho (1994) related the presence of natural fractures to the magnitude of pressure. They presented pressure data from gas-productive, Permian carbonate reservoirs showing that gradients greater than 0.63 psi/ft (14.2 kPa/m) are indicative of relatively small fields.

The association of hydrocarbon accumulations and abnormal pressure is even more evident in unconventional gas accumulations. For example, coalbed methane, shale gas, basin-centered gas, and low-permeability microbial gas are nearly always associated with abnormal pressures.

Gas in shale and coal are self-sourced reservoirs that are commonly abnormally pressured. In the Appalachian Basin, oil and gas are produced from organically-rich, Devonian shale (de Witt, 1984; Reeves et al., 1996). In some productive regions in the Appalachian Basin, oil and gas are produced from fractured, underpressured shale (Hunter, 1962; de Witt, 1984). Some

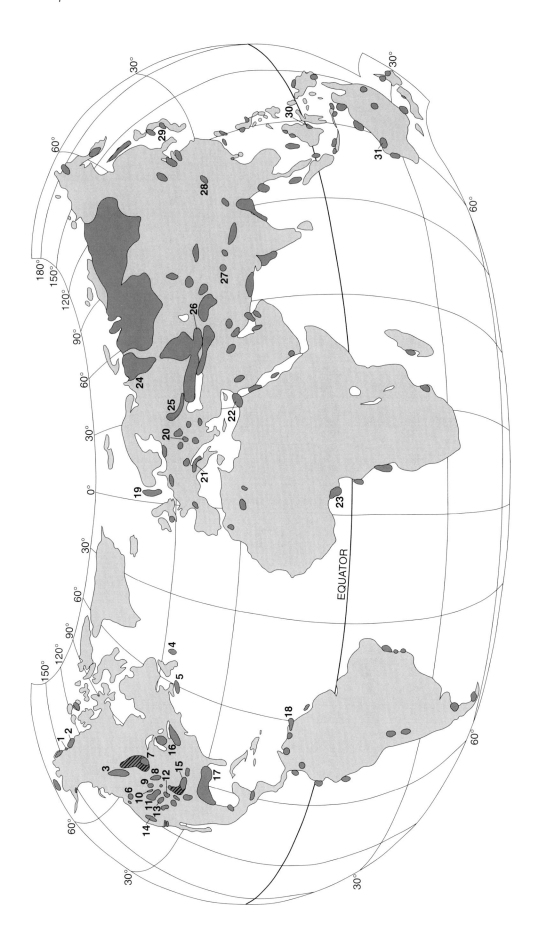

Figure 1. Map showing the global distribution of abnormal pressures. Heavier shaded, diagonally ruled patterns are used to avoid masking of darker patterned areas listed on Table 1. Index numbers adjacent to selected abnormally pressured areas refer to additional data provided in Table 1.

Table 1. Selected attributes of abnormally pressured regions of the world. Locations of the regions are shown on Figure 1 and are linked by the Index Number of the region.

Region	Index No.	Age of Abnormally Pressured Rocks	Depositional System	Structural Setting	Depth to Top of Abnormal Pressure ft (m)	Maximum or Minimum Pressure Gradients psi/ft (kPa/m)	Seal	Temperature Top of Abnormal Pressure °F (°C)	Thermal Maturity Top of Abnormal Pressure %Ro	Fluid Pressure Phase	Hydrocarbon Source & Age	Associated Hydrocarbon Accumulation Type	Cause(s)	References	Remarks
Alaska National Wildlife Refuge Alaska, U.S.A.	1	Cretaceous	marine nonmarine	foreland deltaic/fluvial	10,000 (3,000)	0.84 (19.4)	–	150°-163° (66°-73°)	0.55	–	–	–	uncertain	Gautier et al. (1987)	Conclusion based on 2 wells.
Beaufort Mackenzie Canada	2	Paleozoic & Miocene	deltaic	passive margin	6,550-16,400 (2,000-5,000)	0.85 (19.4)	faults	highly variable	0.75	water/gas	–	structure	–	Hitchon et al. (1990)	
Alberta Deep Basin Canada	3	Triassic, Jurassic Cretaceous	deltaic marginal marine	foreland	>3,000 (>1,000)	–	water block*	–	0.7-1.0	gas	Cretaceous Type III OM	basin-center gas	HC	Masters (1979, 1984) Welte et al. (1984)	Under & overpressured. See Masters (1984) Fig. 18. Very little pressure data, largely interpretive.
Jeanne d'Arc Basin Canada	4	Jurassic	marine	failed rift	9,500-15,100 (2,900-4,600)	0.99 (21.3)	shale	–	–	water/gas*	Jurassic	structure	uncertain	Rogers & Yassir (1993)	
Scotian Shelf Canada	5	Jurassic	deltaic	rift	>14,750 (>4,500)	0.87 (20.1)	–	>248° (>120°)	0.8	–	Jurassic-Cretaceous	structure	CD/HC	Mudford & Best (1989) Rogers & Yassir (1993)	
Columbia Basin U.S.A.	6	Tertiary	fluvial	foreland	9,000-10,000 (2,700-3,000)	>0.80 (>18.5)	water block*	200° (93°)	0.8	gas	Tertiary	basin-center gas	HC	Law et al. (1994)	Overpressured sequence covered with thick basalts.
Williston Basin U.S.A.	7	Mississippian Devonian	marine	cratonic basin	9,000 (2,740)	0.73 (16.9)	–	190° (88°)	0.4-0.5	oil	Mississippian Devonian	basin-center oil	HC	Meissner (1978)	Type I organic material.
Powder River U.S.A.	8	Cretaceous	dominantly marine	foreland	10,000 (3,000)	0.8 (18.5)	unconformity water block*	200° (96°)	0.5	gas & condensate	Cretaceous	stratigraphic	HC	Spencer (1987) Surdam et al. (1994)	
Big Horn Basin U.S.A.	9	Cretaceous	fluvial & marine	foreland	11,000 (3,350)	>0.6 ? (13.6 ?)	water block*	200° (96°)	>1.2	gas	Cretaceous	–	–	Spencer (1987)	Experienced cooling associated with uplift and erosion.
Wind River Basin U.S.A.	10	Mississippian Cretaceous Tertiary	marine fluvial	foreland	variable	0.8 ? (18.5 ?)	water block* shale	–	1.1	gas	Cretaceous	basin-center gas	HC	Bilyeu (1978) Johnson et al. (1996)	
Greater Green River Basin, U.S.A.	11	Tertiary Cretaceous	marine-nonmarine deltaic/fluvial	foreland	8,000 (2,440)	0.9 (20.8)	water block	180°-200° (82°-96°)	0.75-0.85	gas	Cretaceous	basin-center gas	HC	McPeek (1981) Law (1984) Spencer (1987)	Underpressured gas accumulation at shallow depths above deeper, overpressured rocks.
Piceance Basin U.S.A.	12	Mississippian Cretaceous	fluvial to marine	foreland	6,000-8,000 (1,830-2,440)	0.82 (18.9)	water block	165°-200° (74°-96°)	>0.8	gas	Cretaceous	basin-center gas	HC	Spencer (1987) Johnson et al. (1987)	Locally overpressured CO_2 in Mississippian carbonates, see Wilson et al. (this volume).
Uinta Basin U.S.A.	13	Tertiary	lacustrine fluvial	foreland	10,000 (3,000)	0.83 (19.2)	water block	210° (98°)	–	oil	Tertiary Type I OM	basin-center oil	HC	Lucas & Drexler (1976) Spencer (1987)	
Sacramento Basin U.S.A.	14	Cretaceous Tertiary	fluvial-deltaic	forearc	3,900-10,000 (1,190-3,050)	0.85 (19.7)	–	–	–	gas*	–	–	TC/AE	Berry (1973) Lico & Kharaka (1983)	Diminished shale gas production when pressure gradient >0.71 psi/ft (>16.3 kPa/m).
Anadarko Basin U.S.A.	15	Mississippian Pennsylvanian	fluvial fluvial deltaic marine, deltaic	foreland	9,000-10,000 (2,740-3,050)	0.9 (20.8)	diagenetic	140°-155° (60°-66°)	0.81-0.89	gas	Devonian Mississippian Pennsylvanian	stratigraphic basin-center*	HC/O/CD	Breeze (1970) Al-Shaieb et al. (1994)	Types II & III kerogen, Burrus & Hatch (1989).
Appalachian Basin U.S.A.	16	Silurian	fluvial deltaic marine	foreland	>2,500 (>760)	0.36* (8.37*)	water block*	–	–	gas	Ordovician ?	basin-center	HC	Russell (1972), Davis (1984), Zagorski (1988) Law & Spencer (1993)	Underpressured, locally overpressured at depths >10,000 ft (>3,000 m).
Gulf Coast U.S.A.	17	Jurassic Cretaceous Tertiary	marine deltaic	passive margin	6,000-16,000 (1,800-4,875)	near lithostatic	shale	–	–	oil/gas/water	–	structure	CD/HC	Dickinson (1953) Bradley (1975)	Numerous overpressured stratigraphic units. The world's most extensively studied abnormally pressured region.

Table 1 (continued). Selected attributes of abnormally pressured regions of the world. Locations of the regions are shown on Figure 1 and are linked by the Index Number of the region.

Region	Index No.	Age of Abnormally Pressured Rocks	Depositional System	Structural Setting	Depth to Top of Abnormal Pressure ft (m)	Maximum or Minimum Pressure Gradients psi/ft (kPa/m)	Seal	Temperature Top of Abnormal Pressure °F (°C)	Thermal Maturity Top of Abnormal Pressure %R_o	Fluid Pressure Phase	Hydrocarbon Source & Age	Associated Hydrocarbon Accumulation Type	Cause(s)	References	Remarks
Trinidad West Indies	18	Cretaceous to Tertiary	marine to deltaic	active margin	1,500-12,000 (460-3,650)	0.95 (21.9)	faults shale	–	–	water/gas/oil	Upper Cretaceous	structure	CD/HC?/ AE?/TC?	Heppard et al. (this volume)	Mud diapirs and mud volcanoes. Max. pressure gradient in productive fields is 0.73 psi/ft (16.9 kPa/m).
North Sea	19	Triassic Jurassic Cretaceous Paleocene	fluvial paralic marine	failed rift	>5,900 (>1,800)	0.87 (20.1)	marls shale faults claystone	–	–	oil/gas/water	Jurassic	structure	CD/HC/TC/AE	Buhrig (1989) Leonard (1993) Gaarenstroom (1993) Holm (this volume)	Examples of pore pressure compartment.
Bekes Basin Hungary	20	Neogene	fluvial deltaic lacustrine	back-arc rift	>5,900 (>1,800)	0.86 (20.0)	–	257° (125°)	0.45	gas/water/ condensate	Neogene Cretaceous ?	structure stratigraphic	CD/HC/CO$_2$	Spencer et al. (1994)	High heat flow. Rapid deposition.
Adriatic Basin Italy	21	Pliocene	turbidite	foreland	variable, hard pressure at >7,500 (>2,300)	near lithostatic	faults/ shale	–	–	water/gas/oil	Lower Triassic for oil	structure	CD	Carlin & Dainelli (this volume)	Biogenic gas to 13,100 ft (4,000 m). Also thermogenic gas.
Nile Delta/North Sinai Basins Egypt	22	Jurassic-Pliocene	pre-Tertiary-marine, fluvial Tertiary - deltaic marine	passive margin	1,700-12,140 (520-3,700)	0.89 (20.6)	evaporites shales faults	185° (85°)	–	gas/water/ condensate/ oil	Jurassic Oligocene-Miocene	structure	CD/AE/HC	Nashaat (this volume)	Temperature & pressure limits of oil & gas production indicated in Nashaat (this volume).
Niger Delta Nigeria	23	Tertiary	deltaic	passive margin	>9,000* (>2,740')	–	shale	200° (96°)	0.6*	water plus dissolved gas*	Eocene	structure	CD	Evamy et al. (1978) Chukwu (1991) Ejedawe et al. (1986)	Nearly all productive fields at or near top of overpressure.
Timan-Pechora Basin Russia	24	Devonian Carboniferous Permian	marine deltaic fluvial	foreland	highly variable >1,000 (>300)	0.38 Permian (8.8)	shale/ evaporites water block	–	0.8-0.9	oil/gas/water	Permian for basin-center	structure reefs, basin-center in Permian	HC in Permian	Law et al. (1996)	Overpressured in Devonian & Carboniferous. Underpressured in Permian.
Dnieper-Donets-Donbas Ukraine	25	Carboniferous	marine to paludal	rift	1,500-14,750 (450-4,500)	0.79 (18.3)	water block* diagenetic evaporites	212°-230° (100°-110°)	0.9	gas	Carboniferous	basin-center*	HC	Polutranko (this volume) Law et al. (1997)	Overpressured in Dnieper-Donetsk & underpressured in Donbas region.
South Caspian Basin Kazakhstan	26	Tertiary	flysch & molasse	foreland	>3,000 (>1,000)	0.85 (19.6)	shale*	–	–	water & dissolved gas	–	structure*	CD	Bredehoeft et al. (1988) Gurevich & Chilingar (1995)	Contains numerous mud volcanoes see Durmish'yan (1972) for discussion of origin & relationship of mud volcanoes to HC accumulation.
Potwar Plateau Pakistan	27	Paleozoic to Neogene	marine fluvial deltaic	foreland	extremely variable 200-1,000 (60-300)	near lithostatic in Neogene & 0.7 (15.8) in pre-Neogene	uncertain	110° (43°) in Neogene & 180° (82°) in older rocks	0.8-0.9 in pre-Neogene	water in Neogene & oil/gas & water in pre-Neogene	Eocene ? Eocambrian ?	structure	CD in Neogene HC/AE/TC in pre-Neogene	Law et al. (this volume)	Two pressure regimes - Neogene & pre-Neogene.
Sichuan Basin China	28	Permian Triassic	marine	foreland	–	0.93 (21.4)	carbonates	–	–	gas	Permian	structure	TC/O/TC	Da-jun & Yun-ho (1994)	Over- and underpressured carbonates.
Nagaoka Plain Japan	29	Miocene	volcanics	active margin	>4,600 (>1,400)	0.69 (16.0)	mudstone	–	–	oil/gas/water	Miocene	structure	CD	Magara (1968)	Overpressured rock are composed of lava, tuff breccia & agglomerates.
Mahakam Delta Indonesia	30	Miocene	marine deltaic	active margin	11,500-13,100 (3,500-4,000)	0.87 (20.1)	shale	200°-248° (100°-120°)	0.6	water/gas	Miocene	structure	CD	Burrus et al. (1992) Oudin & Picard (1982) Burrus (this volume)	Coal is hydrocarbon source rock.
Dampier sub-Basin Australia	31	Jurassic Cretaceous	–	passive margin	>2,870 (>875)	0.85 (19.7)	–	–	–	–	–	–	CD/MT	Nyein et al. (1977)	Only one well used in study.

Cause(s) codes used: CD = Compaction Disequilibrium, HC = Hydrocarbon Generation, AE = Aquathermal Expansion, TC = Tectonic Compression, O = Osmosis, MT = Mineral Transformations, CO$_2$ = Carbon Dioxide Generation
* = Authors Interpretation

evidence also exists of locally occurring, overpressured Devonian shale in the Appalachian Basin (Milici, personal. communication., 1996). The Cretaceous Barnett Shale in Texas is another example of a gas-productive abnormally pressured shale (Reeves et al., 1996).

Another major self-sourced reservoir is coal. Abnormally low and high pressures have been described in coal-gas reservoirs in the Upper Cretaceous Fruitland Formation of New Mexico and Colorado (Meissner, 1984; Kaiser et al., 1991). In the Powder River Basin of Wyoming, microbial gas is produced from thick (65–100 ft, 20–30 m), underpressured coal beds in the Paleocene Tongue River Formation (Law et al., 1991). In both the Timan-Pechora Basin of Russia and the Donbas region of Ukraine, gas is vented to the atmosphere from underpressured Permian and Carboniferous coal beds, respectively.

The presence of abnormally high or low pressures is one of the more important attributes of basin-centered gas accumulations. Examples of abnormally pressured, basin-centered gas accumulations include the Alberta Basin of Canada (Masters, 1979, 1984), the Greater Green River Basin of Wyoming, Colorado, and Utah (Law et al., 1979; McPeek, 1981; Law, 1984; Spencer, 1987; Law et al., 1989), the Piceance Basin of Colorado (Johnson, 1989; Johnson et al., 1987; Spencer, 1987), the San Juan Basin of New Mexico and Colorado (Berry, 1959; Meissner, 1984), and the Appalachian Basin of Ohio, Pennsylvania, New York, and West Virginia (Davis, 1984; Zagorsky, 1988; Law and Spencer, 1993). In countries other than those in North America, the concept of abnormally pressured basin-centered gas accumulations is not well known and very little published information is available. In Russia, a large, underpressured basin-centered gas accumulation has been identified in Permian rocks in the Timan-Pechora Basin. A basin-centered gas accumulation in Carboniferous age rocks of the Dnieper-Donets Basin of Ukraine has also recently been identified (Law et al., 1997). In South America, a probable basin-centered gas accumulation has been identified in Devonian rocks in the Chaco Basin of Bolivia by Williams et al., (1995). In Jordan, in the Middle East, gas is produced from underpressured, Ordovician sandstone reservoirs (Ahlbrandt et al., 1996, 1997). And there are undoubtedly many more unidentified abnormally pressured, basin-centered gas accumulations distributed around the world.

Low-permeability, shallow, underpressured, microbial gas accumulations in the northern Great Plains of the United States and Canada have been described by Rice and Schurr (1980). Shallow, underpressured gas accumulations in Cretaceous reservoirs are also known to occur in eastern Colorado, and western Kansas.

Curiously, the fluid phase of nearly all abnormally pressured hydrocarbon accumulations is gas. Notable exceptions include the organic-rich Mississippian and Devonian Bakken Shale in the Williston Basin of North Dakota and Montana (Meissner, 1978) and the Pale-

ocene and Eocene Wasatch, Colton, and Green River Formations in the Uinta Basin of Utah (Lucas and Drexler, 1976; Spencer, 1987; Fouch et al., 1992). These two overpressured systems are basin-centered oil accumulations. The reason for the disproportionately few occurrences of basin-centered oil accumulations is not known. We suggest that in abnormally pressured, thermally over-mature reservoirs, originally trapped oil might be expected to have been thermally cracked to gas. This explanation may partially account for the small number of basin-centered oil accumulations.

SUMMARY

Abnormally pressured rocks are globally distributed in a wide range of geologic conditions. An evaluation of causal mechanisms cited in the literature indicates that compaction disequilibrium is the most commonly cited mechanism, followed closely by hydrocarbon generation. In young, rapidly deposited sediments, compaction disequilbrium is most commonly cited as the principle cause of abnormally high pressure, while in older rocks, the most commonly cited overpressure mechanism is hydrocarbon generation. In thermally mature, underpressured systems, the pressures most likely evolved from an originally overpressured system due to gas loss, and gas volume contraction associated with uplift, erosion, and cooling.

There is a strong association of abnormal pressures and conventional and unconventional hydrocarbon accumulations. A general decrease in the size and frequency of oil and gas fields with increasing pressure is common, with the bulk of production coming from reservoirs with pressure gradients less than 0.75 psi/ft (17.4 kPa/m). The threshold for economic oil and gas production in conventionally trapped accumulations is approximately 0.85 psi/ft (19.6 kPa/m). Unconventional gas accumulations are commonly associated with abnormally high or low reservoir pressures.

ACKNOWLEDGEMENTS *The authors gratefully acknowledge F. Meissner and V.I Slavin for assisting in the task of providing locations of some abnormally pressured systems. Illustrations for this manuscript, as well as several other manuscripts in this book, were graciously prepared by Carol Holtgrewe. The manuscript benefited greatly from the comments of T.D. Dyman, R.C. Johnson, M.D. Lewan, and L.C. Price.*

REFERENCES CITED

Ahlbrandt, T.S., O.A. Okasheh, and M.D. Lewan, 1996, A middle east basin center hydrocarbon accumulation in Paleozoic rocks, eastern Jordan, western Iraq and surrounding regions, *in* S. Longacre, B. Katz, R. Slatt, and M. Bowman, convenors, Compartmentalized reservoirs: their detection, characterization and management: AAPG/EAGE Research Symposium, October 20–23, 1996, 3p.

Ahlbrandt, T.S., O.A. Okasheh, and M.D. Lewan, 1997, A middle east basin center hydrocarbon accumulation in Paleozoic rocks, eastern Jordan, western Iraq and surrounding regions, 1997 AAPG International Conference and Exhibition, Vienna, Austria, Sept. 7–10, 1997, p. A1–A2.

Al-Shaieb, Z., J.O. Puckett, A.A. Abdulla, and P.B. Ely, 1994, Megacompartment complex in the Anadarko Basin: A completely sealed overpressured phenomenon, *in* Ortoleva, P.J., ed., Basin compartments and seals: AAPG Memoir 61, p. 55–68

Belonin, M.D. and V.I. Slavin, 1998, Abnormally high formation pressures in petroleum regions of Russia and other countries of the Commonwealth of Independent States (CIS), *in* Law, B.E., G.F. Ulmishek, and V.I. Slavin eds., Abnormal pressures in hydrocarbon environments: AAPG Memoir 70, p. 115–121.

Berry, F.A.F., 1959, Hydrodynamics and geochemistry of the Jurassic and Cretaceous Systems in the San Juan basin, northeastern New Mexico and southwestern Colorado: Unpublished Ph.D. Thesis, Stanford University, 1959, 192 p.

Berry, F.A.F., 1973, High fluid potentials in California Coast Ranges and their tectonic significance: AAPG Bulletin, v. 56, p. 1219–1249.

Bilyeu, B.D., 1978, Deep drilling practices - Wind River basin of Wyoming, *in* Thirteenth Annual Field Conference Guidebook: Wyoming Geological Association, p. 13–24.

Bradley, J.S., 1975, Abnormal formation pressure: AAPG Bulletin, v. 59, p. 957–973.

Bredehoeft, J.B., R.D. Djevanshir, and K.R. Belitz, 1988, Lateral fluid flow in a compacting sand-shale sequence: South Caspian Basin: AAPG Bulletin, v. 72, p. 416–424.

Breeze, A.F., 1970, Abnormal - subnormal pressure relationships in the Morrow Sands of northwestern Oklahoma: Unpublished Univ. Oklahoma M.Sc. Thesis, 122 p.

Buhrig, C., 1989, Geopressured Jurassic reservoirs in the Viking Graben - Modeling and geological significance: Marine and Petroleum Geology, v. 6, p. 31–48.

Burrus, J., 1998, Overpressures models for clastic rocks: their relation to hydrocarbon expulsion: a critical reevaluation, *in* Law, B.E., G.F. Ulmishek, and V.I. Slavin eds., Abnormal pressures in hydrocarbon environments: AAPG Memoir 70, p. 35–63.

Burrus, J., E. Brosse, G. C. de Janvry, and J. Oudin, 1992, Basin modeling in the Mahakam delta based on the integrated 2D model TEMISPACK, Proceedings Indonesian Petroleum Association, 21st Annual. Convention, p. 23–43.

Burst, J.F., 1969, Diagenesis of Gulf Coast clayey sediments and its possible relationship to petroleum migration: AAPG Bulletin., v. 53, p. 73–93.

Carlin, S. and J. Dainelli, 1998, Pressure regimes and pressure systems in the Adriatic foredeep (Italy), *in* Law, B.E., G.F. Ulmishek, and V.I. Slavin eds., Abnormal pressures in hydrocarbon environments: AAPG Memoir 70, p. 145–160.

Chapman, R.E., 1994, Abnormal pore pressures: essential theory, possible causes, and sliding, *in* W.H. Fertl, R.E. Chapman, and R.F. Hotz, eds., Studies in abnormal pressures: Developments in petroleum science 38, Elsevier, p. 51–91.

Chukwu, G.A., 1991, The Niger Delta complex basin: Stratigraphy, structure and hydrocarbon potential: Journal of Petroleum Geology, v. 14, p. 211–220.

Da-jun, P. and L. Yun-ho, 1994, Genetic mechanism of abnormal pressure, pressure seals and natural gas accumulations in carbonate reservoirs, Sichuan Basin, *in* Law, B.E., G. Ulmishek, and V.I. Slavin, eds., Abnormal pressures in hydrocarbon environments: AAPG Hedberg Research Conference, Golden, Colorado, June 8–10, 1994, unpaginated.

Davis, T.B., 1984, Subsurface pressure profiles in gas-saturated basins, *in* Masters, J.A., ed., Elmworth - Case study of a deep basin gas field: AAPG Memoir 38, p. 189–203.

de Witt, W., 1984, Devonian gas-bearing shales in the Appalachian Basin, *in* Spencer, C.W. and R.F. Mast, eds., Geology of tight gas reservoirs: AAPG Studies in Geology 24, p.1–8.

Dickey, P.A. and W.C. Cox, 1977, Oil and gas in reservoirs with subnormal pressures: AAPG Bulletin, v. 61, p. 2134–2142.

Dickinson, G, 1953, Geological aspects of abnormal reservoir pressures in Gulf Coast Louisiana: AAPG Bulletin., v.37, p.410–432.

Doré, A.G. and L.N. Jensen, 1996, The impact of late Cenozoic uplift and erosion on hydrocarbon exploration: offshore Norway and some other uplifted basins: Global and Planetary Change, v. 12, p. 415–436.

Dow, W.G., 1984, Oil source beds and oil prospect generation in the upper Tertiary of the Gulf Coast: Transactions Gulf Coast Association of Geological Societies, p. 329–339.

Durmish'yan, A.G., 1972, Role of anomalously high formation pressures (AHFP) in development of traps for, and accumulations of oil and gas in the southern Caspian basin: International Geology Review, v. 15, no. 5, p. 508–516.

Ejedawe, J.E., 1986, The expulsion criterion in the evaluation of the petroleum source beds of the Tertiary Niger Delta: Journal of Petroleum Geology, v. 9, p.439–450.

Evamy, B., J. Maremboure, P. Kamerling, W.A. Knapp, G. Malloy, and P. Rowlands, 1978, Hydrocarbon habitat of the of the Tertiary Niger delta: AAPG Bulletin, v. 62, p. 1–39.

Fertl, W.H., 1976, Abnormal formation pressures: Elsevier Scientific Publishing Company, Amsterdam, 382 p.

Fertl, W.H., R.E. Chapman, and R.F. Hotz, 1994, Studies in abnormal pressures: New York, Elsevier, 454 p.

Fertl, W.H., and W.G. Leach, 1990, Formation temperature and formation pressure affect the oil and gas distribution in Tertiary Gulf Coast sediments: Transactions - Gulf Coast Association of Geological Societies, v. XL, p.205–216.

Fouch, T.D., V.F. Nuccio, J.C. Osmond, L. MacMillan, W.B. Cashion, and C.J. Wandrey, 1992, Oil and gas in uppermost Cretaceous and Tertiary rock, Uinta Basin, Utah, *in* Fouch, T.D., V.F. Nuccio, T.C. Chidsey Jr., eds., Hydrocarbon and mineral resources of the Uinta Basin, Utah and Colorado: Utah Geological Association Guidebook 20, p.9–48.

Gaarenstroom, L., Tromp, R.A.J., de Jong, M.C., and Brandenburg, A.M., 1993, Overpressures in the Central North Sea - Implications for trap integrity and drilling safety, Petroleum geology of Northwest Europe: Proceedings of the 4th Conference: The Geological Society, p. 1305–1313.

Gautier, D.L., K.J. Bird, and V.A. Colten-Bradley, 1987, Relationship of clay mineralogy, thermal maturity, and geopressure in wells of the Point Thomson area, *in* Bird, K.J. and Magoon, L.B., eds., Petroleum geology of the northern part of the Arctic National Wildlife Refuge, northeastern Alaska: U.S. Geological Survey Bulletin 1778, p.

Gurevich, A.E., and G.V. Chilingar, 1995, Abnormal pressures in Azerbaijan: a brief critical review and recommendations: Journal of Petroleum Science and Engineering, v. 13, p. 125–135.

Heppard, P.D., H.S. Cander, and E.B. Eggertson, 1998, Abnormal pressure and the occurrence of hydrocarbons in offshore eastern Trinidad, West Indies, *in* Law, B.E., G.F. Ulmishek, and V.I. Slavin eds., Abnormal pressures in hydrocarbon environments: AAPG Memoir 70, p. 215–246.

Hitchon, B., J.R. Underschultz, S. Bachu, and C.M. Sauveplane, 1990, Hydrology, geopressures and hydrocarbon occurrences, Beaufort-Mackenzie Basin: Bulletin of Canadian Petroleum Geology, v. 38, p. 215– 235.

Holm, G.M., 1998, Distribution and origin of overpressure in the Central Graben of the North Sea, *in* Law, B.E., G.F. Ulmishek, and V.I. Slavin eds., Abnormal pressures in hydrocarbon environments: AAPG Memoir 70, p. 123–144.

Hunt, J.M., 1990, Generation and migration of petroleum from abnormally pressured compartments: AAPG Bulletin, v. 74, p. 1–12.

Hunt, J.M., J.K. Whelan, L.B. Eglinton, and L.M. Cathles, III, 1994, Gas generation - A major cause of deep Gulf Coast overpressures: Oil and Gas Journal, July 18, 1994, p. 59–63.

Hunt, J.M., J.K. Whelan, L.B. Eglinton, and L.M. Cathles, III, 1998, Relation of shale porosities, gas generation, and compaction to deep overpressures in the U.S. Gulf Coast, *in* Law, B.E., G.F. Ulmishek, and V.I. Slavin eds., Abnormal pressures in hydrocarbon

environments: AAPG Memoir 70, p. 87–104.

Hunter, C.D., 1962, Economics influence east Kentucky gas future: Oil and Gas Journal, July 9, 1962, p.170–174.

Johnson, R.C., 1989, Geologic history and hydrocarbon potential of Late Cretaceous-age, low-permeability reservoirs, Piceance Basin, western Colorado: U.S. Geological Survey Bulletin 1787-E, 51 p.

Johnson, R.C., R.A. Crovelli, C.W. Spencer, and R.F. Mast, 1987, An assessment of gas resources in low-permeability sandstones of the Upper Cretaceous Mesaverde Group, Piceance Basin, Colorado: U.S. Geological Survey Open-File Report 87-357, 21 p.

Johnson, R.C., T.M. Finn, R.A. Crovelli, and R.H. Balay, 1996, An assessment of in-place gas resources in low-permeability Upper Cretaceous and lower Tertiary sandstone reservoirs, Wind River Basin, Wyoming: U.S. Geological Survey Open-File Report 96-264, 67 p.

Kaiser, W.R., T.E. Swartz, and G.J. Hawkins, 1991, Hydrology of the Fruitland Formation, San Juan Basin, *in* Geologic and hydrologic controls on the occurrence and producibility of coalbed methane, Fruitland Formation, San Juan Basin: Gas Research Institute, Topical Report GRI-91/0072.

Law, B.E., 1984, Relationships of source-rock, thermal maturity, and overpressuring to gas generation and occurrence in low-permeability Upper Cretaceous and lower Tertiary rocks, Greater Green River Basin, Wyoming, Colorado, and Utah, *in* Woodward, J., Meissner, F.F., and Clayton, J.L., eds., Hydrocarbon source rocks of the greater Rocky Mountain region: Rocky Mountain Association of Geologists, p. 469–490.

Law, B.E., and W. Dickinson, 1985, Conceptual model for origin of abnormally pressured gas accumulations in low-permeability reservoirs: AAPG Bulletin, v. 69, p. 1295–1304.

Law, B.E., B.P. Kabyshev, A.Yu. Polutranko, G.F. Ulmishek, and T.M. Prigarina, (1997), Basin-centered gas accumulation, Dnieper- Donets Basin and Donbas region, Ukraine: 1997 AAPG International Conference, Vienna, Austria, September 7–10, 1997, p. 36.

Law, B.E., D.D. Rice, and R.M. Flores, 1991, Coal-bed gas accumulations in the Paleocene Fort Union Formation, Powder River Basin, Wyoming, *in* Schwochow, S.D., ed., Coalbed methane of western North America: Rocky Mountain Association of Geologists Guidebook, Fall Conference and Field Trip, Glenwood Springs, Colo., September 17–20, 1991, p. 179–190.

Law, B.E., S.H.A. Shah, and M.A. Malik, 1998, Abnormally high formation pressures, Potwar Plateau, Pakistan, *in* Law, B.E., G.F. Ulmishek, and V.I. Slavin eds., Abnormal pressures in hydrocarbon environments: AAPG Memoir 70, p. 247–258.

Law, B.E., and C.W. Spencer, 1993, Gas in tight reservoirs - An emerging major source of energy, *in*

Howell, D.G., ed., The future of energy gases: U.S. Geological Survey Professional Paper 1570, p. 233–252.

Law, B.E., C.W. Spencer, and N.H. Bostick, 1979, Preliminary results of organic maturation, temperature, and pressure studies in the Pacific Creek area, Sublette County, Wyoming, *in* 5th DOE Symposium on Enhanced Oil and Gas Recovery and Improved Drilling Methods, v. 3, Oil and Gas Recovery: Tulsa, Oklahoma, Publishing Co., p. K-2/1- K2/13.

Law, B.E., C.W. Spencer, R.R. Charpentier, R.A. Crovelli, R.F. Mast, D.L. Dolton, and C.J. Wandrey, 1989, Estimates of gas resources in overpressured low-permeability Cretaceous and Tertiary sandstone reservoirs, Greater green River Basin, Wyoming, Colorado, and Utah: Wyoming Geological Association Fortieth Field Conference Guidebook, p. 39–61.

Law, B.E., Tennyson, M.E., and Johnson, S.Y., 1994, Basin-centered gas accumulations in the Pacific Northwest - A potentially large source of energy: AAPG Annual Meeting, Denver, Colorado, June 12–15, 1994.

Leach, W.G., 1993a, New exploration enhancements in south Louisiana Tertiary sediments: Oil and Gas Journal, Mar. 1, 1993, p. 83–87.

Leach, W.G., 1993b, Fluid migration, hydrocarbon concentration in south Louisiana Tertiary sands: Oil and Gas Journal, Mar. 15, 1993, p. 71–74.

Leach, W.G., 1993c, Maximum hydrocarbon window determination in south Louisiana: Oil and Gas Journal, Mar. 29, 1993, p. 81–84.

Leonard, R.C., 1993, Distribution of sub-surface pressure in the Norwegian Central Graben and applications for exploration, Petroleum geology of Northwest Europe: Proceedings of the 4th Conference, The Geological Society, p. 1295–1303.

Lico, M.S., and Y.K. Kharaka, 1983, Subsurface pressure and temperature distributions, *in* Hester, R.L. and D.E. Hallinger, eds., Selected Papers AAPG Pacific Section 1983 Annual Meeting: Sacramento , California, p. 57–75.

Lucas, P.T., and J.M. Drexler, 1976, Altamont-Bluebell - A major, naturally fractured stratigraphic trap, Uinta Basin, Utah, *in* Braunstein, J. ed., North American oil and gas fields: AAPG Memoir 24, p. 121–135.

Magara, K., 1968, Compaction and migration of fluids in Miocene mudstone, Nagaoka Plain, Japan: AAPG Bulletin, v. 52, p. 2466–2501.

Masters, J.A., 1979, Deep basin gas trap, western Canada: AAPG Bulletin, v. 63, p. 152–181.

Masters, J.A., 1984, ed., Elmworth - Case study of a deep basin gas field: AAPG Memoir 38, 316 p.

McPeek, L.A., 1981, Eastern Green River Basin: a developing giant gas supply from deep overpressured Upper Cretaceous sandstones: AAPG Bulletin, v. 65, p. 1078–1098.

Meissner, F.F., 1978, Petroleum geology of the Bakken Formation, Williston basin, north Dakota and Montana, *in* D. Rehrig, ed., Williston Basin Symposium Guidebook: Montana Geological Society, p. 207–227.

Meissner, F.F., 1984, Cretaceous and lower Tertiary coals as sources for gas accumulation in the Rocky Mountain area, *in* Woodward, J. Meissner, F.F., and Clayton, J.L., eds., Hydrocarbon source rocks of the greater Rocky mountain region: Rocky Mountain Association of Geologists, p. 401–431.

Mouchet, J.P., and A. Mitchell, 1989, Abnormal pressures while drilling: Manuels Techniques Elf Aquitaine, v. 2, 264 p.

Mudford, B.S., and Best, M.E., 1989, Venture Gas Field, offshore Nova Scotia: Case study of overpressuring in region of low sedimentation rate: AAPG Bulletin, v. 73, no. 11, p. 1383–1396.

Nashaat, M., 1998, Abnormally high fluid pressure and seals impacts on hydrocarbon accumulations in the Nile Delta and North Sinai Basins, Egypt, *in* Law, B.E., G.F. Ulmishek, and V.I. Slavin eds., Abnormal pressures in hydrocarbon environments: AAPG Memoir 70, p. 161–180.

Nyein, R.K., L. MacLean, and B.J. Warris, 1977, Occurrence, prediction and control of geopressures on the northwest shelf of Australia: APEA Journal, p 64–72.

Oudin, J.L. and Picard, P.F., 1982, Genesis of hydrocarbons in the Mahakam Delta and the relationship between their distribution and the over pressured zones: Proceedings of the Indonesian Petroleum Association, Eleventh Annual Convention, v. 1, p. 181–202.

Polutranko, A.J., 1998, Causes of formation and distribution of abnormally high formation pressure in petroleum basins of Ukraine, *in* Law, B.E., G.F. Ulmishek, and V.I. Slavin eds., Abnormal pressures in hydrocarbon environments: AAPG Memoir 70, p. 181–194.

Reeves, S.R., V.A. Kuuskraa, and D.G. Hill, 1996, New basins invigorate U.S. gas shales play: Oil and Gas Journal, Jan. 22, 1996, p. 53–58.

Rice, D.D., and G.W. Schurr, 1980, Shallow, low-permeability reservoirs of northern Great Plains - Assessment of their natural gas resources: AAPG Bulletin, v. 64, p. 969–987.

Rogers, A. L., and N.A. Yassir, 1993, Hydrodynamics and overpressuring in the Jeanne d'Arc basin, offshore Newfoundland, Canada: possible implications for hydrocarbon exploration: Bulletin of Canadian Petroleum Geology, v. 41, p. 275–289.

Russell, W.L., 1972, Pressure-depth relations, Appalachian region: AAPG Bulletin, v. 59, p. 528–536.

Schaar, G., 1976, The occurrence of hydrocarbons in overpressured reservoirs of the Baram delta (Offshore Sarawak, Malaysia): Proceedings Indonesian Petroleum Association, Fifth Annual. Conference, p.163–169.

Spencer, C.W., 1987, Hydrocarbon generation as a mechanism for overpressuring in Rocky Mountain region: AAPG Bulletin, v. 71, p. 368–388.

Spencer, C.W., A. Szalay, and E. Tatar, 1994, Abnormal pressure and hydrocarbon migration in the Bekes

Basin, *in* Teleki, P.G. et al., eds., Basin analysis in Petroleum Exploration: Kluwer Academic Publishers, Netherlands, p. 201–219.

Surdam, R.C., Z.S. Jiao, and R.S. Martinsen, 1994, The regional pressure regime in Cretaceous Sandstones and Shales in the Powder River Basin: AAPG Memoir 61, p. 213–233.

Swarbrick, R.E. and M.J. Osborne, 1998, Mechanisms that generate abnormal pressures: an overview, *in* Law, B.E., G.F. Ulmishek, and V.I. Slavin eds., Abnormal pressures in hydrocarbon environments: AAPG Memoir 70, p. 13–34.

Timko, D.J. and W.H. Fertl, 1971, Hydrocarbon accumulation and geopressure relationships and prediction of well economics with log calculated geopressures: Journal of Petroleum Technology, v. 23, p. 923–933.

Welte, D.H., R.G. Schaefer, W. Stoessinger, and M. Radke, 1984, Gas generation and migration in the Deep Basin of western Alberta, *in* Masters, J.A., ed., Elmworth - Case study of a deep basin gas field: AAPG Memoir 38, p. 35–47.

Williams, K.E., B.J. Radovich, and J.W. Brett, 1995, Exploration for deep gas in the Devonian Chaco Basin of southern Bolivia: Sequence stratigraphy, prediction, and well results: 1995 AAPG Annual Convention abstracts, p. 104A.

Zagorski, W.A., 1988, Exploration concepts and methodology for deep Medina Sandstone reservoirs in northwestern Pennsylvania: AAPG Bulletin, v. 72, p. 976.

Swarbrick, R.E. and M.J. Osborne, 1998, Mechanisms that generate abnormal pressures: an overview, *in* Law, B.E., G.F. Ulmishek, and V.I. Slavin eds., Abnormal pressures in hydrocarbon environments: AAPG Memoir 70, p. 13–34.

Chapter 2

Mechanisms that Generate Abnormal Pressures: an Overview

Richard E. Swarbrick
Mark J. Osborne[1]
Department of Geological Sciences
University of Durham
Durham, United Kingdom

Abstract

Normally pressured reservoirs have pore pressures which are the same as a continuous column of static water from the surface. Abnormal pressures occur where the pore pressures are significantly greater than normal (overpressure) or less than normal (underpressure). Overpressured sediments are found in the subsurface of both young basins from about 1.0 to 2.0 km downwards, and in older basins, in thick sections of fine-grained sediments. The main mechanisms considered responsible for most overpressure conditions can be grouped into three broad categories, based on the processes involved: (1) ineffective volume reduction due to imposed stress (vertical loading during burial, lateral tectonic processes) leading to disequilibrium compaction, (2) volume expansion, including porosity increases due to changes in the solid:liquid ratios of the rock, and (3) hydraulic head and hydrocarbon buoyancy. The principal mechanisms which result in large magnitude overpressure are disequilibrium compaction and fluid volume expansion during gas generation. Disequilibrium compaction results from rapid burial (high sedimentation rates) of low-permeability rocks such as shales, and is characterized on pressure vs. depth plots by a fluid retention depth where overpressure commences, and increases downwards along a gradient which can closely follow the lithostatic (overburden) gradient. Disequilibrium compaction is typical in basins with a high sedimentation rate, including Tertiary deltas and some intracratonic basins. In older basins, disequilibrium compaction generated earlier in the basin history may be preserved only in thick, fine-grained sequences, but lost by vertical/lateral leakage from rocks with relatively high permeabilities. Gas generation from secondary maturation reactions, and oil cracking in the deeper parts of sedimentary basins, can result in large fluid volume increases, although the magnitudes are uncertain. In addition, the effect of increased pressures on the reactions involved is unknown. We doubt that any of the other mechanisms involving volume change can contribute significant regional overpressure, except in very unusual conditions. Hydraulic head and hydrocarbon buoyancy are mechanisms whose contributions are generally small; however, they can be easily assessed and may be important when additive to other mechanisms. The effects of transference of overpressure generated elsewhere should always be considered, since the present pressure distribution will be strongly affected by the ability of fluids to move along lateral and vertical conduits. Naturally underpressured reservoirs (as opposed to underpressure during depletion) have not been as widely recognized, being restricted mainly to interior basins which have undergone uplift and temperature reduction. The likely principal causes are hydraulic discharge, rock dilation during erosional unroofing, and gas migration during uplift.

INTRODUCTION

Abnormally pressured rocks are typical of many sedimentary basins worldwide. The pressures are mostly recorded during drilling of deep boreholes (e.g., by wireline pressure tools such as the Repeat Formation Tester, or during a production test) in permeable sedimentary units, and also can be inferred from drilling parameters (such as mud weight) in some conditions. The pore pressure of low-permeability rocks, such as mudstones and shales, is rarely known, and must be inferred from adjacent permeable rocks, or from the interpretation of wireline logs (for example, by comparison of measured and expected porosity for a given depth, the porosity data usually being derived from the sonic, density, neutron or resistivity/conduc-

[1] *Present Affiliation: BP Exploration, Shared Petrotechnical Resource, Basin Modeling and Geochemistry, Middlesex, United Kingdom*

tivity logs (Hottman and Johnson, 1965; Mouchet and Mitchell, 1989)).

In one of the first studies of abnormal pressures, using data from the Gulf of Mexico, Dickinson (1953) suggested that the high pressures in the clastic sequences there could be explained by incomplete dewatering of the sediments. Other explanations were later suggested for the high pressures in these rocks, including thermal effects (Barker, 1972), clay mineral changes (Powers, 1967; Burst, 1969; Bruce, 1984) and osmosis (Marine and Fritz, 1981). In other basins, the influence of hydrocarbon generation/maturation has been proposed to explain overpressured reservoirs (Spencer, 1987), and gas generation as a mechanism has been invoked for the Gulf of Mexico situation (Hunt et al., 1994; 1998-this volume). Some authors suggest one dominant mechanism, e.g., disequilibrium compaction (Bredehoeft and Hanshaw, 1968; Summa et al., 1993), or hydrocarbon generation (Meissner, 1978a, b), whilst others have attempted to ascribe the relative proportion of each mechanism (Hart et al., 1995). Most papers dealing with abnormal pressures provide a brief overview of the range of possible mechanisms, but there are few papers which deal specifically with each, and the conditions required for them to create abnormal pressures. A review of the mechanisms, especially as applied to drilling overpressured reservoirs, is provided in Fertl (1976), Mouchet and Mitchell (1989) and Fertl et al. (1994). Hall's (1993) review of overpressure mechanisms explains each in terms of the physical and chemical reactions which govern rock behavior. A comprehensive list of the mechanisms and the literature pertaining to each is also available in Martinsen (1994).

In this paper we re-examine the basis for the inclusion of each abnormal pressure mechanism on a list of possibilities. We have become aware that there is strong empirical evidence for some of the popular mechanisms (e.g., smectite-illite transformation, hydrocarbon generation), but great uncertainty about the exact processes involved and hence their ability to create abnormal pressures under normal burial conditions in sedimentary environments. We conclude that for many of the mechanisms further research is required to verify the conditions under which they can produce abnormal pressures of the magnitude and distribution observed in sedimentary basins around the world.

Overpressure

Overpressure is defined as "any pressure which exceeds the hydrostatic pressure of a column of water or formation brine" (Dickinson, 1953 and Figure 1). Another way to view overpressure, in terms of the dynamics of subsurface fluid flow, is the inability of formation fluids to escape at a rate which allows equilibration with hydrostatic pressure, calculated from a pressure gradient which varies from 0.433 psi/ft (9.71 kPa/m) for fresh water to about 0.51 psi/ft (11.44 kPa/m) for saturated brine. Hence any measured pres-

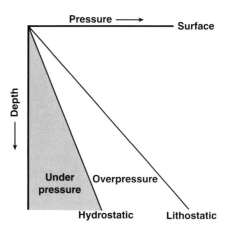

Figure 1. Pressure vs. depth plot. Rocks whose pressures plot between the hydrostatic and lithostatic gradients are termed "overpressured" and those whose pressures are less than hydrostatic are "underpressured".

sure in the subsurface can be compared with the pressure of a column of formation water existing from the surface to the same depth. Overpressure is a disequilibrium state resulting from fluid retention, and one of the primary controls on the presence and distribution of overpressure is therefore permeability, the rock attribute which controls seal behavior.

Many of the world's basins contain overpressured reservoirs; Hunt (1990) cites 180 basins, including basins from the Americas, Africa, Europe, Middle East, Far East, and Australasia, where overpressure has been recognized. The location of some of these, plus tabular data, appear in Law and Spencer (1998-this volume). The age of rocks in which overpressures have been recognized varies from Pleistocene to Cambrian. Overpressure is found in carbonate and clastic reservoirs, and in rocks deposited in the full range of sedimentary environments. Hydrocarbons are often associated with overpressures, but not exclusively so, and there does not seem to be a universal relationship between overpressure and hydrocarbon traps. For example in the Sable Basin, offshore Nova Scotia, gas is found above the top of the overpressure in the shallower traps in the southwest, but largely below the top in the deeper traps to the northeast (Williamson and Smyth, 1992). In the Northern and Central North Sea, shallow oil and gas condensate occur in normally pressured reservoirs, above deeper and highly overpressured reservoirs, which contain a mixture of oil and gas condensate. Fluid inclusion data from the Northern North Sea have been used to argue for hydrocarbon migration when the reservoirs were normally pressured, and prior to their current overpressured condition (Swarbrick, 1994).

Underpressure

Underpressure exists when pore pressure is significantly lower than the hydrostatic pressure (Figure 1). Fewer examples of underpressure are documented, in comparison with overpressure, a factor which Hunt (1990) attributes to the difficulty of recognizing underpressures during conventional drilling operations. Underpressure results from depletion during oil and gas production, but **naturally** underpressured reservoirs have been described from a number of basins, particularly in Canada and U.S.A., including West Canada Basin, Alberta (Gies, 1984; Davies, 1984; Grigg, 1994; Rostron and Toth, 1994); Silurian Clinton sand, eastern Ohio (Davies, 1984); San Juan Basin, New Mexico and Colorado (Berry, 1959; Meissner, 1978a; Kaiser and Scott, 1994); and Red Desert and Green River Basins, Wyoming (Davies, 1984). In each of the above case histories, the basin has been uplifted and contains gas-bearing reservoirs, which have experienced reducing temperatures. Law and Dickinson (1985) and Grigg (1994) believe underpressured reservoirs today have been overpressured in the past.

OVERPRESSURE—THE SYSTEM

In order to understand overpressure in terms of fluid retention we need to know about four principal aspects of the rock and fluid conditions (Figure 2):

A. Causal Mechanism
B. Rock Permeability as it Relates to Seal Behavior
C. Timing (and Rate)
D. Fluid Type

Causal Mechanism

The amount of overpressure and the rate at which it can build up will relate directly to the mechanism which is generating the excess fluid. A wide variety of mechanisms have been proposed for generating over-

pressure in subsurface rocks. The processes which create overpressure allow the mechanisms to be grouped into three categories (Figure 3):

1. Stress-related, (i.e., compression leading to pore volume reduction)
    Mechanisms:
        Disequilibrium compaction (vertical loading stress)
        Tectonic stress (lateral compressive stress)

2. Fluid volume increase
    Mechanisms:
        Temperature increase (aquathermal)
        Water release due to mineral transformation
        Hydrocarbon generation
        Cracking of oil to gas

3. Fluid movement and buoyancy
    Mechanisms:
        Osmosis
        Hydraulic head
        Buoyancy due to density contrasts

In addition, overpressure can result from redistribution of overpressured fluids from elsewhere, sometimes referred to as "transference". Although not a mechanism in itself, transference may exert a strong influence on many of the pore pressure profiles seen in the subsurface, and may mask recognition of the underlying causal mechanism (Swarbrick and Osborne, 1996).

Permeability as it Relates to Seal Behavior

Permeability of a rock is a constant (k) defined by the equation which expresses Darcy's law and relates

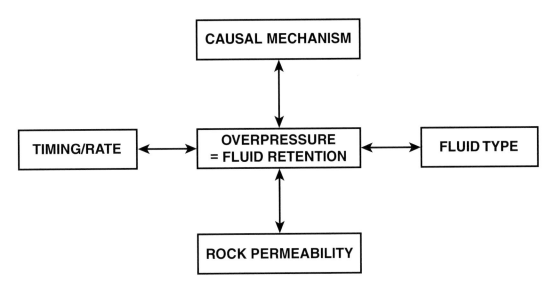

Figure 2. Factors which control overpressure and the inability of rocks to maintain equilibrium at hydrostatic pressures. Each of the factors is interlinked. Permeability of a rock is directed related to its ability to act as a seal.

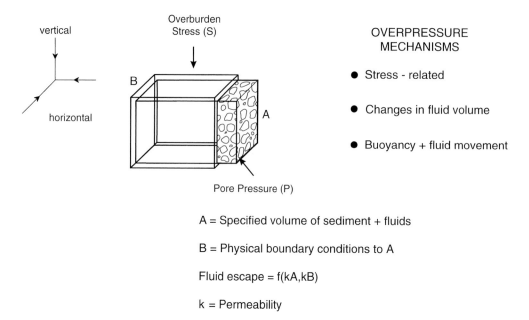

A = Specified volume of sediment + fluids

B = Physical boundary conditions to A

Fluid escape = f(kA,kB)

k = Permeability

Figure 3. The principal classes of mechanisms include increased stress (vertical loading and horizontal compression), changes in fluid volume (mineral dehydration and transformation reactions) and hydraulic flow and buoyancy. The permeability of both the overpressured rock and its boundaries (seals) will contribute to the overpressure observed.

flow rate of a given fluid per unit of time, along a known flow path across which there is a specified pressure drop:

$$Q = \frac{-kA\Delta P}{\mu L}$$

where:
Q = volume rate of flow
k = permeability
L = Length scale
ΔP = pressure drop across L
μ = fluid viscosity
A = area across which flow takes place

Hence, permeability is an intrinsic property of the rock which can be measured. Permeability is controlled by the properties of the rock (e.g., grain size, grain shape, grain tortuosity) and the properties of its fluid content (e.g., viscosity, density), plus the capillary effects where hydrocarbons are involved. In high-permeability rocks, such as aquifers and hydrocarbon reservoirs, permeability can be routinely measured. However, fluid retention leading to overpressure is largely controlled by the low-permeability, non-reservoir rocks, e.g., shales, evaporites and well-cemented carbonates (often referred to as "seals" in the petroleum literature). Measurement of fluid flow in very low-permeability rocks is difficult and prone to large inaccuracies.

Permeability is the property which controls the ability of a rock to act as a seal or barrier to flow. The term **seal** is defined by Watts (1987) as a rock which prevents natural buoyancy-related upwards migration of hydrocarbons. In this paper we are concerned with the ability of rocks to retain all fluids, in particular water throughout the sedimentary column, since overpres-

sured rocks include the lower water-bearing section as well as the upper hydrocarbon-bearing section. (A notable exception is the tight, regionally overpressured low-permeability reservoirs of the Rocky Mountains which produce only hydrocarbons—Law, 1984; Spencer, 1987; Surdam et al., 1994). Seals in the sense of Watts (1987) can be a permanent barrier to flow, where oil and gas are unable to flow across a membrane if the minimum displacement (or entry) pressure of the caprock, controlled by its capillary properties, is not reached (Watts, 1987). Hunt (1990) extended the definition of a seal to any rock which is capable of preventing all pore fluid movement (oil, gas and water) over substantial periods of geologic time, as this was critical to his model of static pressure compartments. Deming (1994) argues that rocks are not capable of sustaining zero effective permeability to water over extended periods of geological time. In this paper we subscribe to the view that pressure build up and dissipation are continuous processes modifying the pore pressures of abnormally pressured rocks through time (see below). However, since seals are viewed by some as the absence of flow over time, we will not over-emphasize the term, beyond its implication for a restriction to flow which assists in the creation of abnormally pressured rocks.

Overpressure dissipation, or leakage, can be accomplished not only by porous media flow through the rock, but also by fracturing. In tectonically active areas, fault reactivation may be the release mechanism (Byerlee, 1993). However, if pore pressures reach the fracture pressure of the rock during overpressure generation, the rock will hydraulically fracture, potentially releasing large volumes of fluid, and rapidly dissipating the excess pressure until the fractures reseal (Engelder, 1993). A close balance between the pore pressures measured in the Jurassic and Triassic reservoirs, and the overlying Kimmeridge Clay Formation organic-rich shales in the Central North Sea (inferred from Leak Off Test data), suggests that these reservoirs are near fracture pressure, and may have already released the entrapped hydrocarbons from some valid traps (Gaarenstroom et al., 1993; Holm, 1998-this volume).

Timing and Rate

Darcy's equation above defines **flow rate,** which is time dependent. Overpressure is a disequilibrium state and will therefore change with time depending on the evolution of the system, unless zero effective permeability is achieved, which is very difficult to maintain in the water phase over geological time (Deming, 1994). The distribution and magnitude of overpressure will therefore change through time, both during the "build-up phase" when the generating mechanism is active, and afterwards during the "dissipation phase" as leakage continues. Bredehoeft et al. (1994) have contrasted their *dynamic* model of overpressure with the *static* model preferred by Hunt (1990). We also emphasize the ephemeral nature of overpressured basins, certainly within the permeable reservoir units. We are looking today merely at a "snap-shot" of the stress state of the overpressured system; pore pressures may have been higher, or lower, in the past.

Fluid Type

Throughout any basin the most common type of fluid is water, varying from near fresh water to high salinity brines. The total dissolved solids (dominantly salts) determine the pressure gradient, through their effect on density, and to a very small degree exert a control on the flow properties, through their effect on viscosity. Where oil and gas are present (almost always in the presence of water) the fluid and flow properties will be dependent on the composition of the hydrocarbons, temperature, hydrocarbon saturation, and rock properties (including relative permeabilities). Fluid properties of hydrocarbons have particular significance to overpressure on account of their buoyancy (controlled by density contrast) and also the capillary pressure effects controlling relative permeability and entry pressure, and hence the effective sealing capacity of the rocks in which they occur. Buoyancy is inversely related to fluid density. Gas is the most buoyant fluid, becoming more dense and decreasing its buoyancy at elevated pressures. Increases and decreases in pressure through time will have an influence on the composition of petroleum in the basin. For example, falling pressure during overpressure dissipation causes gas exsolution if the pressure falls below the bubble point; similarly more gas can remain in solution when pressure rises.

MECHANISMS FOR GENERATING OVERPRESSURE

Stress-Related Mechanisms

Disequilibrium Compaction (Vertical Loading Stress)

In a sedimentary basin, the weight of the overlying rocks at a given depth, known as overburden stress, S_v, is a function of the thickness (Z) and density (ρ_b) of the overlying rocks, and gravity (g):

$$S_v = Z \rho_b g$$

The average bulk density can be determined in a borehole if a density log has been acquired. Bulk density (ρ_b) is determined from the rock matrix density (ρ_{ma}), the fluid density (ρ_{fl}), and the porosity (ϕ), such that:

$$\rho_b = \rho_{ma}(1 - \phi) + \rho_{fl}(\phi)$$

Some of the overburden stress is borne by the fluid, the pore pressure (P), and the remainder is distributed to the contacts between the rock particles, known as the effective stress (σ). The relationship between effective stress and overburden stress is given by Terzaghi's (1923) equation:

$$\sigma = S_v - P$$

In a normally pressured rock, the effective stress at a given depth is the difference between overburden stress and hydrostatic pressure. The overburden stress can be expressed in terms of pressure (sometimes referred to as "lithostatic" pressure) if the average bulk density is known to the depth of interest. At the surface, where the porosity of the fine-grained sediments is high (say 60–70%), the average density is about 1.5 g/cm^3 (0.7 psi/ft), whereas at about 1.0 km the sediment density is typically about 2.3 g/cm^3 (1.0 psi/ft). Although the lithostatic gradient varies with depth, most pressure vs. depth plots use a default gradient of 2.3 g/cm^3 (1.0 psi/ft), based on average sediment density (Mouchet and Mitchell, 1989).

Increases in effective stress, due to loading of sediment during burial, normally cause rocks to compact, reducing the pore volume and forcing out the formation fluids (Plumley, 1980). The rate of porosity loss varies with rock-type. Each rock-type will have a lower limit beyond which no further mechanical compaction is possible and porosity loss is thereafter due to chemical compaction. Sandstones compact at a relatively slow rate from a starting point of about 40–45% porosity to about 20–30% porosity due to rearrangement of the sand grains and some dissolution at grain contacts (McBride et al., 1991 and Figure 4). At depths of only 1.5 to 2.5 km there is little potential for significant further reduction in primary porosity due to mechanical

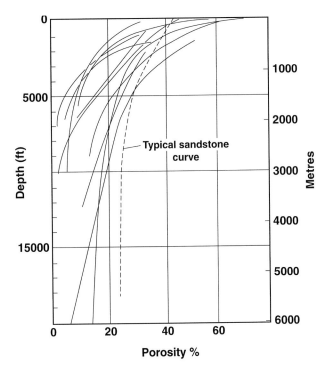

Figure 4. Range of typical porosity-depth curves for mudstones (solid lines), based on published empirical measurements (after Dzevanshir et al., 1986; Aplin et al., 1995). The variability of curves (20% porosity range at 5,000 m) results from the control exerted by different mudstone mineralogy, plus variable overpressure in some data areas. A porosity vs. depth curve (dashed line) for the mechanical compaction of a typical sandstone is shown for comparison.

compaction involving bed thickness reduction, and further porosity reduction is primarily a function of diagenetic cementation (McBride et al., 1991). By contrast, clays have a typical porosity at the time of deposition of 65–80% and compact more quickly than sands (Figure 4). Clays will continue to compact, by grain rearrangement and ductility, to great depths (typically 4–6 km) where the porosity can be reduced to 5–10% of the rock volume (Katsube and Williamson, 1994; Hunt et al., 1994). However, not all clays behave in the same way, which explains some of the variability in porosity vs. depth curves for clay-rich rocks (Figure 4). Aplin et al. (1995) show that the variability is largely controlled by mudstone lithology, although there is also the likelihood that some porosity vs. depth data in Figure 4 were derived from overpressured mudstones in which the rocks are not fully compacted relative to their overburden.

Under conditions of slow burial, the equilibrium between overburden stress and the reduction of pore fluid volume due to compaction can be most easily maintained. Rapid burial, however, leads to faster expulsion of fluids in response to rapidly increasing overburden stress. Where the fluids cannot be expelled fast enough the pressure of the pore fluids increases—a condition known as **disequilibrium compaction.**

Overpressure due to disequilibrium compaction is often recognized by higher porosity than expected at a given depth. Porosity can be considered as a function of the overburden stress and the effective stress. If all the fluid is retained the porosity and effective stress remain constant with depth. Conditions which favor disequilibrium compaction are rapid burial, and low-permeability rocks. Disequilibrium compaction is therefore likely to be found commonly in thick clay and shale successions during continuous rapid burial (England et al., 1987). Overpressure in adjacent, high-permeability reservoir rocks will result from isolation of the reservoir within the low-permeability section.

In addition to Terzaghi's (1923) laboratory experiments, evidence cited for disequilibrium compaction is the anomalous high porosity estimates for low-permeability sections as derived on borehole porosity tools (e.g., sonic log, density log). The sonic vs. depth plot from a North Sea well (Figure 5) shows a departure from the normal trend line at about 1.0 km (3,500 ft) towards anomalously high values of sonic travel time (interpreted as high porosity). This point is interpreted as the top of an overpressure zone. Critical to the interpretation is the validity of a "normal" trend line for porosity (or sonic travel time) and the inference that higher sonic values equate to higher porosity. Other porosity logs (e.g., density, neutron) should show the

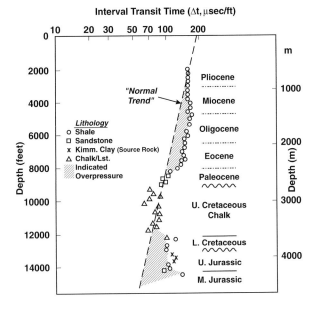

Figure 5. Interval travel time vs. depth plot from the UK 21/20a-1 well, drilled in the Central North Sea. The shale values from the top of Miocene to the base of the Eocene section are interpreted as overpressured due to the high travel times relative to the "normal trend" (see Hottman and Johnson, 1965). Shale values plot close to the normal trend adjacent to the normally pressured Paleocene sands. Deeper, Jurassic shales are also interpreted as overpressured, confirmed by pressure data from the adjacent sands. The cause of overpressure in the Tertiary section is disequilibrium compaction.

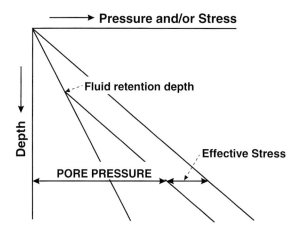

Figure 6. Pressure (and/or stress) vs. depth plot for overpressure generated by disequilibrium compaction. Fluid retention occurs at a depth where the permeability and sedimentation rate combine to prevent complete dewatering. Below the fluid retention depth the profile rapidly changes to almost constant effective stress downwards, i.e., a pressure profile almost parallel to the lithostatic pressure gradient.

same trend to have confidence in this technique. Critical assessment of the rocks is required to verify whether departure from the normal line relates to changing lithologies or mineralogy, or a real manifestation of overpressure and undercompaction (Hermanrud et al., 1998-this volume). The concept of using a "normal trend line" has been valuable in young basins, such as Tertiary deltas (Hottman and Johnson, 1965), but may not be a valid technique everywhere, particularly in older rocks (Wensaas et al., 1994).

Several approaches to modeling disequilibrium compaction are now published (see Mann and Mackenzie, 1990; Audet and McConnell, 1992), and form the basis for fluid flow in basin modeling software packages. The results from the models are critically dependent on the choice of permeability function, especially at low porosity and high stresses. The permeability of mudrocks is poorly known because it is difficult to measure in low-permeability lithologies (see Neuzil, 1994). In addition, changes in permeability with increasing overburden stress are also poorly constrained. Consequently, basin modeling can show the potential for building up overpressures in the subsurface but the actual values are difficult to validate.

The lateral and vertical variability in permeability of a sedimentary sequence (e.g., mudrocks, silts and sands in a clastic sequence) is critically important. A groundwater flow model of up to 2.5 km of stratified high-permeability sand and low-permeability shale (Bethke, 1985) shows complete dewatering by a combination of vertical and lateral flow. Sedimentation rates up to 100 m/m.y. were modeled, and showed that the flow at shallow depth is concentrated upwards in response to relatively high permeabilities in the mudrocks. At depths greater than 1.0 km, flow is

focused laterally to the edge of the basin, with increasing rates of flow towards the basin periphery. At all depths flow was sufficiently slow to allow thermal equilibration, but fast enough to allow no overpressure in the model (Bethke, 1985).

In a section dominated by mudrocks and reservoirs of only restricted lateral extent, the ability to move fluids laterally along carrier beds is lost (Magara, 1974) and overpressures might be expected at depths greater than about 1.0 to 2.0 km (3,300 to 6,600 feet). Mann and Mackenzie (1990) have documented several field examples where pressures in isolated reservoir units show increasing amounts of overpressure with depth in mudrock-dominated sequences. These examples come from Tertiary deltas, such as the Nile and Mississippi, and the Tertiary sediments of the intracratonic North Sea basin. Additional examples include the Malay Basin (Yusoff and Swarbrick, 1994) and the Caspian Sea (Bredehoeft et al., 1988). In each case, disequilibrium compaction commences typically about 1.0 to 2.0 km (3,300–6,600 ft) at the "Fluid Retention Depth" (Figure 6) below which the increase in pressure remains nearly parallel to the lithostatic pressure gradient (and the effective stress remains approximately constant). The depth at which the fluid retention depth occurs is dependent on the sedimentation rate and the rate of permeability reduction during compaction (Mann and Mackenzie, 1990).

When rate of burial slows down or stops, the overpressure generated during rapid burial will dissipate. The rate of dissipation depends primarily on the permeability of the sediments, controlling both vertical and lateral flow to permit pressures to return to equilibrium conditions, i.e., hydrostatic pore pressures. Overpressure will be reduced most rapidly in shallow, permeable units, and most slowly in thick, low-permeability units. Interbeds of contrasting lithologies with high and low-permeability will be differentially drawn down, leading to maximum overpressure in the center of the low-permeability units, e.g., shale, and decreasing amounts of overpressure towards the contact with the high-permeability units (Magara, 1974). The typical shape of the pressure transition zones resulting from disequilibrium compaction (Figure 6) is modified after cessation of burial, and is likely to have a modified transition zone from normally pressured downwards into the overpressure, since the overpressure is preferentially lost from the top (Swarbrick and Osborne, 1996). Luo et al. (1994) describe a thick, overpressured rock unit from the Eastern Delaware Basin, Texas and New Mexico, dominated by low-permeability shales, with the maximum pressures in the core. Pressures in reservoir-dominated units above and below are normally pressured. Although the origin of the overpressure is uncertain, one possibility is that rapid burial 250 Ma created high overpressure by disequilibrium compaction, and that the overpressure today is the residual of a long period of slow dissipation involving an evaporite seal (Castille Formation), possibly augmented by gas generation from deeply buried source rocks (Swarbrick, 1995).

Tectonic (Lateral Compressive Stress)

The same principles of overpressure generation due to compaction and incomplete dewatering apply when reducing pore volume by horizontal tectonic compression. Overpressuring due to lateral stress is reported along major fault zones, both within the fault and in the adjacent porous wall-rock (Byerlee, 1993). Episodic release of overpressured fluids is associated with earthquakes and fault rupture (Byerlee, 1990). In addition, overpressure is a characteristic of accretionary sedimentary prisms, where horizontal compression coupled with loading and underplating is caused by subduction (Davis et al., 1983; Neuzil, 1995). The amounts of horizontal compression are most likely small in passive continental margin and intracratonic areas, but may contribute to overpressure in undeformed sedimentary sequences where the rock neither buckles nor faults. Changes in intra-plate stress over geological time, and their magnitude, are not well documented and are an active research area at the present time (Van Balen and Cloetingh, 1993). High fluid pressures encountered in deep boreholes in the California Coast Ranges are attributed to lateral tectonic compressive forces (Berry, 1973).

Fluid Volume Increase Mechanisms

Several mechanisms have been proposed which can create overpressure, if there is an increase in fluid volume. The main mechanisms proposed include (a) temperature increase (aquathermal expansion), (b) mineral transformation, (c) hydrocarbon generation, and (d) oil to gas cracking.

Temperature Increase (Aquathermal Expansion)

The principle which governs aquathermal expansion as an overpressure mechanism is the thermal expansion of water when heated above 4°C. If the body of water is contained in a **sealed** vessel the pressure rises rapidly. For example, Barker (1972) shows a pressure rise of 8,000 psi (55.1 MPa) in water heated from 54.4° to 93.3°C caused by a volume increase of only 1.65%. This pressure increase over about 1.0–1.5 km would lead to a sharp transition zone at the top of the overpressured section. The critical observation which must be met to satisfy this mechanism is that the environment must be **completely** isolated, with no change in pore volume. Several authors (e.g., Daines, 1982; Luo and Vasseur, 1992) have shown that the conditions for aquathermal pressuring will rarely be met. In particular when water is heated its viscosity reduces and facilitates more rapid expulsion, even at low permeabilities. An additional objection to aquathermal expansion is that in many overpressured rocks there is a gradual transition zone to high amounts of overpressure. This implies permeability and hence the section is

not fully sealed and cannot fulfill the requirements for aquathermal pressuring (Daines, 1982). Numerical modeling of aquathermal expansion (Luo and Vasseur, 1992) shows negligible overpressure development in mudrocks with permeabilities as low as 3 x 10^{-27} m² (3 x 10^{-14} md), permeabilities several orders of magnitude lower that the measured permeabilities of real mudrocks (Deming, 1994).

Aquathermal pressuring is only feasible if a seal with permeability close to zero can be proven to exist. Hunt (1990) suggested that there are diagenetic seals with highly effective sealing properties, typically at depths of about 3.0 km, which are laterally extensive. Although banded cements have been described where the top of the overpressure is observed (Tigert and Al-shaieb, 1990), the origin of these diagenetic seals and their ability to form regional impermeable barriers remains, in our view, open to question.

Mineral Transformation—Water Release Due to Mineral Diagenesis

Several common mineral transformations in sediments involve the release of bound water. The most common of these involves the dehydration of smectite, a multi-layered, mixed-layer clay commonly found in mudrocks. Smectite also transforms to a new mineral, illite, involving the release of water. Other dehydration reactions include gypsum to anhydrite in evaporitic sediments, and coalification (Law et al., 1983).

Smectite Dehydration

Several authors have proposed that smectite dehydration is staged in two (Powers, 1967) or three pulses (Burst, 1969), and that these pulses of released water were instrumental in driving hydrocarbons from source rocks to traps. The overall volume change accompanying the complex smectite-illite reaction is not currently known (Hall, 1993). Our own calculations indicate a total increase in volume of 4.0%, occurring in three pulses of water release. The first two are likely to take place within the top 1.0 km of burial, with only the last pulse at depths where significant amounts of overpressure are measured. However, the volume of water released is only about 1.4%, and will not create significant overpressure unless the rock is completely sealed. Colton-Bradley (1987) has suggested that overpressure would inhibit the dehydration reaction, since the dehydration temperatures are elevated with increasing pore-fluid pressure. The smectite dehydration reaction is therefore thought to be a secondary rather than a major cause of overpressure, but may be additive to overpressure created by disequilibrium compaction.

Gypsum to Anhydrite Dehydration

The temperature-controlled reaction of gypsum transforming to anhydrite results in loss of bound water, and is thought to be an important mechanism to generate overpressure in evaporite sections. The reaction occurs at 40°–60°C at ambient pressure and can potentially generate a fluid pressure significantly in excess of overburden pressure at 1.0 km (Jowett et al., 1993). High fluid pressures in Permian Zechstein carbonate-evaporite sequences in the North Sea and adjacent areas are attributed to this mechanism, as well as in the Buckner Formation of Mississippi, U.S.A. (Mouchet and Mitchell, 1989).

Smectite-Illite Transformation

Clays such as smectite can adsorb water due to an imbalance in their ionic charge. During burial, smectite alters chemically by addition of Al and K ions and the release of Na, Ca, Mg, Fe, and Si ions plus water, to produce illite, which does not have the same capacity to adsorb water. The reaction is kinetically controlled and is dependent on the combined effects of time and temperature, as well as the influence of mineral fabric and permeability (Hall, 1993). In several mud-dominated basins a gradual and systematic change from smectite to illite downwards in the stratigraphic section is observed, broadly coincident with the transition to high amounts of overpressure (Bruce, 1984). The transition occurs over a temperature range of 70°–150°C and appears to be independent of sediment age and burial depth. By contrast, in the highly overpressured Caspian Sea basin, there is no change in smectite to illite ratio to a depth of 6.0 km and temperature of 96°C (Bredehoeft et al., 1988).

The overall volume change involved with the smectite transformation is not well known, in part because the exact chemistry of the reaction(s) is not known. Conversion of one volume of smectite to illite would release 0.36 volumes of water according to the reaction proposed by Hower et al. (1976), leading to an overall volume decrease of about 23% if all the reactions occur in a closed system. Boles and Franks (1979) claim a release of 0.56 volumes of water, which would create a volume increase of about 25%. Hence the origin of overpressure by this mechanism is far from conclusive. However, the coincidence of overpressure at the same stratigraphic levels as smectite to illite transformation may be related to the ensuing changes in the rock fabric, trapping excess fluids generated by another mechanism, e.g., disequilibrium compaction.

The transformation of smectite to illite clay is accompanied by changes in the physical characteristics of the sediments. Firstly, the collapse of the smectite clay framework and release of bound water influences the compressibility of the sediment. If the rock compressibility factor is increased, the overburden induces additional compaction requiring expulsion of the newly released water from the rock to achieve equilibrium. If the low-permeability of the rock acts to retain the fluid, then overpressure will result, i.e., disequilibrium compaction induced by mineral dehydration.

Another consequence of the mineral transformation from smectite to illite is the release of silica. Foster and Custard (1980) proposed that diagenetic silica reduces the permeability of the shales to produce a hydraulic

seal and hence potentially a transition into the overpressured section below. Freed and Peacor (1989), however, have argued that the coincidence of overpressure near the depth of the smectite-illite transformation results from a reduction in the permeability of the shales, not due to silica cementation promoted by the reaction, but by reordering of the illite into packets. Reduction of permeability by mineral reordering would help to retain fluids and hence contribute to the preservation of overpressure, but would not be responsible for its initiation. Silica cementation could involve some volume change but the rate is likely to be too slow to create overpressure.

Hydrocarbon Generation

Generation of liquid and gaseous hydrocarbons from kerogen maturation is kinetically controlled and dependent on a combination of time and temperature. The kinetics have been broadly described for each of the main kerogen Types I, II, II-Sulfur, and III; but the kinetics of each of the many individual reactions involved is not well known (Tissot et al., 1987). Each of

(a) *After Meissner, 1978b*

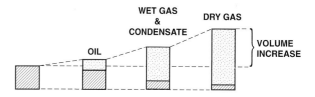

(b) *After Ungerer et al. 1983*

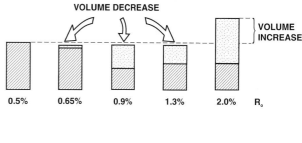

Figure 7. (a) Estimation of volume change when Type II kerogen in the Bakken shale, Williston Basin, matures to produce oil, then wet gas and condensate, and finally dry gas (Meissner, 1978b). Note the increase in volume at all stages of thermal maturity of the kerogen. (b) Estimation of volume change when Type II kerogen in the Toarcian Black shale, Paris Basin, France, matures to oil, then gas (Ungerer et al., 1983). Early oil maturity corresponds to 0.65% R_o (vitrinite reflectance) and peak gas generation occurs at a thermal maturity of about 2.0% R_o. Note the volume decrease during oil generation, and volume increase during gas generation.

the compositional changes during kerogen maturation has implications for the total volume of the reactant and products with potential for both volume increases and decreases. The two main reactions involved with the generation of oil and gas from petroleum source rocks are: (1) kerogen maturation to produce oil and/or gas, and (2) oil and bitumen cracking to gas. These reactions typically occur at depths of 2.0 to 4.0 km and at temperatures in the range 70°–120°C for kerogen maturation (Tissot et al., 1987), and 3.0–5.5 km and 90°–150°C for oil cracking to gas (Barker, 1990).

Kerogen Maturation—Oil Generation

There are two parallel and generally coeval processes involving volume change that take place during kerogen maturation/hydrocarbon generation. First is the creation of mobile generated fluids (mainly oil and hydrocarbon gases, but also CO_2 and water) from an original solid immobile kerogen; and second is the creation of porosity volumes that are not in equilibrium with normal overburden compaction until fluids have been expelled.

The coincidence of overpressure and hydrocarbon generation was given early prominence by the study of the Type II Bakken shale in the Williston Basin, Montana and North Dakota, U.S.A. (Meissner, 1978a, b). The abnormal pressure was attributed by Meissner (1978b) to two processes: (1) increased volume of hydrocarbons and residue relative to unaltered organic material, and (2) inhibited structural collapse of the rock framework as overburden-supporting solid organic matter is converted to hydrocarbon pore fluid. The proposed increase in volume has since received more attention in the literature as a possible mechanism, rather than the collapse of the rock framework. The fluid volume increase was estimated by Meissner (1978a) at about 25%, with even greater increases in volume when maturation proceeds from oil to wet gas, and later to dry gas (Figure 7a). Burrus et al. (1996; 1998-this volume) remodeled the overpressure and maturation histories of the Bakken shales in the Williston Basin using TEMISPACK, a 2D basin model. Their model produces a close match between the overpressure recorded in the Bakken siltstones within the organic-rich shales and volume change during oil generation, although it is not clear how the overpressure is created within their model. Their analysis shows no potential contribution from disequilibrium compaction generated by the burial history of the rocks in this part of the succession.

Spencer (1987) extended the link between overpressure and volume increase during oil generation to most of the deeper parts of the Rocky Mountain basins. Bredehoeft et al., (1994) concluded that maturation of the Type I Green River shales can account for the high overpressures in the Altamont-Bluebell field, Uinta Basin, Utah. Sweeney et al., (1995) model volume increases from organic maturation during oil generation, leading to 25% of the total overpressure found in

the La Luna Formation source rocks in the Maracaibo Basin, Venezuela. The remainder of the overpressure is due to disequilibrium compaction within the maturing source rocks.

In our view, there is considerable uncertainty about the volume change (i.e., comparison between volume of starting reactants compared with total volume of products and residue) when kerogen matures to oil, plus associated gas, residue and by-products (mainly CO_2, H_2S and H_2O), at least at the maturation stage up to $R_o = 1.2\%$. In contrast to Meissner's (1978a) volume increase, Ungerer et al., (1983) show a small volume decrease when modeling the composition changes from kerogen maturation (up to $R_o = 1.3\%$) in Type II Toarcian black shale in the Paris Basin, France (Figure 7b). Part of the explanation for differences in calculated volume change relate to assumptions about the gases generated (e.g., CO_2 and H_2S) and to the volume and density of the residual kerogen/coke. We consider that it is premature to assume that there is an overall fluid volume increase in all cases of kerogen maturation to oil, and additional research (included in the GeoPOP project, see Acknowledgements) is needed to resolve the volumetric change with burial and maturation for the range of kerogens found in petroliferous source rocks.

It is generally accepted that high pressures are necessary to drive expulsion (primary migration) of petroleum from low-permeability source rocks into carrier beds (England et al., 1987, who believe that volume change during oil maturation is negligible). The pore pressures in the source rocks must be sufficient to force the oil and gas out of the micropores and/or to initiate microfractures to release the petroleum (Palciauskas and Domenico, 1980). If there is no fluid volume change during the early maturation of source rocks to oil, how do hydrocarbons migrate out of these source rocks? Can primary migration be achieved without volume change? The second main change during kerogen maturation, the creation of additional pore volume, may provide an answer.

Kerogen maturation involves increased pore volume as a consequence of transforming solid immobile kerogen into mobile fluids. Reduction of the solid fraction as the kerogen transforms to liquids, gases and residue, alters the distribution of the overburden stress between solid rock and pore fluids. For example, take a source rock with a porosity of 13% and a further 10% by volume of kerogen, immediately prior to maturation. If half of the kerogen by volume is transformed to liquid hydrocarbons the total porosity is increased from 13% to 18% as the solid fraction is reduced from 87% to 82% of the original volume, assuming no loss in hydrocarbons due to primary migration. Where the kerogen sustains part of the overburden stress, that stress will be transferred to the liquid phase. If the liquid cannot escape, there will be an increase in pore pressure. This change is illustrated in Figure 8, where the amount of overpressure is related to the effective stress vs. porosity relationship for the modified sediment (i.e., the disequilibrium compaction when the solid:liquid ratio goes down, and the fluids cannot escape). In the example the kerogen is instantaneously transformed at Point A into liquids. The pressure increases by 1,100 psi (Point B), a maximum determined by the effective stress which is in equilibrium with the rock with a porosity of 18% (i.e., a burial depth at Point C). The calculation of overpressure in Figure 8 assumes an even distribution of kerogen throughout the rock, a maximum amount of overpressure, and no expulsion of fluids. In practice, the magnitude of overpressure resulting from this mechanism will vary throughout the source rock depending on the distribution of the original kerogen–lean laminae will experience little change, whereas almost continuous organic-rich laminae could experience pressures as high as fracture pressure (e.g., >85% lithostatic), if almost all of the overburden is supported by the fluid phase. The richer the source rock, the greater the increase in average pressure, assuming the same proportion of kerogen is transformed. We believe this mechanism may be largely responsible for primary migration, but further research is required.

In any given basin, the distribution of overpressure created by this process is controlled by the availability of maturing source rocks within the stratigraphic succession, and we are doubtful that this mechanism will create regional overpressure, except under circumstances of close linkage between source rocks and reservoirs. The overpressures in the Bakken Shale section, measured within siltstones enveloped within maturing organic-rich shales (Burrus et al., 1993; 1996), may be an example of local overpressure sourced by this mechanism, and where there has been minimal escape of hydrocarbons.

Kerogen Maturation—Gas Generation

The maturation of a **gas-prone** source rock (e.g., Type III kerogen) results in significant increase in fluid volume during maturation. Law (1984) observed overpressures in low-permeability Upper Cretaceous and Tertiary coal and carbonaceous sediments, which he attributes to volume change during thermogenic gas generation from these Type III source rocks. Both Meissner (1978a, b) and Ungerer et al. (1983) calculate large volume expansion (between 50% and 100%) relative to the initial volume of the kerogen, when a Type II source rock reaches higher levels of thermal maturity (e.g., $R_o = 2.0$) at which the main hydrocarbon products are gas (Figure 7). Hence where source rocks are generating significant amounts of gaseous hydrocarbons, and especially where the hydrocarbon phase is either gas condensate or dry gas, there is the potential for overpressure to result from the large volume changes which occur. The amounts of volume change, their rates, and the influence of the changing pressure conditions remain to be determined. However, the distribution of overpressure resulting from this mecha-

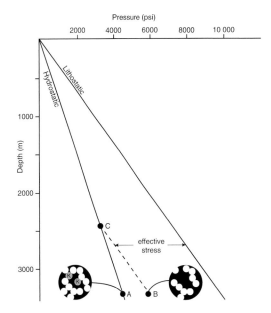

Figure 8. Estimation of the overpressure created by disequilibrium compaction in a maturing source rock. Half of the initial kerogen (10% by volume) is transformed into liquid products, thereby increasing the porosity from 13 to 18%. The effect is to transfer the part of the overburden supported by the original kerogen onto the pore fluid.

nism alone will reflect the depth and temperatures at which the necessary maturation levels are reached, and the location of source rocks in which the volume change is taking place (Swarbrick and Osborne, 1996).

We do not know how a build-up of overpressure during gas generation influences the reaction rates which govern kerogen maturation. It is conceivable that although the transformation of kerogen to hydrocarbons is a kinetically controlled reaction, the build-up of high pressures may act to retard the reaction (Price and Wenger, 1992). Fang et al., (1995) report anomalous immature source rocks in overpressured basins in China, and similar data from Siberian oil and gas basins are reported by Neruchev and Gildeeva (1994). Regardless of any volume change, the presence of oil and gas as separate phases (from water), especially within fine-grained rocks including source rocks, will severely reduce the effective permeability of the rocks. This phenomenon will be very important to the retention capacity of these rocks and their ability to seal excess fluids created by any of the overpressure mechanisms. For example, gas in the Greater Green River Basin, Wyoming, Colorado and Utah, originally generated from Type III kerogen in coals and carbonaceous source rocks, and now in tight overpressured reservoirs, has been trapped by reduced permeability of the rock to gas (Law, 1984).

Oil and Bitumen to Gas Cracking

At high temperatures oil converts to lighter hydrocarbons and ultimately to methane. Thermal cracking is initiated generally at temperatures of 120°–140°C with almost complete cracking to gaseous hydrocarbons (mainly methane) at temperatures in excess of 180°C (Mackenzie and Quigley, 1988). At standard temperatures and pressures, one volume of standard crude oil cracks to 534.3 volumes of gas, plus a small volume of graphite residue. This observation led Barker (1990) to suggest that when the system is effectively isolated (i.e., has a perfect seal) there is an immediate and dramatic increase in pressure as oil cracks to gas. In fact his calculations show that only 1% cracking of oil is required for the pressure to reach lithostatic and thereafter further cracking should lead to fracturing and subsequent leakage. A much smaller volume increase is observed at typical depths of burial (3.0 to 5.0 km), where the compressibility of gas must be considered as

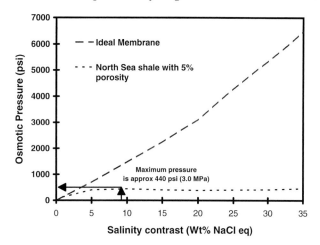

Figure 9. Graph illustrating the osmotic pressure created by a typical North Sea shale, with 5% porosity and 55% clay content, acting as a semi-permeable membrane for a range of salinity contrasts up to 35 Wt.% NaCl eq. Note that pressure reaches a maximum at about 10 Wt.% NaCl eq., and the amount of pressure is much lower than for the theoretical values for an ideal membrane, calculated using the osmotic efficiency equations in Fritz (1986).

well as its solubility in brine. Large volume increases are still likely, however, with the potential for generating overpressure.

The effect of increased pressure as an influence on the rate of reaction of oil to gas cracking is poorly known. Low temperature pyrolysis experiments at increasing pressures (Domine and Enguehard, 1992) show that rate of cracking of C_{20} to C_{5-3} decreased with increasing pressure, as expected from Le Chatelier's principle. As stated earlier, the presence of gas as a separate phase within the fluid will reduce the permeability of the rocks and contribute to their retention capacity and ability to maintain overpressure generated by another mechanism.

There are several basins where the distribution of overpressure is coincident with the deeper parts of the basin where oil to gas cracking is assumed to be occurring. Law et al. (1980) note the active generation of

large amounts of wet gas and the development of overpressure in Upper Cretaceous rocks in the northern Green River Basin. Hunt et al. (1994) observe a strong coincidence of the top of overpressure and peak gas generation in the Gulf of Mexico. Other examples include Jurassic and Triassic reservoirs in the Northern and Central North Sea, the Jurassic Smackover reservoir of Mississippi/Alabama, Lower Pennsylvanian of the Anadarko Basin, Oklahoma, and the Lower Cretaceous of the Powder River Basin, Wyoming (Meissner, personal communication). In the North Sea the highest overpressures are found where the Kimmeridge Clay is most deeply buried and presently mature for gas (Cayley, 1987; Holm, 1998-this volume), but the Jurassic and Triassic reservoirs are everywhere overpressured (Gaarenstroom et al., 1993), even at depths well above active source rock maturation. The origin of the overpressure here remains controversial.

Fluid Movement and Buoyancy Mechanisms

Osmosis

Large contrasts in the brine concentrations of formation fluids, from dilute to saltier water, across a semipermeable membrane can induce transfer of fluids across the membrane. Marine and Fritz (1981) suggested that osmotic pressure could be an explanation for some overpressured sections. We have examined the osmotic behavior of a typical North Sea shale composition in contrast with near perfect membrane behavior modeled elsewhere (see Fritz, 1986). Our calculations (shown in Figure 9) indicate osmotic pressure in typical North Sea rocks is only about 3.0 MPa (435 psi), even with salinity contrasts as high as 35 Wt% NaCl equivalent. If a shale contains microfractures osmosis is impossible.

A further argument against this mechanism having anything but very local importance is the requirement for recharge of the more saline, and discharge of the originally less saline, waters to maintain the pressure. In practice we believe this is unlikely to happen during normal compaction processes. In addition, it has been observed that brines in overpressured zones tend to be of a lower salinity than adjacent normally pressured brine, which would act to reduce the pressure in the overpressured zones.

Hydraulic Head

The hydraulic or potentiometric head resulting from elevation of the water table in highland regions (charge areas) exerts a pressure in the subsurface if the reservoir/aquifer is overlain by a seal (Figure 10). Wells drilled into the overpressured aquifer are known as artesian wells and will produce water flow at the surface due to the excess pressure. The potentiometric head is measured either as the vertical height of the water rise above datum (the practice in hydrogeology) or as the height converted to pressure with knowledge

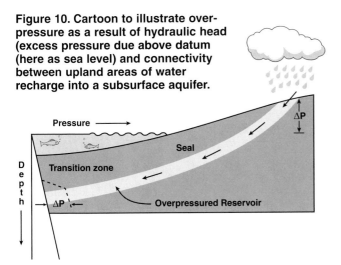

of the formation fluid density. Neuzil (1995) refers to this situation as "equilibrium overpressure" in recognition of the direct relationship between elevation of water table above a reference datum and the amount of overpressure in the sediments. Lateral continuity of reservoirs beneath a continuous seal is required for this mechanism to operate. In many interior foreland basins (e.g., Alberta Basin; Bachu, 1995) there is scope for the generation of significant amounts of overpressure by this mechanism. The amount of overpressure, measured as a potentiometric head (Figure 10), cannot exceed the height of the elevated water table above the potentiometric surface, but in many basins the pressures measured in the subsurface far exceed this value, and an alternative explanation is required.

Hydrocarbon Buoyancy (Density Contrasts)

All gases and most oils have a lower density than the associated formation waters and hence have a lower pressure gradient. Since overpressure is the excess pressure above hydrostatic for a given depth, there is always some amount of overpressure wherever a column of oil or gas is present. This mechanism is restricted to structural and stratigraphic traps of hydrocarbons, and cannot cause regional overpressure. The amount of overpressure within the hydrocarbon accumulation is a function of the pressure gradients of oil, gas and water and the height of the hydrocarbon column (Figure 11). The excess pressure increases from the water contact upwards, and is calculated by multiplying the column of any one fluid by the difference in pressure gradient between the overlying hydrocarbon and underlying water. In the North Sea the maximum overpressure attributed to this mechanism is only 6.0 MPa (600 psi), calculated by adding the effect of the longest oil and gas columns together. In practice, thinner hydrocarbon columns lead to the overpressure due to buoyancy in most fields being less than 1.0 MPa (145 psi). Buoyancy-driven excess pressure is not regarded as "abnormal". The overpressure created by this mechanism is in addition to any regional overpressure in the

water zone, but the additional overpressure may be sufficient to influence the sealing capability of the caprock seals.

There is another potential link between overpressure and hydrocarbon buoyancy. When gas rises and decreases in temperature, the volume of the gas increases and its density decreases. However, in a totally sealed container (e.g., in a drill pipe), gas is unable to expand because of the incompressibility of the surrounding fluid and the pipe, and consequently the pressure of the gas and the fluid rises. The potential for this mechanism to increase the fluid pressure in natural rock systems has not been fully evaluated. In particular, the permeability requirements are not known, and conditions for the generation of free gas are dependent on pressure as well as temperature. If the mechanism did create increases of pressure, the gas, existing as a separate phase, may be forced back into solution at elevated pressure, such that the system may be self-limiting, and resultant overpressures are therefore difficult to predict. In addition, the creation of pressure will depend on the compressibility of the fluids and the rocks (generally considered small).

TRANSFERENCE OF PRESSURE

Transference is the redistribution of excess pore pressure in the subsurface. Although not a primary mechanism to create overpressure within a sedimenta-

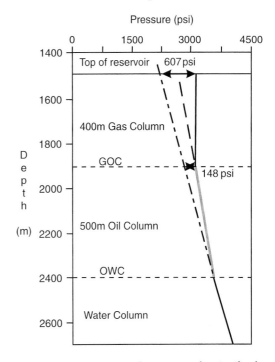

Figure 11. Hydrocarbon buoyancy due to the lower densities of light-medium oils and gases relative to water leads to overpressure, illustrated by a long gas-cap above a medium-light oil, based on data from North Sea fields.

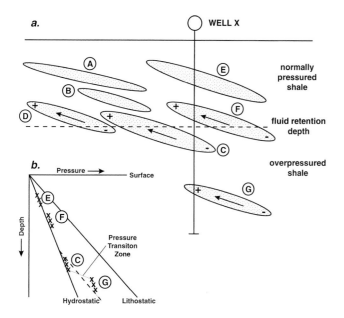

Figure 12. a) Schematic cross-section of a series of iso-lated sandbodies, encased in shales, and tilted uniform-ly. The fluid retention depth (see Figure 6) is indicated by a dashed line. The arrows indicate pressure transfer-ence to equilibrate higher shale pressures at the base, and lower pressures at the crest of the sandbodies. b) Pressure vs. Depth plot for Well X drilled through sand-bodies C, E, F and G. Transference is taking place in all sandbodies at or below the fluid retention depth.

ry basin, transference can be the principal control on the distribution of overpressure found there. Fluid movement is driven by differences in excess pressure and controlled by the permeability of the rocks. High-permeability reservoir sections will be the most effec-tive rocks in redistributing excess pore pressure when it is created, whereas long periods of geological time

will be necessary to equilibrate the overpressure pro-files of low-permeability rocks. The effects of transfer-ence are illustrated in Figure 12, in which the only mechanism creating overpressure is disequilibrium compaction, in the shales below the fluid retention depth. There are a series of tilted sandbodies, A through G, of which four (C, E, F, and G) have been penetrated by Well X. Sand bodies A, B, and E are enveloped within normally pressured shales, and pres-sures within the sandbodies are normal, confirmed by the pressure data from E (Figure 12b). Well X has drilled sandbodies F and G in their updip position, such that the pressures within them are influenced by higher pressures deeper down (since overpressure increases downwards beneath the fluid retention depth; Figure 6). Pressure data from F show a small amount of overpressure at the well location, despite its position above the fluid retention depth, due to trans-ference. The pressures in G are also elevated by trans-ference, above the pressures predicted by disequilibri-um compaction in the shales (Figure 12b). Finally, the pressures in C are lower than anticipated, since Well X penetrated this sand in a downdip position. Note that all the pressures within each sandbody show a pres-sure gradient for water, i.e., parallel to hydrostatic. However, a prominent pressure transition zone (Swar-brick and Osborne, 1996) exists between sandbodies C and G though the shales (Figure 12b).

Examples of overpressure distribution strongly influenced by transference include the Paleocene sand-stone reservoir of the North Sea and the Miocene sedi-ments of the Mahakam Delta, eastern Kalimantan, Indonesia. The Central North Sea Paleocene sand-stones were deposited as an extensive axially-located, turbidite-dominated submarine fan extending laterally into silts and muds, and overlain by up to 3.0 km of fine silts and muds. The high sedimentation rate and

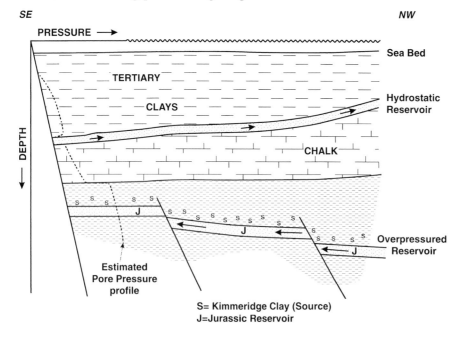

Figure 13. Schematic cross-section from the Central North Sea, with a typical pressure vs. depth plot on the left. Tertiary clays are overpressured (see Figure 5) above a regionally extensive, subcropping, Paleocene reservoir. Fluid escape allows equilibration to hydrostatic pressure in the reservoir due to trans-ference. In the deeper Jurassic shales trans-ference of pressure leads to higher than expected pressures at the crest of each confined fault block.

the fine-grained nature of the post-Paleocene sediments have produced overpressure in the shales, except in the top 1.0 km or so (Ward et al., 1994). The mechanism for overpressure generation is, in this case, disequilibrium compaction. The Paleocene sands beneath are only overpressured at the extreme limits of the sand sheet, where the sandstones are isolated within shales. Elsewhere, there is active transference of excess pore pressures from the basin up the regional slope to the northwest (Figure 13), and eventually to near the sea-floor in the Moray Firth area (Cayley, 1987), a distance of about 250 kilometers. Pressures are at, or close to, hydrostatic wherever the sand sheet is continuous. In the Miocene sediments of the Mahakam Delta there is hydrological continuity from the delta slope to the delta top, with transference from the overpressured mud-dominated section in the east to the sand-dominated, normally pressured section to the west (Burrus et al., 1994; Grosjean et al., 1994).

Transference can also take place vertically, most frequently associated with active faulting (Burley et al., 1989). Leakage of fluids due to fracturing at the crest of structures is an effective way of transferring the excess pore pressures from deeper in a basin to shallower levels. Examples documented in the literature include a Southeast Asia basin where hydrofractures are believed to be permitting upward transfer of hot fluids to shallower reservoirs (Grauls and Cassignol, 1993), and the Gullfaks field, whose pressures are close to the fracture pressure, where a gas chimney is located above the crest of the shallowest of a number of pressure-connected oil fields (Caillet, 1993).

MECHANISMS FOR CREATING UNDERPRESSURE

Observations

From published studies of underpressured rocks it is possible to make some generalizations about the geological settings in which underpressuring occurs. Underpressure commonly occurs in relatively shallowly buried (0.6–3.0 km) permeable rocks which are frequently isolated within, or interbedded with, low-permeability mudrock sections (e.g., Dickey and Cox, 1977). In many instances underpressured rocks have also been uplifted in the geological past (e.g., Bachu and Underschultz, 1995).

A variety of mechanisms have been proposed for the creation of underpressure (Figure 14) and these will be summarized below. The controlling parameters influencing pressure in an underpressured rock remain the same as for an overpressured rock—i.e., the magnitude of the pressure depends on the mechanism, the permeability related to seal integrity of the rocks, the rate at which the mechanism operates and the timing relative to onset/completion of the mechanism, and fluid type.

Mechanisms

Differential Discharge—Groundwater Flow

Underpressure can occur in a topographically-driven flow system where there are very low-permeability rocks in the recharge area, but high permeability in the outflow area (Belitz and Bredehoeft, 1988) (Figure 14). For example, high-permeability Mesozoic and Paleozoic rocks in the Denver basin, U.S.A., are being recharged from the Rocky Mountain uplift to the west, but are discharging fluids along the eastern side of the basin. However, the permeable rocks are isolated from their meteoric recharge area by thick, low-permeability shales. This means that the rate of fluid discharge from the high-permeability units is greater than the rate of recharge through the low-permeability units (Belitz and Bredehoeft, 1988). As more fluid is leaving the permeable units than entering them, there is no continuous fluid column through the rocks from their highest to lowest elevation. This means that pore pressures are less than hydrostatic values, and in the Denver basin, permeable aquifers are often 1,000–1,500 psi (7–10 MPa) underpressured at 0.6–3.0 km depth of burial (Belitz and Bredehoeft, 1988). Underpressuring due to steady-state regional ground water flow is possible in any subaerial, topographically tilted basin which is capped by a thick sequence of low-permeability rocks. The mechanism may also operate where low-permeability barriers exist in the subsurface, effectively disconnecting a high-permeability rock in the deep basin, from its subaerially exposed counterpart at a higher topographic level (Belitz and Bredehoeft, 1988).

Differential Gas Flow

A model for differential hydrocarbon flow was used by Law and Dickinson (1985) as partial explanation for the underpressures observed in several uplifted basins in North America, including the San Juan, Piceance, Appalachian, and Alberta Basins. During uplift gas in overpressured, saturated reservoirs comes out of solution as the temperature and confining pressure are reduced. This exsolved gas migrates out of the lower permeability reservoirs at a greater rate than continued gas generation in the source rock. The imbalance between gas migration and generation leads to a reduction in the overpressuring (Figure 14) and, depending on the magnitude of the temperature reduction and gas loss from the system, pressures below regional hydrostatic may result (Law and Dickinson, 1985).

Rock Dilatancy

During erosion of a shallow-buried, clay-rich lithology, dilation of the pores can occur (Figure 14). The increase in pore volume may facilitate the dissipation of overpressure, and possibly produce underpressure, depending on the amount of dilation, rate of removal, and the permeability of the rock (Luo and Vasseur, 1995; Neuzil and Pollock, 1983). Permeable units may

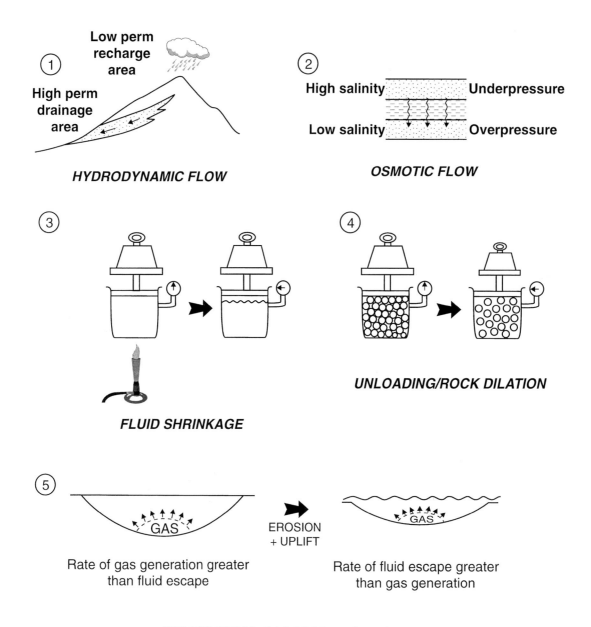

Figure 14. Summary diagram for the major mechanisms thought to be responsible for generating underpressure. For an explanation of each see text.

also become underpressured, if they are isolated within mudrocks which are undergoing dilation, because water is drawn from the sandstones into the shales in response to the expansion in the shale pore volume. Bachu and Underschultz (1995) describe the process as being "analogous to water suction by a previously squeezed sponge."

The amount of dilation following erosion is at present uncertain, because the rheology of various mudrocks is poorly known. Neuzil and Pollock (1983) assumed that mechanical compaction was reversible, hence in their models, rock dilatancy has a major effect on pore pressure. However, this may not be a realistic assumption in all instances, because many experimen-

tal studies indicate that compaction is nearly irreversible (e.g., Karig and Hou, 1992). There is a further problem in that laboratory studies are necessarily of short duration, hence slow unloading processes, such as clay swelling, which occur over thousands of years, may not be reproduced on a laboratory time scale (Peterson, 1958). Such clay swelling is most likely in rocks which are rich in smectitic clays such as montmorillonite. The amount of smectite present in mudrocks generally decreases with depth of burial, due to the conversion of smectite to illite and other minerals (Perry and Hower, 1972). Hence rocks which have been buried to temperatures greater than 70°–100°C will often contain less smectite and be less prone to swelling upon uplift. In addition, deeply buried mudrocks are generally more compacted and cemented, which will again reduce the amount of dilatancy during uplift. Therefore, it seems likely that the effect of pore dilatancy will be most important in smectitic mudrocks which have never been deeply buried (<2.0 km), and which have been uplifted.

Underpressuring due to unloading is also critically dependent upon the sediment permeability. When the mudrock has a moderate permeability and is slightly overpressured before uplift, unloading may result in sub-hydrostatic pressures of up to 600 psi (4 MPa) (Luo and Vasseur, 1995). When mudrock permeabilities are extremely low and the rock is highly overpressured before uplift, rapid unloading will decrease pressures along a path similar to that of the overburden load, and there will be no underpressure (Luo and Vasseur, 1995).

Computational modeling suggests that unloading is a major cause of abnormal pressure in the Alberta Basin of western Canada, in which reservoirs/aquifers are about 440 psi (3 MPa) underpressured and where the uplift can be as much as 800 m in the Cenozoic (Corbet and Bethke, 1992). Pore dilation may also help to explain underpressure in lenticular sandstones of the Mid-Continent of the U.S.A., which are isolated within mudrocks which have been uplifted (Dickey and Cox, 1977). These sandbodies are 500–700 psi (3.4–4.8 MPa) underpressured, are oil and/or gas bearing, but contain virtually no water. This lack of water can be explained if water (but not hydrocarbons) has been drawn into the surrounding shales due to pore dilation (Dickey and Cox, 1977). The lack of hydrocarbon movement may have been due to capillary effects.

Osmosis

As outlined earlier, osmotic pressure can be generated across a shale membrane if fluids of differing salinity exist in formations on either side of the shale (Figure 14). Water will flow from the high salinity side of the membrane to the low salinity side, while salts will tend to be excluded. On the high salinity side of the membrane, underpressures could be produced due to loss of water, while on the low salinity side of the membrane, overpressure could be produced due to increase in fluid volume (Hitchon, 1969). However, most shales are not efficient membranes, because their porosities are too high (>5%) and their cation exchange capacities are generally too low (<30 meq/100g). Hence the amount of underpressure generated by osmosis is likely to be small (<400 psi, <3 MPa). As shales are non-ideal membranes, some ions do pass across the membrane, though in the opposite direction to the movement of water. This means that the salinity contrast will slowly equilibrate due to movement of ions and fluid across the membrane, thus the osmotic potential of the cell will ultimately diminish through time. Underpressure is normally found in fine-grained sediments which are shallowly buried, but at shallow depths the porosity of the shales will be too high for osmotic behavior. In addition, in most underpressured basins, the salinity contrasts required for extensive osmotic flow are not present (e.g., Dickey and Cox, 1977).

Thermal Effects

If water in a completely sealed container is cooled, the fluid will decrease in density. In a completely isolated container, this will result in a reduction in fluid volume, producing underpressuring if the fluid pressure was initially hydrostatic (Figure 14). This mechanism is thought to generate underpressure in reservoirs which possess good lateral and vertical seals (Barker, 1972). The objections we have raised against water expansion as a mechanism to generate overpressures, also apply to this reverse process of water shrinkage. The process will only be an effective cause of abnormal pressure if the rock is perfectly sealed, because the change in water volume during shrinkage is extremely small (<2%). For example, computational modeling of pore fluid cooling during uplift of the Western Canadian Basin, indicated that where the pore fluid is water, fluid shrinkage is a negligible cause of underpressure (<5% of the magnitude) (Corbet and Bethke, 1992). The majority of the underpressure is produced by the effect of rebound and dilation of pores (Corbet and Bethke, 1992).

The volume change in hydrocarbons, due to temperature reduction during uplift, may be greater than in water, due to the higher compressibility of both oil and gas. In most basins the distribution of hydrocarbons is restricted to traps, source beds, and migration pathways, and is volumetrically small in comparison with water-bearing strata. The regional effect on pore pressures is likely to be small. However, in the San Juan and Piceance Basins a pervasive gas phase is reported in the deep, axial portions of uplifted basins, where low-permeability reservoirs are both overpressured and underpressured (Law and Dickinson, 1985; Surdam et al., 1994). The contribution to the underpressuring from volume reduction in hydrocarbons due to temperature reduction requires evaluation, particularly with respect to the magnitude of expansion related to reduced confining pressure. However, despite this uncertainty, it is clear that many instances of underpressure recorded in the literature can be adequately explained by mechanisms other than fluid expansion/contraction. As in the case of aquathermal pressuring, mechanical effects are likely to be far more important than thermal effects in generating abnormal pressure.

CONCLUSIONS

1. There are several mechanisms capable of creating large overpressures in sedimentary basins. The most likely mechanisms for most basin settings, however, are disequilibrium compaction due to rapid loading in fine-grained sediments, and the volume expansion associated with gas generation.

2. Disequilibrium compaction dominates as an active mechanism in young basins experiencing rapid sedimentation, including Tertiary deltas (e.g., Baram, Nile, Niger, Mississippi, Mackenzie), and young, intra-cratonic basins (Caspian Sea, North Sea, Malay, South China Sea). The magnitude of overpressure generated by disequilibrium compaction is limited to the additional overburden stress. Higher amounts of overpressure in these basins is most likely caused by transference, especially where there are laterally continuous but confined reservoirs with considerable vertical relief, or by some other mechanism.

3. Disequilibrium compaction as a consequence of mineral dehydration and transformation may create overpressure in rocks during these alterations. Where part of the rock matrix is converted into free fluid (for example, during kerogen maturation or smectite dehydration), the pore fluid will tend to assume that portion of the overburden previously carried by the rock matrix.

4. Gas generation seems to be a likely mechanism to create regional overpressure, although the volume changes are not yet determined. Both gas generation from gas-prone source rocks and oil to gas cracking are responsible. Possible examples of overpressure due to fluid volume increase with gas generation include the deeper reservoirs of the North Sea, Gulf of Mexico, and the Anadarko Basin. The amount of overpressure generated could be considerable, due to the large potential increases in volume. There is uncertainty about the influence of increasing pressure on the reaction rate, however. Relative permeability effects due to two or three fluid phases may also be important.

5. Other mechanisms which rely on increased volume change are unlikely to create significant overpressures. These mechanisms include smectite to illite transformation, oil generation, and gypsum to anhydrite and smectite dehydration reactions. However, the changes in the physical properties of the rocks undergoing these changes may be significant in producing local overpressure by disequilibrium compaction as the rock:fluid ratio is modified in each case. This local effect may be additive to other synchronous mechanisms, e.g., disequilibrium compaction.

6. Osmosis can only create minor amounts of overpressure in favorable conditions. Shales do not act as ideal membranes and the amount of overpressure is limited, even at high salinity contrasts.

7. Overpressure in oil and gas accumulations due to hydrocarbon buoyancy can be assessed if the densities of the fluids are known. Similarly if the potentiometric (hydraulic) head of water in an upland area is known, its impact in generating overpressure in subsurface aquifers can be calculated. Neither of these mechanisms generates sufficient overpressure to explain high overpressures in many basins.

8. Modeling overpressure is dependent on a full understanding of the geological setting of the basin, and the overpressure mechanism, plus the permeabilities of the fine-grained rocks in which it is created, and retained (by seals). Further research is required to determine the appropriate permeability values to assign to such rocks.

9. Underpressure as a geological phenomenon is less well known than overpressure. Underpressured reservoirs have been recognized mainly in low-permeability, gas-charged reservoirs in uplifted basins. The mechanisms for generating naturally underpressured reservoirs relate to changes in the rock and fluid properties during erosional uplift, e.g., reducing temperature, rock dilatancy, gas solubility and groundwater discharge.

ACKNOWLEDGEMENTS *The authors wish to thank the sponsors of the Overpressure Research Project (GeoPOP) at the Universities of Durham, Newcastle, and Heriot-Watt for their support: Amerada Hess, Agip, Amoco, ARCO, Chevron, Conoco, Elf, Enterprise, Mobil, Norsk Hydro, Phillips Petroleum Company Ltd., Statoil and Total. The authors accept responsibility for the content of the paper and opinions expressed. The manuscript has benefited from reviews by Neil Goulty, Ben Law and Fred Meissner.*

REFERENCES CITED

Aplin, A.C., Y. Yang, and S. Hansen, 1995, Assessment of β, the compression coefficient of mudstones and its relationship with detailed lithology: Marine and Petroleum Geology, v. 12, p. 955–963.

Audet, D.M., and J.D.C. McConnell, 1992, Forward modeling of porosity and pore pressure evolution in sedimentary basins: Basin Research, v. 4, p. 147–162.

Bachu, S., 1995, Synthesis and model of formation-water flow, Alberta Basin, Canada: AAPG Bulletin, v. 79, p. 1159–1178.

Bachu, S., and J.R. Underschultz, 1995, Large-scale underpressuring in the Mississippian-Cretaceous succession, Southwestern Alberta Basin: AAPG Bulletin, v. 79, p. 989–1004.

Barker, C., 1972, Aquathermal pressuring - role of temperature in development of abnormal pressure zones: AAPG Bulletin, v. 56, p. 2068–2071.

Barker, C., 1990, Calculated volume and pressure changes during the thermal cracking of oil to gas in reservoirs: AAPG Bulletin, v. 74, p. 1254–1261.

Belitz, K., and J.D. Bredehoeft, 1988, Hydrodynamics of the Denver basin: an explanation of subnormal pressures: AAPG Bulletin, v. 72, p. 1334–1359.

Berry, F.A.F., 1959, Hydrodynamics and geochemistry of the Jurassic and Cretaceous Systems of the San Juan basin, northwestern New Mexico and southwestern Colorado: Ph.D. dissertation., Stanford Univ., 192p.

Berry, F.A.F., 1973, High fluid potentials in California Coast Ranges and their tectonic significance: AAPG Bulletin, v. 57, p. 1219–1245.

Bethke, C.M., 1985, A numerical model of compaction-driven groundwater flow and heat transfer and its application to the paleohydrology of intracratonic sedimentary basins: Journal of Geophysical Research, v. 80, p. 6817–6828.

Boles, J.R., and S.G. Franks, 1979, Clay diagenesis in the Wilcox sandstones of southwest Texas: Implications of smectite diagenesis on sandstone cementation: Journal of Sedimentary Petrology, v. 49, p. 55–70.

Bredehoeft, J. D., and B.B. Hanshaw, 1968, On the maintenance of anomalous fluid pressures, I: Thick sedimentary sequences: Geological Society of America Bulletin, v. 79, p. 1097–1106.

Bredehoeft, J. D., R. D. Djevanshir, and K. R. Belitz, 1988, Lateral fluid flow in a compacting sand - shale sequence, South Caspian Sea: AAPG Bulletin, v. 72, p. 416–424.

Bredehoeft, J. D., J.B. Wesley, and T.D. Fouch, 1994, Simulations of the origin of fluid pressure, fracture generation and the movement of fluids in the Uinta Basin, Utah: AAPG Bulletin, v. 78, p. 1729–1747.

Bruce, C. H., 1984, Smectite dehydration - its relation to structural development and hydrocarbon accumulation in northern Gulf of Mexico basin: AAPG Bulletin, v. 68, p. 673–683.

Burley, S.D., J. Mullis, and A. Matter, 1989, Timing diagenesis in the Tartan reservoir (U.K. North Sea) - constraints from cathodoluminescence microscopy and fluid inclusion studies: Marine and Petroleum Geology, v. 6, p. 98–120.

Burrus, J., E. Brosse, J.D. Choppin, and Y. Grosjean, 1994, Interactions between tectonism, thermal history, and paleohydrology in the Mahakam Delta, Indonesia: Model results, petroleum consequences (Abs): AAPG Bulletin, v. 78, p. 1136.

Burrus, J., K. Osadetz, J.M. Gautier, E. Brosse, B. Doligez, G. Choppin de Janvry, J. Barlier, and K. Visser, 1993, Source rock permeability and petroleum expulsion efficiency: modeling examples from the Mahakam delta, the Williston Basin and the Paris Basin, *in* J.R. Parker, ed., Petroleum Geology of Northwest Europe: Proceedings of the 4th Conference, The Geological Society, London, p. 1317–1332

Burrus, J. K., K.G. Osadetz, S. Wolf, B. Doligez, K. Visser, and D. Dearborn, 1996, A two-dimensional regional basin model of Williston Basin hydrocarbon systems: AAPG Bulletin, v. 80, p. 265–291.

Burrus, J., 1998, Overpressure models for clastic rocks: their relation to hydrocarbon expulsion: a critical reevaluation, *in* Law, B.E., G.F. Ulmishek, and V.I. Slavin eds., Abnormal pressures in hydrocarbon environments: AAPG Memoir 70, p. 35–63.

Burst, J.F., 1969, Diagenesis of Gulf Coast clayey sediments and its possible relation to petroleum migration: AAPG Bulletin, v. 53, p. 73–93.

Byerlee, J., 1990. Friction, overpressure and fault normal compression: Geophysical Research Letters, v. 17, p. 2109–2112.

Byerlee, J., 1993, Model for episodic flow of high-pressure water in fault zones before earthquakes: Geology, v. 21, p. 303–306.

Caillet, G., 1993, The caprock of the Snorre Field, Norway: a possible leakage by hydraulic fracturing: Marine and Petroleum Geology, v. 10, p. 42–50.

Cayley, G.T., 1987, Hydrocarbon migration in the Central North Sea, *in* J. Brooks and K. Glennie, eds., Petroleum Geology of North West Europe, Graham and Trotman, London, p. 549–555.

Colton-Bradley, V.A.C., 1987, Role of pressure in smectite dehydration - effects on geopressure and smectite to illite transition: AAPG Bulletin, v. 71, p. 1414–1427.

Corbet, T.F., and C.M. Bethke, 1992, Disequilibrium fluid pressures and groundwater flow in the Western Canada sedimentary basin: Journal of Geophysical Research, v. 97, B5, p. 7203–7217.

Daines, S.R., 1982, Aquathermal pressuring and geopressure evaluation: AAPG Bull. , v. 66, p. 931–939.

Davis, D.M., J. Suppe, and F.A. Dahlen, 1983, Mechanics of fold-and-thrust belts and accretionary wedges: Journal Geophysical Research, B, v. 88, p. 1153–1172.

Davies, T.B., 1984, Subsurface pressure profiles in gas-saturated basins, *in* J.A. Masters, ed., Elmworth - case study of a deep basin gas field: AAPG Memoir 38, p. 189–203.

Deming, D., 1994, Factors necessary to define a pressure seal: AAPG Bulletin, v. 78, p. 1005–1009.

Dzevanshir, R.D., L.A. Buryakovsskiy, and G.V. Chilingarian, 1986, Simple quantitative evaluation of porosity of argillaceous sediments at various depths of burial: Sedimentary Geology, v. 46, p. 169–175.

Dickinson, G., 1953, Geological aspects of abnormal reservoir pressures in Gulf Coast, Louisiana: AAPG Bulletin, v. 37, p. 410–432.

Dickey, P.A., and W.C. Cox, 1977, Oil and gas in reservoirs with subnormal pressures: AAPG Bulletin, v. 61, p. 2134–2142.

Domine, F., and F. Enguehard, 1992, Kinetics of hexane pyrolysis at very high pressures, application to geochemical modeling: Organic Geochemistry, v. 18, p. 41-50.

Engelder, T., 1993, Stress regimes in the lithosphere. Princeton University Press, Princeton, New Jersey. 457 pp.

England, W.A., A.S. Mackenzie, D.M. Mann, and T.M. Quigley, 1987, The movement and entrapment of petroleum fluids in the subsurface: Journal of the Geological Society, London, v. 144, p. 327–347.

Fang, H., S. Yongchaun, L. Sitian, and Z. Qiming, 1995,

Overpressure retardation of organic matter and petroleum generation: a case study from the Yinggehai and Qwiongdongnan basins, South China Sea: AAPG Bulletin, v. 79, p. 551–562.

Fertl, W.H., 1976, Abnormal formation pressures: Developments in Petroleum Science 2, Amsterdam, Elsevier, 382 p.

Fertl, W.H., R.E. Chapman, and R.F. Hotz, 1994, Studies in abnormal pressures: New York, Elsevier, 454 p.

Foster, W.R., and H.C. Custard, 1980, Smectite-illite transformation - role in generating and maintaining geopressure: AAPG Bulletin (Abs), v. 64, p. 708.

Freed, R. L., and D. R. Peacor, 1989, Geopressured shale and sealing effect of smectite to illite transition: AAPG Bulletin, v. 73, p. 1223–1232.

Fritz, S.J., 1986, Ideality of clay membrane in osmotic pressure: a review: Clays and Clay Minerals, v. 34, p. 214–223.

Gaarenstroom, L., R. A. J. Tromp, M.C. de Jong, and A.M. Bradenberg, 1993, Overpressures in the Central North Sea: Implications for trap integrity and drilling safety: *in* J.R. Parker,ed., Petroleum Geology of NW Europe, Proceeding of the 4th Conference Geological Society of London, p. 1305–1313.

Gies, R.M., 1984, Case history for a major Alberta deep basin gas trap - the Cadomin Formation, *in* J.A. Masters, ed., Elmworth - case study of a deep basin gas field: AAPG Memoir 38, p. 115–140.

Grauls, D. J., and C. Cassignol, 1993, Identification of a zone of fluid pressure-induced fractures from log and seismic data - a case history: First Break, v. 11, p. 59–68.

Grigg, M., 1994, Geographic distribution of abnormal pressure boundaries in the Western Canadian Basin - the significance to exploration, *in* Law, B.E., G. Ulmishek, and V.I. Slavin, convenors, Abnormal pressures in hydrocarbon environments (Abstracts): AAPG Hedberg Research Conf., Golden, Colorado, June 8–10, 1994, unpaginated and unpublished.

Grosjean, Y., M. Bois, L. de Pazzis, and J. Burrus, 1994, Evaluation and detection of overpressures in a deltaic basin: the Sisi Field case history, Offshore Mahakam, Kutei Basin, Indonesia (Abs): AAPG Bulletin v. 78, p. 1143.

Hall, P.L., 1993, Mechanisms of overpressuring—an overview, *in* D.A.C. Manning, P.L. Hall and C.R. Hughes, eds., Geochemistry of clay-pore fluid interactions, Chapman and Hall, London, p. 265–315

Hart, B.S., P.B. Flemings and A. Deshpande, 1995, Porosity and pressure: Role of compaction disequilibrium in the development of geopressures in a Gulf Coast Pleistocene basin: Geology, v. 23, p. 45–48

Hermanrud, C., L. Wensaas, G.M.G. Teige, E. Vik, H.M.N. Bolås, and S. Hansen, 1998, Shale porosities from well logs on Haltenbanken (offshore Mid-Norway) show no influence of overpressuring, *in* Law, B.E., G.F. Ulmishek, and V.I. Slavin eds., Abnormal pressures in hydrocarbon environments: AAPG Memoir 70, p. 64–85.

Hitchon, B., 1969, Fluid flow in the Western Canada sedimentary basin: 2, effect of geology: Water Resources Research, v. 5, p. 460–469.

Holm, G.M., 1998, Distribution and origin of overpressure in the Central Graben of the North Sea, *in* Law, B.E., G.F. Ulmishek, and V.I. Slavin eds., Abnormal pressures in hydrocarbon environments: AAPG Memoir 70, p. 123–144.

Hottmann, C.E., and R.K. Johnson, 1965, Estimation of formation pressures from log-derived shale properties: Journal of Petroleum Technology, June 1965, p. 717–722.

Hower, J., E.V. Eslinger, M.E. Hower, and E.A. Perry, 1976, Mechanism of burial metamorphism of argillaceous sediment: 1. Mineralogical and chemical evidence: Bulletin of Geological Society of America, v. 87, p. 725–737.

Hunt, J.M., 1990, Generation and migration of petroleum from abnormally pressured fluid compartments: AAPG Bulletin, v. 74, p. 1–12.

Hunt, J.M., J.K. Whelan, L.B. Eglinton, and L.M. Cathles, 1994, Gas generation - a major cause of deep Gulf Coast overpressure: Oil and Gas Journal, v. 92, p. 59–63.

Hunt, J.M., J.K. Whelan, L.B. Eglinton, and L.M. Cathles, III, 1998, Relation of shale porosities, gas generation, and compaction to deep overpressures in the U.S. Gulf Coast, *in* Law, B.E., G.F. Ulmishek, and V.I. Slavin eds., Abnormal pressures in hydrocarbon environments: AAPG Memoir 70, p. 87–104.

Jowett, E. C., L.M. Cathles III, and B.W. Davis, 1993, Predicting depths of gypsum dehydration in evaporitic sedimentary basins: AAPG Bulletin, v. 77, p. 402–413.

Kaiser, W.R., and A.R. Scott, 1994, Abnormal pressure in the San Juan and Greater Green River Basins: Artesian versus hydrocarbon overpressure, *in* Law, B.E., G. Ulmishek, and V. Slavin, convenors, Abnormal pressures in hydrocarbon environments (Abstracts): AAPG Hedberg Research Conference, Golden, Colorado, June 8–10, 1994, unpaginated.

Karig, D.E. and G. Hou, 1992, High-stress consolidation experiments and their geologic implications: Journal of Geophysical Research, v. 97, p.289–300.

Katsube, T.J., and M.A. Williamson, 1994, Effects of diagenesis on shale nano-pore structure and implications for sealing capacity: Clay Mineral, v.29, p.451–461.

Law, B.E., 1984, Relationships of source-rock, thermal maturity, and overpressuring to gas generation and occurrence in low-permeability Upper Cretaceous and lower Tertiary rocks, Greater Green River basin, Wyoming, Colorado and Utah, *in* J. Woodward, F.F. Meissner, and J.L. Clayton, eds., Hydrocarbon source rocks of the greater Rocky Mountain region: Rocky Mountain Association of Geologists, p. 469–490.

Law, B.E., and W.W. Dickinson, 1985, Conceptual model for origin of abnormally pressured gas accumulations in low-permeability reservoirs: AAPG Bulletin, v. 69, p. 1295–1304.

Law, B.E., J.R. Hatch, G.C. Kukal, and C.W. Keighin, 1983, Geological implications of coal dewatering: AAPG Bulletin, v. 67, p. 2250–2260.

Law, B.E., C.W. Spencer, and N.H. Bostick, 1980, Evaluation of organic matter, subsurface temperature and pressure with regard to gas generation in low-permeability Upper Cretaceous and Lower Tertiary sandstones in Pacific Creek area, Sublette and Sweetwater counties, Wyoming: Mountain Geologist, v. 17, p. 23–35.

Luo, M., M.R. Baker, and D.V. Lemone, 1994, Distribution and generation of the overpressure system, eastern Delaware Basin, western Texas and southern New Mexico: AAPG Bulletin, v. 78, p. 1386–1405.

Luo, M., and G. Vasseur, 1992, Contributions of compaction and aquathermal pressuring to geopressure and the influence of environmental conditions: AAPG Bulletin, v. 76, p. 1550–1559.

Luo, M. and G. Vasseur, 1995, Modeling of pore pressure evolution associated with sedimentation and uplift in sedimentary basins: Basin Research, v. 7, p. 35–52.

Mackenzie, A.S., and T.M. Quigley, 1988, Principles of geochemical prospect appraisal: AAPG Bulletin, v. 72, p. 399–415.

Magara, K., 1974, Compaction, ion filtration, and osmosis in shale and their significance in primary migration: AAPG Bulletin, v. 59, p. 2037–2045.

Marine, I. W., and S.J. Fritz, 1981, Osmotic model to explain anomalous hydraulic heads: Water Resources Research, v. 17, p. 73–82.

Mann, D.M., and A.S. Mackenzie, 1990, Prediction of pore fluid pressures in sedimentary basins: Marine and Petroleum Geology, v. 7, p. 55–65.

Martinsen, R.E., 1994, Summary of published literature on anomalous pressures: implications for the study of pressure compartments, *in* Basin Compartments and Seals, ed., P.J. Ortoleva, AAPG Memoir 61, p. 27–38.

McBride, E.F., T.N. Diggs, and J.C. Wilson, 1991, Compaction of Wilcox and Carrizo sandstones (Paleocene-Eocene) to 4420m, Texas Gulf Coast: Journal Sedimentary Petrology, v. 61, p. 73–85.

Meissner, F.F., 1978a, Patterns of source rock maturity in non-marine source rocks of some typical Western interior basins, *in* Nonmarine Tertiary and Upper Cretaceous source rocks and the occurrence of oil and gas in west-central U.S.: Rocky Mountain Association of Geologists Continuing Education Lecture Series, p. 1–37.

Meissner, F. F., 1978b, Petroleum geology of the Bakken Formation, Williston Basin, North Dakota and Montana, *in* 24th Annual conference, Williston Basin symposium: Montana Geological Society, p. 207–227.

Mouchet, J.P., and A. Mitchell, 1989, Abnormal pressures while drilling. Manuels Techniques Elf Aquitaine, v. 2, 264p.

Neruchev, S.G., and I.M. Gildeeva, 1994, Abnormally high reservoir pressure and the principal phase of oil generation, *in* Law, B.E., G. Ulmishek, and V.I. Slavin, convenors, Abnormal pressures in hydrocarbon environments (Abstracts): AAPG Hedberg Research Conference, Golden, Colorado, June 8–10, 1994, unpaginated and unpublished.

Neuzil, C.E., 1994, How permeable are clays and shales?: Water Resources Research, v. 30, p.145–150.

Neuzil, C.E., 1995, Abnormal pressures as hydrodynamic phenomena. American Journal of Science, v. 295, p. 742–786.

Neuzil, C.E., and D.W. Pollock, 1983, Erosional unloading and fluid pressures in hydraulically tight rocks: Journal of Geology, v. 91, p. 179–193.

Palciauskas, V.V., and P.A. Domenico, 1980, Microfracture development in compacting sediments: relation to hydrocarbon-maturation kinetics: AAPG Bulletin, v. 64, p. 927–937.

Perry, E.A. and J. Hower, 1972, Late stage dehydration in deeply buried pelitic sediments: AAPG Bulletin, v. 56, p. 2013–2021.

Peterson, R., 1958, Rebound in the Bearspaw shale, western Canada: Geological Society of America Bulletin, v. 69, p. 1113–1124.

Plumley, W.J., 1980, Abnormally high fluid pressure: survey of some basic principles: AAPG Bulletin, v. 64, p. 414–430.

Powers, M.C., 1967, Fluid release mechanisms in compacting marine mudrocks and their importance in oil exploration: AAPG Bulletin, v. 51, p. 1240–1254.

Price, L. C., and L. M. Wenger, 1992, The influence of pressure on petroleum generation and maturation as suggested by aqueous pyrolysis: Organic Geochemistry, v. 19, p. 141–159.

Rostron, B.J., and J. Toth, 1994, Widespread underpressures in the Cretaceous formations of west central Canada, *in* Law, B.E., G. Ulmishek, and V.I. Slavin, convenors, Abnormal pressures in hydrocarbon environments (Abstracts): AAPG Hedberg Research Conference, Golden, Colorado, June 8–10, 1994, unpaginated and unpublished.

Spencer, C. W., 1987, Hydrocarbon generation as a mechanism for overpressuring in Rocky-Mountain region: AAPG Bulletin, v. 71, p. 368–388.

Summa, L.L., R.J. Pottorf, T.F. Schwarzer, and W.J. Harrison, 1993, Paleohydrology of the Gulf of Mexico Basin: development of compactional overpressure and timing of hydrocarbon migration relative to cementation, *in* A.G. Doré et al., Basin Modeling: Advances and applications, Norwegian Petroleum Society (NPF) Special Publication, v. 3, p. 641–656.

Surdam, R.C., S.J. Zun, J. Liu, and H.Q., Zhao, 1994, Thermal maturation, diagenesis, and abnormal pressure in Cretaceous shales in the Laramide Basins of Wyoming, *in* Law, B.E., G. Ulmishek, and V.I. Slavin, convenors, Abnormal pressures in hydrocarbon environments (Abstracts): AAPG Hedberg Research Conference, Golden, Colorado, June 8–10, 1994, unpaginated and unpublished.

Swarbrick, R.E., 1994, Reservoir diagenesis and hydrocarbon migration under hydrostatic paleopressure conditions: Clay Minerals, 29, 463–473.

Swarbrick, R.E., 1995, Distribution and generation of the overpressure system, eastern Delaware Basin, western Texas and southern New Mexico: Discussion: AAPG Bulletin, v. 79, p. 1817–1821.

Swarbrick, R.E., and M.J. Osborne, 1996, The nature and diversity of pressure transition zones: Marine and Petroleum Geology, v. 2., p. 111-116.

Sweeney, J.J., R.L. Braun, A.K. Burnham, S. Talukdar, and C. Vallejos, 1995, Chemical kinetic model of hydrocarbon generation, expulsion, and destruction applied to the Maracaibo Basin, Venezuela: AAPG Bulletin, v. 79, p. 1515–1532.

Terzaghi, K., 1923, Die Berechnung der Durchlass igkeitsziffer des Tones aws dem Verlanf der Hydrodynamischen Spannungserscheinungen: Sb. Akad. Wiss. Wien, p. 132–135.

Tigert, V., and Al-shaieb, Z., 1990, Pressure seals - their diagenetic banding-patterns: Earth-Science Reviews, v. 29, p. 227–240.

Tissot, B.P., R. Pelet, and P.H. Ungerer, 1987, Thermal history of sedimentary basins, maturation indices, and kinetics of oil and gas generation: AAPG Bulletin, v. 71, p. 1445–1466.

Ungerer, P., E. Behar, and D. Discamps, 1983, Tentative calculation of the overall volume expansion of organic matter during hydrocarbon genesis from geochemistry data. Implications for primary migration, in: Advances *in* Organic Geochemistry, John Wiley, p. 129–135.

Van Balen, R.T., and S.A.P.L. Cloetingh, 1993, Stress-induced fluid flow in rifted basin, *in* A.D. Horbury and A.G. Robinson, eds., Diagenesis and Basin Development, AAPG Studies in Geology, v. 36, p. 87–98.

Ward, C.D., K. Coghill, and M.D. Broussard, 1994, The application of petrophysical data to improve pore and fracture pressure determination in North Sea Central Graben HPHT wells: SPE paper 28297, SPE 69th Annual Tech. Conf., New Orleans, U.S.A., Sept. 25–28, 1994.

Watts, N.L., 1987, Theoretical aspects of cap-rock and fault seals for single- and two-phase hydrocarbon columns: Marine and Petroleum Geology, v. 4, p. 274–307.

Wensaas L., H.F. Shaw, K. Gibbons, P. Aargaard, and H. Dypvik, 1994, Nature and causes of overpressured mudrocks of the Gullfaks area, North Sea: Clay Minerals, v. 29, p. 439–449.

Williamson, M.A., and C. Smyth, 1992, Timing of gas and overpressure generation in the Sable Basin offshore Nova Scotia: implications for gas migration dynamics: Bulletin Canadian Petroleum Geology, v. 40, p. 151–169.

Yusoff, W.I., and R.E. Swarbrick, 1994, Thermal and pressure histories of the Malay Basin, offshore Malaysia (Abs): AAPG Bulletin, v. 78, p. 1171.

Burrus, J., 1998, Overpressure models for clastic rocks, their relation to hydro-
carbon expulsion: a critical reevaluation, *in* Law, B.E., G.F. Ulmishek, and
V.I. Slavin eds., Abnormal pressures in hydrocarbon environments: AAPG
Memoir 70, p. 35–63.

Chapter 3

Overpressure Models for Clastic Rocks, Their Relation to Hydrocarbon Expulsion: a Critical Reevaluation

Jean Burrus[1]
IFP
Rueil Malmaison, France

Abstract

The purpose of this paper is to review advances made in our understanding of the origin of over-pressures in clastic rocks and examine the relationship between overpressuring and hydrocarbon expul-sion. This study uses numerical simulations to examine overpressure models in clastic rocks. It is based on a review of previous regional overpressure modeling studies in rapidly subsiding basins (the Mahakam Delta, Indonesia, and the Gulf Coast, U.S.A.), and in slowly subsiding basins (the Williston Basin, U.S.A.-Canada and the Paris Basin, France). We show that compaction models based on effective stress-porosity relations satisfactorily explain overpressures in rapidly subsiding basins. Overpressures appear primarily controlled by the vertical permeability of the shaly facies where they are observed. Ver-tical permeabilities required to model overpressures in the Gulf Coast and Mahakam basins differ little, they are around 1–10 nanodarcies. Geological evidence and models suggest other causes of overpressure such as aquathermal pressuring or clay diagenesis to be generally small compared with compaction dis-equilibrium. Hydrocarbon (HC) generation can be a minor additional cause of overpressures in rich, mature source rocks. Shale permeabilities calibrated against observed overpressures appear consistent with direct measurements. Specific surface areas of mineral grains and relationships between effective stress/permeability implied by model calibrations agree with independent experimental determination. The main weakness of mechanical compaction models is that they overestimate the porosity of thick overpressured shales. Unlike in previous studies, we suggest that this mismatch is not caused by fluid generation inside overpressured shales. Instead, we infer that it is a consequence of an inappropriate definition of effective stress. If effective stress is defined as $S - \alpha P$, instead of $S - P$, then with α around 0.65–0.85, porosity reversals predicted in overpressured shales are much reduced, and better in agree-ment with observations. Alpha (α) is known in poro-elasticity as the Biot coefficient. We show that the non-linear distribution of horizontal stress often observed in overpressured shale sequences confirms values of the Biot coefficient in the range indicate above.

In slowly subsiding basins, there is no compaction disequilibrium. Pressures are regionally controlled by the surface topography. The persistence of high overpressures in thin (few meters thick), mature source rocks in the HC window implies uncommon conditions : a very rich source interval (total organ-ic carbon content, TOC, >10%), a very low permeability (100–1,000 times smaller than for the Mahakam Delta or Gulf Coast shales), and, possibly, a very low porosity (2–3%). The examples examined suggest that permeability of shales in Paleozoic–Mesozoic, slowly subsiding basins are significantly more vari-able than in Cenozoic rapidly subsiding basins. More complex tectonic and diagenetic histories could explain this greater variability. Our study suggests that, at least at the regional scale considered, diage-netic processes do not need to be invoked in young rapidly subsiding basins. This does not exclude the possibility that locally permeabilities can be decreased if cementation takes place, resulting in an increase of overpressures. It is probable that more mature basins with intermediate sedimentation rates and ages, have mixed chemical and mechanical compaction mechanisms.

[1] *Present Affiliation: Beicip-Franlab, Rueil Malmaison, France*

INTRODUCTION: HISTORICAL PERSPECTIVE

Overpressures have always been an important issue in hydrocarbon (HC) exploration. Most basins contain some overpressured rocks at sufficient depth (e.g., Fertl and Timko, 1970). Still the origin of geopressures is controversial. Dickinson (1953), followed by Dickey et al. (1968) and many others, presented the first evidence that overpressures were controlled by lithofacies distribution, not stratigraphic age, temperature or depth. They found that the overpressures in the Gulf Coast were associated with strata rich in shales, suggesting that compaction disequilibrium could be the cause of overpressures. This conclusion was consistent with the compaction model presented by Hubbert and Rubey (1959). This model, based on Terzaghi's soil consolidation theory (Terzaghi, 1923), assumes that the linear relation observed in soils between void index (the ratio between porosity ϕ and $1 - \phi$) and effective stress σ_e (the difference between total stress S and pore pressure P), also describes the compaction of rocks (Figure 1). The relation means that undercompaction and overpressures are genetically linked. According to Hubbert's model, the shape and magnitude of Gulf Coast overpressures could be explained by the high rate of sedimentation, if permeabilities of shales were below one nanodarcy (1 nanodarcy, 10^{-21} m^2). The likelihood of such low permeabilities appeared, in the absence of convincing experimental evidence, doubtful to many. Owing to measurement difficulties, it was not until the

mid 1980s that reliable shale permeability measurements became available and confirmed such low values. The validity of the Darcy equation applied by Hubbert and Rubey to represent water flow in shales was also disputed. Some argued that most water in shales being adsorbed and crystalline in nature, cannot be displaced at usual hydraulic gradients, and that hydraulic gradients are not linearly proportional to flow rates in shales (e.g., Lutz and Kemper, 1959; Miller and Low, 1963). Such conclusions were shown to be biased by experimental errors and artifices (Olsen, 1965, 1966). But the idea remained that water flow in shales had a different nature than in other rocks. Explanations other than compaction disequilibrium were thus proposed. Shale membrane properties and osmotic characteristics were investigated. Since water expelled out of shales is less saline than remaining water (e.g., Overton and Timko, 1969), it was proposed that downward water flow directed towards areas with increasing salinity could counterbalance upward compactional Darcy flow and this counterflow could explain overpressures (McKelvey et al., 1962; Jones, 1969; Olsen, 1971; Magara, 1974; Hinch, 1978). Such ideas dominated overpressure explanations in the 1960s and 1970s, although the experimental proof was limited and quantification for natural conditions was not attempted.

The theoretical relationship between effective stress and porosity assumed by Hubbert (Figure 1) found in the meantime some practical support. Deviation of porosity-related logs like resistivity or sonic velocity

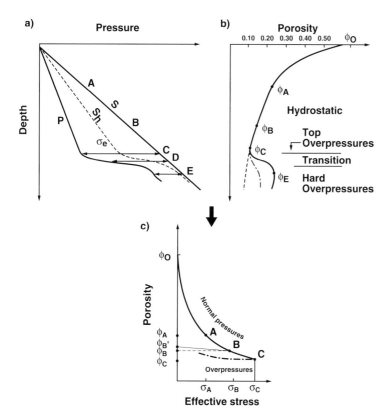

Figure 1. Schematic pressure (a) and porosity (b) profiles in a rapidly subsiding basin, and construction of a porosity/effective stress function (c). S = lithostatic stress; S_h = minimum horizontal stress; P = pore pressure. Effective stresses at depths B and D (or at depths A and E) are equal, so porosities at depths B and D (or at depths A and E) should be equal. The heavy porosity profile in (b) represents such a relation. In reality, porosities in the overpressured section often follow the mixed dashed line, the porosity reversal is significantly less than with the theoretical heavy line. As a consequence, for a given effective stress, the porosity is lower in the overpressured section than in the normally pressured section. Two distinct porosity/effective stress curves are thus obtained in (c) in the normally pressured section (heavy line) and in the overpressured section (mixed dashed line). In the case of stress unloading, there is a hysteresis —only a slight elastic decompaction is allowed. If stress decreases at point B in (c), then the path BB′ is followed, instead of B to A. The S_h profile indicated by the short dashed line in (a) increases markedly at the top of the overpressures. We show in this study how this increase, the existence of the two porosity/effective stress curves in (c) and the excess theoretical porosity in (b) are related.

logs from their normal trends confirmed that effective stresses and large porosities are often linked in overpressured zones (e.g., Hottmann and Johnson, 1965; Magara, 1968; Eaton, 1975). This method is widely applied in exploration to detect overpressures.

Mechanical compaction theory appeared further challenged when illitization profiles observed in illite/smectite (I/S) shale sequences were found to mimic overpressure profiles, like those in the Gulf Coast (Powers, 1967; Burst, 1969; Perry and Hower, 1972; Magara, 1975a; Hower et al., 1976). Conceptual models and experimental work (e.g., Van Olphen, 1963) showed that the progressive burial of clays, was accompanied by their transformation into thermally more stable phases, and the release of bound water. This additional fluid would cause overpressures to develop. When the strength of the rock is reached, microfracturing can take place, leading to episodic leakage. Although the volume and the density of bound water released in the process were a matter of discussion, the illitization theory became popular, mostly because clay dehydration was reported to take place in the Gulf Coast at temperatures (70°–140 °C) or at a depth that roughly coincide with the oil window. This lead to the idea that water released during clay dehydration could help "flush" hydrocarbons out of mature overpressured source rocks and drive oil expulsion (Burst, 1969). The effect of illitization on overpressures was however not quantified, and the data set used to build the concept was limited. Hydrocarbon generation and in particular gas generation (e.g., Tissot and Pelet, 1971; Hedberg, 1974) were also proposed as a mechanism responsible for the development of overpressures in mature source rocks, potentially leading to the opening of microfissures.

Whereas the U.S. Gulf Coast was the dominant source of information on overpressures and compaction profiles until the early 1980s, data from other petroleum provinces became available in the subsequent decade. Additional data showed that basins that have rapidly subsided in the Cenozoic like the Mahakam Delta (Durand and Oudin, 1979; Oudin and Picard, 1982; Burrus et al., 1992), the Beaufort-McKenzie Delta (Issler, 1992) or the Nile Delta (Zein El Din et al., 1988) have overpressure and undercompaction characteristics that are similar to the Gulf Coast. A sharp transition zone marks the boundary between hydrostatic, normally compacted zones and the hard overpressured zones. Some undercompaction is generally observed across the overpressured shales. Transition zones often coincide with porosity reversals (Figure 1). In the North Sea, overpressures are associated with the deposition of Cretaceous–Tertiary shale dominated clastics (e.g., Buhrig, 1989), but, in contrast with the Gulf Coast, porosity profiles do not show significant undercompaction in overpressured series (Chiarelli and Duffaud, 1980; Lindberg et al., 1980; Carstens and Dypvik, 1981; Hermanrud et al., 1998-this volume). In offshore Eastern Canada, significant over-

pressures develop across Jurassic/Cretaceous clastic sequences (Mudford and Best, 1989), but shales hardly show any undercompaction. Enhanced porosity can be observed in overpressured sandstones, but it is attributed to the dissolution of carbonate cements, not to delayed compaction (Jansa and Urea, 1990). Such a conflicting data set suggests that at least two types of processes control overpressure development: (1) mechanical processes, described by the consolidation theory, that preserve porosity, and (2) physical-chemical processes such as viscous creep, pressure solution or diagenesis, in which porosity can be destroyed over geologic time while overpressures are developed.

Finally, the last decade has seen a considerable development of numerical models that simulate compaction and overpressuring. Some models attempt to simulate the functioning of the whole petroleum system, from the source rock to the reservoir (e.g., Ungerer et al., 1987; 1990; Nakayama and Lerche, 1987; Novelli et al., 1988; Burrus et al., 1991, 1992), while other models address solute transport and fluid-rock interactions (e.g., Bethke et al., 1988; Sonnenthal and Ortoleva, 1995). All these models help put quantitative constraints on overpressure mechanisms.

The purpose of this paper is to review advances made in our understanding of the origin of overpressures in clastic rocks and examine the relationship between overpressuring and hydrocarbon expulsion. We present conclusions for the origin of overpressures in several petroleum provinces using IFP's two dimensional numerical model TEMISPACK. The first part discusses "normal" compaction curves and emphasizes the difference between mechanical compaction and chemical compaction. The second part reviews previous compaction/overpressure numerical models. It emphasizes the differences between elastic models based on experimental rock compressibility or specific storage coefficient, and inelastic models that use a "geological" effective stress/porosity relationship. The third part of this paper summarizes results obtained in the modeling of overpressures using TEMISPACK in basins that cover a variety of situations: rapidly subsiding Cenozoic basins like the Mahakam Delta (Indonesia) and the Gulf Coast (U.S.A.), and slowly subsiding, Paleozoic and Mesozoic basins like the Williston Basin (Canada-U.S.A.) and the Paris Basin (France). Implications for hydrocarbon expulsion will be presented. The fourth part discusses the validity of mechanical compaction models. This paper does not address new algorithms or equations. All equations shown are justified in the references mentioned in the text.The list of symbols is given separately.

TYPES OF NORMAL POROSITY PROFILES FOR CLASTIC ROCKS

Normal porosity profiles (NPP) are expected when rocks are normally consolidated and are uniform in lithology, and normally pressured. NPP were introduced in basin models to compute sedimentation rates and decompact layers (Perrier and Quiblier, 1974). According to the literature, concave downwards or exponentially decreasing NPP are considered to result from mechanical compaction processes driven by gravity forces (Korvin, 1984), like in Hubbert's model (Magara, 1980). In contrast, concave upward NPP result from pressure solution models or viscous creep models, in which the rate of porosity loss at grain contact accelerates with temperature, hence with depth (Angevine and Turcotte, 1983, Schneider et al., 1994a). We review below normal compaction profiles for shales, shaly sandstones and clean sandstones.

Normal Porosity Profiles for Shales

NPP are not easy to reconstruct in shales, and the scatter of published curves is large (e.g., Rieke and Chilingarian, 1974). They can be derived from logs, but converting log measurements in shales into porosity values is not straightforward. Porosities are linearly related to densities, but density logs are often perturbed by cavings, and are sensitive to lithology changes. The Neutron log is also sensitive to mineralogy changes. Sonic logs are not linearly related to porosities. Sonic derived porosities using a linear relation differ from density log derived porosities by up to a factor of 2 (Luo et al., 1991). Non-linear sonic velocity/porosity relations have been proposed (Issler, 1992), but they require detailed calibration against laboratory measured porosities which are not usually available.

Athy's (1930) original NPP (Figure 2a) shows a regular exponential porosity decrease between 0.50 at the surface and 0.05 at 2.3 km (7,000 ft):

$$\phi = \phi_0 \cdot e^{-cz} \qquad (1)$$

where c (m^{-1}) is the compaction coefficient, ϕ_0 the surface porosity. This relation has been widely applied, because it is mathematically simple, and many shale NPP generally show a concave downward curvature, like in Equation (1) (e.g., Weller, 1959; Perrier and Quiblier, 1974; Rieke and Chilingarian, 1974; Magara, 1980). There is abundant evidence that Equation (1) is not satisfactory. The surface porosity of 0.50 considered in Athy's original curve is, due to erosion, significantly less than the surface porosity of shales generally found in the range 0.70–0.80 (e.g., Meade, 1966). The curvature of NPP in the first km is actually more than suggested by (1). Following Hedberg (1936), shale NPP are often divided in three parts. The first part is characterized by a very rapid decrease of porosity down to 0.35 in the first few hundred meters. It is followed by a quasi-linear portion down to a porosity of around 0.10 at depth around 2.5–3.5 km, and an even steeper slope below. Issler's (1992) NPP shape for Cenozoic shales in the Beaufort-Mackenzie Delta (Figure 2a) decreases also linearly from 0.35 at a depth of 500 m down to 0.05

at a depth of 3700 m; the curvature is large only in the first 500 m. Similar results were found in the Norwegian Shelf by Hansen (1996). The overall profiles have a downward curvature, suggesting that compaction of shales has a dominant mechanical origin. The change of curvature at a depth of 500 m probably indicates that mechanical compaction processes change below a depth of 0.5–1 km (Heling, 1970). An exponential NPP can either fit the high curvature observed in the first kilometer using a large c value, or it can fit the low curvature below 1 km by using a low c value, it cannot fit both. This misfit was not recognized by Athy in his original work.

Shaly Sandstones

Sandstones containing at least 10–20% of shale material also have concave downward NPPs, hence their compaction probably has a mechanical origin, without much influence of cementation or pressure solution (e.g., Galloway, 1974; Nagtegaal, 1978; Burns and Etheridge, 1979). This has been confirmed in various published compaction experiments that satisfactorily mimic natural trends (Rittenhouse, 1971; Pittman and Larese, 1991).

Clean Sandstones

According to the literature, clean sandstones having less than 10–20% shales or cements do not compact mechanically but do so by pressure solution, except in the first kilometer where mechanical compaction probably dominates pressure-solution (e.g., Tada and Siever, 1989). Unlike shaly sandstones, clean sandstones do not loose much of their porosity under increased effective stresses as demonstrated by laboratory experiments (Pittman and Larese, 1991). The usual surface porosity of clean sandstones, around 0.40, is slightly lower than the maximum theoretical porosity (0.48) of perfect spheres with uniform size (Houseknecht, 1987). The minimum porosity that can be reached by dense rearrangement of spheres with uniform size is 0.26, a value far above the porosity of deeply buried sandstones. Natural observations suggest that pressure solution begins generally at a porosity around 0.30 or depth around 1–1.5 km and is well developed below this depth (Houseknecht, 1987). Porosity can be reduced by pressure solution to very small values (<0.03) at depths of 3–5 km, resulting in quasi-linear or concave upwards NPPs. Numerical pressure solution (e.g., Angevine and Turcotte, 1983) or viscous creep models (Schneider et al., 1994a) and observations (e.g., Galloway, 1974; Scherer, 1987) suggest that clean sandstones loose their porosity more rapidly in high geothermal regions than in cold regions. This dependency can be explained by the fact that diffusion coefficients at grain boundaries (or rock viscosity in creep models) increase with temperature. This increase is however poorly understood. Strain rates obtained in the labora-

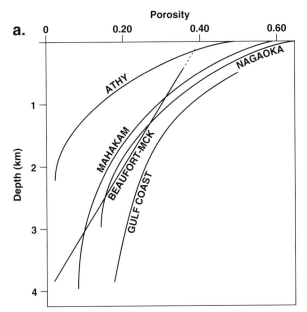

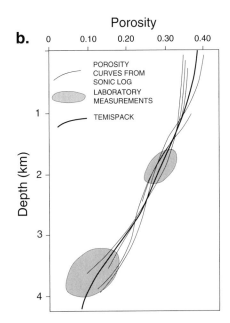

Figure 2. (a) Normal porosity depth profile derived from density logs for the Mahakam shales, compared with Athy's original profile (Athy, 1930), and with shale compaction profiles reported in the Nagaoka Basin (Magara, 1968), in the Gulf Coast (Ham, 1966) and in the Beaufort-Mackenzie Delta (Issler, 1992). Porosities of the Beaufort-Mackenzie, Mahakam and Nagaoka basins agree within 5% between depths of 1 and 3.5 km. **(b)** Porosity vs. depth profile for clean sandstones in the Mahakam Delta, derived from density logs (light lines). Modeled profile used in TEMISPACK is shown as a heavy line. The mean profile is concave upwards, suggesting a strong influence of pressure solution.

tory with pressure solution are 5 to 10 orders of magnitude lower than in nature at equivalent conditions (Gratz, 1991). This discrepancy makes current pressure solution models far from being predictive. Empirical correlations of clean sandstone porosity with time temperature indices have been attempted (Schmoker and Gautier, 1988, 1989), but the scatter is so large that no physical law can be inferred easily (Deming, 1990).

PREVIOUS MECHANICAL COMPACTION/OVERPRESSURE MODELS; THE TEMISPACK MODEL

Previous Mechanical Compaction/Overpressure Models

Both elastic and inelastic models have been used to simulate clastic rock compaction and overpressure development. Elastic models consider specific storage or compressibility coefficients, while inelastic models consider effective stress-porosity relations.

Elastic Models

Elastic compaction models are based on the diffusion equation used in hydrology (Gibson, 1958):

$$\frac{K}{\mu} \cdot \frac{\partial^2 u}{\partial z^2} + \frac{(\rho_s - \rho_w)}{\rho_w} \cdot \omega = S_s \frac{\partial u}{\partial t} \qquad (2)$$

see List of Symbols Used for definition of terms.

With elastic models, the magnitude of overpressures is controlled by the ratio $S_s\omega/K$ (Bishop, 1979)–the development of high overpressure requires a high sedimentation rate, highly compressible rocks (high S_s) and a low permeability. Determining relevant shale compressibility and permeability characteristics is therefore a major goal in overpressure studies.

Shale storage coefficients or compressibilities can be determined in the laboratory, from hydraulic tests in the field or fitted against observed NPPs. If the compressibility of solid grains can be neglected, S_s is related to the elastic compressibility of the rock and of the fluid by (Jacob, 1940):

$$S_s = (\beta_f + \phi \cdot \beta_b) \cdot \rho_w \cdot g \qquad (3)$$

see List of Symbols Used for definition of terms.

As noted by Palciauskas and Domenico (1989), the assumption that elastic properties determined in the laboratory at very large strain rates (typically around 10^{-4} s^{-1}) or in hydraulic tests at intermediate strain rates should apply at the much lower strain rates (typically 10^{-14} to 10^{-16} s^{-1}) observed in nature, is not justified. At geological times, irreversible deformations can be much larger than at laboratory scale, even in the absence of pressure solution. A difference between laboratory and geological compressibilities can be inferred from two series of observations. First, bulk compressibilities of shales determined in the laboratory by oedometer tests (ductile shales: 10^{-8} Pa^{-1} at 5 MPa or 10^3 psi effective stress to 10^{-9} Pa^{-1} at 50 MPa or 10^4 psi effective stress; stiff shales: 10 times lower values) and corresponding specific storage coefficients (ductile shales: 10^{-4} m^{-1} ; stiff shales: 10^{-5} m^{-1}) are typically 100 times less than suggested by hydraulic tests in the field (ductile shales: 10^{-2} m^{-1} ; stiff shales: 10^{-3} m^{-1} [Domenico and Mifflin, 1965; Rieke and Chilingarian, 1974]). Second, during drained oedometer compaction in the laboratory, the minimum porosity reached by shales at effective stresses between 7 MPa and 70 MPa (10^3–10^4 psi) is several tens of percent larger than in nature at equivalent stress (Weller, 1959; Neglia, 1979; Rieke and Chilingarian, 1974; Vasseur et al., 1995). Such a deviation does not seem to be observed for effective stresses less than approximately 7 MPa (1500 psi), which typically correspond to equivalent burial around 700 m (Skempton, 1970; Burland, 1990). This suggests that shales could compact more easily in nature than in the laboratory, at least beyond an equivalent burial around 700 m. Extrapolating directly compressibilities or specific storage coefficients calibrated in the laboratory to the natural system is thus questionable. The coefficients must rather be fitted against observed NPPs. If porosity varies exponentially with depth, then the vertical effective stress, assumed to be $\sigma_e = S - P$, and the bulk compressibility are related to porosity through Athy's compaction coefficient c (e.g., Hubbert and Rubey, 1959; Shi and Wang, 1986; Bethke and Corbett, 1988; Luo and Vasseur, 1992):

$$\sigma = \sigma_0 \cdot \exp\left(-\frac{c \cdot \phi}{(\rho_b - \rho_w) \cdot g \cdot (1-\phi)}\right) \qquad (4)$$

$$\beta_b = \frac{c \cdot \phi}{(\rho_b - \rho_w) \cdot g \cdot (1-\phi)} \qquad (5)$$

Combining Equations (3) and (5) and considering an exponential NPP with a low curvature (c value of 2 x 10^{-4} m^{-1}), we obtain from Equation (5) "geological" S_s values which decrease from 10^{-3} (at 0.50 porosity) to 10^{-5} m^{-1} (at 0.05 porosity). These values are not significantly changed if an exponential NPP with a higher curvature (c value of 8 x 10^{-4} m^{-1}) is considered. These "geological" S_s values correspond to the range of variation of specific storage coefficients indicated (ref. above) by oedometer tests for very ductile shales at large porosity and stiff shales at low porosity. "Geological" S_s values appear thus 10 to 100 times smaller than the S_s value derived from hydraulic tests in the field. Given the different time and space scales implied, it may appear surprising that the "geological" S_s values are in better agreement with oedometer measurements in laboratory than with values determined in the field from pumping tests. Whether or not this coincidence has a physical meaning remains to be investigated. Early linear compaction models assumed constant S_s values. They used S_s values for shales that were derived from hydraulic tests in the field, not from oedometer tests in the laboratory. Hence, according to

the previous discussion, they over estimated S_s by up to two orders of magnitude. As mentioned above, predicted overpressures are controlled by the S_s/K ratio. Since S_s values were overestimated, K was also overestimated by roughly the same factor, compared with models that consider variable, more realistic S_s values. Early models predicted significant overpressures with shale permeabilities around 1–10 microdarcies (Bredehoeft and Hanshaw, 1968; Sharp and Domenico, 1976; Bishop, 1979). In contrast, models that follow Equations (3) and (5) find significant overpressures to be associated with permeabilities in the range 1–10 nanodarcy (e.g., Keith and Rimstidt, 1985; Shi and Wang, 1986; Bethke and Corbett, 1988; Luo and Vasseur, 1992; see the case studies below).

Elastic models also helped clarify ideas about aquathermal pressuring. Thermal expansion of water has been presented as a major cause of overpressuring (e.g., Barker ,1972; Bradley, 1975). Aquathermal pressuring is a prominent mechanism if rocks are perfectly incompressible ($\beta_b = 0$) and perfect seals (K = 0). Whether actual shale permeability and compressibility are small enough to allow aquathermal pressuring to be significant has been a matter of considerable debate (Magara, 1975b; Daines, 1982; Sharp, 1983; Bethke, 1986). Previous models showed that aquathermal pressuring can be neglected compared with the effect of compaction disequilibrium (e.g., Keith and Rimstidt, 1985; Bethke, 1986; Shi and Wang, 1986; Luo and Vasseur, 1992, 1993) if realistic permeabilities (in the range 1 to 10 nanodarcy) and compressibilities (in the range 10^{-8} to 10^{-10} Pa^{-1}) were considered. It has been argued however that aquathermal pressuring would lead to a decline in effective stress with time, even if sedimentation proceeds (Miller and Luk, 1993). In this case, the 100 times lower unloaded specific storage should replace S_s in Equation (2), and aquathermal pressuring becomes a more important source of overpressure compared with compaction disequilibrium (Miller and Luk, 1993). Erosional episodes are examples of geological unloading. But during erosion, rock temperatures decline rather than increase, and even a small elastic rebound causes underpressured rather than overpressured conditions (Neuzil and Pollock, 1983). We will argue in subsequent discussions that, based on modeling, except for very organically rich and very tight source rocks actively generating hydrocarbons, and excluding erosional situations, effective stresses generally do not decline during sedimentation with time. We conclude that aquathermal pressuring can be neglected.

Inelastic Models

Non-linear models represented by Equations (4) and (5) are more adapted than earlier linear elastic models. As indicated above, however, exponential compaction curves cannot simultaneously account for both the high curvature of shale porosity profiles in the first 1.5 km, and for the more linear shape below. In compaction models where the derivative of porosity with depth or stress is a critical parameter, the error can be significant. Inelastic models account for this limitation. They follow the same principle as models obeying Equations (4) and (5), except that the $\phi(\sigma_e)$ relationship is defined point by point (following Figure 1) instead of being defined analytically as in Equation (4). Smith (1971) was the first to show how Hubbert's mechanical model could be generalized for any discrete porosity/depth relation. His numerical model can consider any NPP shape. In particular, the concave upwards clean sandstones NPP, which cannot be described with Equation (1), can be accounted for. Since deformation is irreversible, a distinct pathway is followed during unloading (Figure 1) in case of decreasing effective stress. The slope of this curve is equivalent to an elastic Young's modulus. The maximum vertical effective stress reached in the past, which controls whether the loading curve or the unloading curve is used, is equivalent to a plasticity criteria (Schneider et al., 1994b).

TEMISPACK Model

Principles

TEMISPACK (Doligez et al., 1987; Burrus and Audebert, 1990; Ungerer et al., 1990) is a finite volume model that solves compaction, thermal, water flow, hydrocarbon generation, and migration equations in a deformable Lagrangian mesh. It contains a compaction law of the type presented above. In the studies shown below, fluid and grain matrix densities were assumed to be constant. The permeability is defined for each lithology. It follows a Kozeny- Carman equation:

$$K(\phi) = \frac{0.2 \cdot \phi'^3}{S_0^2 \cdot (1 - \phi')^2} \quad \text{if } \phi > 0.10 \qquad (6)$$

$$K(\phi) = \frac{20 \cdot \phi'^5}{S_0^2 \cdot (1 - \phi')^2} \quad \text{if } \phi < 0.10$$

where ϕ' is the porosity corrected for adsorbed water considered immobile (see List of Symbols Used for definition of other terms). ϕ' is defined as:

$$\phi' = \phi - 3.1 \cdot 10^{-10} \cdot S_0 \qquad (7)$$

Equation (7) assumes that between one and two molecular water layers (with a thickness between 2.5 and 5×10^{-10} m) are adsorbed on the mineral grains (Van Olphen, 1963). This correction is sensitive only for very low porosity (below 0.05) and very large specific surfaces (like in shales). The higher slope for porosities below 0.10 was introduced to better account for experimental data (Jacquin, 1965). Permeability is a tensor, and two permeabilities K_x and K_z are distinguished: $K_x = k_x \cdot K(\phi)$ and $K_z = k_z \cdot K(\phi)$. Both k_x and k_z have no units. They can be viewed as related to the tortuosity τ_x

and τ_z of flow path in the x and z direction (Gueguen and Palciauskas, 1992):

$$k_x = \frac{1}{\tau_x^2}$$

$$\qquad (8)$$

$$k_z = \frac{1}{\tau_z^2}$$

Compacted shales have experimental τ_z values in the range 2–10 (e.g., Katsube et al., 1991), so k_z can be expected to vary between 0.5 and 10^{-2} if Equation (8) is correct. Values of τ_x are closer to 1.

In case of rebound, the maximum porosity rebound, which occurs when σ_e returns to zero (Figure 1), is taken in TEMISPACK as ten percent of the current porosity value–this corresponds to a Young's modulus E of 300 to 350 MPa following the relation $d\phi = 1/E \, d\sigma_e$. This value falls within the plastic clay values (70–500 MPa) (Domenico and Mifflin, 1965). Our model slightly overestimates rebound if one considers the stiff clay Young modulus range of variation (500–1,000 MPa) and the corresponding rebound (3–5% of the current porosity value).

Other model characteristics are as follows. Hydrocarbon generation is described by a first order Arrhenius equation. Hydrocarbon transport uses relative permeabilities, capillary pressures, and a generalized multiphase Darcy equation. Water and hydrocarbon viscosities are functions of temperature. Hydrocarbon generation is accounted for as a source term in the pressure equation. It is assumed that generation of hydrocarbons does not increase the porosity of the source rock, neither does it modify the source rock's compressibility characteristics.

Modeling Procedure

The TEMISPACK modeling procedure used in the case studies presented below is as follows. The first step consists in data preparation. The model section is selected along suspected hydraulic flow lines. It is is divided into chronostratigraphic events (usually between 30 and 60) and in a roughly equal number of columns, forming a grid of typically between 2,000 and 5,000 cells. Paleobathymetry, paleo-uplift and paleo-erosion profiles are also evaluated. Typically, 10 to 20 different lithologies are defined along the section, based on seismic, log, and core data. Each cell in the mesh is attributed to one of the model lithologies. Then petrophysical inputs are defined for each lithology. NPPs are reconstructed from original logs or taken by default. Sand/shale ratios help evaluate matrix density and thermal properties. Porosity and effective stress (ϕ_i, σ_i) inputs are constructed from the NPPs and density information for each model lithology, as indicated in Figure 1. Permeability inputs are set to an initial value, which is inferred from both stratigraphy and lithology. For tight shales, we expect permeabilities around 1–10 nanodarcies (see above). This requires S_0 values around 10^8 m^{-1}, using Equation (8). In the case of permeable sandstones with K around 10–100 millidarcies, K/ϕ plots usually indicate S_0 values in the range 10^5 to 10^6 m^{-1}. The k_x/k_z ratio can be evaluated from the layering of each lithology. It is usually equal or above 10, because rocks always have some anisotropy. In case of laterally continuous sand/shale interbedding, the ratio can be set to values as high as 10^4 or 10^5.

In the second step, overpressures history is simulated and overpressures calculated at present day are compared with observations. Free parameters are adjusted to fit observed pressures. This means modifying the k_z parameter for the tight facies that control overpressure development. The k_x, S_0, or the NPP are usually not modified because convergence is more difficult (S_0), or because the sensitivity is too low (k_x). In most cases, changing k_z by a factor 10–50 from the default value satisfactorily reproduces the observed overpressures. All important parameters, in particular k_z, need to be investigated through a sensitivity analysis. Usually several scenarios can explain observed overpressures. We do not consider relative permeabilities as fitting parameters. We assume relative permeabilities are "X"-shaped concave functions of saturations (Burrus et al., 1996).

OVERPRESSURE MODELING CASE STUDIES

We present below overpressure studies that cover a variety of pressure regimes found in different basins. Only summarized results will be given. Details will be found in separate articles. The Mahakam Delta and Gulf Coast examples correspond to common cases of Cenozoic deltaic basins with very high sedimentation rates (>1,000 m/m.y.). The Williston and the Paris Basins represent examples of much older basins (post-Ordovician, post-Triassic in age, respectively), with moderate thickness (< 3.5, 2.5 km respectively), very low sedimentation rate (<10 m/m.y.), and where com-

paction disequilibrium is absent. Key input parameters used in the studies are summarized in Table 1.

Mahakam Delta Overpressures (Indonesia)

Overpressures generally appear in the Mahakam Delta at a depth of about 3–4 km. They coincide with thick marine shale sequences of post mid-Miocene age (Figure 3). Sonic or density logs indicate that overpressures are associated with some undercompaction (Figure 4a, left). A porosity reversal is generally observed across the transition zone (Duval et al., 1992; Burrus et al., 1992).

To model the Mahakam Delta overpressures, the inputs and assumptions were as follows. Gamma density logs observed in shales interbedded with normally pressured sandstones were converted into porosity using a linear equation. The actual Mahakam Delta clay mineralogy was taken from Rinckenbach (1988). We discriminated illite/smectite, kaolinite, illite, chlorite, quartz, organic matter, pyrite and siderite and progressive illitization. The mineralogy was related to the porosity as follows. We assumed that that the matrix density of mixed mineralogy shales was a linear combination of individual minerals matrix density. The matrix density of individual minerals was taken from reference manuals (e.g., Serra, 1986). The bulk density of mixed mineralogy shales was assumed to be a linear function of the matrix density and of water density, weighted by porosity. The resulting porosity depth curve (Figure 2a) shows a prominent downward curvature. It is similar to the shale NPP reported by Magara (1968) in the Neogene Nagaoka Basin in Japan and mimics the NPP derived by Ham (1966) for normally pressured Gulf Coast shales (Figure 2a).

An effective stress/porosity relation was derived from density log readings and pressure measurements (RFT or DST) (Figure 5). The $\phi(\sigma_e)$ data found in the hydrostatic zone appears distinct from the data observed in the overpressured section. At 10 MPa (1,400 psi) vertical effective stress, porosity is 0.15 less in the overpressured shales than in the normally pressured shales. At 25 MPa (3,500 psi) effective stress, porosity in the overpressured shales is 0.06 less. Similar porosity discrepancies at vertical effective stress 7–8 MPa (1,000–1,100 psi) were already noted in the Gulf Coast (Plumley, 1980; Browers, 1994; Hart et al., 1995). TEMISPACK does not consider distinct $\phi(\sigma_e)$ input data in the hydrostatic and overpressured shales. We therefore used the $\phi(\sigma_e)$ data found in the hydrostatic section to model the development of overpressures.

Results of overpressure models were as follows. Compaction disequilibrium is, in the model, the dominant cause of overpressures. The effect of hydrocarbon generation on overpressures was found to be negligible, less than 1% of the total overpressure. Calculated and observed overpressures at Sisi (location in Figure 3) are shown in Figure 4b, while corresponding porosities are in Figure 4a (right). The fit of model overpres-

Table 1. Summary of critical petrophysical parameters of shales used in overpressure models.

	Permeability Characteristics			Source Rock Initial Potential
	S_0 (m^{-1})	k_z (no unit)	K_z at present-day in overpressured section (10^{-9}D)	kg HC/t* rock
Mahakam Delta	10^8	0.05–0.5	7–70	2 (marine shales)
US Gulf Coast	10^8	0.01	30	5–15 (Tertiary) 10–20 (Cretaceous) 10–30 (Jurassic)
Bakken Shales	10^8	0.01–0.1	0.001	70
Lower Liassic	3×10^7	1	700	15–30

*metric ton

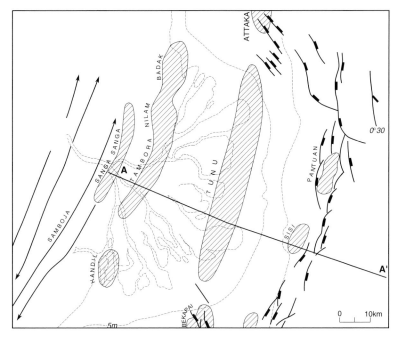

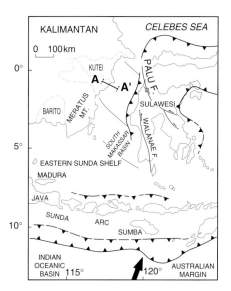

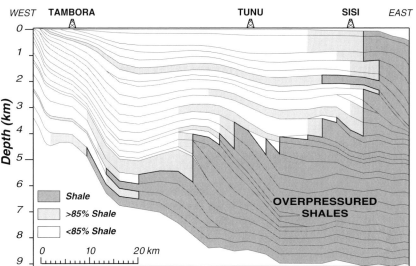

Figure 3. Location map and cross section along the Mahakam Delta. The hydrocarbon bearing structures are indicated with hachures on the map. Shale facies are indicated in light grey and dark grey colors (after Duval et al., 1992). Light grey facies have a slightly higher sand content (15 % instead of 5 %) than dark grey facies. Both facies are overpressured. The hard overpressures coincide approximately with the dark grey facies.

sures indicated by the solid line, and even more the dashed line, appears very good. It corresponds to a S_0 value of 10^8 m^{-1} and k_z value of 0.5 and 0.1, respectively. The model accounts for the position of the top of the overpressures and for the sharp gradient across the transition zone. The lesser gradient of overpressures below 4 km depth (Figure 4b) is known in other deep wells in the region (Burrus et al., 1992). The fit is not so good for the dotted profile, obtained with a S_0 value of 10^8 m^{-1} and k_z equal to 5, which results in 10 times higher permeabilities than with the solid profile. Overpressures, therefore, appear sensitive to shale permeabilities variations of a factor 10 only. The marked reversal shown by model porosity profiles (Figure 4a, right) across the transition zone at Sisi is qualitatively consistent with the reversal shown by the sonic log (Figure 4a, left). When the match of the pressures is

good, like with the solid or the dashed profiles, the porosity reversal appears, however, exaggerated compared with the sonic log. When the match of the porosity reversal seems appropriate, the overpressures appear underestimated (dotted profile). This mismatch is a clear consequence of the existence of distinct effective stress/porosity curves (Figure 5).

For the purpose of comparison of Mahakam Delta and Gulf Coast shales permeabilities (next section), it is important to note that either the solid or the dashed overpressure profiles shown in Figure 4b slightly underestimate the actual overpressures gradient in the transition zone.

The ratio between pore pressure and lithostatic stress P/S gives an indication of hydraulic fracturing (e.g., Du Rouchet, 1981). In areas of regional compression like the Mahakam Delta (Duval et al., 1992),

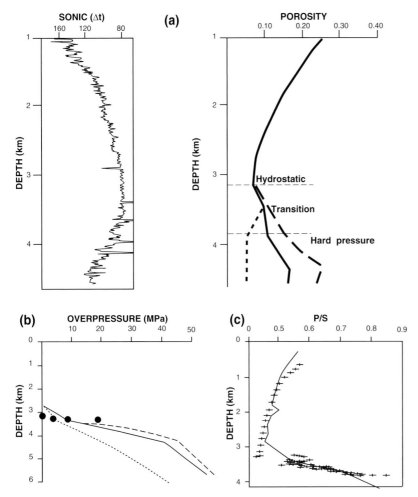

Figure 4. (a) Sonic log at Bekapai (left) and porosity prediction at Sisi (right) in the Mahakam Delta, in regions with similar stratigraphic and overpressures profiles (see location in Figure 3). The k_z parameter of shales is 0.5 for the solid profile, 0.1 for the long dashed profile and 5 for the short dashed profile. The latter profile underestimates shale porosity in the overpressured section, while the former profile overestimates porosity (after Burrus et al., 1992). (b) Observed (solid circles, RFT and DST) and predicted overpressure profiles in the Mahakam Delta (at Sisi). The dotted profile predicts insufficient overpressures. The solid and the dashed profiles predict similar overpressures; the agreement with observations is good, especially for the dashed profile (refer to text). (c) Calculated P/S profile at Sisi with the mean permeability scenario (solid line; see explanation in Figure 4a) and observations derived from logs (crosses). The ratio between pore pressure and lithostatic stress P/S increases sharply across the transition zone, but remains always below 0.85 in the overpressured shales. In a compressional basin like the Mahakam Delta, these values fall below the presumed fracture threshold (after Burrus et al., 1992).

Gulf Coast—Gulf of Mexico Overpressures

A regional model of overpressures and petroleum systems in South Louisiana and offshore Louisiana has been constructed. The lithostratigraphic model distinguished two main facies in the overpressured section. A "shale" facies is found at depth and basinward downdip of any strata, in the hard pressures zone. It is overlain by mixed sand/shale facies deposited on the downthrown blocks of growth faults, while faults were active (Dickey et al., 1968; Dutta, 1987). The top of the overpressures generally occurs within this facies. The mixed facies has been subdivided into a lower part, enriched in shales and less permeable; and an upper part, more rich in sand and slightly more permeable. Above the mixed facies, delta plain and fluvial facies are normally pressured.

Due to the analogy with the Mahakam Delta shale facies, the NPP and effective stress/porosity inputs defined for Mahakam Delta shales were assumed to be valid in the Gulf Coast. Shales and overpressured mixed sand/shale facies were given initial permeability inputs similar to the shales in the Mahakam Delta: $S_0 = 10^8$ m^{-1} and $k_z = 0.5$. These were then adjusted to fit the observed overpressures. We also tested the contribution of hydrocarbon generation to overpressures

hydraulic fracturing conditions are met when this ratio is above a threshold presumably around 0.90–0.95. The predicted P/S ratio (Figure 4c) increases in the Mahakam Delta between 0.55 and 0.80 across the transition zone, and remains below 0.85 in the hard overpressured zone, hence below the presumed threshold of hydraulic fracturing.

Mahakam Delta expulsion/migration models are presented in Figure 6. Hydrocarbon saturations calculated at present day with the previous shales permeabilities are consistent with known hydrocarbon distribution, in particular at Tunu and Tambora (Burrus et al., 1992; Duval et al., 1992). Comparing Figure 6a and 6b suggests that a negligible amount of gas manages to migrate out of the deep overpressured shales. Coals interbedded with hydrostatic sandstones and located higher in the sequence are the actual source of oil and gas accumulations in our model (Burrus et al., 1992), a conclusion supported by geologic and geochemical evidence (see the review in Duval et al., 1992). The contribution of the overpressured shales is comparatively small. This is a consequence of the very low permeabilities implied for the overpressured shales, and of their small initial hydrocarbon generation potential (Table 1).

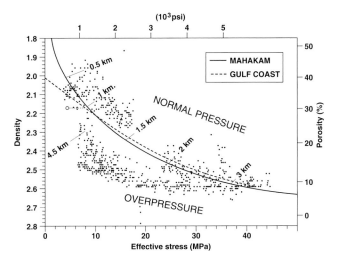

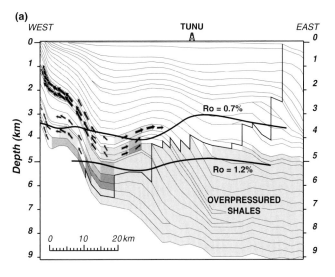

Figure 5. Effective stress/porosity relation in the normally pressured and in the overpressured Mahakam shales, derived from the integration of density logs. Effective stress is defined as S − P. Porosity observed in the normally pressured shale intervals is systematically higher than in the overpressured shales at equivalent effective stress. The approximate depth of burial is indicated in km. The solid line is an approximate mean curve adjusted against the points observed in the normally pressured section. It is equivalent to curve ABC in Figure 1c. The dashed line represents by comparison the effective stress/porosity relation found by Hart et al. (1995) in normally pressured shales in offshore Louisiana (U.S.A., location in Figure 8). The agreement with the Mahakam curve is excellent. The open circle at ~30% porosity and stress of 4 MPa represents measurements in overpressured shales in offshore Louisiana. The circle does not fall on the hydrostatic dashed curve, indicating the existence of distinct porosity/effective stress relations in the normally pressured and in overpressured shales in the Gulf Coast, as in the Mahakam Delta.

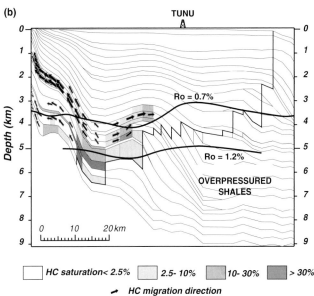

HC saturation< 2.5%   2.5- 10%   10- 30%   > 30%

↗ HC migration direction

Figure 6. Calculated distribution of HC saturations indicated by different grey patterns at present day for two source rock distribution scenarios in the Mahakam Delta (location of section, see Figure 3): (a) a source potential is attributed to the interbedded sand/coal sequences that lie *above* the overpressured shales, and to the overpressured shales (the top of which is underlined by the heavy solid line), (b) the source rock potential of the overpressured shales is set to zero. The top of the oil window (vitrinite reflectance of 0.6%) is indicated (after Duval et al., 1992). The predicted HC distributions at Tunu are similar in both scenarios. They qualitatively agree with actual HC distribution characteristics in Tunu (Duval et al., 1992): the main zone which carries HC is between roughly 3 and 4 km depth, and the east flank of the accumulation is steeper than the west flank. The comparison suggests overpressured shales are not effective sources of Mahakam Delta HC accumulations (after Burrus et al., 1992). Arrows = calculated direction of HC migration at present day. Vitrinite reflectance contours are from Duval et al. (1992).

from the four regional source intervals (Sassen, 1990), using petroleum potentials reconstructed by EXXON as indicated in Table 1.

The results are as follows. After calibration of k_z against observations, overpressures profiles computed at wells B, C, D, and E coincide well with observations (solid line, Figure 7). Best fitting permeability parameters for the shales and lower mixed facies are $S_0 = 10^8$ m^{-1} and $k_z = 0.01$. The upper, mixed facies is around 10 times more permeable than the lower portion. Similar to the Mahakam Delta, the depth to the top of the overpressures, the overpressure gradient in the transition zone, and the kink in the gradients below the transition zone are correctly accounted for. The fit of model overpressures with observations becomes poor when k_z is increased (dashed line) or decreased (mixed line) ten times in the shales and lower mixed facies, confirming the high sensitivity of overpressure to permeability. The contribution of hydrocarbon generation from Tertiary and Cretaceous source rocks to the development of overpressure is negligible (see well D, Figure 7) . Its

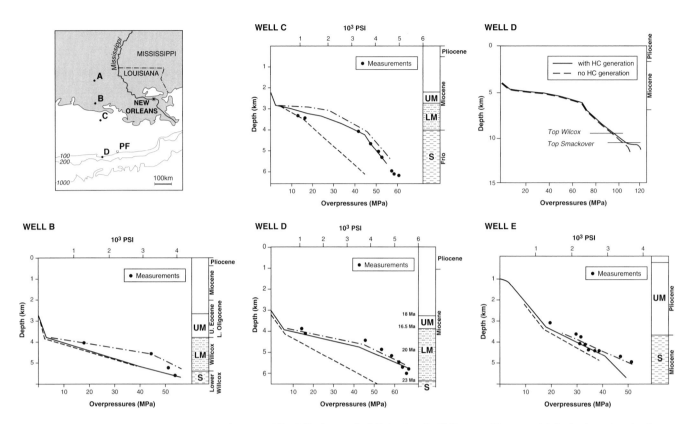

Figure 7. Location map of Wells A, B, C, D, and Pathfinder well, PF, in the Gulf Coast. Observed (circles) and calculated overpressure profiles are shown for Wells B, C, D and E (located 120 km south of Well D). The solid line shows good consistency with observations. The dashed line is ten times more permeable, the mixed dash line is ten times less permeable. Also shown for Well D is the influence of HC generation on overpressures. Except in the Upper Jurassic Smackover source rock interval, this influence is negligible. Facies: UM = upper mixed sand-shale, LM = lower mixed sand-shale, S = shales (see text). Stratigraphy is as indicated.

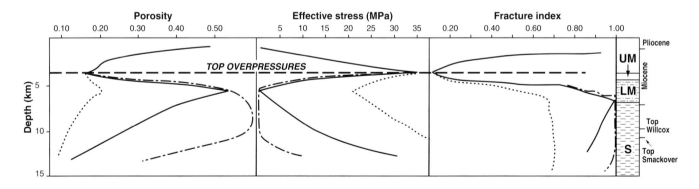

Figure 8. Influence of the vertical permeability of shales on theoretical porosity, effective stress and pore pressure to lithostatic stress P/S ratio predictions in the Gulf Coast (near Well C, Figure 7). Continuous line: theoretical profiles computed with the best fitting permeabilities that match overpressures in Figure 7 (solid line on Figure 7). Dotted profile: ten times higher permeability for shales and lower mixed (LM) facies compared with solid profile. Mixed dash profile: ten times lower permeability in shales and LM facies compared with solid profile. The region below the top of overpressures is predicted to be affected by hydraulic fracturing (P/S = 1 and zero effective stress) if the best fitting pressure profile is considered (solid line). The porosity reversals indicated on the solid and mixed profiles are exaggerated compared with actual porosity profile from Mello et al. (1995). Facies LM, UM, S: see Figure 7.

contribution is moderate for the Smackover source rock (Jurassic), the richest source rock considered. However, it only represents about 10% of the compaction driven overpressures.

The evolution of pressure, porosity, effective stress and the P/S ratio with depth is shown in Figure 8 for Well C for the different k_z values tested above. As in the Mahakam Delta, porosity profiles (Figure 8 left) show considerable reversal across the transition zone. The mixed profiles ($k_z = 0.001$) predict the region between 5 and 10 km depth to be fractured (P/S = 1) and to have no strength ($\sigma_e = 0$). In contrast, the dotted curves ($k_z = 0.1$) indicates no fracturing (P/S <0.70), little porosity reversal, and overpressure that is much below observations. Both profiles do not account correctly for observed overpressures (Figure 7, Well C). The best fitting solid line profile ($k_z = 0.01$) indicates a sharp contrast between a highly compacted region just above the onset of overpressures at 4,200 m, where effective stress is the highest (around 35 MPa, 5,000 psi), and a region just below the onset of overpressure characterized by poor compaction (ϕ >0.40), low effective stress (<5 MPa, 700 psi) and probably affected by hydraulic fracturing (P/S = 1; Figure 8 right). Weedman et al. (1992) provide observations which qualitatively support these results. They report from a core study of a south Louisiana well that compaction fabric of interbedded sand/shale rocks changes very rapidly across the transition zone over a distance less than 100 m. Above the top of the overpressures, abundant stylolithes, crushed fragments, sutured quartz contacts observed in sandstones suggest a high effective stress. In contrast, just below the top of the overpressures, sutured contacts, stylolithes and grain deformations are absent. This absence suggests a much lower effective stress, like in our model.

As in the Mahakam Delta, the model predicts too high porosities in the overpressured section. The modeled porosity reversal indicated by the solid line profile reaches 0.30 (ϕ increases from 0.20 to 0.50 in Figure 8 between depths of 4–5 km), while density and sonic log reversals in south Louisiana or Gulf Coast generally indicate porosity reversals commonly in the range 0.10–0.15 only (e.g., Stuart, 1970; Hinch, 1978; Plumley, 1980; Mello et al., 1994; Browers, 1994; Hart et al., 1995).

Our hydraulic fracturing predictions seem qualitatively consistent with observations of microfractures made in overpressured Gulf Coast shales (e.g., Capuano, 1993). Our model is also consistent with detailed pressure and effective stress measurements acquired in offshore Louisiana at the Pathfinder well, located 90 km east of well D, in the Pleistocene at depth between 2,000 and 2,400 m (see location in Figure 8; Anderson et al., 1994; Hart et al., 1995). These measurements indicate P/S ratio of 0.95 and very low effective stress (4 MPa, 600 psi). The effective stress and P/S ratio predicted by our compaction model are around 5 MPa (700 psi) and slightly above 0.95 respectively at equivalent position. This good agreement with observations suggests that compaction disequilibrium is the main cause of hydrofracturing processes in the Gulf Coast. It indirectly supports the shale permeability model used in our study.

Burrus et al. (1997a) show that, like the Mahakam Delta, the Gulf Coast overpressured shales have a permeability that is too low to allow significant upward hydrocarbon migration into the Tertiary sandy reservoirs. Modeled vertical migration pathways are along listric faults, probably made episodically more permeable by hydraulic fracturing, driven, as indicated above, by compaction disequilibrium. These model results are compatible with previous conceptual hydrocarbon migration models (Sassen, 1990), previous oil source correlations (Wenger and Schwarzer, 1993) and regional evidence of episodic oil flow along listric faults (Anderson et al., 1994). All these processes are the result of compaction disequilibrium.

Williston Basin (Canada-U.S.A.) Bakken Shale Overpressures

The Upper Devonian-Lower Mississippian Bakken organic shales provide an example where hydrocarbon generation, not compaction disequilibrium, is the main source of overpressures (Meissner, 1978). Previously, it was thought that overpressuring in the Bakken would result in the development of hydrofractures, a process that would cause oil expulsion and its migration towards the Mississippian reservoirs, 500 m higher in the sequence (Dow, 1974; Meissner, 1978; Price et al., 1984). Numerical models of Bakken oil generation and migration (Burrus et al., 1996a) confirmed that oil generation was the cause of Bakken overpressures (Figure 9a), but showed that, in contrast to the previous views, hydraulic fracturing thresholds were probably never reached by mature Bakken sources (Figure 9b), except locally, in areas of preexisting faults and locally extensional stress. A good fit of observed Bakken overpressures can be obtained with a k_z value between 0.1 and 0.01 and a S_0 value of 10^8 m^{-1} (Burrus et al., 1996b). Modeled Bakken overpressures are very sensitive to Bakken permeability and organic content. If the Bakken shale vertical permeability is increased by a factor 10, the overpressure peak is halved. If the TOC of the Bakken is 3% instead of 10%, then overpressure becomes negligible (Figure 9a). These TOC values correspond to initial petroleum potentials of 20 and 60 kg hydrocarbon/metric ton of rock, respectively. This great sensitivity has important practical implications. It would imply that hydrocarbon generation plays a dominant role in the origin of overpressure only for source rock intervals with initial petroleum potential well above 20 kg HC/t. Such source rocks are uncommon. A source rock is considered "good" if its initial petroleum potential is above 6 kg HC/t (Tissot and Welte, 1984).

Despite the low permeability of the Bakken, expulsion models suggest that 90% of oil generated in thermally mature Bakken has been expelled (Burrus et al., 1996b). This proportion is in good agreement with observed residual oil content (ibid.). High expulsion efficiency is probably a result of the low storage capacity of the Bakken shale pores (ibid.), which represent less than 3% of the total rock volume (Ropertz, 1994).

Normally Pressured Paris Basin (France)

Typical of shallow intracratonic basins, the Paris Basin (France) shows little overpressuring. Moderate hydraulic heads, typically in the range 0.5 to 2 MPa (70–300 psi), are found in the Triassic and Dogger aquifers at depths down to 2.5 km, due to the 200–350 m uplift observed to the east of the basin (Figure 10; Wei et al., 1990). These hydraulic heads are controlled by the vertical permeability of shaly aquitards. TEMIS-PACK models indicate that observed hydraulic heads imply aquitard permeabilities around a microdarcy (Figure 10b), not a nanodarcy or less like in the previous examples. This value derived from regional flow and detailed temperature profile models (Gaulier and Burrus, 1994) is consistent with regional rare gas distribution models (Castro, 1995). We showed elsewhere

(Burrus et al., 1997b) that upward vertical oil migration from Lower Liassic source rocks across the Liassic aquitards is not impeded over geological time scale when aquitards have permeabilities around a microdarcy. In contrast, if nanodarcy permeabilities are considered, expulsion and upward oil migration past the shaly intervals that separate the Lower Liassic source rocks from the overlying reservoirs is not possible within the time available since oil generation started (around 10–50 Ma). The relatively "high" permeability of the Liassic shales seems therefore to be a key element in the effectiveness of Paris Basin petroleum systems. It differs by a factor of at least 10,000 compared to the Williston Basin Bakken shale permeability.

DISCUSSION

The modeled examples provided in this study show that mechanical compaction can explain overpressures in a variety of situations. In each case shown, overpressures were linked with a low-permeability shale interval. Natural systems contain many non-shale overpressured lithologies like tight sandstones.

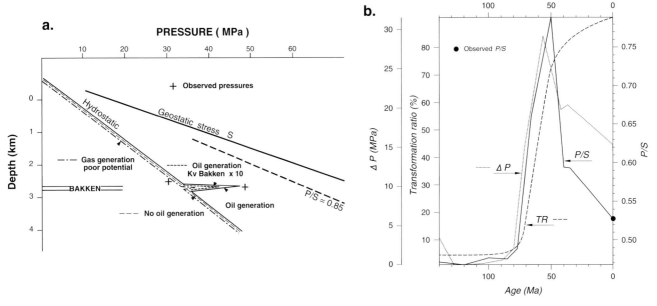

Figure 9. (a) Calculated pressure profiles in the center of the Williston Basin for different assumptions. Continuous line: oil generation in the Bakken is taken into account, Bakken permeability is adjusted to fit the pressure peak observed at the Antelope field (Meissner, 1978). Predicted overpressure peak is halved if Bakken permeability is increased by a factor ten. Dashed line (labeled "no oil generation"): oil generation in the Bakken is not included, no overpressure develops. Mixed dash profile (labeled "gas generation, poor potential"): Bakken TOC is 3% instead of 10%, as for the solid profile (Burrus et al., 1996), and generated HC is assumed to be a compressed gas, in order to maximize impact on overpressure. Still, Bakken overpressure is insignificant. The presumed fracture gradient profile (labeled "P/S=0.85") is far above the current overpressure peak, indicating no hydrofracturing possibilities at present day. (b) Temporal evolution of theoretical overpressure (dotted line), transformation ratio (maturity, dashed line), and pore pressure to lithostatic stress P/S ratio (solid line) in the Bakken shale at the center of the Williston Basin with the best fitting heat flow scenario and best fitting Bakken Shale permeability (details in Burrus et al., 1996). The development of overpressures between 75 and 50 Ma parallels the increase of maturity–this is because oil generation is the cause of overpressures. The decrease of overpressure after 50 Ma is the consequence of regional erosion. Despite the overpressure in the Bakken, the ratio P/S has always been below 0.85-0.90, the presumed hydrofracture threshold (Burrus et al., 1996). Current actual P/S ratio is low, below 0.55.

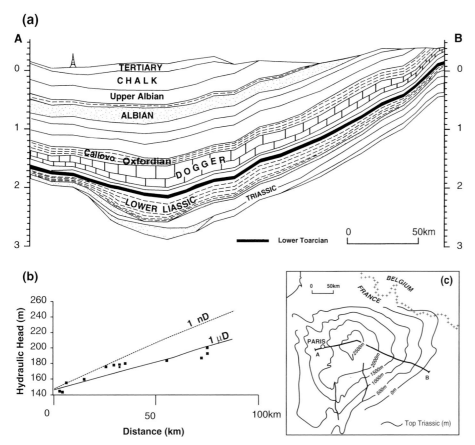

Figure 10. East-west cross section in the Paris Basin (a), actual and theoretical hydraulic head distributions in the Dogger interval (b) and location map (c). The Dogger is an important geothermal and HC reservoir. Main aquifers are indicated by carbonate and sandstone patterns, aquitards are represented by shale pattern in (a). The observed hydraulic head distribution in the Dogger in the west of the section is indicated by solid squares in (b). The actual distribution is compared in (b) with theoretical distributions obtained if aquitards' permeabilities have vertical permeability around a microdarcy (solid line) and a nanodarcy (dotted line). The solid line shows good consistency with observed hydraulic heads, while the dotted line is inconsistent, suggesting that regional scale permeability of Paris Basin shales layers is around 1 microdarcy rather than around 1 nanodarcy as in the Mahakam Delta or Gulf Coast. This is confirmed by rare gas distribution data and models (see text).

Although not addressed in this study, generation of overpressures in these rocks obeys the same physical principles which apply to shale—an overpressure develops when the rate of leakage lags behind the effect of loading and/or fluid generation or expansion inside the system. Whatever the lithology, the development of overpressure due to burial requires a low permeability and a high compressibility. The development of overpressure due to fluid generation or fluid expansion is also facilitated by a low permeability but the compressibility needs be low rather than high (Miller and Luk, 1993; Luo and Vasseur, 1993). Compacted shales have permeabilities generally significantly lower (by typically at least three orders of magnitude) than tight sandstones or carbonates, while compressibilies of shales are generally higher than in carbonates or sandstones (Neuzil and Pollock, 1983). This explains why shales are particularly affected by burial-driven overpressure development.

In subsequent discussions, we address the origin of overpressures and the validity of the shale permeability and effective stress/porosity characteristics used to model overpressures, in rapidly subsiding and in slowly subsiding basins. We believe most of the discussion applies also to overpressure generation in non-shale lithologies, but we acknowledge that additional complications can be expected in such rocks. For instance, diagenetic effects can alter the evolution of permeability and compressibility both through space and time,

making the evolution of overpressures more complicated to analyze.

Young, Rapidly Subsiding Basins

In young (Cenozoic), rapidly subsiding basins like the U.S. Gulf Coast and the Mahakam Delta, we found that compaction disequilibrium is the dominant source of overpressures, far exceeding hydrocarbon generation. The influence of hydrocarbon generation on overpressures is found to be small in rich source rocks (like in the Smackover Formation) or negligible in average to poor source rocks (Mahakam shales, Gulf Coast Cretaceous-Tertiary source rocks). The shale permeability calibrated against overpressure profiles seems to differ by a factor less than 10 from one basin to the other, despite the very different scales of the sections studied in the Mahakam Delta (60 km) and Gulf Coast (800 km). This is an indication of the robustness of the model.

In both the Mahakam Delta and the Gulf Coast, overpressures observed in the shaly intervals imply permeabilities around 0.1–10 nanodarcies (Figure 12). These permeabilities are so low that there is little possibility of hydrocarbon migration during the period elapsed since oil has been generated (few tens of Ma to less than 1 Ma), even if oil columns exceed capillary pressures. More generally, we find that hydrocarbon migration by multiphase Darcy flow through thick overpressured shale sections (with thicknesses of the order of one to several kilometers) is in the absence of hydrofractures or fractures with enhanced permeability along them, ineffective, even over tens or hundreds of million years. Our model ignores hydrocarbon transport by molecular diffusion, but we believe this limitation is unimportant at the scale considered. There is a consensus that diffusion does not affect much oil migration. For gas, the situation is more complex. Although diffusion coefficients of gas in tight shales are uncertain, experimental values generally lead to diffusive flow rates comparable with convective flow rates over distances less than 100 m (Kroos, 1992). Gas diffusion across cap rock can lead to losses of actual gas accumulations in a few million years if cap rock thickness is below 50 m (e.g., Nelson and Simmons, 1995). This process implies that present-day gas accumulations with thin seals have been actively fed until present-day. For much larger shale thickness, like the several kilometer thick overpressured Mahakam and Gulf Coast shales, the efficiency of diffusion (which decreases with the square of distance) becomes negligible (e.g., Kroos, 1992). Our model relies on other assumptions that need further examination. In particular, we need additional constraints on the relative permeabilities used to model hydrocarbon migration across shales. We discuss elsewhere (Burrus, 1997) phenomenological considerations and experimental evidence (e.g., Galle, 1997) that tend to support the shape of relative permeability considered in this study. We also need to

know more about the quantities of hydrocarbon that are retained in the kerogen, and cannot be expelled from source rocks by Darcy flow.

In the Mahakam Delta, the modeled and the observed P/S ratios never exceed 0.85 and there are no listric faults. In a compressional region, this means that the development of pervasive hydrofracturing in the overpressured section has a low probability. In contrast, the overpressured Gulf Coast shales and sandstones reach hydraulic fracturing thresholds (P/S = 1) in the first few kilometers below the top of the overpressures, at depths that generally coincide with the hydrocarbon window. This creates instabilities along the listric normal faults, and increases, probably episodically, the permeability along the fault planes. This mechanism provides the vertical migration pathways along which hydrocarbons can migrate from the source rocks into the reservoirs than are located several kilometers above them through overpressured shales. Overpressures thus influence hydrocarbon migration through hydraulic fracturing and enhancement of regional fault permeabilities.

Other causes of overpressures are probably unimportant, at least in shales. Aquathermal pressuring can be an additional source of overpressures in the case of burial associated with stress unloading in any rock type. Model effective stress history for Mahakam Delta and Gulf Coast shales however does not indicate any decrease of effective stress (Figure 11). Tests not shown indicate that an effective stress decrease may occur in organically rich source rocks (several tens of kilograms hydrocarbon/metric ton rock initial petroleum potential) in the oil window, and at a very low permeability (below 1 nanodarcy). This is the case for the Bakken shale. Except for these particular examples, effective stress increases with burial, so aquathermal pressuring plays a negligible role, confirming earlier conclusions (Luo and Vasseur, 1993). We have not considered the case of carbonates and sandstones in detail. It remains to be investigated if during diagenetic events, cementation or dissolution processes can modify pore structure, rock permeability and compressibility such that the effective stress can decrease through time while sedimentation and burial proceed.

Illitization reactions are also unlikely to be an important source of overpressures in the examples shown, for several reasons. First, in the Mahakam Delta, illitization starts at least 1 km above the top of the overpressured zone and is completed when overpressures appear, so it cannot be a source of overpressure (Rinckenbach, 1988; Furlan, 1993). If illitization was an important source of overpressure in the Gulf Coast and not in the Mahakam Delta, it would be surprising that similar compaction models account simultaneously for the overpressures observed in the Mahakam Delta and in the Gulf Coast. Secondly, earlier models which included clay dehydration have failed to show a significant effect on overpressures (e.g., Bethke, 1986; Dutta, 1987). Thirdly, the development of

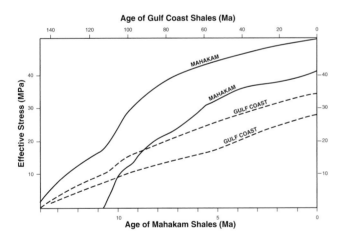

Figure 11. Temporal evolution of effective stress vs. time for overpressured Mahakam shale (bounded by the solid lines; location at Sisi, see Figure 3) and for Gulf Coast Smackover shale (bounded by the dashed lines; location between wells A and D, Figure 7). Permeabilities that allowed the best fitting of present-day overpressures were used in both cases. Despite the fact that hydrocarbon generation was taken into account in both cases, it is important to note that the effective stresses do not decrease through time. Their increase is the consequence of burial and sedimentation. The increase conflicts with recent proposals that "unloading" conditions are met in basins undergoing solid sedimentation and burial.

illitization and its relationship to temperatures are more complex than thought in the past (Elliot et al., 1991; Velde and Vasseur, 1992); insufficient potassium availability seems to be a frequent limitation to illitization reactions. The development of overpressures could even inhibit dehydration reactions (Colten-Bradley, 1987). An indirect effect of illitization on overpressures could be the reduction of permeability, as quartz cement is a by-product of illitization reactions (Hower et al., 1976; Bethke, 1986). Since overpressures are sensitive to small changes in permeabilities, this effect could be locally important.

Slowly Subsiding Petroleum Systems

Slowly subsiding Paleozoic or Mesozoic basins like the Paris or the Williston Basins, in which sedimentation rates were low and where carbonates often represent a significant part of the sedimentary volume (and help drain the basin), show no compaction disequilibrium.

In the Paris Basin, hydraulic heads are regionally controlled by the present day topography and by the regional permeability structure of aquifers and aquitards. The regional distribution of heads imply vertical permeability of the shaly aquitards of about a microdarcy, a value 100–1,000 times higher than in the Mahakam or Gulf Coast shales, and 10,000 times more than the Bakken shales. In the laboratory, however, Lower Toarcian shales have permeabilities that are about 10 nanodarcies (see below). This difference could be the result of pervasive vertical fracturing tentatively related to the episodic reactivation of the faulted Paris Basin basement (Mégnien, 1980).

A different situation prevails in the Bakken Shale. Here hydrocarbon generation is the cause of overpressures that are confined within a thin source rock. Uncommon conditions allow this unique situation–an extremely organic rich source rock (TOC 10%, hydrogen index >600), a very low porosity (<0.03) and very low inferred vertical permeability (10^{-2} to 10^{-3} nanodarcies) due, in our model, to the very low porosity. Bakken Shale permeabilities have not been extensively studied. Their inferred modeled permeabilities are close to the lowest measured permeability reported in the literature (Lin, 1978).

Hydrocarbon generation has been reported to be a dominant cause of overpressuring in several basins, like those in the US Rocky Mountain region, generally because depth of oil windows and overpressured rocks are similar and gas is the pressuring phase (e.g., Law, 1984; Law and Dickinson, 1985; Spencer, 1987). But these basins have average organic content far below the Bakken Shale, and their high proportion of non-shale lithologies (sandstone, siltstone, marls or coal) suggests an average permeability much above that of the Bakken Shale. These basins have generally been affected by considerable recent erosion, a process that develops underpressures rather than overpressures (Neuzil and Pollock, 1983). None of these characteristics seem favorable to the development of overpressures by the processes discussed in this paper. It would be thus be interesting to examine whether a mechanical compaction model can explain the development of overpressures in these basins, and whether the impact of hydrocarbon generation on overpressures in these basins could be significantly more important than the effect of compaction or erosion.

Although pressure regimes in the Lower Toarcian of the Paris Basin and in the Bakken Shale are very different, both source rocks have an equivalent expulsion efficiency of around 80–90% (Espitalié et al., 1987; Burrus et al., 1996). For the Lower Toarcian, the high expulsion efficiency can be related to its "high" permeability–multiphase Darcy models indicate that microdarcy permeability is high enough to allow efficient expulsion over geological time (Burrus et al., 1993b). For the Bakken Shale, the reason seems to be related to the low porosity. The low porosity of Bakken Shale can contain only 10% of the total oil produced in the Bakken (Burrus et al., 1996b), so oil produced at a rate imposed by the thermal history is forcibly expelled from the source despite its very low-permeability. These examples show that relations between overpressures and expulsion efficiency are not straightforward.

Our study suggests that shale permeabilities appear more difficult to predict in old (Paleozoic–Mesozoic), slowly subsiding basins compared with young (Cenozoic) rapidly subsiding basins. They can be either significantly (100–1,000 times) lower or larger than in rapidly subsiding basins. More complex structural and diagenetic histories are probably the cause of this greater variability.

Discussion of Shale Permeability and Mechanical Characteristics

Mahakam, U.S. Gulf Coast and Williston Basin overpressures were modeled with the same shale specific surface area S_0 of 10^8 m^{-1} and a variable vertical k_z of 0.5 to 0.05 in the Mahakam Delta, 0.01 in the Gulf Coast and 0.1 in the Williston Basin. Overpressured shales are predicted to be associated with porosities around 0.15–0.20 in the Mahakam Delta (Figure 4a), 0.30–0.50 in the Gulf Coast (Figure 8) and 0.03 or less in the Williston Basin. Modeled permeabilities across the overpressured formations are thus 1–5 nanodarcies in the Mahakam Delta, 0.1–2 nanodarcies in the Gulf Coast, and 0.01–0.001 nanodarcy in the Williston Basin.

Permeabilities required to define pressure seals are controversial. Using simple considerations, Deming (1994) argues permeabilities of 0.01 to 1 nanodarcy are required for a geologic unit to confine anomalous pressure over geologic time scales and 0.01 to 0.0001 nanodarcy if no significant loading has occurred for 100 m.y. or more. He and Corrigan (1995) argue these permeabilities are too low by a factor 10 to 100. Our analysis, which uses a more complex formulation and a more exact numerical treatment than both previous authors, leads to intermediate results. These results depend markedly on burial rates.

In the following discussion, we examine how modeled parameters compare with direct measurements.

Specific Surface Area

Modeled S_0 values can be compared with experimental surface areas based on nitrogen adsorption techniques. These measurements are usually expressed in g/m². The modeled S_0 value of 10^8 m^{-1} corresponds to a value of 38 g/m² taking an average matrix shale density of 2,650 kg/m³. This value falls within the range 20–60 g/m² reported by Borst (1982) for mud-

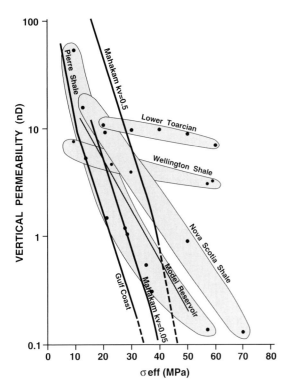

Figure 12. Comparison of theoretical (heavy lines) and experimental permeability/stress relations (solid circles) for Cretaceous Pierre Shale, Wellington Shale, Upper Jurassic Kimmeridge Bay shale, Jurassic Nova Scotia shale and Lower Toarcian shale of the Paris Basin (references in the text). The curve labeled 'Model Reservoir' has been obtained by matching pressure depletion in a Gulf Coast reservoir, at time and space scales intermediate between laboratory and basin models (reference in the text). The solid profiles represent the modeled permeability/effective stress relation used to simulate the Mahakam and Gulf Coast overpressures in Figures 4 and 7. Two Mahakam curves are shown: the one labeled "kv = 0.5" corresponds to the solid profile of Figure 4b; the one labeled "kv = 0.05" yields the best fit with detailed pressure profiles not shown. Agreement of Mahakam Delta and Gulf Coast profiles with Nova Scotia shale or Pierre Shale is good. The Lower Toarcian shale or Wellington Shale curves do not agree with theoretical curves, maybe owing to a high cement content.

stones buried between two and four km and coincides well with the value found between 30 and 40 g/m² for North Sea mudstones buried between 1.3 and 2.2 km (Aplin et al., 1995). Vasseur et al. (1995) report an S_0 value of 40 g/m² for kaolinite and found little variation during compaction experiments up to 50 MPa (7,000 psi) effective stress. All these values are consistent with our inferences.

Vertical Tortuosity

The k_z is, in our study, the free parameter which allows matching calculated and observed overpressures in the shaly successions studied. We found that it was possible to account for the different pressure situations studied with k_z varying between 0.5 and 0.01. This range of variation corresponds to variations of the vertical tortuosity of shale between 1.5 and 10, following Equation (8). Interestingly, this range coincides roughly with the experimental variability at the core scale, as indicated above. Although detailed petrophysical characteristics of the shales studied are not available, this consistency suggests that that the k_z parameter has a physical meaning. The variability of k_z probably reflects the variability of the tortuosity at pore scale or of the microfabric of shales.

Shale Surface Permeability and Tortuosity Evolution

Our modeling procedure implies that k_z is calibrated against overpressures observed at depth at present day. The best fitting k_z value is assumed to be constant with time and depth. It is possible to examine this assumption by comparing permeability predicted and observed at the sea floor. Permeabilities measured on deep sea cores vary between 0.1 and 1 millidarcy, within a factor of 10 (e.g., Pearson and Lister, 1973; Abbott et al., 1981). Following Equation (6), when k_z varies between 0.5 and 0.01 as implied by overpressure models in the different basins studied, the predicted vertical permeability at an assumed surface porosity of 0.75 is between 1 and 100 microdarcies. The second value is in good agreement with experimental determinations. The first value seems too small by a factor of 100. We can conclude that k_z probably decreases by a factor 100 between the surface and 1–2 km depth, instead of being constant as in our model. Such a decrease is consistent with the observation that during burial between 0 and 1 km depth, clay particles become increasingly parallel oriented, thus increasing the vertical tortuosity of the shales (Heling, 1970; Vasseur et al., 1995).

Experimental Permeability/Porosity Relations in Shale

Available compacted shale permeability measurements fall often in the range of 1–10 nanodarcies (e.g., Gondouin and Scala, 1958; McKelvey and Milne, 1962; Young et al., 1964; Luffel and Guidry, 1989). Some are in the range 0.1–0.01 nanodarcy (Young et al., 1964; Lin, 1978). These measurements are however often difficult to compare–the mineralogy, the porosity, the experimental conditions are generally not specified. Experimental permeability/porosity relations in shales have not been studied so far. The Kozeny-Carman equation would correctly predict the magnitude of kaolinite permeability (found between 30 and 100 nanodarcies), but would underestimate the rate of decrease of the permeability when porosity decreases from 0.50 and 0.30 (Vasseur et al., 1995). It is uncertain whether such a conclusion, based on experiments, can be extrapolated to nature. Chenevert and Sharma (1991) studied experimentally the effect of small porosity variations (due to increasing stress) on permeability of Wellington shales. Permeabilities were in the nanodarcy range. They report a permeability decrease by a factor of 1.8 when porosity decreases from 0.109 to 0.107 and by a factor of 1.4 when porosity further decreases from 0.107 to 0.103. Using Equation (6), we find with the S_0 value of 10^8 m^{-1} used in this study a decrease of permeability by a factor 1.2 and 1.3, respectively. A Kozeny-Carman type relation seems also to underestimate the experimental permeability reduction. The experimental porosity variation is, however, narrow. It is thus difficult to derive firm conclusions as to how experimental consolidation or oedometric tests support the Kozeny-Carman relation.

Experimental Permeability/Stress Relations

Recently published (see references below) permeability/effective stress relations provide more insights. Figure 12 represents measured permeability/effective stress relations for shale samples that have permeabilities around or below 1–10 nanodarcies. Two groups can be distinguished. The first group shows a steep slope on the log(K)/σ_e plot. These samples include the Cretaceous Pierre Shale and Wellington Shale (Chenevert and Sharma, 1991), the Upper Jurassic Kimmeridge Bay Shale (Swann et al., 1989) and the Jurassic Nova Scotia shale (Katsube et al., 1991). The second group shows a distinct lesser slope. These include the Lower Toarcian Shale of the Paris Basin (Anonymous, 1989) and the Wellington Shale (Chenevert and Sharma, 1991). The first group shows a decrease of permeability by at least a factor of 100 when effective stress is increased from 10 to 50 MPa. We superposed on these experimental points the permeability/effective stress relation used to model the Mahakam and Gulf Coast overpressures. These relations were obtained by eliminating ϕ from the K(ϕ) and $\phi(\sigma_e)$ relationships. The fit with the experimental plot appears very good at least down to an effective stress of around 40 MPa and permeability of 1 to 0.1 nanodarcy. The Mahakam and the Gulf Coast modeled curves coincide with the experimental trends quite well. Interestingly, data obtained by matching depletion in Gulf Coast reservoirs (Bourgoyne, 1990) also agree with the general trend of group one. These points represent a "loading" experiment with intermediate time (few years) and space (few hundred meters) scales compared with laboratory and basin models. The good linearity of the log(K)/σ_e

curve can be an indication that the reduction of permeability of shales with increasing stress is due to the progressive closure of microcracks (Gueguen et al., 1987; Katsube et al., 1991). The second group shows hardly a factor 2 permeability decrease when effective stress is increased up to 50 MPa. Chenevert and Sharma (1991) suggest a higher cement content might explain the more "rigid" behavior of Wellington and Pierre Shales. This explanation remains to be investigated, but it would provide an interesting geological guide.

The Kozeny-Carman relation is a pipe-type permeability model. There is no reason why this model should apply particularly well to shales. However it appears that the consequences of the Kozeny-Carman model are reasonable when model parameters (specific surface area, tortuosity, $\phi(\sigma_e)$ relations) are compared with experimental constraints, at least in rapidly subsiding basins. Although additional controls are needed, these results are encouraging. We also found that the $\phi(\sigma_e)$ model is not universal. Permeability/stress relations and porosity can be altered by cementation and fracturing, as is possibly the case, in the Williston and Paris Basins.

Discussion of the Causes of Distinct Porosity/Vertical Effective Stress Relations

Compaction models predict porosities that are systematically overestimated in the overpressured section of rapidly subsiding basins. This is a consequence of the fact that actual $\phi(\sigma_e)$ functions are different in the hydrostatic zone and in the overpressured zone. This observation is controversial.

It has been proposed previously that the curve observed in overpressured rocks is an unloading curve. It would be caused by a decrease of effective stress through time, due to fluid generation inside the overpressured rocks (Browers, 1994; Hart et al., 1995). This explanation is conceivable if a significant internal source of fluids exists at a depth coincident with the overpressures. This inference raises difficulties in the Mahakam and Gulf Coast shales. Mahakam or Gulf Coast shales have in general a poor hydrocarbon generation potential (<2 kg HC/t). Models that include hydrocarbon generation do not predict stress decline (Figure 11) or significant overpressure due to fluid generation (Figure 7), but they still show the distinct "unloading" curve in the overpressured section. A distinct effective stress curve is observed in the Gulf Coast Pathfinder well (location in Figure 7) at a depth of 2,200 m (Hart et al., 1995). This depth is at least 2 km above the top of the oil window in the region, so significant hydrocarbon generation is unlikely. A water source due to clay dehydration also appears to be unlikely. In the Mahakam Delta, illitization is already completed when overpressures are encountered.

In a previous paper (Schneider et al., 1993), we proposed an alternative explanation based on a different expression for effective stress. Following poro-elastici-

Figure 13. (a) Modification of effective stress/porosity functions for the Mahakam Delta shales if a Biot coefficient of 0.77 (dashed line) and 0.87 (dotted line) instead of 1 (solid line) is introduced (original data shown in Figure 5). AB = data in normally pressured shale, BC = data in overpressured shale (α = 1). AB′ = data in normally pressured shale, B′C′ = data in overpressured shale (α = 0.87). AB″ = data in overpressured shale, B″C″ = data in overpressured shale (α = 0.77). The two curves found in the normally pressured and in the overpressured sections nearly coincide if the Biot coefficient is 0.77 or 0.87 instead of 1. **(b)** Effective stress/porosity functions in Gulf Coast shales for different values of the Biot coefficient: 0.67 (dashed line), 0.77 (dotted line) and, 1 (solid line). Original data from Browers (1994). The 'normally pressured' curve AB (α = 1) is transformed into AB′ (α = 0.77); the overpressured curve BCDE (α = 1) is transformed into B′C′D′E′ (α = 0.67). The two inserts show the location of points A to E on Browers original data. Both curves coincide if the Biot coefficient is 0.67 in the transition zone (point C) and 0.77 in the hard overpressure zone (point D and E). Browers sonic velocities were transformed into porosity using Issler's relation: $\phi = 1 - (Dt_{ma} - Dt)^{1/x}$, with Dt_{ma} = 220 ms/m and x = 2.19 (Issler, 1992). **(c)** Effective stress/porosity curves in Gulf Coast shales at the PF well (Hart et al., 1995, see location in Figure 7) for different values of the Biot coefficient: 0.67 (dashed line), 0.85 (dotted line) and, 1 (solid line). AB = normally pressured shale data, BC = overpressured shale data. Both curves coincide for a Biot coefficient of 0.85.

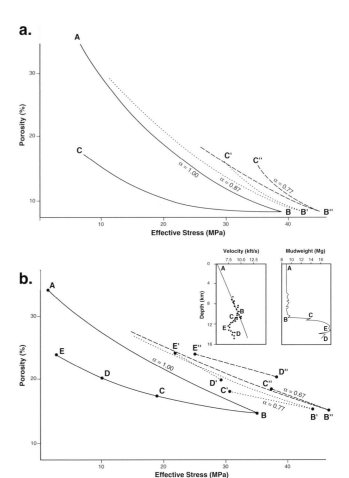

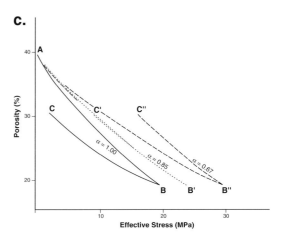

ty (Biot, 1941; Nur and Byerlee, 1971), effective stress can be defined as:

$$\sigma_e = S - \alpha \cdot P \qquad (9)$$

We show that for α equal to 0.87, the two curves found in the Mahakam Delta can be nearly superposed (Figure 13a). The present study confirms this explanation. Figure 13b shows that if α is between 0.67 and 0.77, then the curve observed by Browers (1994) in the hydrostatic section in Gulf Coast shales also becomes coincident with the curve found in the overpressured section. Figure 13c indicates that a value of 0.85 allows superposing the two curves observed by Hart et al. (1995) in the Gulf Coast. The parameter α, known in poro-elasticity as Biot's coefficient, can be related to the solid grain compressibility, β_g, and to the bulk rock compressibility, β_b, by:

$$\alpha = 1 - \frac{\beta_g}{\beta_b} \qquad (10)$$

Following Hubbert and Rubey (1959) and soil mechanics, previous overpressure models have generally assumed an α value of 1. In the laboratory, experimental α values decrease with porosity from a value of 1 at surface conditions to values around 0.6 to 0.8 at porosity around 0.15–0.20 for carbonates and sand-

stones (e.g., Bouteca and Sarda, 1995). Schneider (1993) and Burrus et al. (1997a) showed that this alternative definition of α leads to significantly lower porosity predictions in the overpressured intervals. Predicted porosities associated with overpressured intervals are halved if α varies with porosity from a value of 1 at the surface to a value around 0.67 at a porosity of 0.20, instead of being constant and equal to 1 (Schneider, 1993). Modeled porosities are no more in excess compared with observations if a variable α is considered.

Shale Biot coefficients are poorly documented. Few oedometric experiments on shales and marls have found Biot coefficients around 0.70 (M. Bouteca, personal communication), but data is limited and extrapolation to geological compaction remains questionable. We present below an original analysis which indirectly supports the idea that Biot coefficients for shale are lower than one. The idea is to derive a "statistical" value for the Biot coefficient from the distribution of horizontal stresses in sedimentary basins. In poro-elasticity, in uniaxial compression, the horizontal stress can be related to the vertical stress by (Engelder and Fisher, 1994):

$$S_h = \frac{v}{1-v} \cdot S + \alpha \cdot \frac{(1-2v)}{(1-v)} \cdot P \qquad (11)$$

The Poisson coefficient, v, is around 0.15 to 0.20 for compacted shales (e.g., Swann et al., 1989). It has been observed in the Gulf Coast (Breckels and Van Eekelen, 1982; Zoback and Peska, 1995), in the North Sea (Breckels and Van Eekelen, 1982; Gaarenstroom et al., 1993) and in the Nova Scotia margin in Canada (Yassir and Bell, 1994) that S_h increases more rapidly with depth in overpressured rocks compared with hydrostatic rocks (Figure 1). The anomalous increase of S_h has been explained as a consequence (not a cause) of the development of overpressures (Engelder and Fisher, 1994)–when overpressure develops, part of the overpressure is "transferred" elastically on the lateral stress, which thus becomes more than normal. The stress transfer is related to the Biot coefficient through Equation (11). A compilation of S_h data from the Gulf Coast shows the following non-linear increase with depth (Breckels and Van Eekelen, 1982):

$$S_h(\text{psi}) = 0.197 \cdot D^{1.145} + 0.46 \cdot (P - P_h) \qquad (12)$$

where D is depth in feet, P_h is the hydrostatic pressure (psi), P is the pore pressure (psi). The precision of this equation can be evaluated by comparing predictions with recent stress determinations acquired in offshore Louisiana (Zoback and Peska, 1995). Observations indicate a horizontal stress of 5,377 psi (37.07 MPa) in an overpressured shaly section, when Equation (12) predicts a very close value of 5,434 psi (37.47 MPa). Equation (12) seems therefore predictive. We can rewrite Equation (12) as follows:

$$S_h = (0.197 \cdot D^{1.145} - 0.46 \cdot P_h) + 0.46 \cdot P \qquad (13)$$

The bracket represents a quasi-linear function of depth, not affected by overpressure development, whereas the second term represents the increase of S_h due to overpressures, which is not linear with respect to depth. Equation (11) has also two terms, the first is a quasi-linear function of depth, the second also represents the increase of S_h due to overpressures. We can therefore identify the second terms of Equations (11)

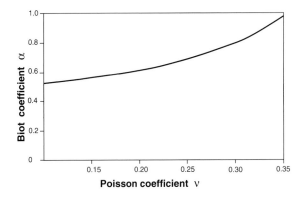

Figure 14 - Variation of the Biot coefficient α versus Poisson's ratio v in overpressured shales. This relation is derived from empirical distribution of horizontal stress observed in the Gulf Coast. It suggests that with realistic v values around 0.20, α is around 0.65-0.70.

and (13), and obtain an empirical approximate expression for α:

$$\alpha = 0.46 \cdot \frac{(1-v)}{(1-2v)} \qquad (14)$$

Equation (14) indicates that, with a reasonable value for v around 0.20, the α value predicted is around 0.65 (Figure 14). It could be higher since the factor 0.46 in Equation (14) seems to be a lower limit; values of 0.49 in Brunei and 0.56 in Venezuela have been reported (Breckels and Van Eekelen, 1982). In this case α values of 0.70 and 0.79, respectively, are obtained. These values coincide or are similar to those that allow superposing the effective stress/porosity curves found in the overpressured and normally pressured portions in the Mahakam Delta or in the Gulf Coast (Figure 13).

This analysis leads to three conclusions. Firstly, geologic data like horizontal stress distribution profiles confirm that Biot coefficient values are different from 1 in compacted shales, despite the lack of direct measurements. Secondly, the distinct $\phi(\sigma_e)$ curve observed in overpressured rocks is probably not an unloading curve. There is a single loading curve in both normally pressured and overpressured rocks if an alternative definition of effective stress is employed. There is no need to invoke problematic fluid generation. Thirdly, this study confirms the earlier explanation of Schneider et al. (1993) concerning the excess modeled porosity found in overpressured rocks–it is an artifact due to the fact that previous overpressure models consider a Biot coefficient equal to 1.

Consequences for Permeability Calibration

The conclusion that effective stress should be defined with a Biot coefficient different from 1 contradicts the assumption upon which the previous case studies were modeled. In fact, previous studies indicate that changing α from 1 to 0.65–0.85 has limited consequences as far as permeability calibration is concerned.

Schneider et al. (1993) compared pressure and porosity predictions in compacting shales using variable α values and a constant value of 1. Similar overpressure predictions were found if permeabilities were increased by a factor between 3 (at porosity 0.20) and 7 (at porosity 0.15) in the first case. This difference in permeability is within the usual uncertainty of permeability calibration and does not significantly alter the previous discussion on permeability/effective stress relations.

More work is needed to confirm the existence and the role of variable Biot coefficients in shales during mechanical compaction. However, recent studies of the evolution of the microfabric of clay particles during compaction tend to support qualitatively this concept. When α equals 1, the mineral skeleton of shales is inferred to have a compressibility which is negligible compared with the bulk rock compressibility. When α equals 0.65, then both compressibilities must be more comparable (Equation 10). We believe recent models of the microfabric of kaolinite and its evolution with consolidation (Vasseur et al., 1995) can help interpret the decrease of α as consolidation proceeds. Kaolinite clays are described as an assemblage of large ellipsoid "particles", typically 2,000–4,000 Å in diameter. Each particle is an aggregate of single crystals, stacked more or less parallel to the particle's main elongation direction. This "stack" encloses some microporosity between the crystals. Each crystal contains around 25 silicate molecular layers and is believed undeformable. Most initial porosity is in the interparticle space. During consolidation the particles reorient themselves in a parallel fashion; while increasing friction at the contact between the particles must be overcome. In this early stage, the internal structure of the particles is not modified. Interparticle friction is small, and the compressibility of the whole rock is large compared with that of each single particle–α is equal to 1. As particles become increasingly reoriented, interparticle friction increases, and it may become more difficult to overcome this friction than to rearrange the crystals within each particle. As a result, the microporosity is affected, the solid compressibility becomes comparable to the bulk compressibility, and α may be much lower than 1.

Comparison with Diagenetic Theories

Mechanical compaction models discussed so far are often opposed to "chemical" compaction models in which overpressures would be solely controlled by the appearance of diagenetic seals cutting across the stratigraphy (e.g., Hunt, 1990). Evidence has indeed been given that in basins like the Anadarko Basin (Al-Shaieb et al., 1995) banded cemented horizons seem to control the extension of overpressures. Reaction Transfer Mechanical Models (e.g., Dewers and Ortoleva, 1988; Sonnenthal and Ortoleva, 1995) explain how these bands can form and how diagenesis and overpressures can interact. Our study of the Gulf Coast and Mahakam overpressures suggests on the other hand that, at least at the regional scale considered, diagenetic processes do not need to be invoked. This does not exclude the possibility that locally permeabilities can be decreased if cementation takes place, resulting in an increase of overpressures. It is possible that more mature basins like the North Sea, with intermediate sedimentation rates and ages, have mixed compaction mechanisms. Mechanical compaction would be responsible for "regional" overpressure development, while mineral diagenesis would alter the local compressibility, permeability, and porosity characteristics of the rocks.

CONCLUSIONS

In this study we have attempted to evaluate the origin of overpressures in different basins applying a 2D mechanical compaction model derived from Hubbert's effective stress/porosity concept. The most important results that emerge from the comparison of different regional studies are as follows:

1. Compaction models based on effective stress-porosity relations calibrated on observed porosity profiles are capable of explaining overpressures in young (Cenozoic), rapidly subsiding basins like the Mahakam Delta and the Gulf Coast. Best fitting shale permeability parameters seem to vary little from basin to basin, an indication of the robustness of the concepts that underlie the model. In rapidly subsiding Cenozoic basins, compaction disequilibrium is found to be the dominant cause of overpressures, compared with aquathermal pressuring, clay dehydration and hydrocarbon generation. With the exception of very organically rich sources in the oil window, effective stresses do not decrease with time during burial. Recent studies (Miller and Luk, 1993) showed that aquathermal pressuring is unimportant, except if unloading conditions are met. We conclude, since unloading does not take place, that aquathermal pressuring is an unimportant process. Oil generation is found to be responsible for a significant or dominant part of overpressures in sources rocks with exceptional characteristics (very high hydrocarbon generation potential; 30–70 kg hydrocarbon/metric ton rock, and permeability much below a nanodarcy).

2. Linear, elastic compaction models that have been used in the past implied that overpressured shales have permeabilities in the range 1–10 microdarcies, in contrast, recent measurements and models agree on values generally in the range 0.1–10 nanodarcies. Most of the discrepancy is due to the incorrect assumptions that "geological" shale compressibility does not vary during compaction and can be deduced from hydraulic tests. For rea-

sons that remain to be elucidated, hydraulic tests yield shale storage coefficients that are significantly (up to a factor of 100) higher than oedometric tests or basin models.

3. Shale permeabilities of 0.1–10 nanodarcies calibrated against observed overpressures seem consistent with independent experimental constraints. Specific surface areas required to model overpressures using Kozeny-Carman permeability models for shales are confirmed by laboratory studies. The relation between vertical effective stress and permeability incorporated in the model appears consistent with experimental results, down to permeabilities around 1 nanodarcy and effective stresses around 40 MPa. Experimental and modeled vertical tortuosities of shales seem also in agreement.

4. Mechanical compaction models overestimate the porosity of overpressured shales. This is due to the existence of two distinct $\phi(\sigma_e)$ curves in the normally pressured and in the overpressured section, a contradiction to Hubbert's assumption of a unique effective stress/porosity relationship. The most likely reason for this discrepancy is not fluid generation inside the rocks, as thought before, but an inappropriate definition of effective stress. If the Biot coefficient is taken equal to 0.65–0.85 instead of 1, then a single curve is obtained and most excess porosity predictions are suppressed. This assumption is shown to be supported by the non-linear distribution of horizontal stress with depth in overpressured basins.

5. In older (Paleozoic–Mesozoic), slowly subsiding basins, shale permeability seems to be considerably more variable than in young (Cenozoic) rapidly subsiding basins. Shale permeabilities observed in layers with similar thickness, depth, and organic content in the Williston Basin and Paris Basin can vary by more than four orders of magnitude. More complex tectonic and diagenetic histories probably explain this greater variability in old slowly subsiding basins. This additional complexity makes the prediction of hydraulic heads more difficult in old intracratonic basins than in young rapidly subsiding basins.

6. Mechanical compaction models do not conflict with diagenetic models. A combination of both models is probably required to account for overpressures and porosity trends in more mature basins where fluid-rock interactions and mineral evolution have altered porosity depth trends and permeabilities and compressibilities of overpressured rock sequences.

7. Modeling the pressure and permeability environment of source rocks is a useful and necessary step in petroleum system analysis. Relations between overpressures and expulsion/migration characteristics are, however, not straightforward. Overpressured and hydrostatic source rocks can have equally good expulsion efficiency (e.g., Paris Basin Lower Toarcian and Williston Basin Bakken). Thick shale sequences with similar overpressure profiles and permeability characteristics act very differently as far as hydrocarbon expulsion and migration are concerned, depending on whether hydraulic fracturing and vertical faults are present (Gulf Coast) or absent (Mahakam Delta). Compaction models facilitate the investigation of the potential of hydraulic fracturing through the evolution of the ratio between pore pressure and lithostatic stress. They allow the integration of pressures distribution and pressure in the investigation of petroleum systems.

ACKNOWLEDGMENTS I thank ELF, Exxon, Petro-Canada and Total for their permission to publish this paper. Regional studies mentioned in this paper benefitted from the participation of J. Barlier (ELF) and J.M. Gaulier (IFP) in the Paris Basin study; F. Schroeder, T. Schwarzer, C. Schaw and R. Lander (Exxon) in the Gulf Coast study; C. Cassaigneau, G. Choppin de Janvry and Y. Grosjean (Total) and E. Brosse (IFP) in the Mahakam study; K. Osadetz (ISPG), D. Dearborn and K. Visser (PetroCanada), B. Doligez and S. Wolf (IFP) in the Williston Basin study. The paper benefitted from discussions with F. Schneider and M. Bouteca (IFP). It was improved by the comments of Indu Meshri, Ben Law, and a third anonymous reviewer

REFERENCES CITED

Abbott, D., W. Menke, M. Hobart, and R. Anderson, 1981, Evidence for excess pore pressures in Southwest Indian Ocean sediments: Journal of Geophysical Research, v. 86, p. 1813–1827.

Al-Shaieb, Z., Puckette, J.O., Abdalla, A.A., and P.B. Ely, 1995, Megacompartment complex in the Anadarko Basin: a completely sealed overpressured phenomenon, *in* P.J. Ortoleva (ed), Basin compartments and seals: AAPG Memoir 61, p. 55–68.

Anderson R.N., P. Flemings, S. Losh, J. Austin, and R. Woodhams, 1994, Gulf of Mexico growth fault drilled seen as oil gas migration pathway: Oil and Gas Journal, June 1994, p. 97–103.

Angevine, C.L., and D.L. Turcotte, 1983, Porosity reduction by pressure solution: a theoretical model for quartz arenites: Geological Society of America Bulletin, v. 94, p. 1129–1134.

Anonymous, 1989, Les propriétés mécaniques des argiles de Grimonvillier 3: Centre National de la Recherche Scientifique, Institut de Physique du Globe (Strasbourg): IFP Contract 12773, 10 p.

Aplin, A.C., Yang S., and Hansen S., 1995, Compaction and characterization of mudstones, Well 8/3-2, Norwegian margin: Marine & Petroleum Geology 8, 3-2.

Athy, L. F., 1930, Density, porosity, and compaction of sedimentary rocks: AAPG Bulletin, v. 14, p. 1–24.

Barker, C., 1972, Aquathermal pressuring: role of temperature in the development of abnormal pressure zone: AAPG Bulletin, v. 56, p. 2068–2071.

Bethke, C.M., 1986, Inverse hydrologic analysis of the distribution and origin of Gulf Coast-type geopressured zones: Journal of Geophysical Research, v. 91, p. 6535–6545.

Bethke, C.M., and T.F. Corbett, 1988, Linear and nonlinear solutions for one-dimensional compaction flow in sedimentary basins: Water Resources Research, v. 24, p. 461–467.

Bethke, C.M., Harrison, W.J., Upson, C., and S.P. Altauer, 1988, Supercomputer analysis of sedimentary basins: Science, v. 239, p. 261–267.

Biot, M.A., 1941, General theory of three-dimensional consolidation. Journal of Applied Physics, v. 12, p. 155–164.

Bishop, R.S., 1979, Calculated compaction states of thick abnormally pressured shales: AAPG Bulletin, v. 63, p. 918–933.

Borst, R.L., 1982, Some effects of compaction and geological time on the pore parameters of argillaceous rocks: Sedimentology, v. 29, p. 291–298.

Bourgoyne, A.T., 1990, Shale water as a pressure support mechanism in gas reservoirs having abnormal formation pressure: Journal of Petroleum Science and Engineering, v. 3, p. 305–319.

Bouteca, M. and J.P. Sarda, 1995, Experimental measurements of thermoporoelastic coefficients, *in* Charlez, P. (ed.), Mechanics of porous media, A.A. Balkema, Rotterdam, p. 31–41.

Bradley, J.S., 1975, Abnormal formation pressure: AAPG Bulletin, v. 59, p. 957–973.

Breckels, I.M., and H.A.M. Van Eekelen, 1982, Relationship between horizontal stress and depth in sedimentary basins: Journal of Petroleum Technology, p. 2191–2199.

Bredehoeft, J. D., and B.B. Hanshaw, 1968, On the maintenance of anomalous fluid pressures: I. Thick sedimentary sequences: Geological Society of America Bulletin, v. 79, p. 1097–1106.

Browers, G.L., 1994, Pore pressure estimation from velocity data: accounting for overpressure mechanism besides undercompaction: Society of Petroleum Engineers , Paper 27488, p. 515–530.

Buhrig, C., 1989, Geopressured Jurassic reservoirs in the Viking Graben: modeling and geological evidence: Marine and Petroleum Geology, v. 6, p. 31–48.

Burland, J.B., 1990, On the compressibility and shear strength of natural clays: Géotechnique, v. 40, p. 329–378.

Burns, L.K., and F.G. Etheridge, 1979, Petrology and diagnostic effects of lithic sandstones: Paleocene and Eocene Umpqua Formation, southwest Oregon, *in* P.A. Scholle and P.R. Schluger (eds), Society Aspects of diagenesis: SEPM Special Publication 26, p.

307–317.

Burrus, J., 1997, Contribution à l'étude du fonctionnement des systèmes pétroliers: apport d'une modélisation bi-demensionnelle, Thèse de Doctorat, Ecole des Mines de Paris, 510 p. *(in French)*.

Burrus, J., and F. Audebert, 1990, Thermal and compaction processes in a rifted basin in the presence of evaporites, Gulf of Lions case study: AAPG Bulletin, v. 74, p. 1420–1440.

Burrus, J., Kuhfuss, A., Doligez, B., and P. Ungerer, 1991, Are numerical models useful in reconstructing the migration of hydrocarbons? A discussion based on the Northern Viking graben, *in* W.A. England, A.J. Fleet (eds), Petroleum Migration: London: The Geological Society, Special Publication v. 59, p. 89–111.

Burrus, J., Brosse, E., Choppin de Janvry, G., Grosjean, Y., and J.L. Oudin, 1992, Basin modeling in the Mahakam Delta based on the integrated 2D model TEMISPACK, *in* Proceedings of the 21st meeting of the Indonesian Petroleum Association: IPA Paper, 92-11-04, p. 23–44.

Burrus, J., Osadetz, K., Gaulier, J.M., Brosse, E., Doligez, B., Choppin de Janvry, G., Barlier, J. and K. Visser, 1993b, Source rock permeability and petroleum expulsion efficiency: modeling examples from the Mahakam Delta, Williston Basin, Paris Basin, *in* J.R. Parker (ed.), Petroleum Geology of NW Europe: Proceedings of the 4th Conference: London, The Geological Society, p. 1317–1332.

Burrus, J., Osadetz, K., Wolf, S., Doligez, S., Visser, K., and O. Dearborn, 1996a, A two dimensional regional model of Williston Basin hydrocarbon systems: AAPG Bulletin 80, 2, p. 265–291.

Burrus, J., Osadetz, K. Wolf, S., and Visser, K, 1996b, Physical and numerical modeling constraints on oil expulsion and accumulation in the Bakken and Lodgepole petroleum systems of the Williston Basin (Canada-U.S.A.), Canadian Society of Petroleum Geologists Bulletin, Special Issue, v. 44, No. 5

Burrus, J., Schwarzer, T.F., Schroeder, F.W., Lander, R., Wenger, L.M., Schneider, F., and Ross, A.H., 1997a, A regional model of Gulf Coast petroleum systems and how they function - north Louisiana to the deep Gulf of Mexico: AAPG Bulletin (submitted).

Burrus, J., Gaulier, J.M., Gable, R., and Barlier, J., 1997b, A regional-scale model of the Paris Basin petroleum systems; validity of Rock Eval Arrherius kinetics and multiphase Darcy equations in reconstructing hydrocarbon generation and migration: Petroleum Geosciences (submitted)

Burst, J. F., 1969, Diagenesis of Gulf coast clayey sediments and its possible relation to petroleum migration: AAPG Bulletin, v. 53, p. 73–93.

Capuano, R.M., 1993, Evidence of fluid flow in microfractures in geopressured shales: AAPG Bulletin, v. 77, p. 1303–1314.

Carstens, H., and H. Dypvik, 1981, Abnormal formation pressure and shale porosity: AAPG Bulletin, v.

65, p. 344–350.

Castro, M.C., 1995, Transfer of rare gases in the waters of sedimentary basins, example of the Paris Basin, Ph.D. thesis, University Pierre and Marie Curie, Paris, 245 p. *(in French)*.

Chenevert, M.E., and A.K. Sharma, 1991, Permeability and effective pore pressure of shales: Society of Petroleum Engineers, paper 21918.

Chiarelli, A., and F. Duffaud, 1980, Pressure origin and distribution in Jurassic of Viking basin (U.K. - Norway): AAPG Bulletin, v. 64, p. 1245–1266.

Colten-Bradley, V.A., 1987, Role of pressure in smectite dehydration-Effects on geopressure and smectite-to-illite transformation: AAPG Bulletin, v. 71, p. 1414–1427.

Daines, S.R., 1982, Aquathermal pressuring and geopressure evaluation: AAPG Bulletin, v. 66, p. 931–939.

Deming, D., 1990, Comment on "Compaction of Basin Sediments: modeling based on time-temperature history" by J.W. Schmoker and D.L. Gautier: Journal of Geophysical Research, v. 95, p. 5153–5154.

Deming, D., 1994, Factors necessary to define a pressure seal, AAPG Bulletin, v. 78, p. 1005–1009.

Dewers, T., and P. Ortoleva, 1988, The role of geochemical self-organization in the migration and trapping of hydrocarbons: Applied Geochemistry, v. 3, p. 287–316.

Dickey, P.A., C.R. Shriram, and W.R. Paine, 1968, Abnormal pressures in deep wells of Southwestern Louisiana: Science, v. 160, p. 609–615.

Dickinson G., 1953, Geological aspects of abnormal reservoir pressures in Gulf Coast Louisiana: AAPG Bulletin, v. 37, p. 410–432.

Doligez, B., Ungerer, P., Chenet, P.Y., Burrus, J., Bessis, F., and G. Bessereau, 1987, Numerical modeling of sedimentation, heat transfer, hydrocarbon formation and fluids migration in the Viking graben, North Sea, *in* Brooks, J., Glennie, K. (eds), Petroleum geology of Northwest Europe, Graham and Trotman, p. 1039–1048.

Domenico, P. A., and M. D. Mifflin, 1965, Water from low-permeability sediments and land subsidence: Water Resources Research, v. 1, p. 563–576.

Dow, W.G., 1974, Application of oil-correlation to source rock data to exploration in Williston Basin: AAPG Bulletin, v. 58, p. 1253–1262.

Durand, B., and J.L. Oudin, 1979, Exemple de migration des hydrocarbures dans une série deltaïque: le Delta de la Mahakam, Kalimantan, Indonésie, Proceedings of the 10th World Petroleum Congress, PDI, 3-111.

Du Rouchet, J., 1981, Stress fields, a key to oil migration: AAPG Bulletin, v. 65, p. 74–85.

Dutta, N.C., 1987, Fluid flow in low permeable porous media, Gulf Coast application, *in* B. Doligez (eds), Migration of hydrocarbons in sedimentary: Technip, p. 567–595.

Duval, B.C., G. Choppin de Janvry, and B. Loiret, 1992,

The Mahakam Delta province: an ever-changing picture and a bright future: Offshore Technological Conference, Paper 6855, p. 393–404.

Eaton, B.A., 1975, The equation for geopressure prediction from well logs: Society of Petroleum Engineers, Paper 5544.

Elliot, W.C., J.L. Aronson, G. Matisoff, and D.L. Gautier, 1991, Kinetics of the smectite to illite transformation in the Denver Basin: clay mineral, K-Ar Data, and mathematical model results: AAPG Bulletin, v. 75, p. 436–462.

Engelder, T., and M.P. Fischer, 1994, Influence of poroelastic behavior on the magnitude of minimum horizontal stress, S_h, in overpressured parts of sedimentary basins: Geology, v. 22, p. 949–952.

Espitalié, J., F. Marquis, and L. Age, 1987, Organic geochemistry of the Paris Basin, *in* Brooks, J., and K. Glennie (eds), Petroleum Geology of Northwest Europe, Graham and Trotman, p. 71–86.

Fertl, W.C., and D.J. Timko, 1970, Occurrence and significance of abnormal-pressure formations: Oil Gas Journal, v. 68, p. 97–108.

Furlan, S., 1993, Les transports de masse dans le delta de la Mahakam: Ph.D. thesis, University of Strasbourg (France), p. 298 *(in French)*.

Gaarenstroom, L., R.A.J. Tromp, M.C. de Jong, and A.M. Brandenburg, 1993, Overpressures in the central North Sea: implications for trap integrity and drilling safety: Proceedings of the 4th Conference on Petroleum Geology of Northwest Europe. p. 1305–1313.

Galle, C, 1997, Migration des gaz et pression de rupture dans une argile compactée destinée à la barrière ouvragée du stockage profond, Bulletin de la Société Géologique de France, (accepted).

Galloway, W.E., 1974, Deposition and diagenetic alteration of sandstone in northeast Pacific arc-related basin: implications for graywacke genesis: Geological Society of America Bulletin, v. 85, p. 379–390.

Gaulier, J.M., and J. Burrus, 1994, Modeling past and present thermal regimes in the Paris Basin: petroleum implications, *in* A. Mascle (ed.), Hydrocarbon and geology of France, Springer-Verlag: Berlin, p. 61–74.

Gaulier, J.M., Burrus, J. and N. Guilhaumou, 1991, Oil generation and migration in the Paris Basin, evaluation by 2D modeling, Third Conference of European Association of Petroleum Geologists, Florence, May 1991 (abstract), p. 26–27.

Gibson, R.E., 1958, The progress of consolidation in a clay layer increasing in thickness with time: Geotechnica, v. 8, p. 171–182.

Gondouin, M., and C. Scala, 1958, Streaming potential and the SP Log: Petroleum Transactions, AIME, v. 213, p. 170–179.

Gratz, A.J., 1991, Solution-transfer compaction of quartzites: progress toward a rate law: Geology, v. 19, p. 901–904.

Gueguen, Y., M. Darot, and M. Reuschlé, 1987, Permeability evolution under stress, *in* B. Doligez (ed), Migration of hydrocarbons in sedimentary basins: Technip, p. 281–295.

Gueguen, Y., and V. Palciauskas, 1992, Introduction à la physique des roches: Hermann (Paris), p. 299 *(in French)*.

Ham, H.H., 1966, New charts help estimate formation pressure: Oil and Gas Journal, v. 64, p. 58–63.

Hansen, S., 1996, A compaction trend for Cretaceous and Tertiary shales on the Norwegian Shelf based on sonic transit times, Petroleum Geoscience, v. 2, p. 159–166.

Hart, B.S., Flemings, P.B., and A. Deshpande, 1995, Porosity and pressure: the role of compaction disequilibrium in the development of geopressures in a Gulf Coast Pleistocene basin: Geology, v. 23, p. 45–48.

He, Z. and J. Corrigan, 1995, Factors necessary to define a pressure seal: discussion, AAPG Bulletin, v. 79, 7, p. 1075–1078

Hedberg, H.D., 1936, Gravitational compaction of clays and shales: American Journal of Sciences, Fifth Series, v. 31, p. 241–287.

Hedberg, H. D., 1974, Relation of methane generation to undercompacted shales, shale diapirs, and mud volcanoes: AAPG Bulletin, v. 58, p. 661–673.

Heling, D., 1970, Micro-fabrics of shales and their rearrangement by compaction: Sedimentology, v. 15, p. 247–260.

Hermanrud, C., L. Wensaas, G.M.G. Teige, E. Vik, H.M.N. Bolås and S. Hansen, 1998, Shale porosities from well logs on Haltenbanken (Offshore Mid-Norway) show no influence of overpressuring, *in* Law, B.E., G.F. Ulmishek, and V.I. Slavin eds., Abnormal pressures in hydrocarbon environments: AAPG Memoir 70, p. 65–85.

Hinch, H.H., 1978, The nature of shales and the dynamics of hydrocarbon expulsion in the Gulf Coast Tertiary section: Continuing Education Course Notes 8: AAPG, Tulsa (Okla), p. 1–31.

Hottmann C. E., and R. K. Johnson, 1965, Estimation of Formation Pressures from log-derived shale properties: Journal of Petroleum Technology, June 1965, p. 717–722.

Houseknecht, D., 1987, Accessing the relative importance of compaction processes and cementation to reduction of porosity in sandstones: AAPG Bulletin, v. 71, p. 633–642.

Hower, J., Eslinger, E.V., and E.A. Perry, 1976, Mechanism of burial metamorphism of argillaceous sediment: mineralogical and chemical evidence: Geological Society of America Bulletin, v. 87, p. 725–737.

Hubbert, K.M., and W.W. Rubey, 1959, Role of fluid pressure in mechanics of overthrust faulting. I. Mechanics of fluid-filled porous solids and its application to overthrust faulting: Bulletin of the Geological Society of America, v. 70, p. 115–166.

Hunt, J.M., 1990, Generation and migration of petroleum from abnormally pressured fluid compartments,
AAPG Bulletin, v. 74, p. 1–12.

Issler, D.R., 1992, A new approach to shale compaction and stratigraphic restoration, Beaufort-Mackenzie Basin and Mackenzie Corridor, Northern Canada: AAPG Bulletin, v. 76, p. 1170–1189.

Jacob, C.E., 1940, On the flow of water in an elastic artesian aquifer: Transaction of the American Geophysical Union, v. 2, p. 574–586.

Jacquin, C., 1965, Interactions entre l'argile et les fluides - Ecoulements à travers les argiles compactées, étude bibliographique: Revue de l'Institut Français du Pétrole, v. 20, p. 1475–1501.

Jansa, L.F., and Urea, 1990, Geology and diagenetic history of overpressured sandstone reservoirs, Venture gas field, offshore Nova Scotia, Canada: AAPG Bulletin, v. 74, p. 1640–1658.

Jones, P. H., 1969, Hydrodynamics of geopressure in the Northern Gulf of Mexico Basin: Journal of Petroleum Technology, July 1969, p. 803–810.

Katsube, T.J., B.S. Mudford, and M.E. Best, 1991, Petrophysical characteristics of shales from the Scotian shelf: Geophysics, v. 56, p. 1681–1689.

Keith, L.A., and J.D. Rimstidt, 1985, A numerical compaction model of overpressuring in shales: Mathematical Geology, v. 17, p. 115–135

Korvin, G., 1984, Shale compaction and statistical physics: Geophysical Journal Royal Astronomical Society, v. 78, p. 35–50.

Kroos, B.M., 1992, Diffusive loss of hydrocarbons through caprock. Experimental studies and theoretical considerations, Erdöl und Kohle-Erdgas-Petrochemie/hydrocarbon Technology, v. 45, p. 387–396.

Law, B.E., 1984, Relationships of source-rock, thermal maturity, and overpressuring to gas generation and occurrence in low-permeability Upper Cretaceous and Lower Tertiary rocks, greater Green River basin, Wyoming, Colorado, and Utah, *in* Woodward, J., Meissner, F.F. and J.L. Clayton (eds), Hydrocarbon source rocks of the greater Rocky Mountain Region: Rocky Mountain Association of Geologists, p. 469–490.

Law, B.E., and W.W. Dickinson, 1985, Conceptual model for origin of abnormally pressured gas accumulations in low-permeability reservoirs: AAPG Bulletin, v. 69, p. 1295–1304.

Lin, W., 1978, Measuring the permeability of Eleana Argillite from Area 17, Nevada Test Site, using the transient method: Lawrence Livermore Laboratory Report, UCRL-52.604.

Lindberg, P., R. Riise, and W.H. Fertl, 1980, Occurrence and distribution of overpressures in the northern North Sea area: Society of Petroleum Engineers. Paper 9339.

Luffel, D.L., and F.K. Guidry, 1989, Reservoir rock properties of Devonian shale from core and log analysis: Conference paper number 8910.

Luo, X., F. Brigaud, and G. Vasseur, 1991, Compaction coefficients of argillaceous sediments: their implications, significance and determination, *in* A.G. Doré (ed) Basin modeling advances and applications, Norwegian Petroleum Society, Special Publication no. 3, p. 321–332.

Luo, X., and G. Vasseur, 1992, Contributions of compaction and aquathermal pressuring to geopressure and the influence of environmental conditions: AAPG Bulletin, v. 76, p. 1550–1559.

Luo, X., and G. Vasseur, 1993, Contributions of compaction and aquathermal pressuring to geopressure and the influence of environmental conditions: Reply: AAPG Bulletin, v. 77, p. 2011–2014.

Lutz, J.F. and W.D. Kemper, 1959, Intrinsic permeability of clays as affected by clay-water intercation: Soil Sciences, 88, 2, p. 83–90.

Magara, K., 1968, Compaction and migration of fluids in Miocene mudstone, Nagaoka Plain, Japan: AAPG Bulletin, v. 52, p. 2466–2501.

Magara, K., 1974, Compaction, ion filtration, and osmosis in shale and their significance in primary migration: AAPG Bulletin, v. 58, p. 283–290.

Magara, K., 1975a, Reevaluation of montmorillonite dehydration as cause of abnormal pressure and hydrocarbon migration: AAPG Bulletin, v. 59, p 292–302.

Magara, K., 1975b, Importance of aquathermal pressuring effect in Gulf Coast: AAPG Bulletin, v. 59, p. 2037– 2045.

Magara, K., 1980, Comparison of porosity depth relationships of shale and sandstone: Journal of Petroleum Geology, v. 3, p 175–185.

McKelvey, J.G., and I.H. Milne, 1962, Flow of salt solutions through compacted clay: Clays and Clay Minerals, v. 11, p. 248–259.

Meade, R. H., 1966, Factors influencing the early stages of the compaction of clays and sands-review: Journal of Sedimentary Petrology, v. 36, p. 1085–1101.

Mégnien, C., 1980, Synthèse géologique du Bassin de Paris: Mémoire du Bureau de Recherche Géologique et Minière, v. 101–103, (Orléans, France) *(in French)*.

Mello, U.T., G.D. Karner, and N. Anderson, 1994, A physical explanation for the positioning of the depth to the top of overpressure in shale-dominated sequences in the Gulf Coast basin, United States: Journal of Geophysical Research, v. 99, p. 2775–2789.

Meissner, F.F., 1978, Petroleum geology of the Bakken Formation Williston Basin, North Dakota and Montana. Williston Basin Symposium: The Montana Geological Society, 24th Annual Conference, p. 207–227.

Miller, R.J. and P.F. Low, 1963, Threshold gradient for water flow in clay systems, Soil Sciences Society of America Proceedings, 27, p. 605–609.

Miller, T.W., and C.H. Luk, 1993, Contributions of compaction and aquathermal pressuring to geopressure and the influence of environmental conditions: discussion: AAPG Bulletin, v. 77, p. 2006–2010.

Mudford, B.S., and M.E. Best, 1989, Venture gas field, offshore Nova Scotia: case study of overpressuring in region of low sedimentation rate: AAPG Bulletin, v. 73, p. 1383–1396.

Nagtegaal, P.J.C., 1978, Sandstone-framework instability as a function of burial diagenesis: Journal of the Geological Society of London, v. 135, p. 101–105.

Nakayama, K., and I. Lerche, 1987, Basin analysis by model simulation: effects of geologic parameters on 1D and 2D fluid flow systems with application to an oil field Gulf Coast: Association of Geological Societies Transactions, v. 37, p. 175–184.

Neglia, S., 1979, Migration of fluids in sedimentary basins: AAPG Bulletin, v. 63, p. 573–597.

Nelson, J.S. and E.C. Simmons, 1995, Diffusion of methane and ethane through the reservoir cap rock: implications for the timing and duration of catagenesis, AAPG Bulletin, v. 79, 7, p. 1064–1074.

Neuzil, C.E., and D.W. Pollock, 1983, Erosional unloading and fluid pressures in hydraulically "tight" rocks: Journal of Geology, v. 91, p. 179–193.

Novelli, L., Welte, D.H., Mattavelli, L., Yalçin, M.N., Cinelli, D., and K.J. Schmitt, 1988, Hydrocarbon generation in southern Sicily. A three dimensional computer aided basin modeling study: Organic Geochemistry, v. 13, p. 153–164.

Nur, A. and J.D. Byerlee, 1971, An exact effective stress law for elastic deformation of rock with fluids. Journal of Geophysical Research, v. 76, 26, p. 6414–6419.

Olsen, H.W., 1965, Deviations from Darcy's Law in saturated clays: Soil Science Society of America Journal, 1965, v. 29, no. 2, p. 135–140;

Olsen, H. W., 1966, Darcy's Law in saturated kaolinite: Water Resources Research, v. 2, p. 287–295.

Olsen, H.W., 1971, Liquid Movement through Kaolinite under hydraulic, electric, and osmotic gradients: AAPG Bulletin, v. 56 p. 2022–2028.

Oudin, J.L., and P.F. Picard, 1982, Relationship between hydrocarbon distribution and the overpressured zones: Proceedings Indonesian Petroleum Association, v. 15, p. 181–202.

Overton, H. L., and D.J. Timko, 1969, The salinity factor: a tectonic stress indicator in marine sands: The Oil and Gas Journal, Oct. 6, 1969, p. 115–124.

Palciauskas, V. and P.A. Domenico, 1989, Fluid pressures in deforming porous rocks, Water Resources Research, v. 25, 2, p. 203–213.

Pearson, W.C., and C.R.B. Lister, 1973, Permeability measurements on a deep-sea core: Journal of Geophysical Research, v. 78, p. 7786–7787.

Perrier, R. and J. Quiblier, 1974, Thickness changes in sedimentary layers during compaction history ; methods for quantitative evaluation: AAPG Bulletin, v. 58, p. 507–520.

Perry, E.A., and J. Hower, 1972, Late-stage dehydration in deeply buried pelitic sediments: AAPG Bulletin, v. 56, p. 2013–2021.

Pittman, E.D., and R.E. Larese, 1991, Compaction of lithic sands: experimental results and applications: AAPG Bulletin, v. 75, p. 1279–1299.

Plumley, W.J. 1980, Abnormal high fluid pressure: survey of some basic principles: AAPG Bulletin, v. 64, p.

414–430.

Powers, M. C., 1967, Fluid-release mechanisms in compacting marine mudrocks and their importance in oil exploration: AAPG Bulletin, v. 51, P. 1240–1254.

Price, L.C., Ging, T., Daws, T., Love, A., Pawlewicz, M., and D. Anders, 1984, Organic metamorphism in the Mississippian-Devonian Bakken shale, North Dakota portion of the Williston Basin, *in* J. Woodward, F.F. Meissner, and J.L. Clayton, (eds), Hydrocarbon source rocks of the greater Rocky Mountain region: Rocky Mountain Association of Geologists, p. 83–134.

Rieke, H.H., and G.V. Chilingarian, 1974, Compaction of argillaceous sediments: Developments in Sedimentology 16, Elsevier, p. 424.

Rinckenbach, T., 1988, Diagénèse minérale des sédiments pétrolifères du delta fossile de la Mahakam: évolution minéralogique et isotopique des composants argileux et histoire thermique: Ph.D. Thesis, University of Strasbourg (France), p. 209 *(in French)*.

Rittenhouse, G., 1971, Mechanical compaction of sands containing different percentages of ductile grains: a theoretical approach: AAPG Bulletin, v. 55, p. 92–96.

Ropertz, O., 1994, Wege der Primäre Migration: Eine Untersuchung über Porennetze, Klüfte und Kerogennetzwerke als Leitbahnen fürden Kohlenwasserstoff-Transport, KFA Report 2875, Jülich, Germany, p. 250 *(in German)*.

Sassen, R. 1990, Lower Tertiary and Upper Cretaceous source rocks in Louisiana and Mississippi, implications to Gulf of Mexico crude oil: AAPG Bulletin, v. 74, p. 857–878.

Scherer, M., 1987, Parameters influencing porosity in sandstones: a model for sandstone porosity prediction: AAPG Bulletin, v. 71, p. 485–491.

Schmoker, J.W., and D.L. Gautier, 1988, Sandstone porosity as a function of thermal maturity: Geology, v. 16, p. 1007–1010.

Schmoker, J.W., and D.L. Gautier, 1989, Compaction of basin sediments: modeling based on time-temperature history: Journal of Geophysical Research, v. 94, p. 7379–7386.

Schneider, F. Burrus, J., and S. Wolf, 1993, Modeling overpressure by effective stress/porosity relationship in low permeability rocks: empirical artifice or physical reality ? *in* Doré, A.G., Auguston, J.H., Hermanrud, C., Stewart, D.J., and O. Sylta (eds), Basin modeling; advances and application: Norwegian Petroleum Society Special Publication, 3, p. 33–341.

Schneider, F., Potdevin, J.L., Wolf, S., and I. Faille, 1994a, Modèle de compaction élasto-plastique et viscoplastique pour simulateur de bassins sédimentaires: Revue de l'Institut Français du Pétrole, v. 49, p. 141–148 *(in French)*.

Schneider, F., Boutéca, M., and G. Vasseur, 1994b, Validity of the effective stress concept in sedimentary basin modeling: First Break, v. 12, p. 321–326.

Serra, O., 1986, Diagraphies différées, Bulletin des Centres Recherches Exploration-Production Elf-Aquitaine, Mémoire 1, 210 p. *(in French)*.

Sharp, J.M., and P.A. Domenico, 1976, Energy transport in thick sequences of compacting sediment: Geological Society of America Bulletin, v. 87, p. 390–400.

Sharp, J.M.Jr., 1983, Permeability controls on aquathermal pressuring: AAPG Bulletin, v. 67, p 2057–2061.

Shi, Y., and C.Y. Wang, 1986, Pore pressure generation in sedimentary basins, overloading versus aquathermal: Journal of Geophysical Research, v. 91, p. 2153–2162.

Skempton, A.W., 1970, The consolidation of clays under gravitational compaction: The Quarterly Journal of the Geological Society of London, v. 125, p. 373–409.

Smith, J.E., 1971, The dynamics of shale compaction and evolution of pore-fluid pressures: Mathematical Geology, v. 3, p. 239–263.

Sonnenthal, E., and P.J. Ortoleva, 1995, Numerical simulations of overpressured compartments in sedimentary basins, *in* Ortoleva, P.J. (ed.), Basin compartments and seals, AAPG Memoir 61, p. 403–416.

Spencer, C.W., 1987, Hydrocarbon Generation as a Mechanism for Overpressuring in Rocky Mountain Region: AAPG Bulletin, v. 71, 4, p. 368–388.

Stuart, C., 1970, Geopressures, Proceedings of the Second Symposium on abnormal subsurface pressures, Louisiana State University, Baton Rouge, p. 121.

Swann, G., J. Cook, S. Bruce, and R. Meehan, 1989, Strain rate effects in Kimmeridge Bay shale: Journal of Mechanical Mining Sciences & Geomechanics, v. 26, p. 135–149.

Tada, R., and R. Siever, 1989, Pressure solution during diagenesis: Annual Review of Earth and Planetary Sciences, v. 17, p. 89–118.

Terzaghi, K. Van, 1923, Die Berechnung der Durchassigkeitsziffer des Tones aus dem Verlauf der hydrodynamischen Spannungerscheinungen, Sitzungbericht Akademie Wissenschaft Wien, Mathematik Naturwissenschaft, Kl. Abteil 2A, v. 132 (3/4), p.125–138 *(in German)*.

Tissot, B., and R. Pelet, 1971, Nouvelles données sur les mécanismes de genèse et de migration du pétrole. Simulation mathématique et application à la prospection: Proceedings of the 8th World Petroleum Congress, v. 2, p. 39–46 *(in French)*.

Tissot, B., and D. Welte, 1984, Petroleum formation and occurrence, 2nd ed. , Springer-Verlag: Berlin, 299 p.

Ungerer, P., B. Doligez, P.Y. Chenet, J. Burrus, F. Bessis, E. Lafargue, G. Ginoir, O. Heum, and S. Eggen, 1987, A 2-D model of basin scale petroleum migration by two-phase fluid flow; application to some case studies, *in* B. Doligez (ed.), Migration of hydrocarbons in sedimentary basins, Technip, p. 415–456.

Ungerer, P., J. Burrus, B. Doligez, P.-Y. Chenet, and F. Bessis, 1990, Basin evaluation by integrated two-dimensional modeling of heat transfer, fluid flow, hydrocarbon generation and migration. American Association of Petroleum Geologists Bulletin, v. 74, p. 309–335.

Van Olphen, H., 1963, Compaction of clay sediments in the range of molecular particle distances, *in* Bradley, W.F. (ed.), Clays and Clay Minerals, v. II, p. 178–187.

Vasseur, G., I. Djeran Maigre, D. Grunberger, D. Tessier, G. Rousset, and B. Velde, 1995, Evolution of structural and physical parameters of clays during experimental compaction. Marine and Petroleum Geology, v. 12, 8, p. 941–954.

Velde, B., and G. Vasseur, 1992, Estimation of the diagenetic smectite to illite-temperature space: American Mineralogists, v. 77, p. 967–976.

Weedman, S.D., S.L. Brantley, and W. Albrecht, 1992, Secondary compaction after secondary porosity: can it form a pressure seal ?: Geology, v. 20, p. 303–306.

Wei, H.F., E. Ledoux, and G. de Marsily, 1990, Regional modeling of groundwater and salt environmental tracers transport in deep aquifers in the Paris Basin, Journal of Hydrology, v. 120, p. 341–358.

Weller, J. M., 1959, Compaction of sediments: AAPG Bulletin, v. 43, p. 273–310.

Wenger, L., and T. Schwarzer, 1993, Secondary migration evaluation of tertiary reservoired hydrocarbons in South Louisiana an application of oil source correlation techniques, 78th AAPG Annual meeting, New-Orleans, 25–28 April 1993 (abstract).

Yassir, N.A., and J.S. Bell, 1994, Relationships between pore pressure, stresses and present-day geodynamics in the Scotian shelf, offshore, eastern Canada: AAPG Bulletin, v. 78, p. 1863–1880.

Young, A., Ph.F. Low, and A. S. McLatche, 1964, Permeability studies of argillaceous rocks: Journal of Geophysical Research, v. 69, p. 4237–4245.

Zein El Din, Y., A. Hegazi, and M. Nashaat, 1988, The use of geopressure data as a new tool for exploration in the Nile Delta, Egypt: 9th exploration and production Conference, Nov. 21, 1988, Cairo, Egypt, p. 13

Zoback, M.D., and P. Peska, 1995, In situ stress and rock strength in the GBRN/DOE Pathfinder well, South Eugene Island, Gulf of Mexico, SPE paper 29233, Journal of Petroleum Technology, July 1995, p. 582–585.

LIST OF SYMBOLS USED

α Biot coefficient (no unit)
β_b Drained bulk compressibility of the rock (Pa^{-1})
β_f Fluid compressibility (Pa^{-1})
β_g Solid grain compressibility (Pa^{-1})
c Athy's compaction coefficient (m^{-1})
ϕ_0 Surface porosity (no unit)
ϕ Porosity (no unit)
ϕ' Porosity corrected for adsorbed water (no unit)
K Permeability (m^2, darcy)
k_x Permeability anisotropy factor in x direction (no unit)
k_z Permeability anisotropy factor in z direction (no unit)
μ Water viscosity (Pa.s)
ν Poisson ratio (no units)
ω Sedimentation rate (ms^{-1})
P Pore pressure (Pa)
P_h Hydrostatic pressure (Pa)
ρ_b Bulk sediment density (kg/m^3)
ρ_w Water density (kg/m^3)
S Lithostatic stress (Pa)
S_h Horizontal stress (Pa)
S_0 Specific surface area (m^2)
S_s Specific storage coefficient (m^{-1})
σ_e Effective stress (Pa)
t Time (s)
τ_x Tortuosity in x direction (no unit)
τ_z Tortuosity in z direction (no unit)
z depth (m)

Chapter 4

SHALE POROSITIES FROM WELL LOGS ON HALTENBANKEN (OFFSHORE MID-NORWAY) SHOW NO INFLUENCE OF OVERPRESSURING

C. Hermanrud
L. Wensaas
G.M.G. Teige
H.M. Nordgård Bolås
S. Hansen
Statoil Research Centre
Trondheim, Norway

E. Vik[1]
Geologisk Institut
Århus Universitet
Denmark

Abstract

Fluid pressure detection and porosity evaluation from well logs are largely based on an assumed relationship between high fluid pressures and high porosities due to undercompaction. However, few data have been presented which demonstrate to what extent porosities are higher in overpressured than in normally pressured shales of similar type, and how this porosity difference is detected by the responses from standard logs. Jurassic intra-reservoir shales on Haltenbanken (offshore mid-Norway) are particularly well-suited for such an investigation because (a) the area is subdivided into two, major, distinctive pressure regimes (one normally pressured, the other highly overpressured) and (b) the lithology, depositional environment and present burial depth do not vary significantly across the area.

Log comparisons reveal that neutron and density responses show no significant porosity difference between the two regimes, whereas sonic and resistivity responses show higher (apparent) porosities in the overpressured area. It is thus suggested that porosity is unaffected by differences in fluid pressures, but that the sonic and resistivity logs are reacting to textural changes induced in the rocks by overpressuring rather than high porosities due to undercompaction.

High fluid pressures in combination with low shale porosities could be explained by pressure unloading (i.e., fluid overpressuring post-dating shale compaction), and this cannot be ruled out from the Haltenbanken data. However, log data from North Sea shales also show that formation density does not significantly vary with fluid overpressuring, whereas sonic log data decreases with depth irrespective of overpressuring. As it is unlikely that fluid overpressuring in all of these formations post-dated compaction, it appears that shale porosity reduction may proceed without significant hindrance by fluid overpressuring.

These findings suggest that standard principles applied to pore pressure evaluation from well logs may not always be valid, thus partly explaining the large degree of uncertainty attached to such work. Furthermore, basin modeling of fluid flow, overpressure buildup, hydrofracturing and hydrocarbon migration appears to rely on equations which give improper descriptions of fluid transport in shales.

INTRODUCTION

The use of wireline logs for pressure analysis of low-permeability formations (shales) dates back to the 1960s and early 1970s. Detection of abnormal pore fluid pressures in open holes is generally based on the assumption that there is a direct relationship between an anomalous log response (porosity) and pore pressure (Figure 1). Several early techniques developed for overpressure quantification are widely used today, for example, those of Hottman and Johnson (1965) and Eaton (1972a, 1972b), and the Equivalent Depth Methods of Rubey and Hubbert (1959) and Ham (1966) —the background and development of which are given elsewhere (Fertl, 1976; Dutta, 1986; Mouchet and Mitchell, 1989; Fertl et al., 1994).

[1] *Present Affiliation: Norsk Hydro, E & P Research Centre, Bergen, Norway*

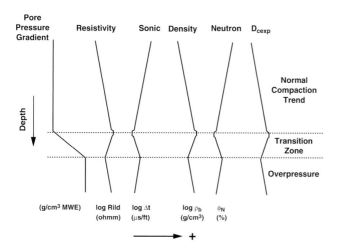

Figure 1. Schematic response of well logging parameters to normal compaction and overpressure by fluid retention (under compaction) (modified from Fertl and Timko, 1972). D$_{cexp}$ is the drilling exponent.

Largely developed from exploration drilling in the young (Tertiary) and rather homogeneous U.S. Gulf Coast deposits, these methods rely on empirical relationships between shale porosity and burial depth throughout a shaly sequence. Overpressures are identified by deviations from a normal compaction trend (Figure 1), with calculated shale pressures increasing as the deviations increase. However, the application of these techniques to the more heterogeneous shales of the Norwegian continental shelf commonly yields inaccurate pressure analyses, which may be related to factors such as: (a) problems in defining a normal compaction trend in the shallow and heterogeneous Tertiary interval (Lindberg et al., 1980) and (b) the lack of any indication of undercompaction within the pre-Tertiary section (Carstens and Dypvik, 1979, 1981; Hunt, 1990; Wensaas et al., 1994). Hunt (1991) argues that compactional disequilibrium in overpressured shaly sequences, both on the Norwegian Shelf and world-wide, is generally far less abundant than hitherto believed. To the best of our knowledge, the occurrence of abnormally high porosities in overpressured shales of the Norwegian continental shelf has not been verified by direct porosity measurements made on equivalent sets of normally pressured and overpressured shales.

More recent methods use an effective stress-porosity approach, where excess pore fluid pressures are determined from the Terzaghi (1923) effective stress relationship:

$$P_{fl} = OB - \sigma_v \qquad (1)$$

where P_{fl} is the calculated fluid pressure, OB represents the overburden load and σ_v the vertical effective stress, which is determined either from a power-law or exponential relationship between effective stress and the petrophysically determined rock porosity (ϕ), or from

related parameters such as the void ratio $\phi/(1 - \phi)$ or solidity $(1 - \phi)$ (Holbrook and Hauck, 1987; Alixant et al., 1989; Alixant and Desbrandes, 1991; Ward et al., 1994, 1995). Various authors have also used the relationship between effective stress and porosity (or another physical parameter, e.g., velocity) when attempting to distinguish between overpressuring caused by fluid retention (undercompaction) and overpressuring caused by fluid expansion mechanisms (Yassir and Bell, 1994; Bowers, 1994; Hart et al., 1995; Ward, 1995).

All of these approaches rely on the common assumption that porosity reduction in shales will not take place if the effective stress is not at its historical maximum value. This assumption also serves as the basis for numerical basin modeling of fluid flow in shales, overpressuring, hydrofracturing and primary hydrocarbon migration. Originating from laboratory experiments conducted under surface conditions (Terzaghi, 1923), this assumption has been universally adopted as a valid principle in sedimentary basins evolving over geological time and reaching temperatures of 200°C or more. However, relationships between porosity and effective stresses in the subsurface may well be different from those derived from laboratory experiments; and by introducing a Biot (1941) coefficient of about 0.65–0.85 in the Terzaghi (1923) equation, similar porosity-effective stress relationships can be obtained for both normally pressured and overpressured sections (Burrus, 1998-this volume).

This paper examines the validity of the Terzaghi-based relationship between porosity and effective stress in shales in the Haltenbanken and North Sea areas (offshore Norway), largely by comparing log responses between overpressured and normally pressured intra-reservoir shales of the same formation and at similar burial depths. In addition, our attempts to quantify pressure transition zones between overpressured cap rocks and normally pressured reservoirs, and our comparisons of log response versus depth in different pressure regimes containing thick shaly sequences in several formations, have proved helpful in supporting conclusions from the intra-reservoir shale investigations. The Haltenbanken area is especially suitable for this as it contains two major, distinctive pressure regimes (Figure 2) located within a restricted area in which the depositional environment, sedimentary properties, and present burial depths do not significantly vary.

THE HALTENBANKEN AREA: GEOLOGICAL SETTING, PORE PRESSURE DISTRIBUTION, AND HYDROCARBON OCCURRENCE

As the geological history of the Haltenbanken area (Figure 2) has been extensively described by other

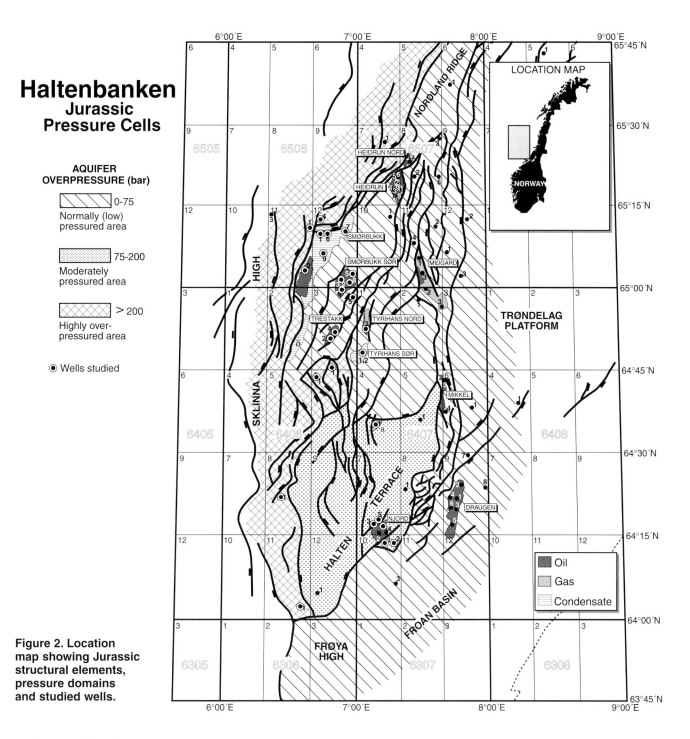

Figure 2. Location map showing Jurassic structural elements, pressure domains and studied wells.

authors (Hollander, 1984; Heum et al., 1986; Ehrenberg et al., 1992; Koch and Heum, 1995), only those features relevant to this paper are outlined below.

Laterally extensive reservoir sandstones of deltaic to shallow marine origin were deposited throughout the Early and Middle Jurassic—the Åre, Tilje, Ile and Garn Formations—and are separated by shaly deposits of the Ror and Not Formations (Figure 3). Alluvial fan deposits of the Tofte Formation are locally present in the northwestern part of the study area, where it

divides the Ror Formation into a lower and upper unit. The Garn Formation is overlain by shales of the Upper Jurassic Melke and Spekk Formations, the latter constituting the main source rock.

Crustal extension formed a faulted terrain in Late Jurassic and Early Cretaceous times. Thermal subsidence created accommodation space for renewed sedimentation in the Early Cretaceous, resulting in deposition of (mainly) shaly sediments throughout the Cretaceous and Tertiary. The Lower to Middle Creta-

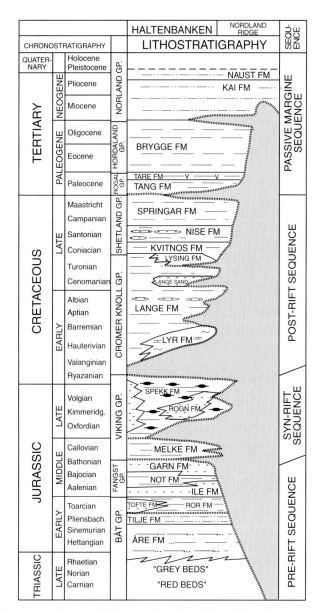

Figure 3. General lithostratigraphy of the Halten Terrace.

ceous Lange Formation and Upper Cretaceous Lysing Formation consist of sandy sequences in which hydrocarbons have been discovered in several wells. Pliocene–Pleistocene uplift of mainland Norway led to deposition of up to 1.5 km of sediments in the Haltenbanken area, which caused a corresponding amount of subsidence of Jurassic rocks and increased source rock maturation and hydrocarbon generation.

Fluid pressures in the study area are close to hydrostatic from the seabed to the Lower Tertiary rocks, where they rise to about 1.6 g/cm³ MWE (mud weight equivalent) at about 2.3 km burial depth and thereafter remain relatively constant down to the Lysing Formation. Renewed pressure buildup occurs in the Lower Cretaceous and Upper Jurassic shales, where in some wells mud weights were in excess of 1.8 g/cm³ MWE (Figure 4).

Fluid pressures of the Jurassic reservoirs vary significantly over the area. Those to the west have high fluid pressures (about 1.8 g/cm³ MWE), while those to the east are close to hydrostatic (low pressured area in Figure 2). The two pressure regimes are separated by a north-south striking fault system. A transitional area (with moderate fluid pressures of about 1.4 g/cm³ MWE) has also been identified in the southern part of Haltenbanken (the Halten Terrace), while no such transition zone seems to exist to the north. Fluid pressures above the Jurassic pressure regimes appear to be identical. Note, however, that determinations of fluid pressures in the shaly sequences are largely based on indirect methods (interpretation of drilling parameters and wireline logs) and are thus highly uncertain. RFT and DST measurements in the Lange and Lysing Formations, and in the Fangst Group, serve as useful control points for calculated fluid pressures as they vary little throughout the study area.

The overpressure seal consists of a 1 km thick section of mainly shaly sediments lying between the Lysing Formation and the Fangst Group reservoirs (Figure 2). The present day temperatures and thermal maturity of the seal ranges from about 115°C and 0.6 %R_o at a depth of 3 km to 150°C and 1.05 %R_o at 4 km (Vik and Hermanrud, 1993). Overpressures to the west have been attributed to rapid Plio-Pleistocene subsidence, possibly assisted by the lateral transfer of high fluid pressures from the deeply buried Møre Basin west of the study area (Koch and Heum, 1995).

All of the 7 exploration wells drilled to Middle Jurassic objectives in the (overpressured) western part of Haltenbanken are water-bearing (pr. 1996), whereas most of the exploration wells to the east (normally pressured) have encountered hydrocarbons. The reason for this relationship between high overpressures and water-bearing wells is still open to debate, but it certainly cannot be explained solely on the basis of

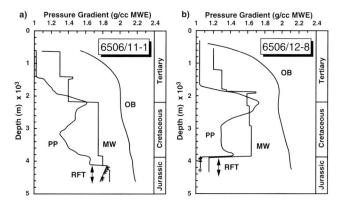

Figure 4. Characteristic pressure profiles in the Jurassic high and low pressure domains. (a) Well 6506/11-1 and (b) Well 6506/12-8. PP = pressure gradient (in mud weight equivalent MWE), OB = overburden gradient, MW = mud weight, (+) represent RFT pressure measurements (made in the reservoirs) from Repeated Formation Tests. Location of wells are shown on Figure 2.

high fluid pressures as even higher pressures are encountered in several hydrocarbon discoveries encountered at similar depths in the North Sea.

DATA SELECTION

Wireline log data were compiled from 28 wells penetrating Jurassic reservoirs: 6 from highly overpressured intervals and 22 from normally pressured or moderately pressured intervals. All of these wells are located close to the pressure regime boundary, which extends from the Smørbukk area in the north to the Njord area in the south (Figure 2). An additional set of 12 shallow wells, located further east on the Halten Terrace, were included to establish a regional porosity versus depth trend for the Not Formation shales. Wherever possible, log data and core measurements were also gathered from the Garn Formation (at burial depths below 3.5 km). Note that the various log data were not always available from all of the wells or studied formations.

As the investigation hinged on our ability to separate the effects of overpressuring from other factors which could influence shale porosity, it was important to identify a single shaly formation which was buried to the same depth in both pressure regimes and showed little lateral variation across the area. The intra-reservoir shales of the Not Formation broadly satisfied these criteria.

Both the Ror and Not Formations (Figure 3) are characterized by overall coarsening upward sequences of fine-grained sediments bounded by erosion surfaces. Their thicknesses vary between 10–120 m and 10–200 m, respectively, and both formations are regionally distributed throughout the *study* area. The Ror Formation generally thins to the northeast and has largely been removed by erosion over the Nordland Ridge (Figure 2); the Not Formation, which is divisible into three sub-units, generally thickens to the southwest and thins to the east on the Trøndelag Platform (Figure 2).

Regional correlations of the Not Formation subunits were based on well log characteristics and core descriptions. The upper subunit consists of a very fine- to fine-grained sandstone which, apart from having overall higher gamma ray readings, cannot easily be distinguished from the overlying Garn Formation sandstones (Figure 5). The middle subunit contains a gradual coarsening upward sequence of shales and silty claystones, characterized by an upward decrease in sonic transit time and a reduction in the separation between the neutron and density logs. The lower subunit consists of fine-grained shales which elicit uniform patterns of high resistivity, neutron and sonic values, and low density values. Regionally, the Not Formation gradually becomes finer-grained to the south and east as the two lowermost subunits gradually thicken and the uppermost silty/sandy unit disappears.

Shale porosity was calculated from averaged log readings (see below) of the lower and middle subunits taken together (the top of the shaly interval being defined by the onset of a downward increase in sonic interval transit time). The lower subunit was investigated separately. As both sets of results are similar, they will no longer be differentiated.

Shale porosity determination for the Ror Formation was based on log readings averaged over all of the individual, minor shale units. The log patterns are rather complex, reflecting interfingering of shale and silty/muddy sands (Figure 6). Because the Ror Formation is separated into upper and lower units towards the north-west (where it interdigitates with sands of the Tofte Formation), the log patterns cannot easily be correlated across the region. The Ror Formation results are simply included here to support those obtained from the Not Formation, and little importance will be attached to them.

Only those log intervals uninfluenced by poor hole conditions (i.e., cavings, washouts, etc. detected from the caliper logs) were used, and the averaging was performed automatically after manual inspection and removal of poor quality signals. The analyses were based on the assumption that the pore fluid pressure in the intra-reservoir shales was equal to that in the intervening sandstone reservoirs.

The cap rock sequences of the Garn Formation were investigated to see if pressure transition zones could be detected. As the onlapping cap rocks consist of shales of different ages and formations, significant deviations were anticipated between wells. Even so, it was expected that the majority of the shales overlying the normally pressured reservoirs would show some indication of porosity decrease close to the reservoirs, whereas those overlying the overpressured reservoirs would not.

Helium porosities (from routine core measurements) were available from 7 wells penetrating the Not Formation shales and 12 wells penetrating the Ror Formation shales (see Figures 5 and 6), all of the wells being located in the low and moderate pressure regimes (see Figure 2). No such measurements were available from the highly overpressured wells. For the Garn Formation sandstones (>3,500 m), the differences between the average total porosities (measured by helium porosimetry) and those calculated from the various wireline logs were used to evaluate the influence of drilling mud composition on log porosity.

POROSITY CALCULATIONS

Shale porosities were calculated from averaged responses of the density, sonic, resistivity, and neutron logs. As the objective was to compare the responses in the three pressure regimes, the calculated porosities for the sonic, resistivity, and neutron logs were adjusted (i.e., by the proper choice of the free parameters [e.g.,

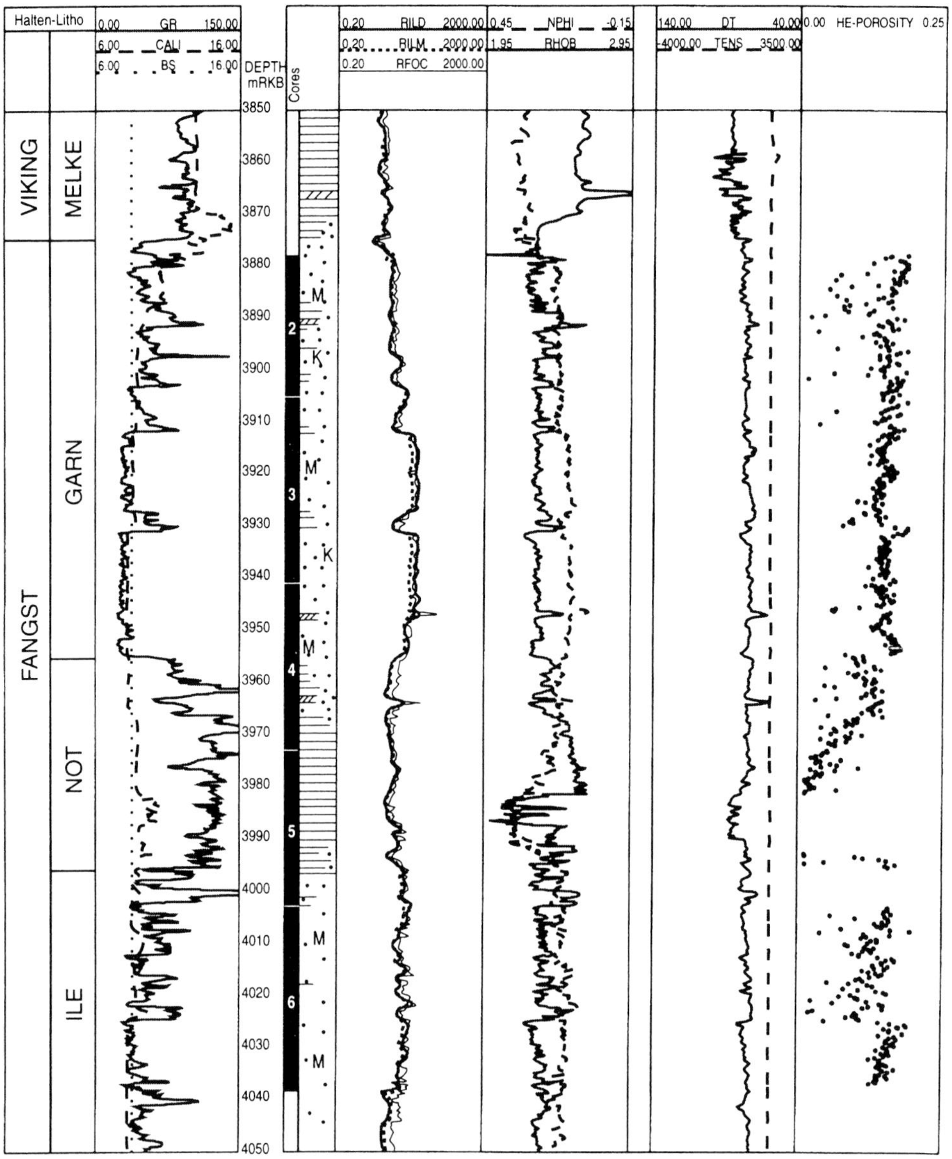

Figure 5. Typical log patterns for the Not Formation (well 6506/12-8 on Figure 2). Average log readings were calculated from the basal shaly interval (3,968–3,993 m). Bulk density readings (RHOB) affected by poor hole conditions between 3,983 and 3,988 m were excluded from the calculations. Also shown are helium porosity measurements from core samples taken from the upper part of the "shaly interval".

formation water resistivity and matrix transit time in the equations]) so that they gave the same average porosity in the Not Formation for a low-pressure reference well (porosity = 7.0% in well 6407/1-2 from the Tyrihans Sør field, Figure 2).

The porosity data were only used to inspect porosity differences between the three pressure regimes. The absolute magnitudes of the calculated shale porosities were thus unimportant as only the relative differences between the data were of interest. Consequently, uncertainties linked to the aforementioned choice of parameters do not influence the conclusions of this paper. Details of the porosity calculations are given in Appendix A.

Additional refinements to those described in Appendix A for sandstone porosity would normally be performed during routine petrophysical log evaluation. Their omission here, however, is inconsequential for the same reasons as explained above—our investigations targeted the difference in porosity between the highly overpressured and the normally pressured pressure regimes at Haltenbanken, which are uninfluenced by the (calculated) absolute porosity values.

RESULTS

Although the averaged log-derived porosities of the Not Formation shaly intervals (Figure 7) show a considerable scatter at any given depth (which may be due to various factors discussed in the next section), the following broad, yet significant observation can be made. The density and neutron log-derived porosities are no higher in the overpressured than in the normally pressured wells, whereas the sonic and resistivity log-derived porosities are significantly higher in the overpressured than in the normally pressured wells. For the overpressured wells, the range in sonic and resistivity data is considerable, even though the rocks with the highest porosities do not contain higher fluid pressures than those with the lowest porosities. Furthermore, the plots for moderately pressured wells (<1.4 g/cm³ MWE) tend to fall amongst those of the normally pressured wells, rather than forming a separate, intermediate cluster. In addition, the range in shale porosity for the normally pressured wells is significantly less for the resistivity log than that for the other log types. Generally similar observations pertain to the Ror Formation data illustrated in Figure 8.

The shale porosities of both formations do not vary significantly with depth between 3.5 to 4.5 km, average log-derived porosities below 3.5 km are given in Table 1. This may mainly be due to scatter in the data. However, the resistivity-derived porosities for the normally

pressured shales suggest a slight reduction. Below 3.5 km, shale porosity in normally pressured areas is generally less than 10%. This is borne out from total helium porosity measurements in normally and moderately pressured Not Formation shales (Figure 9) and also corresponds to the porosity range calculated from the logs (Figure 7). As there are no core measurements available for the overpressured intra-reservoir shales, the abnormal porosities derived from the sonic and resistivity logs cannot be verified.

Results from the analyses of cap rock log responses are shown in Table 2. As expected, shales overlying the overpressured reservoirs show no systematic porosity reduction in the pressure transition zone: two of the five wells show decreasing porosity, two wells show increasing porosity, and one well has a constant porosity throughout. The anticipated decrease in cap rock porosity over the normally pressured reservoirs, how-

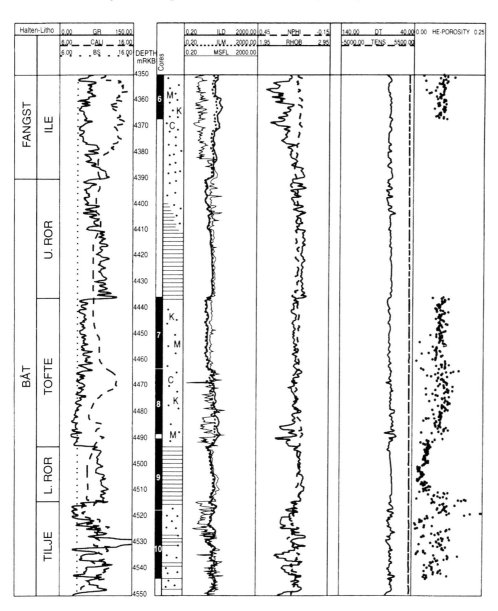

Figure 6. Typical log patterns for the Not Formation in the north-western part of the study area (well 6506/12-6 in Figure 2) . Average log readings were calculated for all shaly intervals (e.g., Upper Ror Formation, 4,410–4,436 m; Lower Ror Formation, 4,505–4,514 m) identified from the logs. Also shown are helium porosity measurements from core samples taken from the Lower Ror Formation.

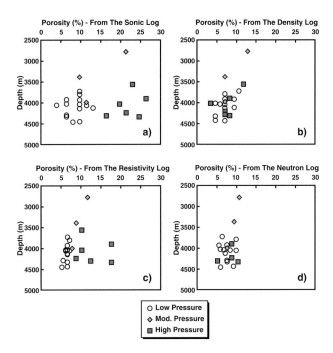

Figure 7. Calculated mean porosity of shaly intervals in the Not Formation: (a) sonic log (DT) porosity; (b) density log (RHOB) porosity; (c) deep resistivity log (RILD) porosity; and (d) neutron log (NPHI) porosity (rescaled to obtain the same mean and standard deviation as the density-derived porosity). See text for various calculation procedures and Figure 2 for definition of pressure regimes. Highly overpressured wells are indicated by squares; moderately overpressured wells by diamonds; low (to normally) pressured wells by (open) circles.

ever, was not apparent from the density log responses. Indeed, the opposite proved to be the case, as 10 out of 12 wells showed increasing porosity. The sonic log responses, however, differed from those of the density log in that 5 of 13 wells showed signs of a pressure transition zone, while in four of the wells sonic-log derived porosity increased.

Averages of helium porosities for the Garn Formation sandstones are shown in Figure 10. Average porosity is about 3–5% higher in the overpressured sandstones than in the equivalent low pressured sandstones. Average porosities from the resistivity logs show a separation between the two major pressure regimes (Figure 11). No such clear relationship emerges from the neutron and density porosities. The large spread in log-derived porosity for the sandstones may result from the fact that routine log analysis refinements, such as variations in hydrocarbon density or water saturation (S_w), were not taken into account.

OVERPRESSURED SHALES—LOW OR HIGH POROSITY?

The fact that the sonic and resistivity logs indicate different shale porosity in the two major pressure regimes; whereas, the neutron and density logs do not suggests that either: (1) porosity is unaffected by overpressuring—the sonic and resistivity logs respond directly to fluid pressure (hypothesis 1); or (2) the overpressured shales are undercompacted but other effects mask the anticipated increase in neutron and density porosity (hypothesis 2).

To decide which of these hypotheses is the most likely, it is necessary to briefly consider possible sources of error, such as borehole conditions, mineralogical variations, and differences in hydrocarbon occurrence and drilling mud composition.

Borehole conditions: As previously mentioned, all of the logs were carefully screened to avoid sections which were affected by poor hole conditions. In the event that such conditions still influenced the data, the outcome would be an overestimate of porosity in caved or washed-out intervals. (Poor hole conditions

Table 1. Average log-derived porosity (%) and number of wells (N) for intra-reservoir shales (Not and Ror Formations) below 3.5 km. LP = low pressured (hydrostatic), OP = overpressured.

Log	Average Log-Derived Porosity Not Formation				Average Log-Derived Porosity Upper Ror Formation				Average Log-Derived Porosity Lower Ror Formation			
	LP wells	N	OP wells	N	LP wells	N	OP wells	N	LP wells	N	OP wells	N
Sonic (DT)	8.6	14	21.8	6	5.8	14	10.8	5	6.0	7	12.1	4
Density (RHOB)	7.3	13	7.6	6	7.7	14	6.9	4	7.5	8	7.3	4
Neutron (NPHI)	7.1	13	7.8	4	6.6	13	8.3	4	6.7	8	7.5	4
Resistivity (RILD)	6.3	14	12.5	6	6.6	14	8.9	4	5.3	8	9.2	3

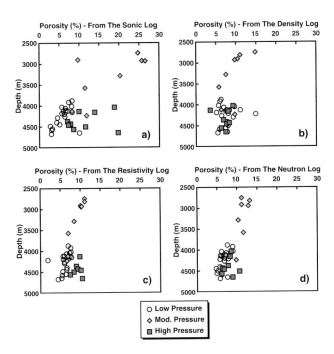

Figure 8. Calculated mean porosity of shaly intervals in the Ror Formation: (a) sonic log (DT) porosity; (b) density log (RHOB) porosity; (c) deep resistivity log (RILD) porosity; and (d) neutron log (NPHI) porosity (rescaled to obtain the same mean and standard deviation as the density-derived porosity). See text for various calculation procedures. Symbols as per Figure 7.

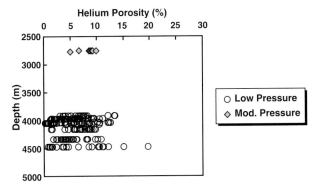

Figure 9. Total porosity of low (circles) and moderately overpressured (diamonds) Not Formation shales measured by helium porosimetry (routine core analysis).

are more frequently encountered in the normally pressured wells, which were drilled with a less dense mud, than in the overpressured wells.) However, poor hole conditions cannot be invoked in favor of hypothesis 2, which demands a suppression of log porosity.

Mineralogy: Any systematic differences in mineralogy between the two areas could possibly lead to a porosity reduction in the western (overpressured) area. This, then, would be cancelled out by enhanced porosity preservation due to undercompaction; i.e., there would be no significant porosity difference between the two pressure regimes. The lateral continuity of the Not Formation, and the short distance between the overpressured and normally pressured wells (especially in the Smørbukk area), suggests that such mineralogical changes are unlikely. The minerals most likely to cause overestimation of neutron and density porosity are opal, coal and zeolites. These minerals are rare to non-existent in the Not Formation, and mineralogical variations cannot therefore be used as an argument in support of hypothesis 2.

Hydrocarbon occurrence: All of the wells in the overpressured area are water-bearing, while 11 of the 17 wells in the normally and moderately pressured areas, which contain Not Formation shales, are oil- or gas-bearing. Rock-eval analysis of the Not Formation shales showed that they contain virtually no free hydrocarbons (S_1 less than 0.5 mg/g rock). Moreover,

the dry wells in the normally pressured area show similar Not Formation log responses to those in the hydrocarbon-bearing wells in the same area. It is therefore safe to conclude that differences in the presence of hydrocarbons between the two pressure regimes cannot lend support to hypothesis 2.

Drilling mud: The influence of drilling mud on the log responses could only favor hypothesis 2 if its composition was significantly different in the two major pressure regimes. As far as we know, the only constituent varying systematically between the regimes is the barite content (see below). Other components (such as KCl) may influence log readings, but these do not seem to correlate with the level of fluid overpressuring.

The invasion of drilling mud filtrate in permeable formations causes a change in the physical properties of the pore-fill. The sonic log measures the transmission of sound waves, which mainly takes place in the rock matrix; any influence of different mud types would be expected to be small. However, as the heavy barite-rich mud is more "solid-like" than the lighter,

Table 2. Sonic (DT) and density (RHOB) log responses in cap rocks. Wells are divided into three categories: those in which the logs indicate (a) an anticipated transition zone (apparent porosity decrease towards the reservoir), (b) no transition zone, and (c) a reverse transition zone (apparent porosity increase towards the reservoir).

Reservoir Type (Log Type)	# Wells with a Transition Zone	# Wells with No Transition Zone	# Wells with a Reverse Transition Zone
Overpressured (DT & RHOB)	2	1	2
Low Pressured (DT)	5	4	4
Low Pressured (RHOB)	0	2	10

less barite-rich mud, the difference in mud composition might possibly lead to an underestimation of porosity in the overpressured wells. The density log could likewise underestimate porosity in the overpressured wells if the formations were invaded by heavy drilling mud. Similarly, the neutron log, responding to the hydrogen density of a formation, could underestimate porosity in wells drilled with (hydrogen-poor) barite-rich mud, compared with those drilled with (hydrogen-rich) bentonite-rich mud. More significantly, the different mud cake compositions may also influence the log readings in a similar way. Of the four logs used in this study, the resistivity is the least likely to be affected by different mud properties in a shaly sequence as it supposedly investigates a rock volume beyond the invaded zone—even in highly permeable sandstones. If there is any barite influence (which is highly unlikely), it should lead to increased resistivity in the overpressured wells and hence an underestimation of porosity.

To gain an impression of just how much different mud compositions can affect log porosity in the study area, an investigation was made of the Garn Formation sandstone reservoirs (in which the influence of drilling mud should be far greater than in the underlying tight shales of the Not Formation for which mudcake build-up will be minimal).

Figure 12 shows the difference between helium core porosities and those calculated from the logs. Only water-bearing wells were used for the resistivity and neutron logs, as these logs are severely influenced by hydrocarbon content and would thus be unsuitable for separating effects ascribable to the different fluid pressure regimes alone. As can be seen (Figure 12), small but systematic differences exist between the porosities in the two regimes. This is what would be expected for the sonic, density and neutron logs. The resistivity log, however, which should be least prone to the influence of drilling mud composition, shows the inverse of what would be expected in the overpressured and normally pressured wells. It is therefore suggested that the

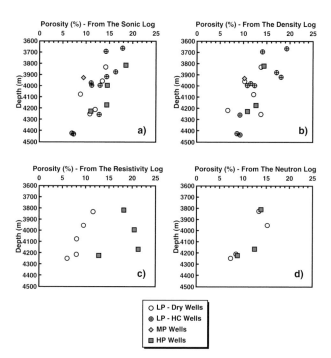

Figure 11. Calculated mean porosity of cored intervals of Garn Formation sandstones: (a) sonic log (DT) porosity; (b) density log (RHOB) porosity; (c) deep resistivity log (RILD) porosity; and (d) neutron log (NPHI) porosity (rescaled to obtain the same mean and standard deviation as the density-derived porosity). See text for various calculation procedures. Symbols as per Figures 7 and 10.

clear distinction between the two pressure regimes shown by the resistivity-log porosities is due to micro-fracturing of the overpressured rocks (see below).

The differences between the measured and the density-derived porosities, tentatively attributed to variations in drilling mud composition, are 3 to 4 porosity units (Table 3). Any influence of the drilling mud on the shaly Not Formation must be significantly less than this. It is therefore concluded that differences in drilling mud composition cannot be used to support hypothesis 2.

Furthermore, if both the neutron and density logs were influenced by a common factor masking high porosities, then a correlation should exist between them. For normally pressured wells, a correlation would also indicate that porosity indeed varies between the wells; conversely, a lack of correlation would indicate that variations in log-derived porosity (as seen in Figures 7 and 8) are mainly due to "random errors". Another possibility is that the log responses for one (or both) of the logs are not proportional to porosity, although we find no supporting arguments for this.

A neutron-density cross-plot for the Not Formation shows that no such correlation exists (Figure 13). This observation, coupled to the lack of a plausible explanation for hypothesis 2, leads us to the conclusion that the (calibrated) neutron and density logs give a valid representation of shale porosity in the two pressure

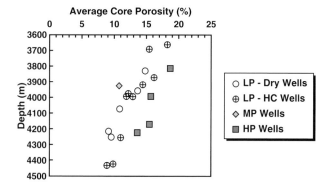

Figure 10. Average helium porosity measurements from cored sections of the Garn Formation. Symbols as per Figure 7. And, where marked, circles (LP - HC Wells) represent hydrocarbon-bearing low (to normally) pressured wells.

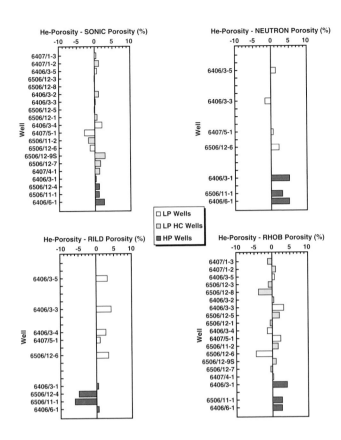

Figure 12. Difference between helium-porosity and log-derived porosity in Garn Formation sandstones. Shaded bars (LP HC Wells) represent low pressured hydrocarbon-bearing reservoirs. Well 6406/3-1 represents a moderately pressured hydrocarbon-bearing reservoir. LP = low pressured; HP = highly overpressured.

regimes; in other words, shale porosity in the two areas can hardly be distinguished. Furthermore, the lack of correlation suggests that the neutron and density logs provide independent estimates of shale porosity. If this is so, estimates of average porosity in the two pressure regimes can be obtained by treating the two data sources as one; the number of porosity estimates in the overpressured and normally pressured areas are thus 10 and 27, respectively. Furthermore, a 95% confidence limit for the porosity difference between the two regimes shows that the average porosity in the overpressured area is at most 1.2% higher than that of the normally pressured area, and that this porosity difference may well be zero or even negative.

These calculations, however, do not account for the possible influence of different drilling muds. Based on the analysis of sandstone log responses in the Garn Formation and the calculation just cited, we suggest that the differences in porosity between the two pressure regimes is at most 3%, probably much less, and possibly zero. This difference is significantly less than the differences suggested by the sonic (13.3%) and resistivity (6.1%) logs. It is thus suggested that hypothesis 1 is the most likely of the two, which means that

the influence of overpressuring on porosities in the Not Formation is small or even non-existent.

WHY DO RESISTIVITY AND SONIC LOGS REACT TO FLUID PRESSURE?

The suggestion that the neutron and density logs provide reliable estimates of differences in shale porosity between the two pressure regimes (and that this difference is almost negligible) strengthens our idea that the sonic and resistivity logs are responding *directly* to fluid pressures rather than indirectly to high shale porosities, the latter being the basis for most pressure prediction methods.

Numerous studies have been described in which the electric and/or acoustic properties of shales have been monitored as increasing stress has been applied (e.g., Johnston, 1987; Fjær et al., 1989; Jing et al., 1990; Nur, 1990; Vernik, 1994). All of these investigations (and many others) find that increased stress influences the sonic and/or resistivity log responses. However, the observed effects are mainly attributed to the closing of micro-cracks which may have formed during the removal of rock samples from in situ to surface conditions. These results cannot therefore be used as evidence of the direct influence of overpressuring on the log responses.

If the sonic and resistivity logs are indeed influenced by fluid overpressuring, and that this is caused by the *same* mechanism, one would anticipate the existence of a strong overall correlation between them. However, if the logs are responding to fluid overpressuring because of *different* mechanisms, no such correlation would be expected within each of the pressure regimes.

This question was investigated by cross-plotting the sonic and resistivity porosities (Figure 14). As both of the logs indicate elevated apparent porosities in the overpressured shales, an overall correlation emerges. The validity of the correlation for each pressure regime, however, is less certain. For example, the normally pressured wells show no apparent correlation, mainly because the resistivity-derived porosities vary by only

Table 3. Calculated averages of differences between helium porosity (from cores) and log-derived porosity for the Garn Formation.

Log Type	Average of Difference, He Porosity - Log Porosity	
	Overpressured Wells	"Normal Pressured" Wells
Sonic	1.27	0.30
Resistivity	-2.31	2.89
Neutron	4.20	0.60
Density	3.21	0.05

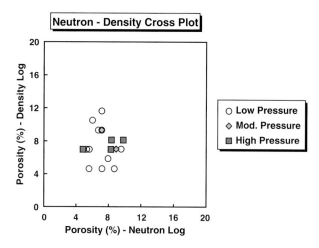

Figure 13. Cross-plot of neutron- vs. density-derived porosities for the Not Formation. The symbols as per Figure 7.

1–3% between them. However, the six overpressured wells (which have both sonic and resistivity log measurements) seem to show a weak positive correlation. Although this might indicate that some common factor, which is modified by fluid pressures, influences both logs, such a thought must remain speculative in light of the scarcity of data and the relative weakness of the correlation.

It should also be recalled that the two logs purportedly reacting to fluid pressure, resistivity and sonic, measure transport properties of the rocks; whereas, the neutron and density logs rely on bulk rock properties. As the conventional sonic log reacts to the rock matrix texture, one might suggest that low fluid pressures and a corresponding high stress at grain contacts would facilitate sound transport between the grains. If this is correct, the sonic log should respond directly to high fluid pressures in moderately consolidated rocks as a reduction in effective stress at grain contacts could cause a reduction in sound transmission. One would not, however, expect the sonic log to overestimate porosities in overpressured and well-cemented sandstones (such as the Garn Formation), as the increased stiffness of the cemented rock framework may be largely unaffected by reductions in effective stress until the level of micro-fracturing is reached. Figure 12, which shows the difference between helium sandstone porosity and the corresponding sonic porosity for the Garn Formation, indicates that log porosities (apart from the resistivity log) are not significantly influenced by fluid overpressures in the western part of Haltenbanken.

As the resistivity log measures a rock's ability to transmit electricity, saline water conducting electricity better than hydrocarbons, good connectivity in the water phase will yield a low resistivity and high log porosity (see Equation A3 in the Appendix). The high apparent resistivity-derived porosities in the overpressured area thus suggest that water connectivity is better than in the normally pressured area, a possible result

of micro-fracturing reducing tortuosity (Miller, 1996).

Figure 11c shows that the resistivity-derived porosity of the Garn Formation sandstones is considerably higher in the overpressured than in the normally pressured rocks, which is contrary to observations of core porosity (Figure 10). This suggests that water connectivity in the former is better, possibly because of a higher frequency of micro fractures.

Whether such a difference in micro fracture intensity can be proven by examining sandstone cores is questionable. This is because cores taken from the normally pressured Garn Formation (which should contain fewer natural fractures in the subsurface than their overpressured counterparts) may experience greater changes in effective stress during exposure to surface conditions. The result of this may be the development of extensive induced fracturing, difficult to distinguish from natural fracturing (without undertaking a major study). For this reason we have not investigated the cores to quantify micro-crack distribution.

Pore pressures can also be detected from the drilling exponent (D_{cexp}), the idea being that the drilling rate increases as high porosity (undercompacted) rocks are penetrated. However, increased drilling rate may result from a reduction in effective stress and need not necessarily be a reliable shale porosity indicator.

In summary, we suggest that the acoustic and electrical transport properties of the rocks are influenced by textural changes induced by overpressuring, and that the sonic and resistivity logs therefore respond directly to fluid overpressuring in shales. The sonic log responds to reduced transport capacity in the matrix of the overpressured rocks, while the resistivity log reacts to increased fluid connectivity in micro fractures.

IMPLICATIONS FOR NUMERICAL BASIN MODELING

Integrated process modeling of heat, fluid flow, hydrocarbon generation, expulsion, and migration has emerged as a major exploration tool during the last decade (Bethke, 1985; Doligez et al., 1986; Hermanrud, 1993; Lerche, 1993). Such modeling also includes shale compaction and overpressuring, and errors made here will influence the simulated timing of hydrocarbon generation and expulsion.

As far as we know, all basin modeling programs (including major commercial software) model shale porosity occlusion on the basis of reductions in effective stress. As a rule, such modeling would produce high porosities in deeply buried overpressured shales, which is contrary to our data from the Not Formation. However, diagenetic processes are known to play a major role in porosity occlusion (at least in reservoir rocks) and have been quantitatively modeled (Dewers and Ortoleva, 1990; Waples and Kamata, 1993). Lower modeled porosities in overpressured shales would be obtained if diagenesis was modeled using temperature as the main controlling factor (rather than effective stress). However, it is beyond the scope of this paper to discuss the various temperature-controlled diagenetic reactions which can lead to shale porosity reduction with burial depth. (For a discussion of possible diagenetic reactions and their temperature dependence, see Pearson and Small (1988), Freed and Peacor (1989a, 1989b), Bjørlykke and Aagaard (1992), Bjørkum (1996), Oelkers et al. (1996).

Schneider et al. (1993) and Burrus (1998-this volume) suggest that for basin modeling purposes Terzaghi's (1923) equation should be replaced by that of Biot (1941):

$$\alpha P_{fl} = OB - \sigma_v \qquad (2)$$

where α, the Biot coefficient, is defined as:

$$\alpha = 1 - \frac{K_{fr}}{K_s} \qquad (3)$$

where K_{fr} is the bulk modulus of the framework and K_s the bulk modulus of the solid. The term $1/K_s$ is often referred to as grain compressibility (Fjær et al., 1992). The consequence of substituting Equation 1 with Equation 2 is that porosity reduction is more controlled by burial depth and less by overpressuring (as indicated by Equation 1). This substitution, with α = 0.67, produced a satisfactory fit between shale density and effective stress in the Mahakam Delta, something which could not be achieved from Equation 1 (Schneider et al., 1993). Values of α are normally reported to be close to 1.0 at shallow burial depths (Fjær et al., 1992), but may well be reduced to values as low as that suggested by Schneider et al. (1993) during burial, at least in tight sandstones (Warpinski and Teufel, 1992). Introducing the (depth-dependent) Biot coefficient (Burrus, 1998-this volume) could provide an alternative approach to modeling shale porosity, which would

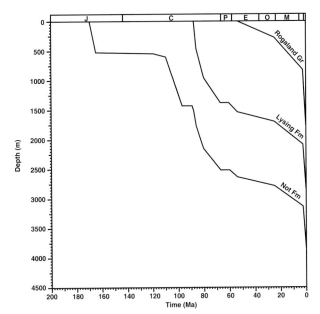

Figure 15. Typical burial history diagram for Halten Terrace wells (6506/11-2). The intra-reservoir shales are presently at maximum burial depth having experienced rapid late Pliocene to Pleistocene burial.

result in a closer match with porosities inferred from density log measurements.

Another explanation for discrepancies between effective stress and shale porosity was suggested by Holbrook and Hauck (1987), and later adopted in the works of Alixant et al. (1989), Alixant and Desbrandes (1991), Ward et al. (1994, 1995), Ward (1995), Yassir and Bell (1994), Bowers (1994) and Hart et al. (1995). These works discriminate between the virgin (loading) curve, which describes the compactional effects on fluid overpressuring, and the unloading curve, which describes overpressuring originating from noncompactional effects such as aquathermal pressuring and hydrocarbon generation. Overpressuring resulting from noncompactional effects can postdate porosity reduction, thus leading to fluid pressures higher than those obtained from Equation 1.

By introducing both the Biot coefficient and the late arrival of overpressuring from noncompactional effects ("unloading"), a smaller difference in porosity between overpressured and normally pressured shales will be obtained than that derived from standard methods. Even so, all of these methods rely on compaction as the driving force, which means that porosities are elevated in overpressured shales unless all overpressuring postdates porosity reduction.

Let us now consider the possibility that overpressuring of the Not Formation postdated porosity reduction. If so, the high fluid pressures in reservoirs to the west could have originated in response to the last period of rapid Pliocene–Pleistocene subsidence (2 Ma, see Figure 15), together with closure of the fault system separating the two main pressure regimes. From Figure 8, it appears that compaction was incomplete until

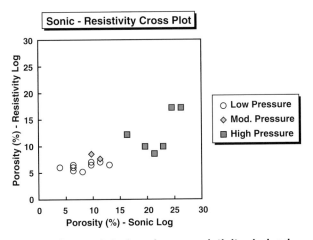

Figure 14. Cross-plot of sonic- vs. resistivity-derived porosities for the Not Formation. The symbols as per Figure 7.

the formation was buried to about 3 km. In this case, overpressuring must have resulted from the last 500 m of subsidence if a diminution in effective stress, due to undercompaction, arrested porosity reduction. However, without precisely knowing the permeability of the Not Formation, it is unclear whether 2 m.y. would have been long enough for the formation to adjust to the reservoir pressures. A lack of adjustment would result in overpressuring (detectable from the sonic and resistivity logs). If overpressuring occurred earlier, the fluid pressure of the Not Formation should be different in the two major pressure regimes.

Basin modeling of cap rock fluid pressures suggests that a fluid pressure transition zone should exist between overpressured cap rocks and normally pressured reservoir rocks. These transition zones should be present irrespective of the mechanism of overpressure development. Furthermore, pressure zones should be accompanied by downward decreasing shale porosities (i.e., from the central parts of the overpressured cap rocks towards the underlying reservoirs), provided that increased overpressuring prevented further compaction.

Recalling that the cap rock did not reveal the presence of porosity transition zones based on log responses, we suggest that vertical variations in rock properties may have more influence on the logs than possible decreases in porosity caused by depletion of fluid overpressures close to the sand-shale boundaries. However, the difference between the sonic and density log responses suggests that fluid pressure transition zones exist in the cap rocks overlying the Garn Formation.

These considerations are consistent with our earlier suggestion; namely, that the density log only responds to porosity while the sonic log can be directly influenced by fluid pressures (and not necessarily high porosities). The density log data, however, cast doubt on the significance of effective stress as a mechanism to reduce Haltenbanken shale porosities at depths below 3.5 km, as lower rather than higher porosities should otherwise be present over the reservoirs. However, our scant knowledge of the cap rocks' heterogeneity prevents us from making a firm conclusion.

Additional information on the mechanisms of shale porosity reduction can be gained by studying log responses of individual shaly formations at different depths and pressures. Such investigations in the North Sea basin are currently being reported by Hansen (1997). The main results to emerge from his work so far are:

1. At depths below 2 km, the density log appears to be unaffected by fluid overpressuring. Some overpressured wells show increased sonic travel times (especially in the tight, tuffaceous Balder Formation); whereas, most of them do not. Overpressures in these shallow formations do not exceed 1.5 g/cm^3 MWE.

2. Both the sonic and density logs indicate high porosities in some (but less than 50%) of the overpressured wells between 2 and 2.5 km in some formations.

It is unclear whether these wells are undercompacted or if the apparent high porosities are the result of lateral lithology variations.

3. At greater depths (below 3–3.5 km), the sonic transit time does not, in general, diminish with depth within overpressured formations, while the density readings suggest normal compaction. These observations agree well with those made in the overpressured Not Formation (Figure 16).

4. Shale porosity, as calculated by either the density or the sonic log, does not correlate with reported fluid pressure (from logs, D_c-exponent, trip gas and connection gas) within a given formation.

While the possible late arrival of overpressures in the Not Formation cannot be discounted, it is less easy to explain how thick (several km) shaly sequences can be substantially overpressured once compaction has ceased; after all, they have very low permeabilities (otherwise they would not be overpressured), which makes it unlikely that substantial fluid volumes could be introduced during a relatively short period of time. Late overpressuring (unloading) is thus not regarded as a viable explanation for the lack of elevated porosity in the deeply buried, overpressured shales of the North Sea.

Taking all of the evidence together—the log responses of the Not and Ror Formations, the cap rocks overlying the Garn Formation reservoirs, and the deeply buried North Sea shales—it appears that while shale porosity is reduced with burial depth, it is only moderately influenced (or entirely uninfluenced) by fluid overpressuring. The sonic log apparently reacts to fluid pressures which are unrelated to shale porosity, at least at depths below 3 km (110°C). Consequently, the results of pressure build-up and fluid dynamics from Terzaghi-based basin modeling should be interpreted with great care, the (only) mechanism for shale porosity reduction being based on increases in effective stress following Equation 1. Furthermore, calibrations of modeled shale porosity to sonic log responses should not be performed, as these will yield overestimates and consequently too low thermal conductivities.

IMPLICATIONS FOR FLUID PRESSURE ANALYSIS

Detection of overpressure zones

This study also provided an opportunity to test the reliability of various logging tools for pressure prediction: as previously mentioned, the detection and quantification of overpressure from logs is based on the assumption that there is a direct relationship between an anomalous shale porosity and a pressure anomaly. As our results indicate that the difference in intra-reservoir shale porosities between the two major pressure regimes varies by less than 1–3%, we require tools that not only respond to porosity variations but

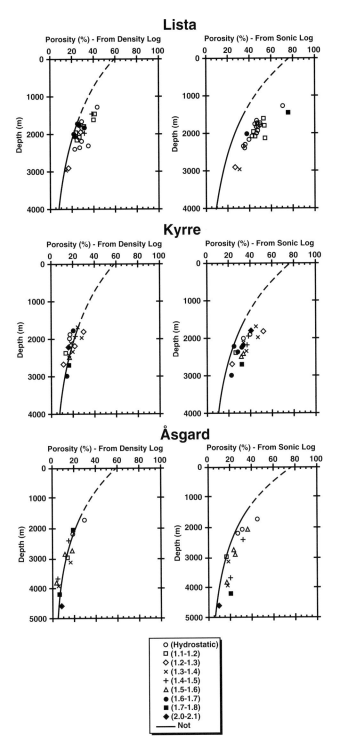

Figure 16. Porosities calculated from bulk density and sonic transit time from three North Sea formations (i.e., the Paleocene Lista Fm., the Late Cretaceous Kyrre Fm., and the Early Cretaceous Åsgard Fm.), with the normal porosity trend established for the Not Formation superimposed. The porosities were calculated using Equations A1 and A2 with free parameter values as used for the Not Formation. The deviations from the Not Formation sonic normal porosity trend (lines) are most likely due to differences in matrix transit time

also, *directly,* to increases in pore fluid pressure in order to separate the two pressure regimes from well logs.

Since the density and neutron logs appear to be unaffected by the present level of overpressure in deeply buried shales, we recommend that they should not be used for this purpose. The sonic and resistivity logs, however, appear to respond to high fluid pressures even when there is no apparent variation in shale porosity. As the moderately pressured rocks cannot be distinguished from the normally pressured rocks, one might speculate that the reduction in transport properties of compressional sound waves and the increased connectivity in the fluid phase (leading to increased electrical conductivity) require a pressure level approaching fracture pressure before their influence on the logs is significant.

Normal trends for calculated pressures (see below) were established for the Not Formation using regression analysis of mean sonic and density log values from normally pressured wells (Figure 17). These calculations also included log data from wells outside the main study area where the Not Formation shales occur at much shallower burial depths (i.e., 1.4–2.5 km). The apparent porosity versus depth trend for the sonic log gives a higher calculated porosity than that of the density log, although the difference becomes small at depths below 4.0 km. (Core measurements of shale porosity were unavailable for calibration.)

Figure 18 shows the calculated porosity from the sonic and density logs versus effective vertical stress (overburden minus pore fluid pressure) for both normally pressured and overpressured Not Formation shales. The present vertical effective stress of the overpressured low porosity shales, at depths between 3.6 and 4.4 km, is similar to the effective stresses of normally pressured and high porosity shales at a present depth of 1.7 and 2.4 km. The cross-plot of density-derived porosity versus effective stress (Figure 18) shows a clear separation between overpressured and normally pressured rocks. This separation may arise from noncompactional (diagenetic) mechanisms for shale porosity occlusion, from unloading, or from a less pronounced influence of overpressuring on effective stress than that derived from the Terzaghi relationship (Schneider et al., 1993).

The cross-plot of sonic-derived porosity versus effective vertical stress (Figure 18) shows that the overpressured wells neither follow the virgin loading curve, as determined by the normally pressured wells, nor the unloading curve of effective stresses of 400–500 bars (as do the density-derived shale porosities). This information can be used to assess the relative influence of porosity and effective stress on the sonic log responses based on the following reasoning: The calculated average sonic porosity for the overpressured wells is 21.8% (Table 1) at effective stresses of about 150 bars. As seen earlier, the porosities of these shales are in fact no higher than those obtained from the low pressured regimes (8.6%, Table 1). The apparent porosity difference, which is due to the direct influence of over-pressuring, is thus about 13%. However, the shallower, normally pressured rocks, with present effective stresses equivalent to those of the overpressured wells (150 bars), have porosities of about 26%, which is about 17% higher than the deeply buried normally pressured rocks. The overpressured wells show an apparent enhanced porosity of about 13/17 (76%) of the porosity difference between wells at the virgin curve between 150 and 450 bars. This suggests that the reduction in interval transit time is mainly due to an increase in effective vertical stress (76%) and, to a lesser extent, on porosity reduction by increased burial (24%).

Generally speaking, it is expected that pressure detection methods based on the identification of normal compaction trends will underestimate levels of overpressure in the Haltenbanken area. This is because they are based on the assumption that the overpressured wells in Figure 18b should fall on the virgin curve, while in fact they have effective stresses of about 70 bars less than that suggested by the curve.

Calculated magnitudes of fluid pressures

Calculations of fluid pressure in the Not Formation from the sonic log were made to investigate the relative merits of different techniques (see Appendix B for details). The results (Table 4) indicate that both the matrix stress (Rubey and Hubbert, 1959; Ham, 1966) and effective stress (Hart et al., 1995) methods tend to yield higher pore fluid pressures than the Eaton (1972a, b) method. However, all of the methods tend to underestimate the level of overpressure. The use of conventional methods on the mean sonic porosity data of the Not Formation may underestimate the pressure gradi-

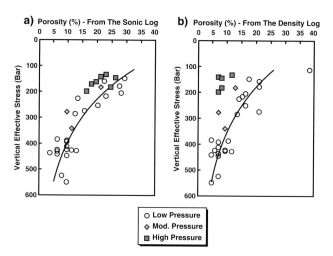

Figure 18. Log-derived porosities for Not Formation shales vs. effective vertical stress (overburden minus pore fluid pressure): (a) bulk density log; (b) sonic log. Exponential regression analyses (curves) indicate: $\phi_{RHOB} = 40.3e^{-0.0039\sigma}$ **and** $\phi_{DT} = 52.3e^{-0.0042\sigma}$**, where** σ **is the vertical effective stress. Symbols as per Figure 7.**

ent by as much as 0.2–0.3 g/cm^3 MWE. Furthermore, the velocity and resistivity anomalies cannot be used to identify or quantify moderate overpressures (below 1.40 g/cm^3 MWE), as observed in the Njord area, because they cannot be distinguished from the hydrostatic pressures.

The differences between the calculated and measured fluid pressure, and the differences between the two methods themselves, probably reflect one or more of the following: (a) the inappropriateness of sonic velocity for porosity-fluid pressure relationships (on which all three methods are based); (b) uncertainty related to the identification of the "normal sonic gradient"; and (c) uncertainty related to determination of the overburden gradient. The variations in overburden gradient in the studied wells is about ±0.05 g/cm^3, and the standard deviation of the sonic normal trend is 7.8 ms/ft. The latter may cause a 1,000 m variation in equivalent depth and a subsequent variation in the estimated pore fluid pressure of ±0.1 g/cm^3. We suspect that these variations are artifacts, and that they introduce errors to the calculated pore pressures.

The calculated pore pressures indicate that both the matrix and effective stress methods correlate better with the observed reservoir pressures than the Eaton method (1972a, b). All three, however, would benefit from empirical calibrations to local data to produce calculated fluid pressures which correspond more closely than they do at present to the reservoir pressures.

Hart et al. (1995) suggested that the difference between the observed pressure and the (lower) pressure calculated from the simple effective stress approach can be used to quantify the relative contributions to pore pressure of various fluid expansion mechanisms. This, however, assumes that an overpressured shale is associated with abnormally high porosity, an

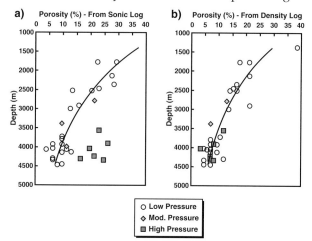

Figure 17. Estimated "normal compaction" trends for Not Formation shales: (a) sonic-derived porosity vs. depth, (b) density-derived porosity vs. depth. Normal compaction trends (curves) determined from exponential regression analysis (based only on data from the low pressured wells) indicate: $\phi_{DT} = 77.07e^{-0.00053z}$ **and** $\phi_{RHOB} = 58.93e^{-0.00051z}$**, where z is the burial depth. Symbols as per Figure 7.**

assumption that is invalid for the Not Formation shales. The Hart et al. (1995) method cannot therefore be applied to the deep Jurassic shales in the Haltenbanken area.

We therefore suggest that shale porosity in the study area should be identified from the density log, while the presence of high pore fluid pressures can be verified from sonic and resistivity logs. The study of the Not Formation has shown that sonic and resistivity logs respond both to variations in porosity and directly to high (above 1.60 g/cm^3 MWE) overpressure. Since present (Terzaghi-based) effective stress versus porosity relationships are based on the assumption that the effective stress is the only factor which causes shale porosity reduction, future improvements should be made to methods employing logs for pressure analysis. Such methods should be based on relationships between fluid pressure and sonic and/or resistivity log responses, which do not rely on relationships between porosity and effective stress as calculated using Terzaghi's principle (Equation 1). Shale porosity, as determined from density logs, could be used to correct the calculated fluid pressures for the effects of porosity variations. Wherever possible, regional relationships between sonic log responses and effective stress should still be determined for individual shale units in order to characterize the variation in shale compaction as a function of effective stress, mineralogy and porosity.

SUMMARY AND CONCLUSIONS

The porosity of Haltenbanken intra-reservoir shales determined from density and neutron logs does not

vary significantly between highly overpressured and normally pressured areas; however, higher *apparent* porosities in the overpressured areas are indicated by the sonic and resistivity logs. An explanation for this is that the sonic and resistivity logs are responding to textural changes in the shales induced by overpressuring rather than to higher porosities.

Porosity transition zones could not be identified in overpressured Haltenbanken cap rocks which overlie normally pressured reservoirs. Several wells, however, appear to contain fluid pressure transition zones, even though there is no decrease in porosity immediately above the boundary between the cap rocks and the reservoirs. This suggests that shale porosity was not reduced by the decreased effective stress experienced by these rocks.

An inspection of density and sonic log data from deeply buried shales in about 30 North Sea wells supports the findings from Haltenbanken. The density logs failed to discriminate between highly overpressured and normally pressured shales, whereas the sonic logs recorded slightly higher interval transit times in the overpressured wells.

These results confirm that the response of the sonic log to decreasing shale porosity with depth (at least in the areas under investigation) is not largely due to compaction following Terzaghi's principle (as calculated by Equation 1), as this would yield lower porosities in normally pressured rocks and higher porosities in their overpressured counterparts. Late overpressuring could possibly explain the low porosities encountered in the overpressured rocks on Haltenbanken, but cannot explain why the thick, shaly, overpressured sequences of the North Sea do not exhibit abnormally high porosities. These observations might be explained

Table 4. Pore fluid pressures calculated from sonic logs in highly overpressured settings of the Not Formation. Average depths are given as total vertical depths in meters below Kelly Bushing (mKB). Pressures are given as mud weight equivalents (g/cm^3 MWE): MWE describes the density of a mud column, extending from the drill floor (Kelly Bushing) to the reservoir, needed to balance the fluid pressure in the formation.

Highly Overpressured Wells	Average Depth of Shaly Interval (mKB)	Observed Reservoir Pressure in MWE (g/cm^3)	Calculated Reservoir Pressure in MWE (Calculation Methods)		
			Matrix Stress Method (g/cm^3)	Eaton Method[1] (g/cm^3)	Effective Stress Concept[2] (g/cm^3)
6406/11-1	3561	1.68	1.52	1.42	1.55
6406/3-1	3901	1.80	1.77	1.59	1.71
6506/12-4	4033	1.72	1.55	1.44	1.56
6506/11-1	4239	1.83	1.68	1.54	1.65
6406/6-1	4305	1.73	1.54	1.42	1.51
6406/8-1	4339	1.78	1.82	1.65	1.74
6506/11-3	4340	1.82	1.63	1.48	1.57

1 Eaton (1972a, b).
2 The effective stress concept suggested by Hart et al. (1995).

from calculations of effective stress using Equation 2 with a depth-dependent Biot coefficient, or simulations of temperature-controlled clastic diagenesis.

To the best of our knowledge, numerical basin modeling of overpressuring, fluid flow, shale compaction and primary migration is always based on Terzaghi-type (Equation 1) compaction as the sole means of reducing shale porosity (and this largely applies to sandstones as well). We therefore suggest that modeling results be interpreted with great care as they will frequently predict excessively high shale porosities in overpressured basins.

We also suggest that efforts should be made to improve predictions of porosity in overpressured rocks in future basin modeling programs.

Methods commonly used to evaluate fluid pressures from well logs are also based on "Terzaghi-compaction" and the assumption that overpressuring results in abnormally high porosities. Although in most cases the sonic log responds to the direct influence of fluid overpressuring, improvements could be made if a relationship involving both sonic and resistivity responses was developed.

ACKNOWLEDGEMENTS *We are grateful to Antony T. Buller, Terje Eidesmo, Arne M. Raaen and Per Arne Bjørkum (of Statoil) and two anonymous referees for constructive criticism of an earlier version of this paper. Antony T. Buller is also thanked for improving the English language. Den norske stats oljeselskap a.s. (Statoil) is acknowledged for granting permission to publish this work.*

REFERENCES CITED

Alixant, J. L., R. Desbrandes, and T. Delahaye, 1989, A new approach to real-time pore pressure evaluation: Society of Petroleum Engineers Inc. SPE paper 19336, presented at the SPE Eastern Regional Meeting, West Virginia, October 24–27, 1989.

Alixant, J. L. and R. Desbrandes, 1991, Explicit pore-pressure evaluation: concept and application: Society of Petroleum Engineers Inc. (SPE) Drilling Engineering, v. 6, Sept. 1991, p. 182–188.

Archie, G. E., 1942, The electrical resistivity log as an aid in determining some reservoir characteristics: Journal Petroleum Technology, v. 5, p. 54–62.

Athy, L. F., 1930, Density, porosity, and compaction of sedimentary rocks: AAPG Bulletin, v. 14, p. 1–22.

Bethke, C. M., 1985, A numerical model of compaction-driven ground water flow and heat transfer and its application the paleohydrology of intracratonic sedimentary basins: Journal of Geophysical Research, v. 90, p. 6817–6828.

Biot, M. A., 1941, General theory of three-dimensional consolidation: Journal of Applied Physics, v. 12, p. 155–164.

Bjørkum, P. A., 1996, How important is pressure in causing dissolution of quartz in sandstones: Journal of Sedimentary Research, v. 66, no. 7, p. 147–154.

Bjørlykke, K., and P. Aagaard, 1992, Clay minerals in North Sea sandstones, *in* Origin, diagenesis, and petrophysics of clay minerals in sandstones, SEPM Special Publication, No. 47, p. 65–80.

Bowers, G. L., 1994, Pore pressure estimation from velocity data; accounting for overpressure mechanisms besides undercompaction: IADC/SPE paper 27488, presented at the 1994 IADC/SPE Drilling Conference, Dallas, Texas, Feb. 15–18, 1994.

Burrus, J., 1998, Overpressure models for clastic rocks; their relation to hydrocarbon expulsion: a critical reevaluation, *in* Law, B.E., G.F. Ulmishek, and V.I. Slavin eds., Abnormal pressures in hydrocarbon environments: AAPG Memoir 70, p.35–64

Carstens, H. and H. Dypvik, 1979, Prediction, detection and evaluation of abnormal formation pressures in virgin areas: some limitations: Norwegian Petroleum Society Paper NSS 13, 14 p.

Carstens, H. and H. Dypvik, 1981, Abnormal formation pressure and shale porosity: AAPG Bulletin, v. 65, p. 344–350.

Dewers, T., and P. J. Ortoleva, 1990, Interaction of reaction, mass transport, and rock deformation during diagenesis: mathematical modeling of intergranular pressure solution, stylolites, and differential compaction/cementation, *in* I. Meshri and P. J. Ortoleva, eds., Prediction of reservoir quality through chemical modeling, AAPG Memoir, 49, p. 147–160.

Doligez, B. F., Bessis, J., Burrus, J., Ungerer, P. and P. Y. Chénet, 1986, Integrated numerical modeling of sedimentation, heat transfer and fluid migration in a sedimentary basin: the THEMIS model, *in* J. Burrus, ed., Thermal Modeling of Sedimentary Basins, Paris, Technip, p. 173–195.

Dutta, N. C., 1986, Shale compaction, burial diagenesis, and geopressures: a dynamic model, solution and some results, *in* J. Burrus, ed., Thermal modeling in sedimentary basins, IFP, Editions Technip 27 Rue Ginoux 75737 Paris Cedex 15, p. 149–172

Eaton, B. A., 1972a, Graphical method predicts geopressures worldwide: World Oil, v. 182, no. 6, p. 51–56.

Eaton, B. A., 1972b, The effects of overburden stress on geopressure predictions from well logs; Journal of Petroleum Technology, Aug. 1972, p. 929–934.

Ehrenberg, S. N., H. M. Gjerstad, and F. Hadler-Jacobsen, 1992, Smørbukk Field - a gas condensate fault trap in the Haltenbanken Province, offshore Mid-Norway, *in* M. T. Halbouty, ed., Giant oil and gas fields of the decade, AAPG Memoir, 54, p. 323–348.

Fertl, W. H., and D. H. Timko, 1972, How down hole temperature, pressure affect drilling. part 3: overpressure detection from wireline methods, World Oil, Aug. 1, 1972, p. 36–66

Fertl, W.H., 1976, Abnormal formation pressures, Developments in Petroleum Science 2, New York, Elsevier, 382 p.

Fertl, W. H., R. E. Chapman, and R. F. Hotz, eds., 1994, Studies in abnormal pressures, New York, Elsevier, 454 p.

Fjær, E., R. M. Holt, and A. M. Raaen, 1989, Rock mechanics and rock acoustics, *in* V. Maury and D. Fourmaintraux, editors, Rocks at Great Depths, v. 1, Rotterdam, Balkama, p. 355–362.

Fjær, E., R. M. Holt, P. Horsrud, A. M. Raaen, and R. Risnes, 1992, Petroleum related rock mechanics: Developments in Petroleum Science, 33, Amsterdam, Elsevier, 338 p.

Freed, R. L., and D. R. Peacor, 1989a, Geopressured shale and sealing effects of smectite to illite transition: AAPG Bulletin, v. 73, no. 10, p. 1223–1232.

Freed, R. L., and D. R. Peacor, 1989b, Variable temperature of the smectite/illite reaction in Gulf Coast sediments: Clay Minerals, v. 24, p. 171–180

Ham, H. H., 1966, A method of estimating formation pressures from Gulf Coast well logs: Transactions - Gulf Coast Association of Geological Societies, 16, p. 185–197.

Hansen, S, 1997, Quantification of sediment compaction with emphasis on shales: Ph.D. Dissertation, Dept. of Geology, Univ. of Århus, Denmark

Hart, B. S., P. B. Flemings, and A. Deshpande, 1995, Porosity and pressure: role of compaction disequilibrium in the development of geopressures in a Gulf Coast Pleistocene basin: Geology, v. 23, p. 45–48.

Hermanrud, C., 1993, Basin modeling techniques - an overview, *in* A. G. Doré, J. H. Augustson, C. Hermanrud, D. J. Stewart, and Ø. Sylta, eds., Basin modeling: advances and applications, Norwegian Petroleum Society (NPF) Special Publication 3, Amsterdam, Elsevier, p. 1–34.

Heum, O. R., A. Dalland, and K. Meisingset, 1986, Habitat of hydrocarbons at Haltenbanken (PVT-modeling as a predictive tool in hydrocarbon exploration), *in* A. M. Spencer et al., eds., Habitat of the Norwegian continental shelf, Norwegian Petroleum Society, London, Graham and Trotman, p. 259–274.

Holbrook, P. W. and M. L. Hauck, 1987, A petrophysical-mechanical math model for real-time wellsite pore pressure/fracture gradient prediction: Society of Petroleum Engineers Inc. (SPE) paper 16666, presented at the 62nd SPE Annual Technical Conference, Dallas, Texas, Sept. 27–30, 1987.

Hollander, N. B., 1984, Geohistory and hydrocarbon evolution of the Haltenbanken area, *in* A.M. Spencer, S.O. Johnsen, A. Moerk, E. Nysaether, P. Songstad, and Å. Spinnangr eds., Petroleum geology of the north European margin, Norwegian Petroleum Society, London, Graham and Trotman, p. 283–384.

Hottman, C. E. and R. K. Johnson, 1965, Estimation of formation pressures from log derived shale properties: Journal of Petroleum Technology, v. 17, p. 717–722.

Hunt, J. M., 1990, Generation and migration of petroleum from abnormally pressured fluid compartments: AAPG Bulletin, v. 74, no. 1, p. 1–12.

Hunt, J. M., 1991, Generation and migration of petroleum from abnormally pressured fluid compartments: Reply, AAPG Bulletin, v. 75, no. 2, p. 328–330.

Jing, X. D., T. S. Daltaban, and J. S. Archer, 1990, Experimental measurements on the effects of pressure and temperature on electric properties of natural and synthetic rocks, *in* V. Maury and D. Fourmaintraux, eds., Rocks at Great Depths, v. 3, Rotterdam, Balkama, p. 1357–1368.

Johnston, D.H., 1987, Physical properties of shales at temperature and pressure: Geophysics, v. 52, no. 10, p. 1391–1401.

Koch, J. O., and O. R. Heum, 1995, Exploration trends of the Halten Terrace, *in* S. Hanslien, ed., 25 years of petroleum exploration in Norway, Norwegian Petroleum Society (NPF), Special Publication 4, Amsterdam, Elsevier, p. 235–251.

Lerche, I., 1993, Theoretical aspects of problems in basin modeling, *in* A. G. Doré, J. H. Augustson, C. Hermanrud, D. J. Stewart and Ø. Sylta, eds., Basin Modeling: Advances and Applications, Norwegian Petroleum Society (NPF) Special Publication 3, Amsterdam, Elsevier, p. 35–65.

Lindberg, P., R. Riise, and W. F. Fertl, 1980, Occurrence and distribution of overpressures in the northern North Sea: Society of Petroleum Engineers Inc. paper SPE 9339, presented at the 55th SPE Annual Fall Technical Conference, Dallas, Texas, Sept. 21–24, 1980.

Miller, T. W., 1996, New insight on natural hydraulic fractures induced by abnormally high pore pressures: AAPG Bulletin, v. 79, no. 7, p. 1005–1018.

Mouchet, J. P., and A. Mitchell, eds., 1989, Abnormal pressures while drilling: Elf Aquitaine Manuels Techniques 2, Boussens, 1989, 264 p.

Nur, A., 1990, General report: seismic rock properties for rock mass description and fluid flow monitoring, *in* V. Maury and D. Fourmaintraux, eds., Rocks at Great Depths, v. 3, Rotterdam, Balkama, p. 1081–1102.

Oelkers, E. H., Bjørkum, P. A. and Murphy, W. M., 1996, A petrographic and computational investigation of quartz cementation and porosity reduction in North Sea sandstone: American Journal of Science, v. 296, p. 420–452.

Pearson, M. J., and J. S. Small, 1988, Illite-smectite diagenesis and paleotemperatures in northern North Sea Quaternary to Mesozoic shale sequence: Clay Minerals, v. 23, p. 109–132.

Rubey, W. W., and M. K. Hubbert, 1959, Role of fluid pressure in mechanics of overthrust faulting: Geological Society of America Bulletin, v. 70, p. 167–206.

Schneider, F., J. Burrus, and S. Wolf, 1993, Modeling overpressures by effective-stress/porosity relationships in low-permeability rocks: empirical artifice or physical reality?, *in* A. G. Doré, J. H. Augustson, C. Hermanrud, D. J. Stewart, and Ø. Sylta, eds., Basin modeling: advances and applications, Norwegian

Petroleum Society (NPF) Special Publication 3, Amsterdam, Elsevier, p. 333–341.

Sclater, J. G. and P. A. B. Christie, 1980, Continental stretching: an explanation of the post-mid-Cretaceous subsidence of the Central North Sea basin: Journal of Geophysical Research., v. 85, p. 3711–3739.

Terzaghi, K., 1923, Die berechnung der durchlässigkeitsziffer des tones as dem verlauf der hydrodynamichen spannungsercheinungen: Sitzungsgbrg. Akad. Wiss. Wien, Math. Naturwiss. K1., IIa, 132 (3/4), p. 125–138.

Vernik, L., 1994, Hydrocarbon-generation induced micro-cracking of source rocks: Geophysics, v. 59, no. 4, p. 555–563.

Vik, E. and C. Hermanrud, 1993, Transient thermal effects of rapid subsidence in the Haltenbanken area, in A. G. Doré, J. H. Augustson, C. Hermanrud, D. J. Stewart, and Ø. Sylta, eds., Basin modeling: advances and applications, Norwegian Petroleum Society (NPF) Special Publication 3, Amsterdam, Elsevier, p. 107–117.

Waples, D.W. and H. Kamata, 1993, Modeling porosity reduction as a series of chemical and physical processes, in A. G. Doré, J. H. Augustson, C. Hermanrud, D. J. Stewart, and Ø. Sylta, eds., Basin modeling: advances and applications, Norwegian Petroleum Society (NPF) Special Publication 3, Amsterdam, Elsevier, p. 107–117.

Ward, C. D., K. Coghill and M. D. Broussard, 1994, The application of petrophysical data to improve pore and fracture pressure determination in North Sea Central Graben HPHT wells: Society of Petroleum Engineers Inc. paper SPE 28297, presented at the SPE 69th Annual Technical Conference, New Orleans, Louisiana, Sept. 25–28, 1994.

Ward, C. D., K. Coghill, and M. D. Broussard, 1995, Brief: Pore- and fracture-pressure determinations: effective-stress approach: SPE paper 30141, Journal of Petroleum Technology, Feb. 1995, p. 123–124.

Ward, C. D., 1995, Evidence for sediment unloading caused by fluid expansion overpressure-generation mechanisms:, in M. Fejerskov and A. M. Myrvang, eds., Proceedings of the workshop Rock Stresses in the North Sea, p. 218–231.

Warpinski, N. R., and L. W. Teufel, 1992, Determination of effective stress law for permeability and deformation in low-permeability rocks: Society of Petroleum Engineers Formation Evaluation, v. 7, June 1992, p. 123–131.

Wensaas, L., H. F. Shaw, K. Gibbons, P. Aagaard, and H. Dypvik, 1994, Nature and causes of overpressuring in mudrocks of the Gullfaks area, North Sea: Clay Minerals, v. 29, p. 439–449.

Wyllie, M. R. J., A. R. Gregory, and G. H. F. Gardner, 1958, An experimental investigation of the factors affecting elastic wave velocities in porous media: Geophysics, v. 23, p. 459–493.

Yassir, N. A., and J. S. Bell, 1994, Abnormally high fluid pressures and associated porosities and stress regimes in sedimentary basins: Proceedings SPE/ISRM EUROCK '94 Conference, Delft, the Netherlands, Aug. 29–31, 1994, Rotterdam, Balkama, p. 879–886.

APPENDIX A—POROSITY CALCULATIONS

The *density log*, which measures the electron density of a formation, responds to variations in total rock porosity. Porosity was calculated using the following relationship by assuming a fluid density (ρ_f) of 1.0 g/cm^3 and a matrix density (ρ_{ma}) of 2.72 g/cm^3 (Sclater and Christie, 1980):

$$\phi_{RHOB} = \frac{\rho_{ma} - \rho_b}{\rho_{ma} - \rho_f} \tag{A1}$$

where ϕ_{RHOB} is the calculated shale porosity and ρ_b is the (log-measured) bulk density of the formation.

The *sonic log* measures the interval transit time of a compressional sound wave within approximately 30 cm of the formation, and responds to variations in intergranular porosity. Shale porosity (ϕ_{DT}) was calculated from average sonic log readings (Δt) using the classic empirical equation of Wyllie et al. (1958):

$$\phi_{DT} = \frac{\Delta t - \Delta t_{ma}}{\Delta t_f - \Delta t_{ma}} \tag{A2}$$

where Δt is the average sonic transit time ($\mu s/ft$) of the shaly intervals; Δt_f is the transit time of the fluid (Δt_f = 189 $\mu s/ft$, 620 $\mu s/m$); and Δt_{ma} is the matrix transit time (Δt_{ma} = 68.8 $\mu s/ft$, 226 $\mu s/m$).

The resistivity and neutron logs are not very well suited to estimating shale porosity. However, differences in porosity between overpressured and normally pressured wells of the same formation can be assessed assuming that the errors are common to both major pressure regimes. The calculated shale porosity from these logs was calibrated to give comparable porosities to those of the density logs. This calibration was performed by parameter determination for the resistivity logs, and by rescaling the neutron logs.

The *resistivity log* measures a rock's ability to transmit an electric current. Resistivity-derived porosities (ϕ_{RILD})were determined using Archie's (1942) porosity-resistivity relationship for water-bearing zones:

$$\phi_{RILD} = \left(\frac{a \dfrac{R_w}{R_t}}{S_w^n} \right)^{\frac{1}{m}} \tag{A3}$$

where a is the tortuosity factor; R_w is the formation water resistivity; R_t is the shale resistivity from the deep laterolog or deep induction log; S_w is the water saturation, n is the saturation exponent; and m is the cementation exponent. For the calculations, we assumed constant values for water saturation (S_w = 1.0), saturation exponent (n = 2), cementation exponent (m = 2), tortuosity factor (a = 1), and formation water resistivity (R_w = 0.0294 ohmm). The value of R_w was calculated from Equation A3 using the average resistivity reading (R_t = 6.0 ohmm) to fit the average density-derived porosity (ϕ_{RHOB} = 0.07) in the calibration well, 6407/1-2.

The *neutron log* measures the concentration of hydrogen ions in a formation. Shale porosity (in % limestone-equivalent units) in the Not and Ror Formations ranges between 24–28% and 8–31%, respectively. The large spread and high apparent porosities suggest that significant amounts of hydrogen are bound in the crystal lattices of the clay particles. The neutron log measurements were rescaled to give the same mean and standard deviation as those for the density log. This was partly done to make the presentation of the neutron log data more appealing (i.e., that the porosities should "look right"), but mainly because a crossplot of the density and sonic log data should define a 45° trend if the porosities from these two sources are correlatable (see below).

The Garn Formation sandstone porosity in fully water-saturated rocks was derived from the density and sonic logs using Equations A1 and A2, a fluid density (ρ_f) of 1.0 g/cm^3, a matrix density (ρ_{ma}) of 2.65 g/cm^3, a fluid transit time (Δt_f) of 189 µs/ft (620 µs/m), and a matrix transit time (Δt_{ma}) of 55.5 µs/ft (182 µs/m). Porosity from the density log in hydrocarbon-bearing formations was adjusted according to the reported water saturation data (S_w) and a constant hydrocarbon density (ρ_{HC}) of 0.5 g/cm^3. The sonic porosity, however, was not adjusted for the presence of hydrocarbons. Neutron sandstone porosity was used directly (water-bearing formations only), while porosity from resistivity logs in dry wells was obtained from Equation A3, values of a, n, and m as for the shales, and a calculated formation water resistivity (R_w = 0.062 ohmm) chosen to match the density-derived porosity (ϕ_{RHOB} = 0.12) in a dry reference well (6406/3-3).

APPENDIX B—CALCULATION OF FLUID PRESSURES

Pore fluid pressure calculations from Not Formation sonic data and the corresponding normal compaction trend (See Figure 17) were performed using three methods (Table 4). This was done to investigate their relative merits and fit to RFT pressures in the surrounding reservoir rocks.

The Equivalent Depth (or the matrix stress) method is based on comparisons between the level of com-

paction in an overpressured and an undercompacted (i.e., high porosity) shale, for which the latter is assumed to correspond to a normally compacted shale located at a shallower equivalent depth. Shale porosity is a function of the net confining pressure (matrix stress), which requires an estimate of the change in overburden gradient:

$$P_i = OB_i - (OB_e - P_e)\frac{D_e}{D_i} \qquad (B1)$$

where
  P_i = pore pressure gradient at the depth of interest
  P_e = pore pressure gradient at an equivalent depth
  D_i = depth of interest
  D_e = equivalent depth
  OB_i = overburden gradient at the depth of interest
  OB_e = overburden gradient at an equivalent depth

The Eaton (1972a, b) or exponential method, which has been used for a variety of logs and drilling parameters, generally assumes that the pore pressure at any depth can be estimated from the observed parameter/normal parameter ratio and the change in overburden gradient:

$$P_i = OB_i - (OB_i - P_n)\left(\frac{dt_n}{dt_i}\right)^3 \qquad (B2)$$

where
  P_i = pore pressure gradient at the depth of interest
  P_n = normal pore pressure gradient
  OB_i = overburden gradient at the depth of interest
  dt_n = normal shale travel time at the depth of interest
  dt_i = shale travel time at the depth of interest

The third method, proposed by Hart et al. (1995), links shale porosity (ϕ) to vertical effective stress (σ_v) according to Athy's (1930) exponential relationship (σ_v given in kPa):

$$\phi = \phi_0 e^{a\sigma_v} \qquad (B3)$$

The porosity at zero effective stress, ϕ_0 = 52.3, and the empirical constant, a = –0.0042 kPa^{-1}, were determined by exponential regression of effective stress versus sonic-derived porosities of normally pressured Not Formation shales (Figure 18).

The fluid pressure was determined by combining Equations 1 (see main text) and Equation B3, whereby:

$$P_{fl} = OB - \frac{\ln\left(\dfrac{\phi_{DT}}{\phi_0}\right)}{a} \qquad (B4)$$

where ϕ_{DT} represents the sonic-derived porosity.

Hunt, J.M., J.K. Whelan, L.B. Eglinton, and L.M. Cathles III, 1998, Relation of shale porosities, gas generation, and compaction to deep overpressures in the U. S. Gulf Coast, *in* Law, B.E., G.F. Ulmishek, and V.I. Slavin eds., Abnormal pressures in hydrocarbon environments: AAPG Memoir 70, p. 87–104.

Chapter 5

RELATION OF SHALE POROSITIES, GAS GENERATION, AND COMPACTION TO DEEP OVERPRESSURES IN THE U.S. GULF COAST

John M. Hunt
Jean K. Whelan
Lorraine Buxton Eglinton
Woods Hole Oceanographic Institution
Woods Hole, Massachusetts, U.S.A.

Lawrence M. Cathles III
Department of Geological Sciences
Cornell University
Ithaca, New York, U.S.A.

Abstract

Direct measurements of porosities from Tertiary and Cretaceous shales in the Texas-Louisiana Gulf Coast show that in many areas shale porosity is either constant or increasing at the depths where high overpressures occur and where hydrocarbons are being generated. In the absence of a decrease in porosity with sediment load (depth), gas generation becomes the principal cause of overpressures and hydrocarbon expulsion.

Gulf Coast shale porosities decrease exponentially in normally compacting shales only down to porosities of about 30%, after which the decrease is linear until a constant porosity is reached. These linear trends are believed to be related to the high quartz content (74%) of the clay-size fraction (<4 microns).

The depths at which shales reach relatively constant porosity values appear to depend on the internal surface areas of the shales. Shales containing minerals with small, internal surface areas, such as fine-grained quartz and carbonates, stop compacting at porosities around 3%, whereas shales containing minerals with large surface areas, such as smectite and illite, stop compacting around 10%. This interval of no compaction usually is reached at depths around 3 to 4 kilometers (temperatures of 85° to 110°C) prior to the development of deep high overpressures and the generation of large quantities of hydrocarbons in the Gulf Coast. Model studies indicate that gas generation is the dominant process creating these deep overpressures.

The porosity-depth profiles that show a linear decrease with depth followed by a constant porosity do not conform to the hypothesized exponential profiles used in many modeling programs today. This means that more direct shale porosity measurements are needed to confirm the type of profiles that actually exist and should be used in any basin modeling program.

INTRODUCTION

Compaction is defined as the loss of porosity due to stress. Shales showing no loss in porosity through thousands of feet of burial indicate no compaction. The objectives of this paper are to show how direct porosity measurements made on both cuttings and cores from many wells in the U.S. Gulf Coast indicate the following three points:

1. Under hydrostatic conditions shale porosities below about 30% tend to decrease linearly (not exponentially) with depth to a point below which there is no further decrease.
2. The cessation of compaction does not appear to be related to overpressuring. This phenomenon occurs with normally pressured shales. The two-stage, linear compaction is thus a "normal" compaction trend.

3. At depths where compaction no longer occurs, gas generation seems to be the major cause of overpressures and the probable cause of hydrocarbon expulsion from source rocks.

Two kinds of porosity-depth curves for shales have been published over the past several decades. Composites of porosity-depth data from several wells over a large area generally show a scatter of data points through which an exponential porosity curve can be drawn. Such curves indicate a rapid decrease in porosity near the surface and a slower decrease with increasing depth. Several examples and their variability are discussed by Rieke and Chilingarian (1974, p. 42), Dickey (1976) and Hunt (1979, p. 202). The second type of curve, which is best recognized from analyses of single wells rather than composites, shows that after an initial exponential decrease to 30–35% porosity, the subsequent porosity decrease follows two straight-line segments in angular end-to-end contact with each other. The second line frequently has no slope, thus indicating no compaction.

Hedberg (1936) was the first to observe this latter phenomenon in undisturbed Tertiary shales of the Eastern Venezuelan Basin (Figure 1). Most of his data were from a single well supplemented with a few shallow and deep data points from two other wells in the same area.

Hedberg described four stages in the compaction of shales, the first three due to mechanical compaction and the last to chemical compaction. These stages were described as mechanical rearrangement for porosity decreasing from 90 to 75%, dewatering from 75 to 35%, mechanical deformation from 35 to 10%, and recrystallization, i.e., chemical compaction for porosity decreases to below 10%. Hedberg observed that, "reduction of pore space below 10% takes place very slowly and only with large increments of pressure. Chemical readjustment becomes the dominant factor in the fourth stage of compaction." Hedberg's division of mechanical compaction and chemical compaction or diagenesis for the Venezuelan shales at around 10% porosity becomes important in our later discussion of a two stage model for the decrease in the porosity of shales with depth.

In 1959, Storer published dry bulk density-depth curves for several wells in the Po Basin of Italy. They showed some of the same linear changes reported by Hedberg. Linear shale compaction trends also have been reported by Korvin (1984) and Wells (1990).

Many of the Tertiary shales from the Gulf Coast discussed in this paper show little change in porosity at depths >3 km whereas sandstones show a wide variation. Loucks et al. (1984) analyzed sandstone porosities in 7,564 cores from the Lower Tertiary of the Texas Gulf Coast. They found that porosities ranged from 3 to 30% at various depths between 9,000 and 15,000 feet (2,440 and 4,570 m). Over half of the total porosity below 10,000 ft (3,050 m) was secondary due to chemical diagenesis. They observed that pore networks in the

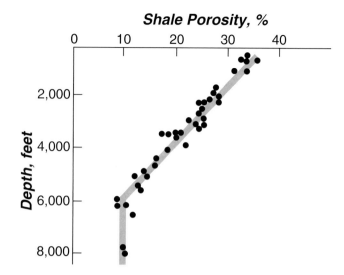

Figure 1. Shale porosity vs. depth for Venezuelan shales illustrating porosity decrease along two discrete straight-line segments. Hedberg (1936) drew a linear porosity line from 35 to 10% porosity. This was the only linear trend shown by Rieke and Chilingarian in their 1974 compilation of published porosity curves.

sandstones were composed predominantly of secondary porosity.

Several published examples of high secondary porosity in sandstones at depths >3 km emphasize that erroneous compaction models (such as assuming undercompaction) can result unless there is clear petrographic evidence that the porosity is primary (Meloche, 1985; Franks and Forester, 1984; Jansa and Noguera Urrea, 1990). Secondary porosity in sandstones is caused mainly by chemical diagenesis, not compaction. Consequently, shale porosities are more useful than sandstone porosities for following regional compaction trends and for recognizing the difference between normally compacted and undercompacted rocks–as will be discussed later. Note that, all porosity data in this report represents only shales, or in a few cases, limestones.

METHODOLOGY

All porosity measurements reported in this paper were carried out by Amoco using the method of Hinch (personal communication. See also, Hinch, 1980). Although shales in several thousand wells were analyzed, few of the data were published (Powley, 1993). Recently, part of Amoco's data bank was released through a cooperative project with the Gas Research Institute. The raw data from about 30 wells were used for the current paper.

Hinch's technique involves measuring the quantity of an organic liquid, Varsol, taken up by a dried, evacuated shale sample. Porosities measured by this procedure were compared with those obtained by Core Laboratories on the same samples using a helium uptake method. The two methods agreed to within ±1.5% of each other. Helium tends to go into smaller pores than Varsol but, the Varsol is adsorbed more strongly on the mineral surfaces thereby forcing open some pores that may not be easily reached by helium. Although the two techniques were comparable, they both were between 5 and 10% lower than the porosity obtained on water saturated conventional cores. The reason for this apparent discrepancy is that a small amount of non-effective porosity (5 to 10% of the total porosity) develops when a sample shrinks on drying. The shrinkage tends to decrease with deeper samples with lower porosity. It is <5% of the porosity at depths >3 km. The interior sections of conventional connate water saturated shale cores give the most accurate porosities but they are too costly to take continuously.

A critical factor in making valid shale porosity measurements is that wet shale samples, including conventional cores containing large amounts of smectite and illite, must be analyzed very soon after collection. If not, they expand by hydration, resulting in porosities much higher than the original in situ values. Hydration also occurs in the sidewalls of a borehole during drilling. Consequently, if sidewall cores are taken at the end of a drilling operation they should not be used for porosity measurements. In this work, all cuttings were washed and dried as they arrived at the surface in order to prevent adsorption of enough water to cause hydration. Replicate experiments gave results within ±0.5% of each other.

Hinch (personal communication) also compared his porosity measurements with those obtained by various logging techniques (Figure 2). The dried cuttings (DC) line represents a moving average of porosity data determined on cuttings by the Hinch method. Porosities were averaged over a 250 ft (76 m) depth interval. The average was recalculated as each deeper sample was added with the shallowest being removed. Sixty-eight samples were analyzed from 4,000 ft (1,220 m) to 15,000 ft (4,575 m). The borehole gravity meter line (BGM) in the upper section of the well represents porosities calculated from wet bulk densities measured with this meter. Gravity is measured at two different depths in the well bore and the difference in gravity is proportional to the density of the rock. The advantage of this instrument is that the densities can be measured laterally in the rock formations through a circle 1,000 ft (300 m) in diameter. This avoids hydration effects around the borehole. Consequently, porosities obtained with this meter are close to the in situ porosities (Hinch, personal communication).

The curve in the lower section of the well in Figure 2 marked "Resistivity + % Illite (R+%I)" represents in situ porosity determined from the short normal resistivity and the percent of expandable mixed layer illite. According to Hinch (personal communication), E.R. Michaelis of Amoco found that the porosity of shales could be estimated from the following equation:

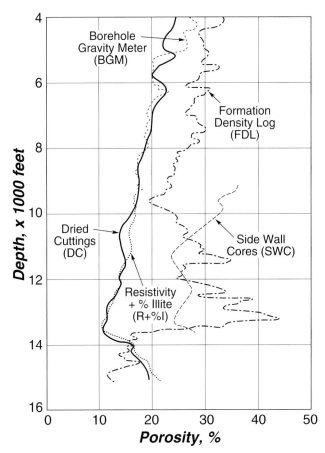

Figure 2. Laboratory porosity measurements on dried cuttings and sidewall cores compared to downhole porosity measurements made with the borehole gravity meter, formation density log, and to the in situ porosity calculated from the short normal resistivity and percent expandable mixed layer illite. This is Miami Corp. well no. 27 drilled by Amoco in Cameron Parish, Louisiana (Hinch, personal communication).

$$\text{Shale Porosity} = 4.88 \left(\frac{I_{EML}^{0.35}}{R_{SN}^{0.47}} \right) \tag{1}$$

where I_{EML} = % expandable mixed-layer illite and R_{SN} = short normal resistivity

The method, which requires X-ray diffraction analysis to determine the percent of expandable mixed layer illite, gives the porosities within ±2.5% of the value measured on a conventional water saturated core sample from the same depth (Hinch, personal communication). The most inaccurate porosities were measured with the formation density log (FDL) and with sidewall cores (SWC) as shown in Figure 2. Hinch believes this is due to extensive hydration of the shales around the well bore.

Exclusive use of either the formation density logs or sidewall cores to measure porosities can lead to the erroneous conclusion that the shales are undercompacted. Shale hydration at the borehole can be mistaken for undercompaction in deltaic sediments. The discussion which follows relies on the Hinch porosity analysis of cores and dried cuttings, thus avoiding these problems.

COMPACTION PROFILES OF NORMALLY COMPACTED SHALES

The first publication of Amoco data showing linear compaction appeared as a profile of density and porosity vs. depth for shales in a West Delta well offshore Louisiana—see well 20 in Figure 13 for porosity curve only (Bradley,1976). The data showed a linear increase in dry bulk density and decrease in porosity down to about 12,000 ft (3,660 m) followed by essentially no change in density or porosity to total depth (16,400 ft, 5,000 m). Bradley also showed other linear compaction profiles which indicated that a complete range of densities can be associated with abnormal pressures. He concluded that overburden stress is not the sole

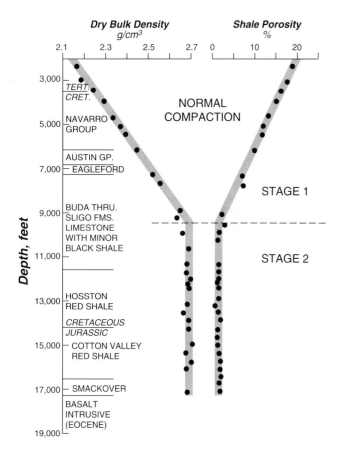

Figure 3. Shale dry bulk density and porosity vs. depth shown as two linear segments in angular end-to-end contact for the normally pressured Amoco Lena Buerger well in Frio County, Texas (Powley, 1993). These segments, divided by the dashed line, are called compaction Stages 1 and 2 by Hinch (1980) and Powley (1993). Black dots are individual sample measurements.

cause of abnormally high pressures because of this lack of any correspondence between overpressures and shale density.

Hinch (1980) and Powley (1993) define the linear porosity change from 35 to 10% as compaction Stage 1 and the constant porosity interval as Stage 2 in normally compacting sediments. These are equivalent to Hedberg's mechanical deformation and recrystallization stages, respectively, which represent the two-stage compaction model in this paper.

For example, Figure 3 contains density and porosity data (black dots) for a normally compacted well, the Amoco Lena Buerger, in Frio County, Texas. Starting at a depth of 2,000 ft. (610 m) in Figure 3, the density increases and porosity decreases along straight line segments until they reach relatively constant values of 3% porosity and 2.7 g/cm³ density at a depth of about 9,500 ft (2,900 m). From here to a basalt intrusive at 17,500 ft (5,335 m) there is no systematic decrease in porosity, indicating no compaction. We define this two stage compaction model as the normal compaction curve for this well. The low porosity of 3% in Stage 2 is due to the rocks containing mainly carbonates and red shales with kaolinite, both of which have very low mineral surface areas compared to smectite and illite.

Some geologists have speculated that the Stage 2 density and porosity lines in Figure 3 are caused by undercompaction. However, undercompacted shales are universally within overpressured compartments and there is no evidence that this well was ever overpressured (Powley, 1993). Overpressures can be recognized by drill stem tests, mud weights and resistivity logs. The Lena Buerger well was drilled with 9 lb/gal. mud to total depth. There was persistent lost circulation in the well and no shift from normal to low resistivities. All these observations indicated normal hydrostatic pressure throughout the well. The change from Stage 1 to 2 is not due to a change in lithology. The entire section through this shift is in Upper Cretaceous rocks with relatively similar lithologies.

This two-stage model can only be recognized by direct porosity and density measurements on samples from single wells. Composites of data from several wells generally show only a scatter of points.

Figure 4 is a map of the wells used in our study. The Lena Buerger well is no. 21, southwest of San Antonio. Table 1 contains the county or parish, state, latitude, longitude, operator, lease, and API well number for the wells cited. We included wells which covered the entire Gulf Coast area from the most southern part of Texas to the most eastern part of Louisiana. In order to limit the geological variables that would affect the porosity we only used wells from continuously sinking areas devoid of uplift and erosion or other structural changes that could cause variability in the porosity profile.

An important concept in this paper concerns the change from a continuous linear decrease in porosity with depth to no apparent decrease (no slope). The depth of this change is shown as a break in the slope of

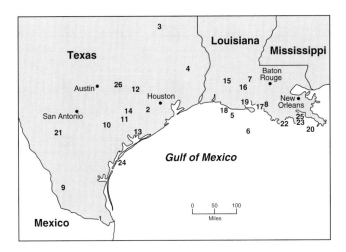

Figure 4. Location map for Gulf Coast wells in this study. Location data appears in Table 1.

the porosity lines in Figures 3, 5, 9, 11, 12, and 13. Four typical normal compaction porosity profiles from Louisiana and Texas (well numbers 5, 7, 8, and 11 in Table 1 and Figure 4) are shown in Figure 5. The vertical depth scales differ depending on the samples available for analysis. Each porosity profile is represented by two linear best fit segments. These correspond to compaction Stages 1 and 2, the second stage having no slope in the porosity vs depth profile. The break between Stages 1 and 2 occurs at a different depth in each of the four areas. This critical change can best be

recognized from individual well data or closely spaced wells in the same oil field. If the data from the four wells in Figure 5 were composited the resulting curve would not be useful for determining the depth of the shift from a linear decrease to no decrease in porosity.

Our study confirms that of Bradley (1976) in showing no correlation between the top of Stage 2 and the onset of overpressures. The approximate tops of overpressures in these wells are: East Cameron, 13,000 ft (3,960 m), St. Landry, deeper than 15,000 ft (4,570 m), St. Mary, 15,500 ft (4,730 m), and Lavaca 10,000 ft (3,050 m). The interval from the top of Stage 2 to the top of the overpressure in these wells ranges between 1,000 and 5,000 ft (300 and 1,525 m).

The East Cameron well in Figure 5 shows a steady decrease in shale porosity until reaching a depth of about 7,800 ft (2,380 m). For the next 3,000 ft (915 m) or more of burial the shale porosity is remarkably uniform, showing no systematic increase or decrease. This contrasts with the Gulf Coast sandstones mentioned earlier by Loucks et al., (1984) which showed sandstone porosities ranging from 3 to 30% at these depths.

The St. Landry and St. Mary wells (7 and 8 in Table 1) show no systematic decrease in shale porosities starting around 12,000 ft (3,660 m) and 14,500 ft (4,420 m) respectively. The porosity of the St. Landry well is 12.1% at the top of Stage 2 and 12.3 % at the bottom of the hole, about 3,000 ft (915 m) deeper.

The well in Lavaca County, Texas (no. 11 in Table 1) is overpressured starting around 10,000 ft (3,050 m)

Table 1. Gulf Coast wells for which porosity profiles were studied. *Lease abbreviations: OCS = Outer Continental Shelf, OCSG = OCS Government land, SL = State Lease and, ST-TR = State Tract.

	County	State	Latitude	Longitude	Operator	Lease*	API Number
1.	Cameron Co.	TX	26.0S	97.7W	Pan American Co.	M.E. Wentz	(420610009700)
2.	Fort Bend Co.	TX	29.7S	95.8W	Mobil Oil Corp.	B.C. Harwood	(421570297100)
3.	Smith Co.	TX	32.3S	95.3W	Fairway O G Co.	Fairway et al.	(424230081800)
4.	Tyler Co.	TX	30.9S	94.4W	Pan American Co.	Long Bel Co.	(424570005700)
5.	E. Cameron Ph.	LA	29.5S	92.6W	Pan American Co.	SL 1186	(177030004800)
6.	South Marsh Is.	LA	28.8S	92.0W	Pan American Co.	OCS 00785	(177070044800)
7.	St. Landry Ph.	LA	29.7S	92.0W	Halbouty M.	K.S. Stelly et al.	(170972004100)
8.	St Mary Ph.	LA	29.7S	91.3W	Pan Am	St. Mary B&T Co.	(171010199600)
9.	Zapata Co.	TX	27.1S	99.0W	Union Prod	Jennings	(425050040600)
10.	De Witt Co.	TX	29.1S	97.4W	Atlantic Richfield	Eliza Smith	(421230033200)
11.	Lavaca Co.	TX	29.3S	96.7W	Magnolia	Simpson Heirs	(422850015500)
12.	Washington Co.	TX	30.2S	96.4W	Shell	Jackson	(424770029142)
13.	Matagorda Co.	TX	28.8S	96.3W	Tennessee Gas	B W Trull Estate	(423210230900)
14.	Colorado Co.	TX	29.5S	96.6W	Shell Oil et al.	Plow Realty Co.	(420890049700)
15.	Allen Ph.	LA	30.5S	92.8W	Pan American	J.A. Bell et al.	(170032000500)
16.	Acadia Ph.	LA	30.3S	92.2W	Continental Oil	Faddy Arceneaux	(170010189600)
17.	St. Mary Ph	LA	29.7S	91.5W	Humble Oil	A. Wettermark	(171010193200)
18.	East Cameron OCS	LA	29.6S	93.0W	Pan American	OCS 1327	(177030023700)
19.	Vermilion Ph.	LA	29.8S	92.0W	Pan American	Vermilion Land	(171132007900)
20.	West Delta	LA	28.9S	89.6W	Pan American	OCSG 01088	(177190124600)
21.	Frio Co.	TX	28.0S	99.0W	Pan American	Lena Buerger	(421630150900)
22.	S. Timbalier	LA	29.1S	90.6W	Placid	SL 2857	(177150091400)
23.	Grande Isle	LA	29.2S	90.0W	Humble Oil	SL 05013	(177170072600)
24.	Mustang Island	TX	27.8S	96.8W	Atlantic	ST TR 726-L	(427020000900)
25.	Plaquemines Ph.	LA	29.3S	89.6W	Amoco	Bastian Bay Fd.	(170750459300)
26.	Lee Co.	TX	30.4S	96.9W	Pan Am	Willie Matejcek	(422870009100)

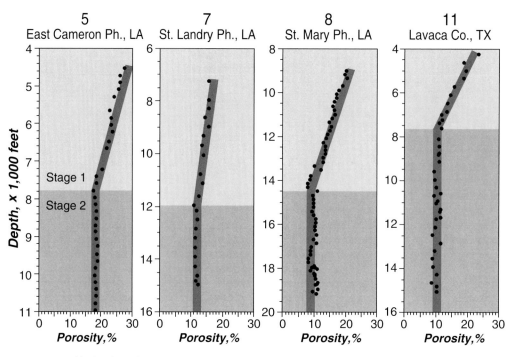

Figure 5. Porosity vs. depth for four wells in Louisiana and Texas (numbers 5, 7, 8 and 11 in Table 1). The vertical scales are different on these graphs due to limitations in the ranges of sampling.

where the pressure gradient is 0.69 psi/ft (16 kPa/m). Porosity data from two nearby wells in the same field were used to construct the curve (for this well) in Figure 5 because of gaps in the data from the individual wells. The linear decrease in porosity of Stage 1 extends to a depth of approximately 7,500 ft (2,290 m). At greater depths, there is no consistent decrease in porosity indicating no compaction through the next 7,500 ft (2,290 m) of burial. There is ~2,500 ft (760 m) of no compaction above the overpressured compartment.

Powley (1985, 1993) reported in his study of over 100 wells that the tops of deep overpressured compartments commonly occur a few thousand feet below the top of Stage 2. Less commonly they occur at the contact between Stages 1 and 2. Least common is when the overpressure top is found in compaction Stage 1. No clear relationship was reported by Powley between the top of Stage 2 and lithology. However, this top does seem to be related to subsurface temperatures. Hinch (1980) plotted the temperatures at the top of Stage 2

versus the age of the rocks for 65 wells in the Gulf Coast (Figure 6). Most of the temperatures were between 90° and 100°C (194° and 212°F) but they were as low as 82°C (180°F) in Cretaceous rocks and as high as 110°C (230°F) in Pleistocene rocks. This suggests that Stage 2 may be initiated by some process which depends on time-temperature conditions.

Statistical Analysis

Most basin models in recent years have assumed an exponential curve of porosity versus depth for their model based on the 1930 study of Athy (Dutta, 1987, Forbes et al., 1992, Ungerer et al., 1990, Waples and Okui, 1992). This assumption has prevailed despite evidence of linear porosity profiles as observed by Hedberg (1936) and discussed previously. Here we have examined the porosity data using statistical analysis. Our objective was to determine (1) whether an exponential or a linear variation in porosity with

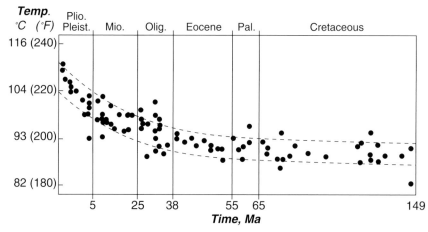

Figure 6. Relation between temperature and time (geologic epoch) of samples at the top of Stage 2 where porosity stops decreasing with depth (Hinch, 1980). The dashed lines enclose 80% of data as of 1980.

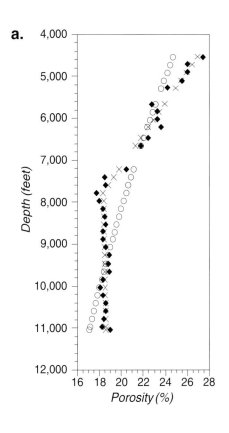

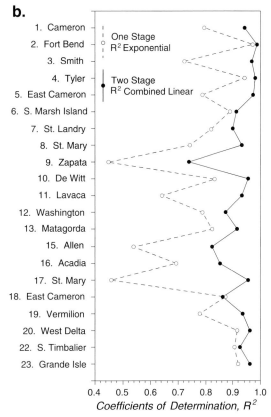

Figure 7. (a) Comparison of predicted porosities with measured porosity for an East Cameron Ph., La. well (no.5, Table 1). Symbols used: ◆ = measured porosity; O = predicted porosity from one-stage exponential curve fit (R^2 = 0.79); and ✕ = predicted porosity from two-stage linear curve fit (R^2 = 0.95). (b) Graph showing the comparison of coefficients of determination (R^2) for a one-stage exponential curve fit and a two-stage linear curve fit for 21 normally compacted shale porosity-depth profiles in the U. S. Gulf Coast.

depth better fits the experimental data and (2) whether the compaction Stage 2 shows a constant porosity or a systematic increase or decrease with increasing depth.

If one considers porosity to decrease exponentially with depth over the entire data range, then correlation coefficients (r) should be indicative of this. However, porosity coefficients show a poor exponential relationship with increasing depth. In contrast, comparison of porosity changes modeled as two lines, either linear or exponential, yield much more convincing correlation parameters. Figure 7a shows an example of the predicted porosity (R^2) for a two-stage linear curve fit and a one-stage exponential curve fit compared to the actual measured porosity for the Louisiana, East Cameron well (no. 5 in Table 1). Clearly, the two-stage linear curve fits the measured porosities much better than the one-stage exponential curve. This also was true of all but one of the wells in this study as shown in Figure 7b (the exception is East Cameron, no. 18). For example, the R^2 exponential coefficient for the Lavaca well (no. 11) in Figure 7B is 0.642 compared to the two-stage linear coefficient of 0.933.

The more important aspect of our model, however, is to determine whether porosity in Stage 2 changes in a systematic way with increasing depth. To address this question, we used predicted porosity values from exponential and linear regression data solely for Stage 2. There is essentially no difference between coefficients of determination (R^2) for the two models in Stage 2 only. Both confirm that, within the errors of determination, there is no slope in the porosity versus depth curve for all wells except S. Marsh Island and Matagorda (no. 6 and 13) which show a slight decreasing and increasing porosity, respectively. We evaluated the upper and lower 95% confidence limits for the porosity versus depth slope determinations (Figure 8). These data confirm that porosity is constant throughout Stage 2 (i.e., there is no systematic change of porosity with increasing depth). The data support our concept that there is no compaction in Stage 2 for most normally compacted shales in the Gulf Coast.

Why Linear Compaction Profiles?

The conclusions in this paper apply to the Gulf Coast only. However, they also indicate that one cannot assume exponential porosity decreases in shales having porosities less than 30%. We believe that these deep linear profiles in the Gulf Coast may be related to mineralogy. It has been known since the 1960s that clay minerals compact exponentially while sandstones compact linearly (Chilingar and Knight, 1960; Atwater and Miller, 1965). Shales are composed largely of clay-sized particles which are assumed to be clay minerals, but what if they are quartz? Bradley (1993, personal communication) evaluated X-ray and elemental analyses of several hundred shales from the U.S. Gulf Coast. The shale sections in the cores were defined as shale by electric logs and visual inspection. The clay-size fraction (<4 μm) was found to consist of 74% quartz and

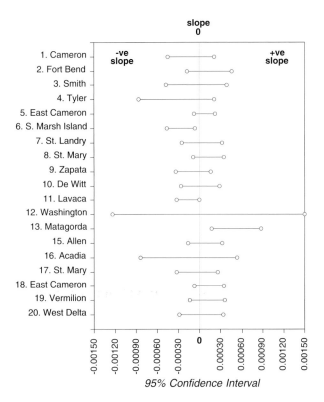

Figure 8. Graph showing Stage 2 porosity slope determinations for the wells in Figure 7b. Porosities of all but two of the wells (no. 6 and 13 in Table 1) have no slope within the 95% confidence interval. No. 6 has a slight negative slope (-ve) and well no. 13 a positive slope (+ve).

26% clay minerals. The average particle size of the quartz was 2 μm, and that of the clay minerals, 0.1 μm. The average Tertiary Gulf Coast shale based on Bradley's study contains 67% quartz, 7% feldspar, 20% clay minerals, 4% carbonate, and 1% organics and other minerals. Possibly this high content of clay-size quartz is causing the linear compaction profiles.

Why does the constant porosity Stage 2 reach a minimum value ranging from 3 to 18% with an average value around 10% (Figures 3, 5, 9, 11, 12, and 13)? This also appears to be related to mineralogy. For example, Figure 9 shows the porosity-depth profile for the Baltimore Canyon B-2 COST well drilled off the East Coast of the U.S.A. This is a normally pressured well with a normal compaction curve. Stage 2 porosities range from 3 to 6%. In Figure 10 the smectite-illite content of Stage 2 shales in this well are plotted against porosity and compared with similar data from a normally pressured Texas well. In this figure, a higher smectite-illite content equates with a higher minimum porosity. The Lena Buerger well in Figure 3, which has no smectite or illite, reaches a minimum porosity of 3% in Stage 2 while the typical Gulf Coast well with 20% smectite-illite averages 10% porosity in Stage 2.

In earlier work, Chilingar and Knight (1960) compacted clay minerals under 200,000 psi (13,800 MPa) pressure. The minimum porosity reached by smectite

was 18% compared to 8% for kaolinite. Because of their enormous surface areas, smectite and illite retain far more water during burial than kaolinite, quartz and calcite. Powers (1967) showed that it takes 80,000 ft (24 km) of overburden pressure to remove the last two monomolecular layers of water from smectite. Thus shale porosities are unable to decrease much below 10% within current drilling depths when appreciable quantities of smectite and illite are present.

In constructing porosity and density profiles by depth, it is very important to plot the well data for shales only. Limestones and sandstones generally have different porosities than shales at the same depth. The differences with sandstones were mentioned in the introduction. Differences with limestones are shown in Table 2 for a well in Lee County, Texas (no. 26 in Table 1). Limestones are interbedded with shales at depths greater than about 12,000 ft (3,660 m) in this well. Porosities of the shales ranged between 10 and 16%. Limestones at the same depths have porosities of 2.6 to 7%. Indirect methods of estimating porosities, such as logging, would average these numbers.

COMPACTION PROFILES OF OVERPRESSURED SHALES

Figure 11 shows the two-stage normal compaction density and porosity profiles and a pressure-depth plot

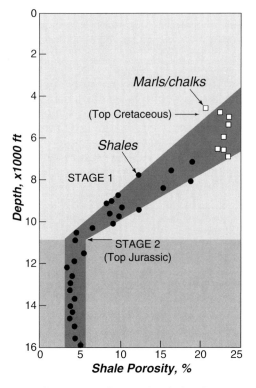

Figure 9. Shale porosity vs. depth for the normally pressured COST B-2 well in Baltimore Canyon, (offshore) New Jersey, U.S.A.

for a well in Bastian Bay Field, Plaquemines Parish, Louisiana (no. 25 in Figure 4 and Table 1). Here the shale reaches a minimum porosity of 10% at about 11,000 ft (3,350 m). The next 5,000 ft (1,525 m) or so represents the no compaction, constant porosity Stage 2 after which the porosity increases within an overpressured fluid compartment. An overpressure is defined as any pressure gradient above 12 kPa/m (0.53 psi/ft) which is the hydrostatic pressure of a saturated brine (Bradley and Powley, 1994). Drill stem tests within the top seal of Figure 11 show overpressures starting around 14,000 ft (4,270 m) within the Stage 2 interval. However, undercompaction, as indicated by an increasing porosity, is only evident below the second seal at about 16,000 ft (4,880 m). This is 2,000 ft (610 m) below the top of the overpressure.

A possible explanation for this is that the undercompaction originally extended to the top seal, but it has partially dewatered since forming. Powley (1993) defines an undercompacted shale as any shale which has not yet dewatered into the characteristic normal porosity-depth profile, in this case 10%. He claims that less than half of the deep (>3 km) overpressured rocks of the Gulf Coast are undercompacted. Examples of currently dewatering Gulf Coast shales are in Cameron Parish, Louisiana (Hinch, 1980) and Mustang Island (Hunt, 1996, p. 293). Additional examples are in Powley (1993).

Overpressures in the Gulf Coast may or may not lead to undercompaction. The Plaquemines Parish well in Figure 11 shows an increase in shale porosity (undercompaction) below 16,000 ft (4,880 m) but the well in Figure 12 in the Sheridan Field of Colorado County, Texas (no. 14 in Figure 4 and Table 1) shows no change in porosity or density with overpressure (Powley, 1993). There is a linear increase in dry bulk density and decrease in shale porosity down to a depth of about 8,700 ft. (2,650 m) at the top of Stage 2 (Figure 12). The top of an overpressured fluid compartment containing about 5,000 psi excess pressure occurs at around 12,000 ft (3,660 m) which is more than 3,000 ft (915 m) below the beginning of the constant porosity interval of Stage 2. There is no evidence of undercompaction occurring with the overpressures. Porosities and densities are essentially the same at 17,000 ft.

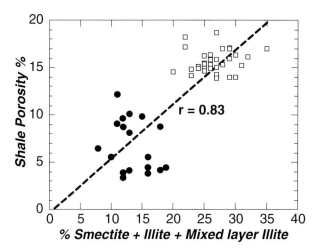

Figure 10. Correlation of shale porosity and smectite-illite content in normally pressured Stage 2 shales. Solid circles are shales from the Baltimore Canyon well (COST B-2), squares are shales from a Gulf Coast well offshore Texas.

(5,180 m) as at 8,700 ft (2,650 m). It is difficult to see how compaction can play any significant role in the development of these overpressures since there is no systematic reduction in porosity or increase in density starting above and extending through the overpressured fluid compartment.

The pressure measurements in Figure 12 (solid circles) are from drill stem tests. There is a suggestion of an increase in dry bulk density and corresponding decrease in porosity right at the seal in Figure 12 as might be expected.

Well numbers 15, 16, and 17 (Figure 4 and Table 1) in Allen, Acadia and St. Mary Parishes, Louisiana also show overpressures within the no compaction Stage 2 interval. For example, in the Allen well (no. 15) the pressure/depth gradient is 0.76 psi/ft (17.1 kPa/m), 1,500 ft (460 m) below the top of Stage 2.

HYDROCARBON GENERATION AS A CAUSE OF OVERPRESSURES

The top of the no compaction Stage 2 usually occurs between 90° and 100°C (194° and 212°F, Figure 6). Some geochemists believe these temperatures are not high enough to generate large quantities of hydrocarbons from Type III kerogen in rapidly depositing deltas. Time-temperature modeling based on the Arrhenius equation has indicated that although petroleum generation may start at temperatures as low as 60°C (140°F), the peak in oil and gas formation is usually at temperatures higher than 95°C (203°F) (Wood, 1988; Mackenzie and Quigley, 1988; Hunt and Hennet, 1992). Arrhenius kinetics for a Type III kerogen indicates a temperature of about 120°C (248°F) is required to initiate oil generation (Hunt, 1996, p. 159). Smith

Table 2. Porosity of limestones and shales at similar depths in a Lee Co.,Texas well

| Depth | | Percent Porosity | |
feet	(meters)	Limestone	Shale
12,940	(3,945)	2.6	10.6
13,850	(4,223)	3.7	12.6
14,900	(4,543)	5.8	15.8
15,260	(4,652)	7.0	14.8
15,620	(4,762)	6.2	13.5
15,950	(4,863)	5.5	13.2

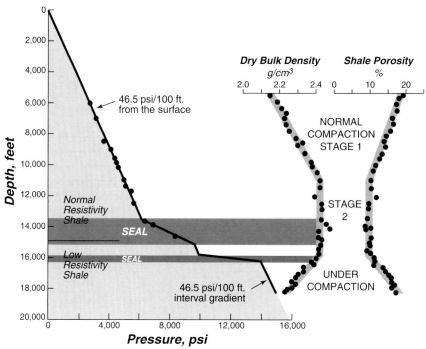

Figure 11. Porosity-density profiles of a well in Plaquemines Parish, Louisiana showing no compaction for 5,000 ft (1,525 m) above an overpressured sealed compartment and undercompaction within the compartment (Powley, 1993).

(1994) reported that <10% of the total gas yield from Type III kerogen is expelled at a vitrinite reflectance (R_o) of 1.3, whereas >60% is expelled at an R_o of 1.8, based on Shell Oil Company data files. These R_o values are roughly equivalent to paleotemperatures of 100° and 170°C, respectively. Because these temperatures are higher than those in Figure 6, it means that significant hydrocarbon generation occurs in the Gulf Coast below the top of compaction Stage 2. Overpressuring often peaks within this hydrocarbon generation window.

Figure 13 shows the porosity profile for an East Cameron well offshore Louisiana (no. 18, Figure 4) that is near a well that has been studied in considerable detail at the Woods Hole Oceanographic Institution. This area has continuously subsided and has not experienced uplift or erosion. The shales of well 18 are normally compacted with the break between compaction Stages 1 and 2 at about 10,700 ft (3,260 m).

The hydrocarbon generation window was determined by using pyrolysis techniques plus headspace analysis of canned cuttings such as were used in a South Padre, Texas well by Huc and Hunt (1980). The beginning of oil generation in the East Cameron well was at about 12,000 ft (3,660 m). The generation peak

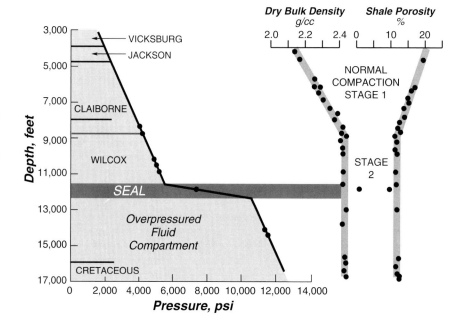

Figure 12. Compaction profile including dry bulk density and porosity for shales with depth for the Sheridan Field in Colorado County, Texas (no. 14 in Table 1). Pressure data points are by drill stem tests. Normal hydrostatic pressure extends from the surface to about 11,000 ft (3,350 m). The overpressure starts 3,000 ft (914 m) below the top of the constant porosity Stage 2. There is no evidence of undercompaction in this well and the porosity is relatively constant through the overpressured section (Powley, 1993).

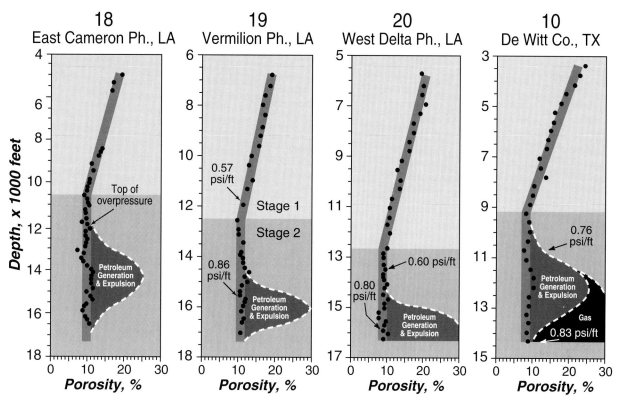

Figure 13. Porosity vs. depth for four Gulf Coast wells, showing the close relationship between the intervals of high overpressures and the petroleum generation and expulsion windows. All of these overpressures are in the intervals of no compaction. These wells are no. 10, 18, 19, and 20 in Table 1.

was in the 14,000 to 15,000 ft range (4,270–4,570 m). The distribution of individual hydrocarbons with depth indicated that considerable hydrocarbon expulsion and migration was occurring at 15,000 ft (4,570 m) (DOE Report no. COO-4392-4, 1981). Headspace analyses showed the C_2–C_8 hydrocarbons peaking at about 14,500 ft (4,420 m). Vitrinite reflectance values ranged from $R_o = 0.65$ to 0.9% through the hydrocarbon generation interval.

A. Yukler carried out a one-dimensional maturation model in this area for us using the model of Yukler and Kokesh (1984). Computed sterane isomerization ratios predicted that equilibrium should have been reached at about 15,000 ft (4,570 m). The computed hydrocarbon generation showed that the active oil generation zone is from 13,000 to 15,000 ft (3,960–4,570 m) with the peak at 15,000 ft corresponding to the equilibrium in the sterane isomerization ratios. The model also shows that the gas/condensate generation zone was from 15,000 ft (4,570 m) to the bottom of the well. The model results showed a good match with the Woods Hole experimental data.

Figure 13 shows the computed maturation interval in the East Cameron area (well no.18). There are about 4,000 ft (1,220 m) of rocks showing no compaction above the peak in oil expulsion and migration. The top of overpressure coincides with the start of petroleum generation at ~12,000 ft (3,660 m). Consequently, in this well, it appears that hydrocarbon generation is related

to the generation of overpressures, but not to the cessation of compaction.

The second porosity profile in Figure 13 is for well no. 19 in the onshore Vermilion area of Louisiana. The top of the no compaction (Stage 2) is at about 12,500 ft (3,810 m) and the top of the overpressure is around 11,000 ft (3,350 m) This is an example of the overpressure starting in Stage 1. Data were available for constructing a burial history curve in this area for the Oligocene and Eocene formations. Using this curve for the base of the Oligocene Vicksburg group along with Arrhenius kinetics for Type III kerogen (Hunt, 1996, p. 159) it was possible to estimate the location of the oil window as shown in Figure 13. Generation was calculated as beginning around 15,000 ft (4,820 m) for the base of the Upper Oligocene with peak generation and expulsion extending about 1,000 ft (305 m) deeper. Although the porosity data for this well end at 17,000 ft (5,180 m) there is no significant decrease in porosity indicating no compaction. The high pressure/depth gradient of 0.86 psi/ft (19.4 kPa/m) is centered in the hydrocarbon generation window.

In 1974, La Plante developed a set of simultaneous equations designed for calculating the amounts of methane, carbon dioxide, water, and nitrogen as thermodynamically stable products generated during the conversion of kerogen to petroleum. He used this model to determine the depths at which petroleum and other volatiles were formed from the kerogen in three wells in the Louisiana Gulf Coast. The porosity/depth plot for his well, in West Delta, Louisiana, is in Figure 13 (well no. 20). He found that the beginning of the oil window occurred around 14,000 ft (4,270 m) at a temperature of 205°F (96°C). At 16,500 ft (5,030 m) his calculations showed that about 10% of the original kerogen had been converted to hydrocarbons, mostly gas. This represented about one-third of the total hydrocarbon generating capability calculated for the kerogen using Arrhenius kinetics. Consequently, La Plante's (1974) model indicated that the oil plus gas window in this well extends deeper than the total depth drilled. There is no apparent compaction occurring in this interval based on the porosity/depth profile. Nevertheless, the pressure/depth gradient reaches 0.80 psi/ft (18 kPa/m) within the oil and gas generation window. No data was available for defining the top of the overpressure.

Yukler and Dow (1990) applied a quantitative basin analysis model to a drilling site within 6 miles (10 km) of the De Witt County Texas well in Figure 13 (well no. 10). They used data on wells in this area to determine the geologic evolution of the basin and to quantify pressure, temperature, organic matter maturation, and hydrocarbon generation histories. The hydrocarbon generation was determined from the kinetic equations given by Tissot and Espitalie (1975) with corrected cracking parameters and computed temperatures (Yukler, 1987). The top of the overpressure in this area is at about 10,000 ft (3,050 m).

Their model showed that the oil generation window for a mixed Type II, III kerogen began at a depth around 10,000 ft (3,050 m) and peaked at a depth around 11,808 ft (3,600 m). The oil generation phased out around 14,760 ft (4,500 m). However, active gas generation continued down to 18,040 ft (5,500 m). The increase in overpressure with depth in the De Witt well in Figure 13 correlates directly with the increase in gas generation computed by the Yukler and Dow model (1990). Thus, gas generation appears to be related to the overpressure. Compaction cannot be causing the overpressure since there is no decrease in porosity through the oil and gas generation window in any of these wells. All of these cases shown in Figure 13 support the concept that gas generation is causing the observed overpressures.

Several petroleum geologists have suggested that hydrocarbon maturation causes overpressuring and primary hydrocarbon migration. For example, Momper (1978) reported that primary hydrocarbon migration on a significant scale from a shale source rock occurs only after the kerogen has generated about 15 bbl of oil per acre-foot of rock (850 ppm). Momper's model calculated that at peak oil generation the conversion of organic matter to liquids and gases can cause a net volume increase of up to 25 percent over the original organic volume. In the restricted pore space of a fine-grained source rock, this would create a localized pressure build-up causing the opening of

existing microfractures or formation of new ones with the expulsion of oil. Overpressures of 0.6 to 0.7 psi/ft (13.6 to 15.8 kPa/m) would be sufficient to reopen closed vertical fractures and possibly form new ones (Momper, 1981). After oil is expelled, the fractures close until the pressure builds up from subsequent generation. This results in the pulsed expulsion of oil until the generating system runs down. Generation *causes* migration. When generation stops, the primary migration stops (Hunt, 1996, p. 316–319).

Meissner (1978) came to a similar conclusion in studying the Bakken shale source rock of the Williston Basin. High overpressures (>0.6 psi/ft, >13.6 kPa/m) in the Bakken shale are confined both stratigraphically and regionally to that part of the shale that is actively generating oil. Pressures above and below the Bakken shale are slightly subnormal and pressures to the west and east in the immature, non-generating, shallower section of the Bakken are normal. Meissner concluded that active high-rate hydrocarbon generation was causing the high fluid formation pressures.

In the Altamont Field of the Uinta Basin of Utah, overpressures up to 0.8 psi/ft (18 kPa/m) are confined to the Upper Wasatch-Green River Black Shale, with normal pressures above and below (Hunt, 1979, p. 245). Based on comparisons of actual data to their models, both Sweeney et al. (1987) and Spencer (1987) concluded that hydrocarbon generation caused overpressures in the Altamont Field of the Uinta Basin.

Law (1984) studied source rocks and overpressures in the Upper Cretaceous and Lower Tertiary rocks of the Greater Green River Basin in Wyoming, Colorado and Utah. He found that abnormally high formation pressures in this basin are always associated with gas-bearing reservoirs, which suggests that the overpressures are caused by the generation of gas. A subsequent study by Law and Dickinson (1985) indicated that overpressured gas accumulations in the Rocky Mountain region are caused by the thermal generation of gas in low-permeability rocks where gas accumulation rates are higher than rates of gas loss.

Barker (1990) modeled the conversion of oil to gas at 12,000 ft (3,660 m) in an isolated system to determine if it could cause overpressures. He found that the conversion of less than 2% of oil to gas would create overpressures exceeding the fracture gradient. Although Barker's model applied to reservoirs, there is over 100 times as much disseminated oil in the source rocks of the world as in reservoirs (Hunt, 1972). Conversion of this residual oil to gas with deeper burial could create small fractures throughout the source rock comparable to those observed in the Green River Shale of the Altamont field, Uinta Basin, the Bakken Shale of the Antelope field, Williston Basin and the Bazhenov Shale of the Salym field, West Siberian Basin. All of these source rocks are highly overpressured and fractured within the oil generation zones almost exclusively. For example, Vernik (1994) reported that bedding-parallel microfractures are pervasive in the deepest part of the

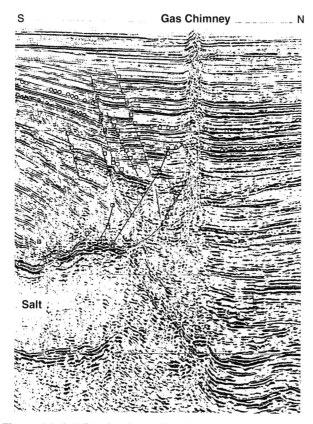

Figure 14. A 3-D seismic profile of a gas chimney rising from depths greater than 15,000 ft (4,570 m) up through Plio-Pleistocene sediments in the South Marsh Island area, offshore Louisiana. The gas plume is adjacent to a 7,000 ft (2,135 m) thick allochthonous Jurassic salt. The straight lines are faults, and boxes are crossline interpretations of horizons and faults (after Hunt, 1996, p.460).

Bakken Shale due to high overpressures caused by hydrocarbon generation. In the Deep Alberta Basin, there are isolated, overpressured carbonate gas reservoirs containing bitumen filled microfractures apparently resulting from the conversion of oil to gas (Marquez and Mountjoy, 1996).

Barker's model was primarily for methane generation but CO_2 is a major component of gases from both organic and inorganic sources in deltaic sediments such as in the Gulf Coast. In Miocene through Jurassic rocks of the Texas Gulf Coast the CO_2 ranges from <1 mole% in reservoirs at 7,000 ft (2,130 m) to 7 mole% in reservoirs at 12,000 ft (3,660 m) according to Franks and Forester (1984). The role of CO_2 in expelling hydrocarbons from shale source rocks is discussed further in our conceptual model of oil and gas expulsion.

Dahl and Yukler (1991) used a basin model to follow the geological and geochemical processes in the Oseberg area of the North Sea. Their computed pressure history showed that abnormal pressures occurred in the Viking group source rocks simultaneously with oil generation. A similar result was obtained in the Gulf Coast calculations discussed above that Yukler per-

formed for us based on Yukler and Kokesh (1984). The computed excess pore pressures in these Gulf Coast calculations reached their highest level at a depth of 12,000 ft (3,660 m) and continued at that level to the bottom of the well.

Finally, Lewan (1987) found that oil and gas are expelled from chunks of source rocks in hydrous pyrolysis experiments where compaction plays no role.

Copious quantities of gas are migrating vertically from depths >3 km in the Gulf Coast. Figure 14 is an example of a three-dimensional seismic profile through Plio–Pleistocene sedimentary rocks in the South Marsh Island area offshore Louisiana. On the left is the edge of an estimated 7,000 ft (2,130 m) thick Jurassic salt contacting the Plio–Pleistocene sediments. On the edge of the salt is a gas chimney extending to the surface. Most gas chimneys in the Gulf Coast are small (about 400 m, 1,312 ft in diameter) and vertically oriented, so they are not usually seen on regional two-dimensional seismic grids. The gas plume in Figure 14 is not going up faults because it is nearly vertical, whereas the faults are at an angle. It is not syndepositional, slow acting, or continuous because there is no evidence of thinning of the sediment layers adjacent to the plume. It looks like a high-pressure gas blowout shooting up like a bullet. Some plumes look like wormholes in that they rotate slightly on the way up. The source of the gas in Figure 14 is difficult to determine. It may be gas spilling over from an accumulation under the salt or from a deeper overpressured compartment with a fractured seal (Hunt, 1996, p. 459).

A CONCEPTUAL MODEL OF COMPACTION AND HYDROCARBON GENERATION AND MIGRATION

The above discussion suggests the conceptual model illustrated in Figure 15. Grain dissolution compaction occurs in hydrostatically pressured shales from the base of mechanical compaction at ~500 m depth to the depth at which strata temperatures reach 85° to 110°C. Pressure dissolution at grain contacts is hypothesized to produce a plastic interpenetration of grain boundaries whose magnitude is a function only of the effective stresses pressing the grains together. Palciauskas and Domenico (1989) have shown that this kind of plastic penetration produces linear compaction. They derived a physical-chemical model for sand compaction from first principles. Remarkably, their model has no arbitrary parameters that need to be fitted. The predicted compaction depends only on the geometrical and physical properties of the load-bearing matrix grains. For a simple grain geometry and appropriate properties of quartz, their model predicts linear compaction of the right magnitude. Since Gulf Coast shales are 74% quartz in the clay-sized fraction (<4 μm) as previously stated, the model should be equally appropriate for describing shale compaction.

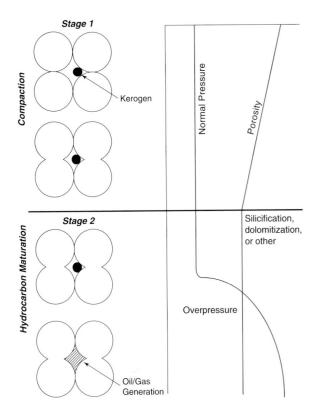

Figure 15. Conceptual model of compaction, overpressuring, and hydrocarbon generation. Linear compaction occurs until minimum adsorbed water and diagenetic reactions arrest further compaction. Overpressures are caused by the generation of hydrocarbon gases.

We suggest that compaction is arrested when only about three monomolecular layers of oriented water are still adsorbed on the mineral surfaces. Removal of the 3rd and 2nd layers from smectite-illite require 20,000 and 40,000 ft (6,100 and 12,200 m) of overburden, respectively (Powers, 1967). There also may be temperature dependent chemical changes involved such as silicification or dolomitization creating a rigid framework in the rock but we have no clear evidence for it.

Overpressures are generated in the organic-lean Gulf Coast rocks when maturation reactions occur whose products are of greater volume than the reactants. The positive volume change forces both the hydrocarbons and pore waters out of the source shales. If the shales have low enough permeability, overpressures are produced.

Figure 16 summarizes estimates of the densities and masses of the reactants and products of Type III kerogen maturation which is typical of the Gulf Coast. Consider the first stage of maturation, in which kerogen decomposes to bitumen, CO_2 and a residue, R. The change in volume of reactants and products depends mainly on the density of the CO_2 phase. Table 3 shows the density of CO_2, CH_4, and C_3H_8 along a hypothetical pressure-temperature depth profile. Temperatures increase at 25°C/km. Pressures increase hydrostatically to 2.9 km depth, then increase rapidly across a seal

to lithostatic levels at 3.0 km depth, and thereafter again increase along a hydrostatic gradient. The density of CO_2 within the overpressured compartment on the high pressure side of the seal is ~0.89 g/cm³. This density was calculated at 90°C and 660 bars using the Redlich Kwong equation of state. With this gas density, Figure 16 shows that the change in volume of the bitumen generation reaction depends on whether CO_2 dissolves or is isolated as a gas phase.

The volume changes shown in Figure 16 were calculated by dividing the mass of each reactant or product by its density and summing all reactants and products with the convention that reactant volumes are negative. The volume change was then converted to the change in volume within each cm³ of sediment by multiplying by the grams of kerogen per cm³ of sediment. The grams of kerogen per cm³ sediment equals 3.6 x 10⁻² for 1.5 wt% kerogen in the sediments. The kerogen is assumed to contain 67% carbon (TOC). The mass per unit volume was calculated from the relation TOC x ρ_g x $(1 − \phi)$, where TOC = 0.01, ϕ is the sediment porosity of 0.1 and ρ_g is the mineral grain density of 2.65 g/cm³.

It is unlikely that any but the most organic-rich sediments will produce a separate CO_2 gas phase because CO_2 is very soluble in water. At 660 bars and 100°C, for example, a simple Henry's Law calculation indicates that water can dissolve ~0.4 grams of CO_2 per gram H_2O. Sediments with 1.0 wt% TOC contain 3.6 x 10⁻² grams of kerogen per cm³ sediment and can produce 7.2 x 10⁻³ g CO_2/cm³. Pore waters can dissolve 4 x 10⁻² g CO_2/cm³ at 10% porosity. If the CO_2 dissolves, and we neglect the small volume change of the water caused by this dissolution, Figure 16 shows that the volume change of the bitumen generation reaction is negative (the products have less volume than the reactants). No overpressures will be generated, and no bitumen will be expelled from the source shales.

If on the other hand the CO_2 does not dissolve, the volume change of the reaction is positive. The products occupy more volume than the reactants, overpressures are produced, and CO_2 and bitumen are expelled from the shales. The overpressures may be very high and can be estimated following the approach of Barker (1990). The volume change of the bitumen-generating reaction, considering only the non-gas phases is 0.11 cm³ per gram kerogen reacted. The reaction produces 4.5 x 10⁻³ moles of CO_2 per gram of kerogen reacted. The compressibility factor, z, for CO_2 at 660 bars and 90°C is 1.09. The gas law equation (below) relates pressure, P, in bars, volume, V, in cm³, the moles of gas, n, temperature in degrees Kelvin (T_K = 373°K), and the gas constant, R = 83.14 cm³ bar/mole °K:

$$P = \frac{83.14 z n T_K}{V} \qquad (2)$$

From this equation the pressure required to contain 4.5 x 10⁻³ moles of CO_2 in 0.11 cm³ is 1,346 bars, or about twice lithostatic. If CO_2 does not dissolve, the

Type III Kerogen

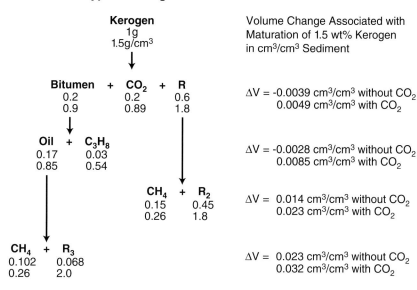

Volume Change Associated with
Maturation of 1.5 wt% Kerogen
in cm^3/cm^3 Sediment

$\Delta V = -0.0039\ cm^3/cm^3$ without CO_2
$0.0049\ cm^3/cm^3$ with CO_2

$\Delta V = -0.0028\ cm^3/cm^3$ without CO_2
$0.0085\ cm^3/cm^3$ with CO_2

$\Delta V = 0.014\ cm^3/cm^3$ without CO_2
$0.023\ cm^3/cm^3$ with CO_2

$\Delta V = 0.023\ cm^3/cm^3$ without CO_2
$0.032\ cm^3/cm^3$ with CO_2

Figure 16. Maturation tree for Type III kerogen showing the grams of product (upper no.) generated from 1 gram of kerogen and the densities (lower no.) these products will have under overpressured conditions (90°C and 660 bars). Volume changes correspond to each of the four reactions shown. R_n = residue for that reaction.

overpressures will fracture the shale and expel CO_2 and bitumen (which is fluid at 90°C at these pressures).

If CO_2 dissolves, the first reaction that generates positive volume change is the methane generation reaction. Methane is much less soluble in water than is CO_2. At lithostatic pressures and realistic basin temperatures, the solubility of methane is about 5,000 ppm (Bonham, 1978). For sediments with porosities of 10%, about 5×10^{-4} g CH_4 can thus be dissolved per cm^3 sediment. For 1.5 wt% kerogen in the sediments this corresponds to 0.014 g CH_4 per gram of kerogen. Only about 10% of the methane generated by the third reaction in Figure 16 can be dissolved in the pore waters of 10% porosity sediment.

Assuming that all CO_2 and 10% of the CH_4 generated dissolves, the pressure produced by methane generation can be estimated by the same methods applied above. The non-gas volume ($R_2 + R_3$) created by the four reactions in Figure 16 is 0.41 cm^3/g kerogen reacted. The reactions generate 9.3 millimoles of methane, of which 0.93 millimoles are dissolved in the pore waters. With a z factor of 1.4 and 90°C temperatures, 845 bars are required to compress 8.4 millimoles of CH_4

into 0.41 cm^3. Pressures well in excess of lithostatic, which is about 660 bars at 3 km, can be generated by methane production.

The amount of hydrocarbon expulsion that can result from methane overpressures is indicated for Type III kerogen in Figure 16. The assumption is that a volume of pore fluids and hydrocarbons expelled equals the overall ΔV of reaction. The expelled volume can be expressed as a percentage of the original porosity if the figures in Figure 16 are divided by 0.1 and multiplied by 100. For 1.5 wt% kerogen in the sediments the first stage of methane generation could thus expel 14% of the bitumen, oil and pore fluids from a 10% porosity sediment.

Expulsion of both bitumen and oil would be greatly assisted by the presence of CO_2. At a subsurface temperature of 90°C (195°F) and pressure of 4,000 psi, about 500 standard cubic feet of CO_2 will dissolve in a barrel of 10° API oil reducing its viscosity by a factor of 30 (Murtada and Hofling, 1987).

Gas generation and the development of overpressures could of course arrest compaction by greatly reducing the effective stress on the mineral grains. This

Table 3. Densities (ρ) and compressibility (z) factors for CO_2, CH_4, and C_3H_8 along a hypothetical basin pressure-temperature profile. Parameters for CH_4 and C_3H_8 were computed from the Behar et al. (1985) equation of state. Parameters for CO_2 were computed from the Redlich Kwong equation of state. Z = depth.

Z (km)	T (°C)	P(bars)	ρ CO_2	z CO_2	ρ CH_4	z CH_4	ρ C_3H_8	z C_3H_8
1.4	50	140	0.60	0.38	0.10	0.88	0.50	0.46
2.4	75	240	0.66	0.56	0.14	0.94	0.49	0.75
2.6	80	260	0.67	0.59	0.15	0.95	0.49	0.81
2.9	87.5	290	0.68	0.63	0.16	0.98	0.49	0.88
3.0	90	660	0.89	1.09	0.26	1.36	0.54	1.8
3.4	100	700	0.89	1.13	0.26	1.39	0.53	1.9
3.8	110	740	0.89	1.17	0.26	1.43	0.54	2.0

mechanism for arresting compaction could be important in some cases, but for most of the Gulf Coast data summarized here, compaction is arrested above rather than at the top of the overpressure. Gas generation could affect permeability through capillary effects and could drive diagenetic reactions by changing pH. It is thus possible that gas generation and arrested compaction are indirectly related.

The most important aspect of this discussion is that gas generation provides an alternative to compaction as a mechanism for producing overpressures and expelling hydrocarbons. Where compaction is arrested, as in much of the deep Gulf Coast, gas generation becomes the most probable cause of overpressuring and hydrocarbon expulsion.

CONCLUSIONS

Direct measurements of porosity and density of shales reported by Bradley (1976), Hinch (1980), and Powley (1985, 1992, and 1993) laid the groundwork for showing that most shale compaction in the U.S. Gulf Coast occurs in two stages for porosities <30%. Powley's early work, reported by Hinch (1980), indicated that there is a time-temperature control of the abrupt change from a systematic decrease in porosity (compaction Stage 1) to no decrease (no compaction, Stage 2).

This paper supports those early conclusions by giving statistical evidence that a two-stage linear plot of shale porosity versus depth at porosities <30% fits the data better than a one-stage exponential plot. We show that in Stage 2 the porosity versus depth line has no slope indicating no compaction through as much as 10,000 ft (3,050 m) of rocks.

The peaks of oil, condensate and gas generation and expulsion and the tops of overpressures were commonly found to occur within intervals of no compaction, thereby, indicating that shale compaction cannot be the major contributor to overpressures or to the expulsion of hydrocarbons from these deep Gulf Coast rocks. The major cause of deep (>3 km) overpressures and the expulsion of hydrocarbons from these rocks appears to be the increase in volume of pore fluids caused by the thermal generation of gas.

The examples in this paper show that many porosity-depth relations in the Gulf Coast at porosities <30% can be described in two stages: (1) a linear decrease with depth, and (2) a deeper stage showing no decrease. This means that *the one stage exponential porosity-depth relation still used in most basin modeling today is not always valid.* Consequently, it is important to make direct shale porosity measurements on individual wells to define the type of porosity profile that actually exists before proceeding with a basin modeling program. The use of hypothetical curves not based on real data or the use of composites of data from several wells in a large area may lead to erroneous conclusions.

ACKNOWLEDGEMENTS *The authors particularly wish to thank the Amoco Production Co. for providing the basic porosity data and Henry H. Hinch, formerly of Amoco, for providing the information on his analytical methods, the smectite-illite data on shales, and for his constructive review of the manuscript. Thanks also are due to K.E. Peters, D.E. Powley, J.S. Bradley, and J.A. Curiale for critically reviewing various revisions of the manuscript. This work was supported by the Gas Research Institute, Contract No. 5091-260-2298. This is the Woods Hole Oceanographic Institution Contribution No. 8573.*

REFERENCES CITED

Athy, L.F., 1930, Compaction and oil migration: Bulletin of the American Association of Petroleum Geologists, v. 14, p. 25–36.

Atwater, G.I. and Miller, E.E. 1965, The effect of decrease in porosity with depth on future development of oil and gas reserves in south Louisiana: AAPG Bull., v 49, p. 334 abstract.

Barker, C.E., 1990, Calculated volume and pressure changes during the thermal cracking of oil to gas in reservoirs: AAPG Bull., v. 74, p. 1254–1261.

Behar, E., R. Simonet, and E. Rauzy, 1985, A new non-cubic equation of state: Fluid Phase Equilibria, v. 21, p.237–255.

Bonham, L.C., 1978, Solubility of methane in water at elevated temperatures and pressures: AAPG Bull., v. 62, p. 2478–2481.

Bradley, J.S., 1976, Abnormal formation pressure: Reply: AAPG Bull., v. 60, p. 1127–1128.

Bradley, J.S. and D.E. Powley, 1994, Pressure compartments in sedimentary basins: A review. *in* P.J. Ortoleva (ed.), Basin Compartments and Seals: AAPG Memoir 61, Tulsa, American Association of Petroleum Geologists: p. 3–26.

Chilingar, G.V., and L. Knight, 1960, Relationship between pressure and moisture contents of kaolinite, illite, and montmorillonite clays: AAPG Bulletin, v. 44, p. 89–94.

Dahl, B., and A. Yukler, 1991, The role of petroleum geochemistry in basin modeling of the Oseberg area, North Sea. *in* R. K. Merrill (ed.), Source and migration processes and evaluation techniques, Treatise of petroleum geology handbook, Tulsa, The American Association of Petroleum Geologists. pp. 65–85.

Dickey, P.A., 1976, Abnormal formation pressure: discussion: AAPG Bull., v. 60, 1124–1128.

DOE Report No. COO-4392-4, 1981, Organic geochemistry of continental margin sediments.

Dutta, N.C., 1987, Fluid flow in low permeable porous media. *in* B Doligez (ed.), Migration of hydrocarbons in sedimentary basins. Collection 45, Paris: Editions Technip, pp. 567–596.

Forbes, P.L., P. Ungerer, and B.S. Mudford, 1992, A two-dimensional model of overpressure development and gas accumulation in Venture Field, Eastern Canada: AAPG Bull., v. 76, p. 318

Franks, S.G., and R.W. Forester, 1984, Relationships among secondary porosity, pore-fluid chemistry and carbon dioxide, Texas Gulf coast *in* D.A. McDonald and R.C. Surdam (eds.) , Clastic diagenesis. AAPG Memoir 37. Tulsa: The American Association of Petroleum Geologists, pp. 63–80.

Hedberg, H.D., 1936, Gravitational compaction of clays and shales: Am. J. Sci., fifth series, 31, (184), 241–287.

Hinch, H.H., 1980, The nature of shales and the dynamics of hydrocarbon expulsion in the Gulf Coast Tertiary section in W.H. Roberts III and R.J. Cordell (eds.), Problems of Petroleum Migration: AAPG Studies in Geology, No. 10: Tulsa, The American Association of Petroleum Geologists, pp. 1–18.

Huc, A.Y. and J.M. Hunt, 1980, Generation and migration of hydrocarbons in offshore South Texas Gulf Coast sediments: Geochim. Cosmochim. Acta, v. 44, p. 1081–1089.

Hunt, J.M., 1972, Distribution of carbon in the crust of the earth: AAPG Bull., v. 56 , p. 2273–2277.

Hunt, J.M., 1979, Petroleum geochemistry and geology. first edition. San Francisco: W. H. Freeman and Company. 617 p.

Hunt, J.M. and R.J-C. Hennet, 1992, Modeling petroleum generation in sedimentary basins *in* J.W. Whelan and J.W. Farrington (eds.), Organic matter: productivity, accumulation and preservation in recent and ancient sediments: New York: Columbia University Press, pp. 24–52.

Hunt, J.M., 1996, Petroleum geochemistry and geology. second edition. New York: Freeman and Company. 743 pp.

Jansa, L.F., and V.H. Noguera Urrea, 1990, Geology and diagenetic history of overpressured sandstone reservoirs, Venture gas field, offshore Nova Scotia, Canada: AAPG Bull., v. 74, p. 1640–1658.

Korvin, G., 1984, Shale compaction and statistical physics: Geophysical Journal of the Royal Astronomical Society, v. 78, p. 35–50.

La Plante, R.E., 1974, Hydrocarbon generation in Gulf Coast Tertiary sediments: AAPG Bulletin, v. 58, p. 1281–1289.

Law, B.E., 1984, Relationships of source-rock, thermal maturity, and overpressuring to gas generation and occurrence in low-permeability Upper Cretaceous and Lower Tertiary rocks, greater Green River Basin, Wyoming, Colorado, and Utah *in* J. Woodward, F.F. Meissner, J.L Clayton (eds.), Hydrocarbon source rocks of the greater Rocky Mountain region. Denver: Rocky Mountain Association of Geologists. pp. 469–490.

Law, B. E. and W. W. Dickinson, 1985, Conceptual model for origin of abnormally pressured gas accumulations in low-permeability reservoirs: AAPG Bull., v. 69, pp. p. 1295–1304.

Lewan, M.D., 1987, Petrographic study of primary petroleum migration in the Woodford Shale and related rock units *in* B. Doligez (ed.), Migration of Hydrocarbons in Sedimentary Basins: Paris: Editions Technip, p.113–130.

Loucks, R.G., M.M. Dodge and W.E. Galloway, 1984, Regional controls on diagenesis and reservoir quality in Lower Tertiary sandstones along with Texas Gulf Coast *in* D.A. McDonald and R.C. Surdam (eds)., Clastic Diagenesis: AAPG Memoir 37. Tulsa Okla.: American Association of Petroleum Geologists, p. 15–45.

Mackenzie, A.S. and T.M. Quigley, 1988, Principles of geochemical prospect appraisal: AAPG Bull., v. 72, p. 399–415.

Marquez, X.M., and E.W. Mountjoy, 1996, Microfractures due to overpressures caused by thermal cracking in well-sealed Upper Devonian reservoirs, Deep Alberta Basin: AAPG Bull., v. 80, p. 570–588.

Meissner, F.F., 1978, Petroleum geology of the Bakken Formation, Williston Basin, North Dakota and Montana. Proceedings of 1978 Williston Basin symposium, "The Economic Geology of the Williston Basin," September 24–17, 1978, Montana Geological Society, Billings, pp. 207–227.

Meloche, J.D., 1985, Diagenetic evolution of overpressured sandstones - Scotian Shelf (abs.). 1985 Annual Convention Geological Association of Canada, Fredericton, N.B.

Momper, J.A., 1978, Oil migration limitations suggested by geological and geochemical considerations *in* Physical and chemical constraints on petroleum migration. Vol. 1. Notes for AAPG short course, April 9, 1978, AAPG National Meeting, Oklahoma City. 60 p.

Momper, J.A., 1981, The petroleum expulsion mechanism - a consequence of the generation process *in* AAPG Geochemistry for Geologists School. Notes for AAPG Short Course , Feb. 23–25, 1981, Denver Colorado

Murtada, H. and B. Hofling, 1987, Feasibility of heavy-oil recovery *in* R.F. Meyer (ed.), Exploration for Heavy Crude oil and Natural Bitumen, AAPG Studies in Geology No. 25, Tulsa, OK: American Association of Petroleum Geologists, pp. 629–643.

Palciauskas, V.V. and P. A. Domenico, 1989, Fluid pressures in deforming porous rocks: Water. Res. Research, v. 25, p. 203–213.

Powley, D. E., 1985, Pressures, normal and abnormal. Lecture Notes, Techniques of Petroleum Exploration II. American Association of Petroleum Geologists School. South Padre Island, Sept 16–19, 1985.

Powley, D.E., 1992, Shale porosity-depth relations in normally compacted shale: Second Symposium on Deep Basin Compartments and Seals, Gas Research Institute Oklahoma State University, Sept. 29–Oct. 1, 1992. Stillwater, Oklahoma.

Powley, D.E., 1993, Shale compaction and its relationship to fluid seals. Section III, Quarterly report, Jan. 1993–Apr. 1993, Oklahoma State University to the Gas Research Institute, G.R.I., Contract 5092-2443.

Powers, M.C., 1967, Fluid-release mechanisms in compacting marine mud rocks and their importance in oil exploration: AAPG Bulletin, v. 51(7), p. 1240–1254.

Rieke III, H.H., and G.V. Chilingarian, 1974, Compaction of Argillaceous Sediments: New York, Elsevier Scientific Publishing Co., pp. 41–43.

Smith, J.T., 1994, Petroleum system logic as an exploration tool in a frontier setting *in* L.B. Magoon, and W.G. Dow (eds.) The petroleum system–from source to trap. AAPG Memoir 60. Tulsa: American Association of Petroleum Geologists. p. 31.

Spencer, C.S., 1987, Hydrocarbon generation as a mechanism for overpressuring in Rocky Mountain Region: AAPG Bull., v. 71, p. 368–388.

Storer, A., 1959, Constipamento dei sedimenti argillosi nel Bacino Padano, Giacimenti Gassiferi dell' Europa Occidentale. Rome: Acad. Nazionale dei Lincei, Rome. pp. 519–544.

Sweeney, J.J., A.K. Burnham, and R.L. Braun, 1987, A model of hydrocarbon generation from Type I kerogen: application to Uinta Basin, Utah: AAPG Bull., v. 71, p. 967–985.

Tissot, B., and J. Espitalie, 1975, L'évolution thermique de la metière organique des sédiments: application d'une simulation mathématique. Rev. Inst. Franc. Petrol., v. 30, p. 743–777.

Ungerer, P., J. Burrus, B. Doligez, P. Y. Chenet, and F. Bessis, 1990, Basin evaluation by integrated two-dimensional modeling of heat transfer, fluid flow, hydrocarbon generation, and migration: AAPG Bull., v. 74, p. 309–335.

Vernik, L., 1994, Hydrocarbon generation induced micro-cracking of source rocks: Geophysics, v. 59, p. 555–563.

Waples, D.W., and A. Okui, 1992, Overpressuring and hydrocarbon expulsion. (Abstract) American Association of Petroleum Geologists Annual Convention, Calgary, June 21–24. p. 137.

Wells, P.E., 1990, Porosities and seismic velocities of mudstones from Wairarapa oil wells of North Island, New Zealand, and their use in determining burial history: New Zealand Journal of Geology and Geophysics, v. 33, p. 29–39.

Wood, D.A., 1988, Relationships between thermal maturity indices calculated using Arrhenius equation and Lopatin Method: Implications for petroleum exploration: AAPG Bull., v. 72, p. 115–134.

Yukler, M.A., 1987, How essential is quantitative basin modeling in petroleum exploration? 7th Biannual Petroleum Congress of Turkey Proc., April 6–12.

Yukler, M.A., and W.G. Dow, 1990, Temperature, pressure and hydrocarbon generation histories in San Marcos Arch area De Witt County, Texas *in* D. Schumacher, and B.F. Perkins (eds.), Gulf Coast oils and gases: their characteristics, origin, distribution, and exploration and production significance: Proceedings of the Ninth Annual Research Conference, Society of Economic Paleontologists and Mineralogists Foundation, pp. 99–104.

Yukler, M.A., and F Kokesh, 1984, Review of models used in petroleum resource estimation and organic geochemistry *in* J. Brooks and D. Welte (eds.), Advances in Petroleum Geochemistry 1984, v. 1. London: Academic Press. pp 69–113.

Slavin, V.I., and E.M. Smirnova, 1998, Abnormally high formation pressures: origin, prediction, hydrocarbon field development, and ecological problems, *in* Law, B.E., G.F. Ulmishek, and V.I. Slavin eds., Abnormal pressures in hydrocarbon environments: AAPG Memoir 70, p. 105–114.

Chapter 6

ABNORMALLY HIGH FORMATION PRESSURES: ORIGIN, PREDICTION, HYDROCARBON FIELD DEVELOPMENT, AND ECOLOGICAL PROBLEMS

V.I. Slavin[1]
E.M. Smirnova
VNIGRI
St. Petersburg, Russia

Abstract

Exploration for and the development of oil and gas fields in zones of abnormally high formation pressures (AHFP) require a good understanding of the origin of AHFP and the development of predictive methods. We classify the causes of AHFP into two genetic groups: (1) a syn-sedimentary group characterized by undercompaction of rocks and (2) a post-sedimentary group characterized by secondary decompaction of rocks. The choice of methods for the prediction and evaluation of AHFP should be based on the origin of abnormal pressure and the lithology of the rocks. The reservoir properties of rocks in AHFP zones suggest that large, in-place resources of hydrocarbons may be present in complex low-permeability clastic and carbonate reservoirs despite low rates of production.

Plastic deformation of reservoir rocks which result from the decrease of reservoir pressure in the course of well testing and production in AHFP zones is dependent on the origin of abnormal pressure. Careful monitoring of the critical limit of formation pressure is necessary to avoid irreversible deformation of reservoir rock. Changes in the stress field in the productive reservoir should be controlled in order to prevent the initiation of induced earthquakes. In the AHFP zones of syn-sedimentary origin, water flooding should be implemented from the start of production in order to prevent subsidence with consequent environmental damage.

INTRODUCTION

Abnormally high formation pressures (AHFP) are known to occur in all types of petroleum basins from shallow to very great depths. The deeper parts of petroleum basins typically possess substantial undiscovered resources of oil and gas; however, widespread abnormally high formation pressures at great depths pose serious technological problems for both the exploration and development of these resources. In a number of cases, studies of abnormal pressure can provide direct indications of deeper hydrocarbon pools. Studies of the mechanism of formation of AHFP help in understanding such important processes as primary and secondary migration of hydrocarbons, formation (and reformation) of hydrocarbon pools, and diagenetic changes of reservoir rocks. These studies are also very

important for increasing the efficiency of development of oil and gas fields.

ORIGIN OF ABNORMALLY HIGH PRESSURE

The origin of abnormally high formation pressure has been widely discussed (Fertl, 1976; Magara, 1978; Mouchet and Mitchell, 1989; Slavin and Bruk, 1987; Swarbrick and Osborne, this volume). Basically, all authors agree that the formation of AHFP is a result of two main factors. The first of these factors is the relative isolation of rocks from the effects of fluid flow. The second factor is the change of either fluid volume and/or pore volume that may occur during the burial history of a given sequence of rocks. In some cases, both of

---

[1] *Present Affiliation: Consulting Petroleum Geologist, Berlin, Germany*

these factors can operate simultaneously. The analysis of models of AHFP formation suggests that three main mechanisms can result in the development of AHFP. These three mechanisms include: 1) sedimentation, 2) hydrocarbon generation, and 3) the vertical migration of fluids.

Sedimentation

The main factor which determines changes of pore space is compaction, or decompaction of rocks under changing load. Experimental studies by Terzaghi (1923) of the mechanics of soils have shown that rocks saturated with liquid under increasing load correctly reflects the present understanding of the normal compaction of sediments. The simplest model of Terzaghi (1923) assumes that mineral particles are incompressible and the compaction of rocks is due exclusively to decreasing pore volumes filled with water. The rate of compaction is defined as the rate of water discharge from the pores. Based on laboratory experiments, Terzaghi suggested that the deformation of a rock is determined by the effective stress, σ_{ef}, which is related to the applied load (P_{geost}) as:

$$\sigma_{ef} = P_{geost} - P_{fm} \qquad (1)$$

where σ_{ef} is the effective stress, P_{geost} is the geostatic load, and P_{fm} is the formation pressure, in MPa. The equation reflecting the decrease of porosity of rocks in the course of subsidence is as follows (Fertl, 1976; Magara, 1978):

$$K_n = k_{no}e^{-\beta_n\sigma_{ef}} \qquad (2)$$

where K_n is the porosity of a sediment at depth H, k_{no} is the porosity of a sediment at the surface, and β_n is a constant, characterizing the level of compaction (reversible and irreversible deformation) of rock. Therefore, if the rate of sedimentation is moderate and formation water can be expelled from the pore space, then the pore pressure in the sediments will be equal to the hydrostatic pressure (Figure 1). At rates of high sedimentation, abnormally high formation pressure will develop because of the inability of the pore water to escape (Figure 1) at a rate commensurate with sediment loading. In this case, a large part of the weight of the overburden is borne by the pore fluids.

Hydrocarbon generation

The generation of hydrocarbons involves the transformation of kerogen in organic matter into liquid and gaseous phases. This transformation results in an increase of the fluid volume and a very small decrease in the volume of the organic matter. The decrease of volume of the solid phase is insignificant. The increase of the fluid volume leads to the growth of formation pressure and related reduction of the effective stress.

Although hydrocarbon generation is a very slow process, the effect on the development of AHFP is significant. This stems from the fact that the appearance in the pore space of originally one phase, and later two or three phases (water, gas, and liquid hydrocarbons), strongly alters the permeability of the rocks. Muskat (1949) calculated that if the pore space contained equal volumes of water and oil, the permeability of the rock decreased threefold. If the pore space is filled by equal volumes of water and gas, the relative permeability (ratio of phase permeability to absolute permeability) for water is only 9 percent and that for gas is 29 percent. If the mixture consists of oil and gas and oil saturation is 60 percent, the relative permeabilities for oil and gas are equal and amount to 10 percent.

Changes of permeability are more complex when a three-phase mixture exists in the pore space. For example, if a mixture consists of 40 percent water, 40 percent oil, and 20 percent gas, the relative permeability for gas is less than 1 percent, that for oil is about 12 percent,

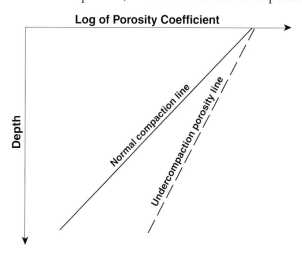

Figure 1. Change of porosity with depth in normally compacted and undercompacted rocks.

and the movement of formation water is virtually precluded. Therefore, the generation of hydrocarbons results in the development of AHFP. This is due to both the transformation of the solid phase of the rock into liquid and gaseous phases with a simultaneous increase of fluid volume in a nearly constant rock volume and a substantial (more than seven times) decrease of rock permeability.

The small increase of porosity associated with the process of hydrocarbon generation is analogous to the porosity increase associated with dilation of a rock matrix accompanying tectonic uplift and erosion. This increase of porosity has been described by a number of researchers (Dobrynin, 1970; Avchyan, 1966; Sitnikov et al., 1971). Their studies show that compaction and decompaction of rocks occur under different relationships between porosity and effective stress (Figure 2). We propose that the change of porosity during decom-

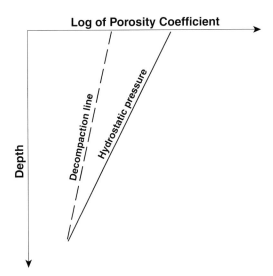

Figure 2. Diagrammatic cross-plot showing porosity loss with depth during sedimentation and compaction (solid Line) and changes of porosity associated with decompaction during uplift and erosion (dashed line).

paction related to the decrease of effective stress may be described by the following equation:

$$K_n = K_{np}e^{-\beta_{np}\sigma_{ef}} \qquad (3)$$

where K_n is the in situ porosity of the rock, K_{np} is the porosity of the rock on the surface after decompaction, β_{np} is a constant characterizing a degree of decompaction (reversible deformation) of a rock, and σ_{ef} is the effective stress in MPa.

The increase of porosity during generation of hydrocarbons is due to the elasticity of the rock matrix, therefore, the amount of this increase is much less than that found under the same AHFP in undercompacted rocks deposited under high sedimentation rates. The increase of porosity and the magnitude of AHFP in this case depends on the generation potential of the rocks.

Vertical migration of fluids

The presence of faults in sedimentary sequences provide a possible fluid-flow pathway for the discharge of AHFP in the lower parts of rock sequences and the development of AHFP in overlying rocks. The increase of porosity in rocks in which AHFP has been developed under these conditions is due to the elastic properties of the rocks. The discharge of AHFP through faults occurs only when gas is present in the composition of the fluid. If the migrating fluid is only water, the pressure decrease is immediate because the water is non compressible and the elastic forces cannot respond quickly enough to alter the volume of pore space. Only the presence of gas, which is compressible and changes its volume with a change of pressure, provides the conditions necessary for long-term vertical migration. Vertically migrating gas may be in solution and, after

reaching a reservoir come out of solution. It is then distributed in the reservoir as a function of buoyancy. Under these conditions, the change of porosity with depth during the formation of AHFP, is shown in Figure 3. In this case, the upper boundary of the abnormally pressured zone is due to the presence of a low-permeability seal and possibly the presence of a relative permeability barrier due to a multiphase fluid.

METHODS OF PREDICTING ABNORMALLY HIGH PRESSURE

Excluding direct measurements of pore pressure, all indirect methods of estimating AHFP can be conditionally divided into the four following groups: (1) methods based on the concept of equivalent depths; (2) methods based on the empirical relationships among geologic, geophysical, and technological parameters and the magnitude of formation pressure; (3) methods based on analogy; and (4) methods based on the hydrodynamic relationship between formation fluids and drilling mud. The first two methods, which are most commonly used are discussed below with proposed changes and innovations in their application.

The method of equivalent depths is based on the theory of compaction of water-saturated, clay-rich sediments. As previously discussed, high sedimentation rates preclude the effective expulsion of pore water and the rock becomes undercompacted. From this relationship, Foster and Whalen (1966) have proposed that

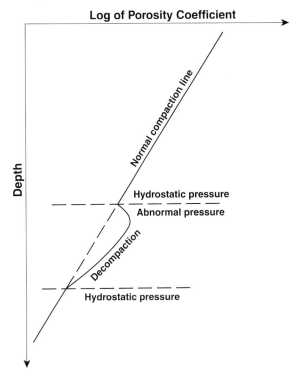

Figure 3. Change of porosity with depth in zone of AHFP related to vertical migration of hydrocarbons.

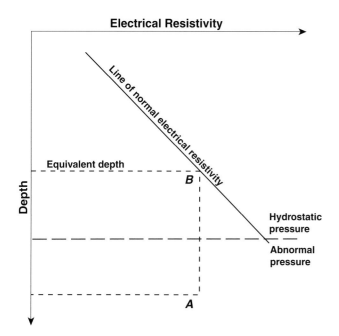

Figure 4. Diagrammatic cross plot illustrating the estimation of AHFP in undercompacted rocks by the method of equivalent depths.

the magnitude of abnormal pressure may be estimated based on the equality of the effective stress (and accordingly porosity) at different depths. The application of the method is as follows.

Beds of carbonate-free shale are defined in the borehole and their specific electric resistivities are measured. The measured resistivities (in logarithmic scale) are plotted against depth (in linear scale). In the zone of normal pressure and normal compaction of the shales, all points are located along a straight line as shown on Figure 4. In the zone of AHFP, where the resistivity values deviate from the normal compaction curve, the magnitude of abnormal pressure at point A (Figure 4) is determined based on the concept of equivalent depths. The resistivity at depth point A, is equal to the resistivity at depth point B along the line of normal compaction (Figure 4). Because the effective stresses at points A and B are equal and the geostatic pressures at these depths are also equal, the AHFP at point A can be calculated from Equation (1).

We propose that the concept of equivalent depths may also be used to estimate AHFP caused by hydrocarbon generation and vertical migration (post-sedimentary mechanisms) (Slavin, 1983; Slavin and Khimich, 1987). For the estimation of AHFP in these cases, values of some geologic, geophysical, or technological parameters are plotted against depth. In the example shown on Figure 5, the line of normal compaction is the line of normal change of electrical resistivity in the zone of normal reservoir pressure. The line of decompaction is drawn from a point located on the downward extension of the line of normal compaction at a depth for which abnormally high pressure is to be

determined (Point C on Figure 5). The magnitude of abnormal pressure at point A is then determined by the method of equivalent depths. From the relationships among formation pressure, effective stress, and porosity given in Equations (1) and (3), point B (Figure 5) is determined and the decompaction line is drawn between points C and B (Figure 5).

A number of approaches to determine the slope of the decompaction line have been proposed. One of them involves measurements of the chosen parameter taken from an equivalent shale bed at the crest and flanks of a post-sedimentary anticlinal structure. The measured values can then be plotted against depth, as shown on Figure 6A. Another approach involves the determination of the selected parameter in cores. The third approach involves direct measurements of the chosen parameter in the AHFP zone. Using any of these approaches the decompaction line for a shale bed can be obtained and the inclination of this line can be extrapolated onto other shale beds.

Application of two methods for determination of the slope of the decompaction line are demonstrated on Figure 6. This example is taken from the Nedzhelin gas field in the Vilyuy Basin of east Siberia, Russia and is controlled by a post-depositional anticlinal trap. In the first method, the Lower Jurassic Sontar shale bed in the field was correlated in seven wells of the field, located at the crest and on the flanks of the anticline. Resistivity measurements of the shales in these wells were plotted and used to draw the decompaction line (Figure 6A). In the second method, the line of decompaction is drawn based on the direct measurement of the AHFP in well 20. From the magnitude of AHFP, the equivalent depth was then calculated using Equation (1) and the method of equivalent depths. The weighted mean

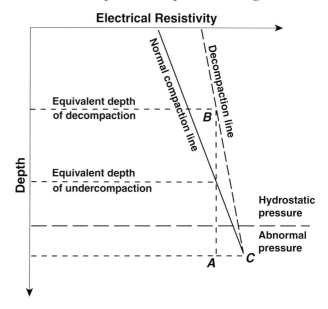

Figure 5. Diagrammatic cross-plot illustrating the method of estimating AHFP in decompacted rocks by the method of equivalent depths.

density of rocks was taken to be equal to 2.3 g/cm³. The line of decompaction is then drawn through points A and B on Figure 6. It can be seen from the plot that the slope of both decompaction lines obtained by different and unrelated methods are very similar. The slope of the decompaction lines were then used to calculate reservoir pressure in the AHFP zone of well 20 (Figure 6B).

From the above discussion it is obvious that the use of the equivalent depths method requires knowledge of the origin of AHFP. Some ways to determine the origin of AHFP have been proposed by Slavin and Khimich (1987). In some cases, formation pressure can be directly measured in the AHFP zone and the results are then compared with calculated values of AHFP for both undercompaction and decompaction AHFP mechanisms. Comparisons of the similarity between the measured and calculated values may provide information on the origin of AHFP. Alternatively, another approach is based on the results of well testing. Because the syn-sedimentary mechanism of AHFP formation involves undercompaction of rocks, a decrease of formation pressure during testing may result in the irreversible reduction of porosity and permeability. Therefore, the results are different during testing in the direct and reverse directions.

Among the methods based on the empirical correlation among geologic, geophysical, or technological parameters and the magnitude of formation pressure, the most precise and express methods use the relationship between drilling parameters and pressure difference at the bottom of a drilling well (ΔP). A close correlation among drilling speed, d-exponent, and differential pressure (ΔP) was described by Vidrine and Bent (1968) and Jorden and Shirley (1966). Based on our studies in northern West Siberia and in the Kobystan-Kura region of the South Caspian Basin

(Azerbaijan) we proposed a similar method for estimating AHFP which we called the method of varying normalized speed of drilling (Slavin and Matus, 1985).

Firstly, the method requires that the mechanical speed of drilling is normalized to technological parameters such as diameter of the bit, speed of bit rotation, load on the bit, and wear of the bit. Therefore, the d-exponent was chosen as the base. After this normalization, the rate of bit penetration depends on only two factors, the degree of compaction and differential pressure at the bottom of the well. The compaction of rocks with increasing depth is described by Equation (2), therefore, the equation for d-exponent is:

$$d = d_n - d_{no}e^{-c\sigma_{ef}} \qquad (4)$$

where c is a constant characterizing the lithology of the rocks, σ_{ef} is the effective stress, and d_n and d_{no} are the d-exponents at depth H and conditionally at the surface, respectively.

Equation (4) normalizes the mechanical speed of drilling to the compaction of rocks with depth. As a result, the dependence of the d-exponent on the pressure difference at the bottom of the well can be obtained and from this, AHFP can be determined.

The application of the described method is shown in Figures 7 and 8. A plot of the d-exponent (in logarithmic scale) against depth (in linear scale) is constructed for the zone of known pressure. Straight lines are drawn connecting points of the d-exponent with equal differential pressures (ΔP) to a chosen depth (2,000 m on Figure 7A). These lines represent the compaction lines of the normalized drilling speed. A graph of the dependence of the d-exponent on differential pressure is then made for the 2,000 m depth (Figure 7B). Finally, a point of d-exponent in the zone of unknown pressure is projected parallel to the lines of compaction, to the

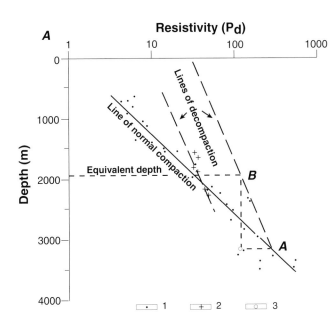

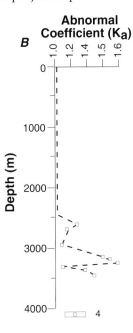

Figure 6. Estimation of the abnormality coefficient (B) in decompacted rocks from a plot of electric resistivity of shales vs. depth (A). Data is from well 20 in the Nedzhelin area. (1 = values of electric resistivity, 2 = values of electric resistivity in shales of the Sontar Formation, 3 = measured formation pressures, 4 = coefficient of abnormality values).

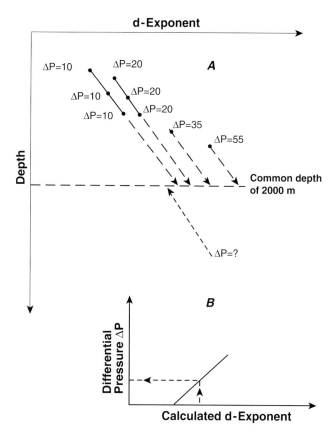

Figure 7. Estimation of AHFP by the method of varying normalized drilling speed. (A) compaction lines based on normalized drilling speed drawn to a common depth of 2,000 m. (B) plot of differential pressure versus d-exponent values reduced to a depth of 2,000 m.

been proposed by Slavin et al. (1985). Decompaction of rocks in the case of post-sedimentary origin of AHFP leads to relatively insignificant increases of porosity. The resulting correction would be well in the range of available precision of the technological parameters.

DEVELOPMENT OF OIL AND GAS FIELDS IN ZONES OF ABNORMALLY HIGH PRESSURE

The development of hydrocarbon fields in zones of AHFP is a difficult task. Reservoir properties of rocks in the AHFP zones are often poor compared to reservoir properties in zones of normal pressure. Diagenetic changes in clastic reservoir rocks within AHFP zones are commonly extensive. Carbonate reservoirs are characterized by primary sub-capillary (< 1 micron), secondary capillary, and supra-capillary pores developed in the matrix. The flow of fluids in both clastic and carbonate rocks occurs mainly through fractures because the permeability of the matrix is very low. Hydrocarbon pools may have large reserves, but daily yields from reservoirs of this type may be low.

Hydrocarbon pools in zones of AHFP are commonly complex. The pools may contain areas filled with formation water and oil-water and gas-water contacts are often poorly defined. Such an irregular distribution of hydrocarbons is probably caused by the highly variable distribution of organic matter and the hydrocarbon-generative potential of these rocks. Part of the pore space is often filled with anomalous, highly viscous oil with a high content of paraffin, asphaltenes, and resins.

Drilling into productive beds in the AHFP zones seldom occurs under optimal conditions because changes of formation pressure in these zones are highly irregular. Potentially productive beds are most commonly subjected to large pressure gradients which have resulted in the plugging of pores and the formation of thick zones of mud filtrate invasion (up to 7.0 m thick). Because the bottom hole pressure during drilling is higher than the formation pressure, the effective stress decreases which results in elastic decompaction of the reservoir rocks and consequent mud invasion into the prospective reservoir. In the course of testing these reservoirs, bottom hole pressures dissipate and become smaller than the original pressure. The reservoir rock compresses and the penetrated mud filtrate forms a crust inside the reservoir, completely plugging it. The mud crust virtually cannot be destroyed either during drilling or during testing of the well. In addition, penetration of mud filtrate into the bottom hole zone of the reservoir creates problems in interpretation of wireline logs which may result in missing productive beds.

Significant technological difficulties arise during testing of productive beds. We believe that the main problem is the creation of conditions favorable for the development of irreversible deformation which may lead to closing of fractures in the bottom hole zone.

chosen depth of 2,000 m (Figure 7A). The differential pressure is then determined from the graph of the d-exponent vs. differential pressure (Figure 7B), and the gradient of the formation pressure is calculated.

Figure 8 is an example of the method using a computerized method (ANOPRESS) for well 204 in the Inzyrey field of the Timan-Pechora Basin, Russia. Figure 8B is a cross plot of the dependence of the normalized d-exponent on differential pressures (ΔP). and is analogous to the plot shown on Figure 7B. In this case the depth of normalization is the surface (0 m). Figure 8A is analogous to Figure 7A showing the zone of study. The lines of ΔP were determined from the cross plot shown on Figure 8B. The calculated results of the method in this well are shown on Figure 8C. The calculated pressure gradient (shown as a dotted line) compare very favorable with the mud density measured during drilling of the well.

The precision of the method depends largely on the presence of a number of d-exponent points with close values of differential pressure in the zone of known formation pressure and on the ability to evaluate undercompaction and decompaction. For AHFP of syn-sedimentary origin, undercompaction should be taken into account. The way to evaluate undercompaction has

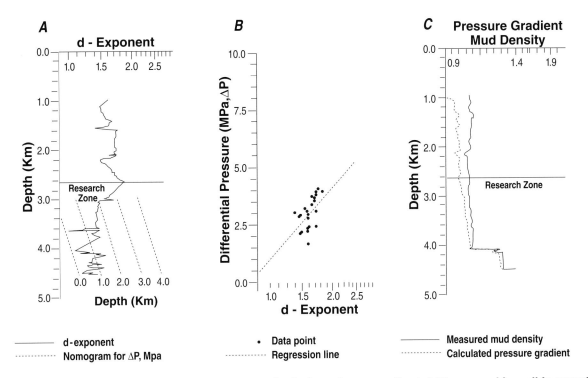

Figure 8. Estimation of formation pressure by the method of varying normalized drilling speed in well Inzyrey 204, Timan-Pechora Basin: (A) plot of d-exponent vs. depth, (B) dependence between differential pressure and d-exponent values reduced to a conditional depth in the Inzyrey 204 well, (C) computer calculation of the pressure gradient (coefficient of abnormality) in the Inzyrey 204 well section as determined by the theoretical model shown in Figure 7.

The two main types of AHFP origin (syn-sedimentary and post-sedimentary) are characterized by different types of fluid flow curves from a well during testing. The first type is characteristic of the syn-sedimentary mechanism of AHFP formation (Figure 9). In this case, any decrease of formation pressure results in irreversible deformation of the reservoir rock. Direct and reverse regimes of testing produce dissimilar results. In the reverse regime, the flow is less because the permeability is irreversibly lower. The second type of fluid flow curve is characteristic of the post-sedimentary mechanism of AHFP formation (Figure 10). The curve has an initial straight segment related to the elastic deformation of rocks. Direct and reverse regimes of testing in the range of this segment produce similar results. New, irreversible deformation appears outside this range.

The problem of irreversible deformation of reservoir rocks complicates production in the AHFP zone of syn-sedimentary origin. The deformation occurs when for-

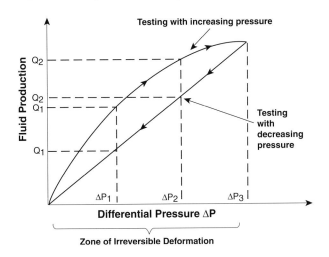

Figure 9. Indicator curve of a test in the AHFP zone of syn-sedimentary origin.

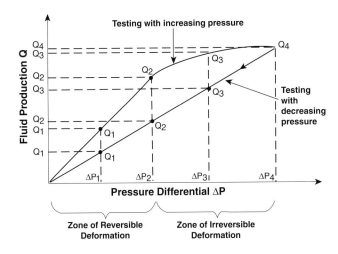

Figure 10. Indicator curve of a test in the AHFP zone of post-sedimentary origin.

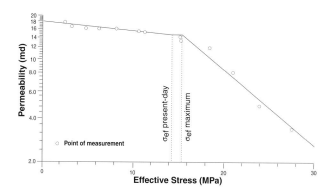

Figure 11. Experimental relationships of permeability to effective rock stress in a sample of Devonian carbonate at a depth of 1,198 m in well Ulyanov 1382, Romashkino field, Volga-Ural Province.

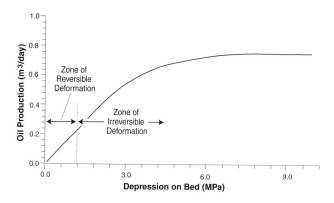

Figure 12. Prediction of daily yields of oil from well 1382 of the Romashkino field (Ulyanov area), Volga-Ural Province, at different draw down pressures. The change of permeability is interpreted from the example shown in Figure 11.

mation pressures decrease during production. The presence of AHFP of post-sedimentary origin does not put additional limits on production as compared with production under normal hydrostatic pressure. The general rule is that during production, the decrease of formation pressure should not exceed the limits of the area of elastic deformation of the reservoir rock.

Each productive bed has its own critical limit of decreasing formation pressure beyond which irreversible deformation occurs. The determination of the critical limit of reversible deformation was proposed by Belonin et al. (1987). Samples of reservoir rocks are subjected to compression from all sides and deformation curves are constructed based on measurements of porosity or permeability. From these curves, the maximum effective stress is determined and the present effective stress is calculated. The difference between the maximum and present effective stresses define the difference between the initial formation pressure in the reservoir and the minimum pressure during production below which irreversible deformation occurs. Figures 11 through 14 show an example of the prediction of production characteristics in the Ulyanov area of the Romashkino field (Volga-Ural Province). The determination of the critical limit for decreasing formation pressure is shown in Figure 11 and the prediction of daily yields of wells is shown in Figure 12. Figures 13 and 14 demonstrate the prediction of irreversible changes of reservoir rocks with distance from the well at a maximum differential pressure of 8 MPa.

SOME ECOLOGICAL PROBLEMS

Drilling and production of hydrocarbons in zones of AHFP, if improperly conducted, may result in significant ecological damage because of possible contamination of underground water with chemical compounds, danger of blowouts, and induced earthquakes. Probably the most dangerous operation is drilling through the transitional zone between zones of normal and

abnormally high pressures. Models of the vertical distribution of pore space in the AHFP zones indicate that changes of formation pressure through a given sequence of rocks are highly irregular (Leonard, 1993; Slavin et al., 1994). The maximum increase of formation pressure gradients occurs in the transitional zone. In this zone, the commonly used method of "balanced" drilling when the mud density is increased according to the growth of overpressure, may result in serious problems. When the mud density is adjusted to overpressure at the well bottom, a significant pressure differences can appear in the upper part of the transitional zone and loss of mud may occur in permeable horizons. This may result in hydrofracturing of the relatively high permeability reservoirs in the upper part of the transition zone, catastrophic loss of drilling mud, and blowouts from the deeper, higher pressured beds in the lower part of the transition zone.

Production of hydrocarbons in the AHFP zone commonly leads to irreversible plastic deformation of reservoir rocks. Reversible deformation disrupts the natural equilibrium and may induce earthquakes. To estimate the magnitude of plastic deformation, data on

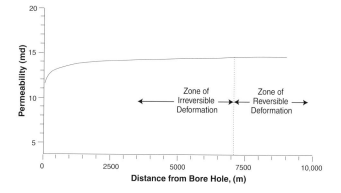

Figure 13. Calculated change of permeability in a producing reservoir as a function of distance from the borehole. Well number 1382 from the Romashkino field (Ulyanov area), Volga-Ural Province.

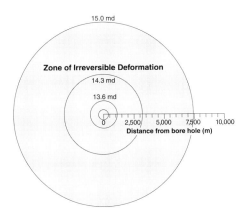

Figure 14. Map of calculated permeability around the bottom of well 1382 of the Romashkino field (Ulyanov area), Volga-Ural province, in the zone of irreversible plastic deformation (see Figures 11 and 13). The calculation is made for the depression radius of 10,000 m and maximum differential pressure of 10 MPa.

two oil pools of the Balakhany-Sabunchi-Romany field in Azerbaijan have been studied. Sandstone reservoirs containing the pools are located in the Pliocene Kirmakin and Podkirmakin Formations. Reservoir data from these two pools (Table 1) show that during 20 years of production, the average porosity of sandstones in the Kirmakin Formation decreased from 26.4 to 22.8 percent and that of the Podkirmakin Formation decreased from 27.7 to 22.2 percent. Thickness of the productive zones in these units decreased by 9.55 and 4.7 m, respectively. Such large changes in reservoir characteristics are dangerous and may induce earthquakes.

Because plastic deformation is a cause of induced earthquakes, the stress field in reservoir rocks should be continuously monitored during production. Changes of porosity in hydrocarbon-saturated beds should be measured and documented from the time of drilling and testing through the time of production. The resulting information file of these producing beds should contain deformational curves on porosity and the results of testing and production (Figure 15).

CONCLUSIONS

1. Although several causal mechanisms leading to AHFP are known, all of them may be divided into

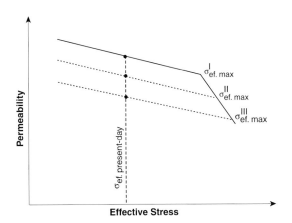

Figure 15. Record of irreversible deformation of productive bed.

two groups—syn- or post-sedimentary. The two groups differ in their effect on the porosity of rocks. At the same magnitude of overpressure, the porosity in an AHFP zone of syn-sedimentary origin (undercompaction) is substantially larger than the porosity in an AHFP zone of post-sedimentary origin (decompaction).

2. The estimation and prediction of AHFP requires the use of a variety of methods. The specific method chosen depends on the origin of AHFP and the lithology of the rocks. Methods based on the concept of equivalent depths are best for AHFP of syn-sedimentary origin. For AHFP of post-sedimentary origin, the best results are obtained using the method of varying normalized speed of drilling.

3. Reservoir properties of overpressured rocks are commonly poorer than those of normally pressured rocks, especially in overpressured rocks of post-sedimentary origin. The fluid flow in overpressured clastic and carbonate rocks occurs dominantly along fractures because the permeability of the matrix is very low. In these low-permeability reservoirs, rates of production are commonly low, even in fields with large, in-place resources.

4. The decrease of formation pressure during testing and production in AHFP zones of syn-sedimenta-

Table 1. Deformation of reservoir rocks of the Balakhany-Sabunchi-Romany oil field, Azerbaijan after 20 years of hydrocarbon production.

Reservoir	Original Thickness (m)	Present Day Thickness (m)	Original Porosity (%)	Present Day Porosity (%)	Porosity % Decrease During Production
Kirmakin	266	256.45	26.4	22.8	13.6
Podkimakin	85.1	80.4	27.7	22.2	19.8

ry origin inevitably results in irreversible plastic deformation of reservoir rocks. Deformation of rocks in the AHFP zone of post-sedimentary origin is similar to that in normally pressured rocks. In general, the lowering of formation pressure during production should not exceed the limits of elastic deformation. Exploitation of hydrocarbon pools in the AHFP zone of syn-sedimentary origin should involve repressuring by water flooding or gas injection from the beginning of production, in order to avoid plastic deformation and the risk of initiating earthquakes. The stress field in the reservoir should be continuously monitored.

REFERENCES CITED

Avchyan, G.M., 1966, Estimation of pressure magnitude affecting rocks: Doklady AN SSSR, v.170, no.2, p.399–401.

Belonin, M.D., Slavin, V.I., and Matus, B.A., 1987, Sposob vozdeystviya na prizaboynuyu zonu skvazhiny (Method of influence on the well bottom), Patent no.1502812.

Dobrynin, V.M., 1970, Deformatsii i izmeneniya fizicheskikh svoystv kollektorov nefti i gaza (Deformation and changes of physical properties of oil and gas reservoirs): Moscow, Nedra, 239 p.

Fertl, W.H., 1976, Abnormal formation pressure, Elsevier Scientific Publishing Company, Amsterdam-Oxford-New York, 398 p.

Foster, J.B., and H.E. Whalen, 1966, Estimation of formation pressure from electrical surveys—offshore Louisiana: Journal of Petroleum Technology, v.18, no.2, p.165–171.

Jorden, J.R., and O.J. Shirley, 1966, Application of drilling performance data to overpressure detection: Journal of Petroleum Technology, v.28, no.11, p.1387–1394.

Leonard, R.C., 1993, Distribution of subsurface pressure in the Norwegian Central Graben and applications for exploration: Petroleum Geology The Geological Society, London, p.1295–1303.

Magara, K., 1978, Compaction and fluid migration: Elsevier Scientific Publishing Company, Amsterdam-Oxford-New York, 296 p.

Muskat, M., 1949, Physical principles of oil production: New York, McGraw Hill Book Co.

Mouchet, J.P., and A. Mitchell, 1989, Abnormal pressures while drilling: Origins, prediction, detection, evaluation: Boussens, Elf Aquitaine, 255 p.

Slavin, V.I., 1983, Peculiarities of the stress state of sedimentary rocks and their effect on improvement of methods of study of the sedimentary section, *in* Problemy metodiki poiska, razvedki i osvoeniya neftyanykh i gazovykh mestorozhdeniy Yakutskoy SSSR (Problems of methods for exploration for and development of oil and gas fields in Yakutsk Republic): Yakutsk, YaOSO AN SSSR, p.16–19.

Slavin, V.I., and B.A. Matus, 1985, Sposob opredeleniya plastovogo davleniya (Method for determination of formation pressure): Author Certificate no.1183670.

Slavin, V.I., V.V. Sheverdyaev, and B.A. Matus, 1985, Determination of abnormally high formation pressure based on technological drilling data: Neftyanoe Khoziaystvo, no.5, p.35–38.

Slavin, V.I., and V.F. Khimich, 1987, Geodynamic models of formation of abnormally high formation pressure and their practical significance, in Study of geologic section and prediction of abnormally high formation pressure: Leningrad, p.42–54.

Slavin, V.I., and L.M. Bruk, 1987, Main hypotheses of development of abnormally high formation pressure and their classification, *in* Izucheniye geologicheskogo razreza i prognozirovaniye AVPD (Study of the geologic section and prediction of abnormally high formation pressure): Leningrad, p.7–22.

Slavin, V.I., O.A. Ulybin, and E.M. Smirnova, 1994, Distribution of abnormally high formation pressures in petroleum-bearing sequences of the Timan-Pechora basin and Barents Sea, *in* M.D. Belonin and V.N. Makarevich, eds., Poiski, razvedka i dobycha nefti i gaza v Timano-Pechorskom basseyne i Barentsevom more (Petroleum exploration and production in the Timan-Pechora basin and Barents Sea): St. Petersburg, VNIGRI, p.137–144.

Sitnikov, M.F., A.L. Volin, and A.K. Kutov, 1971, Rezultaty issledovaniya deformatsii obraztsov porod pri vtorichnom uplotnenii (Investigation results of deformation of rock samples during secondary compaction): RNTS, ser. Burenie, vyp.9, p.33–36.

Swarbrick, R.E. and M.J. Osborne, 1998, Mechanisms which generate abnormal pressures: an overview, *in* Law, B.E., G.F. Ulmishek, and V.I. Slavin eds., Abnormal pressures in hydrocarbon environments: AAPG Memoir 70, p.13–34.

Terzaghi, K., 1923, Die Berechnung der Durchlassigkeitsziffer des Tones aws dem Verlanf der Hydrodynamischen Spannungsercheinungen: Sb. Akad. Wiss., Wein, p.132–135.

Vidrine, D.J., and E.J. Bent, 1968, Field verification of the effect of differential pressure on drilling rate, SPE of AIME 42th Annual Fall Meeting: Houston, Texas, SPE 1859, 11 p.

Belonin, M.D., and V.I. Slavin, 1998, Abnormally high formation pressures in
petroleum regions of Russia and other countries of the C.I.S., in Law, B.E.,
G.F. Ulmishek, and V.I. Slavin eds., Abnormal pressures in hydrocarbon
environments: AAPG Memoir 70, p. 115–121.

Chapter 7

Abnormally High Formation Pressures in Petroleum Regions of Russia and Other Countries of the C.I.S.

M.D. Belonin
V.I. Slavin[1]
VNIGRI
St. Petersburg, Russia

Abstract

Statistical analyses of oil and gas pools in ancient platforms (Precambrian), young platforms (post-Hercynian), and mobile belts (foreland troughs, and intermontane depressions) of the Commonwealth of Independent States (C.I.S.), do not reveal any significant differences in the relationships between commercial production and the magnitude of overpressure. However, pressure data from all three structural provinces indicate 90% of all oil and gas pools are in reservoirs with pressure abnormality coefficients (A_c—the quotient of the measured pressure divided by the hydrostatic pressure) less than 1.8 (0.81 psi/ft, 18.7 kPa/m), suggestive of a commercially practical upper limit for productive reservoirs.

In ancient platforms, more than 90% of hydrocarbon pools are present in reservoirs in which the A_c is less than 1.7 (0.77 psi/ft, 17.8 kPa/m). Oil pools in ancient platform provinces are more common in reservoirs where the abnormality coefficient ranges from 1.1–1.3 (0.50–0.54 psi/ft, 11.5–12.4 kPa/m), whereas most gas pools occur in reservoirs with an A_c ranging from 1.06–1.1 (0.48–0.50 psi/ft, 11–11.5 kPa/m). Very few oil and gas pools in ancient platforms occur in reservoirs with abnormality coefficients greater than 1.6 (0.72 psi/ft, 16.6 kPa/m). On young platforms, the majority of oil pools occur in reservoirs in which the A_c is less than 1.6 (0.72 psi/ft, 16.6 kPa/m). Gas pools in young platforms are found in reservoirs with abnormality coefficients as high as 2.0 (0.9 psi/ft, 20.8 kPa/m). In mobile belt provinces, 90% of oil pools occur in reservoirs with A_c values less than 1.8 (0.81 psi/ft, 18.7 kPa/m). A small number of pools occur in reservoirs where the abnormality coefficient is as high as 2.0 (0.9 psi/ft, 20.8 kPa/m). In mobile belt provinces, 90% of all oil and gas pools occur in reservoirs with A_c values less than 1.8 (0.81 psi/ft, 18.7 kPa/m) and are most common where the abnormality coefficient ranges from 1.2 to 1.3 (0.54–0.59 psi/ft, 12.4–13.6 kPa/m). The statistical data demonstrate that as the A_c increases, the frequency of occurrence of commercial oil and gas accumulations generally decreases. Very few oil and gas pools, regardless of structural setting, occur in reservoirs with abnormality coefficients greater than 1.8 (0.81 psi/ft, 18.7 kPa/m).

INTRODUCTION

The first report of abnormally high formation pressure (AHFP) in the former Soviet Union was recorded on October 14, 1875 in a well drilled by Colonel A.A. Burmeister on the Apsheron Peninsula near Baku, Azerbaijan. Since that time, AHFP has been identified in many areas within the former Soviet Union from a wide range of geologic conditions on ancient (Precambrian) platforms, young (post-Hercynian) platforms, and in mobile belts (geosynclines, foreland troughs, and intermontane depressions). In Russia alone, AHFP has been detected in more than 5,000 wells drilled in

the Timan-Pechora, West Siberian, East Siberian, Volga-Ural, Sakhalin, North Caspian, South Caspian, and North Caucasus petroleum provinces (Figure 1). Although a large amount of undiscovered oil and gas is believed to be present in these provinces, the exploitation of these resources is hindered by the presence of abnormally high formation pressures.

Because of the drilling and completion problems encountered while attempting to exploit undiscovered oil and gas resources in overpressured reservoirs, the purpose of this study is to present the results of statistical analyses on the distribution of oil and gas pools as a function of the magnitude of reservoir pressure in

[1] *Present Affiliation: Consulting Petroleum Geologist, Berlin, Germany*

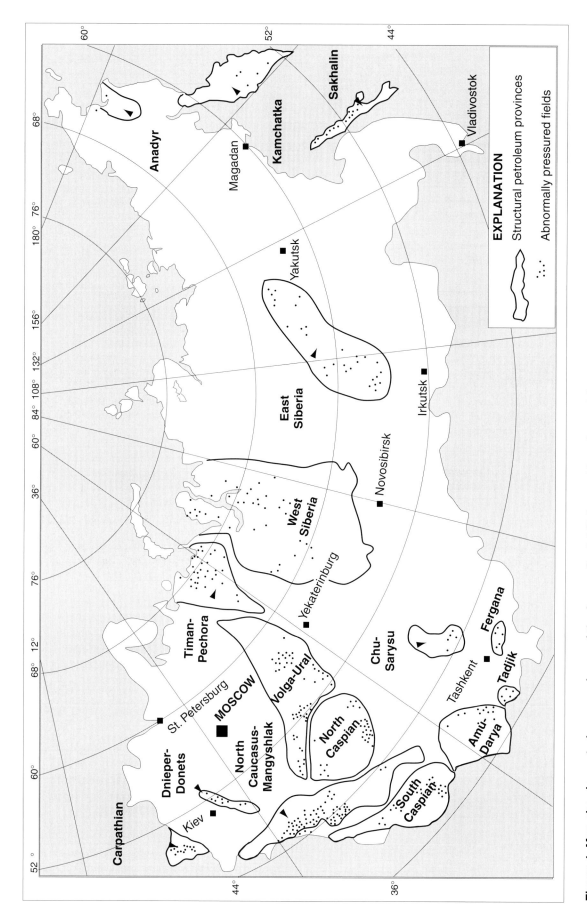

Figure 1. Map showing petroleum provinces of the former Soviet Union and locations of fields with overpressure. Ancient platforms: Timan-Pechora, Volga-Ural, Dnieper-Donets, North Caspian, and East Siberia; Young platforms: North Caucasus-Mangyshlak, Amu-Darya, Chu-Sarysu, and West Siberia; Mobile belts: Carpathian, South Caspian, Tadjik, Fergana, Sakhalin, Kamchatka, and Anadyr.

selected petroleum provinces of the former Soviet Union. The structural provinces used in this study include ancient platforms, young platforms, and mobile belts. Pressures are reported as abnormality coefficients (A_c). The A_c is the quotient of the measured pressure divided by the hydrostatic pressure.

ANCIENT PLATFORM PROVINCES

Timan-Pechora Petroleum Province

Abnormal reservoir pressures in the Timan-Pechora province (Figure 1) are found in reservoirs at depths exceeding 2,500–3,000 m. The abnormally pressured zones in the province are bounded above by either Devonian (Afonin and Kynov-Sargay Horizons) shale seals or anhydrite and dolomite seals which mainly occur in the oil and gas productive Middle Devonian and lower Upper Devonian (lower Frasnian) parts of the sedimentary sequence. The shale seals are regionally extensive, whereas the anhydrite and dolomite seals occur only in the northern part of the province.

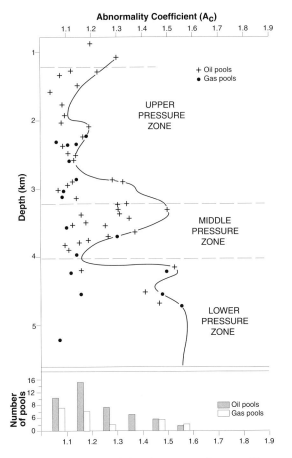

Figure 2. Cross-plot of depth versus abnormality coefficient, showing distribution of oil and gas fields in zones of abnormally high pressure in the Timan-Pechora province. Location of Timan-Pechora province shown on Figure 1.

The cross-plot of the A_c versus depth for 60 oil and gas pools (including gas condensate) in the Timan-Pechora province shows that the entire sedimentary section may be divided into three pressure zones (Figure 2). In the upper zone, sandstones and siltstones are highly permeable and there is fluid communication across the shales. Reservoir pressures in the upper zone are only mildly overpressured with most A_c values less than 1.15 (0.52 psi/ft, 12 kPa/m) and a few values as high as 1.34 (0.60 psi/ft, 13.8 kPa/m). In most pools, however, pressures at the oil-water or gas-water contacts do not exceed hydrostatic pressure.

The middle pressure zone includes Devonian rocks and occurs at depths ranging from 3.2 to 4.0 km. This abnormally pressured zone is bounded above by the Kynov-Sargay regional shale seal. The zone contains a larger number of oil and gas pools and higher pressures than the upper pressure zone. However, pools with pressures at or slightly above hydrostatic are also present in the zone. We suggest that the low pressures in this zone are due to localized leakage through the seals. The A_c of oil pools in this zone reaches 1.5 (0.68 psi/ft, 15.7 kPa/m) and the A_c in gas pools does not exceed 1.4 (0.63 psi/ft, 14.5 kPa/m). In the lower pressure zone, at depths greater than 4 km, fluid pressures are highly variable. Oil and gas pools in this pressure zone have A_c values ranging from 1.1 to 1.6 (0.50–0.72 psi/ft, 11.5–16.6 kPa/m).

The histogram shown on the bottom of Figure 2 indicates that the largest number of oil and gas pools occur in slightly overpressured reservoirs with A_c values less than 1.2 (0.54 psi/ft, 12.4 kPa/m). Further increases in pressure result in a gradual decrease in the number of pools. There are no pools with an A_c greater than 1.6 (0.72 psi/ft, 16.6 kPa/m).

North Caspian Petroleum Province

In the North Caspian province (Figure 1), the A_c in overpressured reservoirs generally increases with increasing depth (Figure 3). However, the difference in the A_c between pools located above and below the regional salt seal is much greater than the difference between upper and lower pressure zones in the Timan-Pechora province. In pools above the seal and in reservoirs within the seal, the A_c commonly does not exceed 1.5 (0.68 psi/ft, 15.7 kPa/m). Below the seal, the A_c may be as high as 2.1 (0.95 psi/ft, 21.9 kPa/m). The pressure differences between the supra- and sub-salt rocks indicate that the sedimentary sequence can easily be divided into two pressure zones separated by the Kungurian (Lower Permian) salt seal (Figure 3). Abnormality coefficients in oil pools in the upper zone gradually increase with depth from 1.1 to 1.5 (0.50–0.68 psi/ft, 11.5–15.7 kPa/m), whereas the abnormality coefficients increase abruptly from 1.5 to 2.1 (0.68–0.95 psi/ft, 15.7–21.9 kPa/m) in the lower zone (Figure 3). The very high A_c values (2.1) in the lower pressure zone probably reflect the high efficiency of the salt seal.

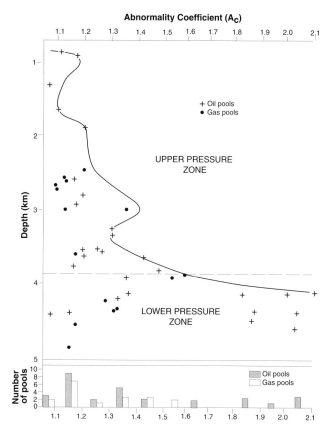

Figure 3. Cross-plot of depth versus abnormality coefficient, showing distribution of oil and gas fields in zones of abnormally high pressure in the North Caspian province. Location of Caspian province shown on Figure 1.

YOUNG PLATFORM PETROLEUM PROVINCES

West Siberian Petroleum Province

In most of the West Siberian province (Figure 1), the distribution of abnormally high reservoir pressures (Figure 4) is determined by the efficiency of the Upper Jurassic-Valanginian shale seal. The top of the abnormally pressured zone is most commonly coincident with the bottom of the seal or within the seal. Isolines of A_c generally parallel structural contour lines of the shale seal.

Abnormally high formation pressures above the Upper Jurassic-Valanginian regional seal have been reported in the Yamal-Taz region. For example, in the Kharasavey field, the A_c at a depth of 1,500 m is 1.15 (0.52 psi/ft, 12 kPa/m) and increases with increasing depth to 1.7 (0.77 psi/ft, 17.8 kPa/m) before decreasing to 1.25 (0.56 psi/ft, 12.9 kPa/m) at even deeper depths. Abnormally high reservoir pressures also occur in the Lower Cretaceous Achimov and Megion Formations in the Urengoy field, above the regional seal. Abnormali-

ty coefficients in the Kharasavey and Urengoy fields range from 1.3 to 1.5 (0.59–0.68 psi/ft, 13.6–15.7 kPa/m). The Kharasavey and Urengoy fields are characterized by hydraulic communication between reservoirs above and below the regional seal.

The vertical distribution of overpressured pools through the sedimentary sequence in the West Siberian province indicates that there are two pressure zones (Figure 4). The upper zone in the province is characterized by pressures slightly above hydrostatic, with abnormality coefficients less than 1.35 (0.61 psi/ft, 14 kPa/m). The lower pressure zone occurs at depths of 2.0 to 3.8 km and consists of the Achimov, Bazhenov, Tyumen, and Vasyugan Formations. The lower zone contains nearly all of the overpressured oil pools in the province. The abnormality coefficients in pools within this zone decrease with depth from 1.95 to 1.7 (0.88–0.77 psi/ft, 20.3–17.8 kPa/m) (Figure 4). The decreasing pressure gradient with increasing depth is a distinctive characteristic of the West Siberian province. The large changes of pressure across the shale seal in this province implies the presence of an effective seal.

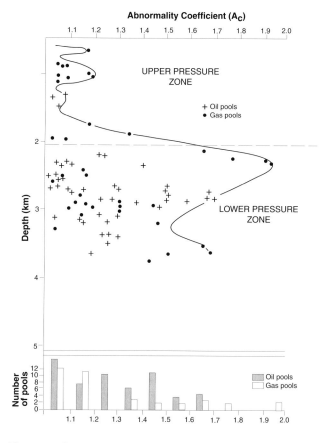

Figure 4. Cross-plot of depth versus abnormality coefficient, showing distribution of oil and gas fields in zones of abnormally high pressure in the West Siberian province. Location of West Siberian province shown on Figure 1.

MOBILE BELT PETROLEUM PROVINCES

South Caspian Petroleum Province

In the South Caspian petroleum province (Figure 1), the principal oil and gas producing reservoirs are in Oligocene and Pliocene orogenic clastic red beds. The sequence is as thick as 4 km and consists of alternating sandstone and shale. This sequence contains 70% of all hydrocarbon pools in the province. Commonly, the fields contain several different reservoirs distributed throughout the sequence. The South Caspian province is also well known for its mud volcanoes which have a bearing on the pressure history of the province.

A plot of the distribution of abnormally pressured pools (Figure 5) show that there are no clearly defined pressure zones in this province and there is an irregular decrease, with depth, in A_c from 1.1 to 2.3 (0.50–1.03 psi/ft, 11.5–21.6 kPa/m). It should be noted that the higher pressures reported here are slightly above lithostatic. About 60 % of the oil and gas pools occur within abnormally high pressured reservoirs with abnormality coefficients ranging from 1.2 to 1.6 (0.54–0.72 psi/ft, 12.4–16.6 kPa/m) (Figure 5).

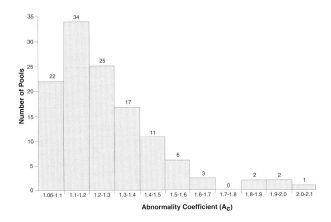

Figure 6. Histogram showing distribution of oil pools in zones of abnormal pressure on ancient platforms.

The apparent absence of a regional seal may provide some information concerning the origin, or cause of overpressure in this province. We suggest that the orogenic Oligocene and Pliocene sediments were deposited rapidly, resulting in a compaction disequilibrium state, where much of the weight of the overburden is borne by water in the pore system. This disequilibrium state may not require a pressure seal if the condition is dynamic. In addition, lenticular mudstone beds may contribute to sealing efficiency.

DISCUSSION

In ancient platform provinces, the A_c in pools increases with increasing depth (Figures 2 and 3). The pressure gradient is variable and is primarily dependent on the efficiency of seals. Hydrocarbon pools on ancient platforms are found chiefly in overpressured rocks with abnormality coefficients ranging from 1.06 to 1.7 (0.48–0.77 psi/ft, 11–17.8 kPa/m) (Figures 6 and 7). The largest number of oil pools in ancient platforms occur within the A_c interval 1.1–1.2 (0.50–0.54 psi/ft, 11.5–12.4 kPa/m) (Figure 6) and the majority of gas pools occur within the A_c interval 1.06–1.1 (0.48–0.50 psi/ft, 11–11.5 kPa/m) (Figure 7). The number of oil pools gradually decreases from 34 in A_c interval 1.1–1.2 (0.50–0.54 psi/ft, 11.5–12.4 kPa/m), to 3 in A_c interval 1.6–1.7 (0.72–0.77 psi/ft, 16.6–17.8 kPa/m) (Figure 6), and the number of gas pools in this province decreases from 25 in A_c interval 1.06–1.1 (0.48–0.54 psi/ft, 11–12.4 kPa/m), to 8 in A_c interval 1.5–1.6 (0.68–0.72 psi/ft, 15.7–16.6 kPa/m) (Figure 7). In ancient platforms, 94% of all oil pools occur in reservoirs with A_c values less than 1.7 (0.77 psi/ft, 17.8 kPa/m) and 92% of all gas pools occur in reservoirs with A_c values less than 1.6 (0.72 psi/ft, 16.6 kPa/m).

On young platforms, 90% of all oil pools are in reservoirs with A_c values less than 1.8 (0.81 psi/ft, 18.7 kPa/m) (Figure 8). However, at abnormality coefficients greater than 1.5 (0.68 psi/ft, 15.7 kPa/m), oil pools are much less common. The maximum number

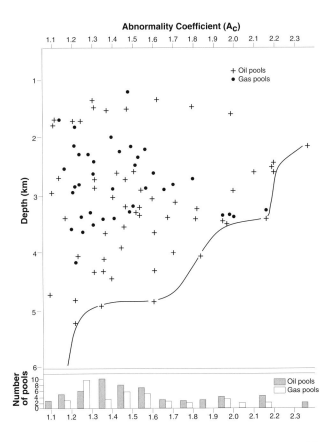

Figure 5. Cross-plot of depth versus abnormality coefficient, showing distribution of oil and gas fields in zones of abnormally high pressure in the South Caspian province. Location of South Caspian province shown on Figure 1.

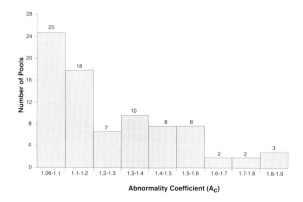

Figure 7. Histogram showing distribution of gas pools in zones of abnormal pressure on ancient platforms.

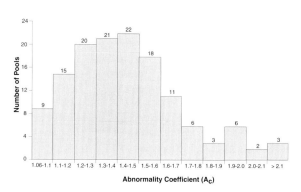

Figure 10. Histogram showing distribution of oil pools in zones of abnormal pressure in mobile belts.

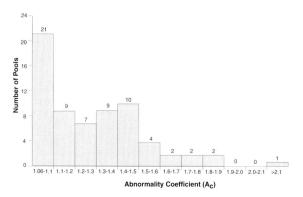

Figure 8. Histogram showing distribution of oil pools in zones of abnormal pressure on young platforms.

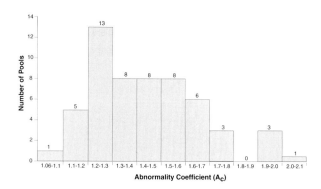

Figure 11. Histogram showing distribution of gas fields in zones of abnormal pressure in mobile belts.

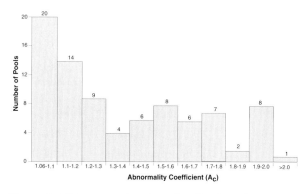

Figure 9. Histogram showing distribution of gas pools in zones of abnormal pressure on young platforms.

of gas pools occur in the A_c interval 1.06–1.1 (0.48–0.50 psi/ft, 11–11.5 kPa/m) (Figure 9). Slightly less than 90% of all gas pools occur in reservoirs with A_c values less than 1.8 (0.81 psi/ft, 18.7 kPa/m).

The distribution of oil and gas pools in mobile belts is similar to the distributions in ancient and young platforms. In mobile belts, 90% of all oil pools are found in reservoirs with A_c values less than 1.8 (0.81

psi/ft, 18.7 kPa/m) (Figure 10) and 93% of all gas pools occur in reservoirs with A_c values less than 1.8. (0.81 psi/ft, 18.7 kPa/m) (Figure 11).

The statistical data from the three structural provinces indicate that the largest number of oil pools occur in reservoirs with abnormality coefficients ranging from 1.06 to 1.8 (0.48–0.81 psi/ft, 11–18.7 kPa/m) in mobile belts, 1.06 to 1.5 (0.48–0.68 psi/ft, 11–15.7 kPa/m) in young platforms, and 1.06 to 1.6 (0.48–0.72 psi/ft, 11–16.6 kPa/m) in ancient platforms (Table 1). The distribution of gas pools in the structural provinces is similar.

There do not appear to be statistically significant differences in the distribution of oil and gas pools among the different structural provinces (Table 1). Most oil and gas pools occur in reservoirs that are only mildly overpressured (1.1–1.3 A_c), and in reservoirs that exceed A_c values of 1.8 (0.81 psi/ft, 18.7 kPa/m), there is a low probability of commercial production. Over 90% of all commercial oil and gas production is from reservoirs with A_c values less than 1.8 (0.81 psi/ft, 18.7 kPa/m) (Table 1). Reservoir pressure differences between provinces may be due to the nature and integrity of the seal, causes of abnormal pressure, or the duration of pressure development.

SUMMARY

These statistical data demonstrate that as the abnormality coefficient increases, the frequency of occurrence of commercial oil and gas accumulations generally decreases. There do not appear to be any significant differences among the three structural provinces between the occurrence of commercial oil and gas pools and magnitudes of overpressure. Most oil and gas pools occur in mildly overpressured reservoirs where the A_c ranges from 1.06 to 1.3 (0.48–0.54 psi/ft, 11–12.4 kPa/m). In all structural provinces examined, an abnormality coefficient value of about 1.8 (0.81 psi/ft, 18.7 kPa/m) appears to be the practical threshold for the occurrence of commercial oil and gas production. Although there are a few oil and gas pools in reservoirs with abnormality coefficient values greater than 1.8 (0.81 psi/ft, 18.7 kPa/m), over 90% of all oil and gas pools in the structural provinces examined occur in reservoirs with A_c values less than 1.8 (0.81 psi/ft, 18.7 kPa/m).

Table 1. Summary of relationships between occurrences of oil and gas pools and overpressure in selected structural provinces of the former Soviet Union. A_c = Pressure Abnormality Coefficient (quotient of measured pressure divided by hydrostatic pressure), * Maximum A_c value at which 90% of production occurs.

Province	Oil Pools Pressure Range, A_c	Gas Pools Pressure Range, A_c	Upper Pressure Limit (A_c) of Commercial Production*
Ancient Platforms	1.06 to 2.1	1.06 to 1.9	1.7
Young Platforms	1.06 to 2.1	1.06 to 2.0	1.8
Mobile Belts	1.06 to 2.1	1.06 to 2.0	1.8

Holm, G.M., 1998, Distribution and origin of overpressure in the Central Graben of the North Sea, *in* Law, B.E., G.F. Ulmishek, and V.I. Slavin eds., Abnormal pressures in hydrocarbon environments: AAPG Memoir 70, p. 123–144.

Chapter 8

Distribution and Origin of Overpressure in the Central Graben of the North Sea

Gordon M. Holm
Hydrocarbon Management International Ltd.,
Edinburgh, Scotland

Abstract

The Central Graben is a major hydrocarbon province. Oil, condensate, and gas are found in a variety of horizons ranging in age from the Devonian to Eocene. The principal economic zones are the Upper Jurassic Fulmar Sandstone, the Upper Cretaceous Chalk Group, and the Paleocene Forties Formation.

The Upper Jurassic sandstones in the Central Graben vary from being normally pressured, 0.01 MPa/m (0.45 psi/ft) near the graben margins, to pressure gradients in excess of 0.02 MPa/m (0.867 psi/ft) in the center of the graben. The Paleocene sandstones consist of sheet sandstones forming a normally pressured regional aquifer, and the Chalk Group, where overlain by these sandstones, is similarly normally pressured. In the southern part of the graben, the Paleocene consists of claystones that act as a seal to both overpressure and hydrocarbons in the Ekofisk Formation of the Chalk Group.

The origin of the overpressure was formerly considered to result from compaction disequilibrium. In recent years it has been recognized that hydrocarbon generation can play a crucial role in generating the extreme overpressures that are present in the Jurassic sandstones. The Kimmeridge Claystone Formation is an excellent source rock with a TOC locally greater than 10%. It varies from sub-mature at the graben margins, to late-stage gas generation within the depocenter of the graben.

In highly overpressured rocks, pore pressures may approach the fracture gradient and the dynamic interplay between pore pressure and fracture gradient is indicative of a dynamic overpressure system. In this type of system, the overpressure is produced by the volumetric expansion associated with the generation of oil and gas. When the pore pressure exceeds the minimum confining stress, episodic fluid expulsion occurs through fractures in the seal.

Recognition that the overpressure is generated from hydrocarbon generation in a dynamic system is important both in modeling the distribution and magnitude of the overpressure and to understanding the relationship between migration and distribution of hydrocarbons in the Central Graben of the North Sea.

INTRODUCTION

The Central Graben of the North Sea (Figure 1) is one of the world's most prolific oil producing provinces. It contains major reserves of oil, gas, and condensate present in reservoirs ranging in age from Devonian to Early Eocene. These include super giant fields like Ekofisk (Figure 2), which produces from overpressured Upper Cretaceous chalk, and the Forties field (Figure 2), which produces from the normally pressured, Paleocene Forties Sandstone. A deep gas-condensate play exists in Jurassic and Triassic reservoirs that exhibit pore pressures approaching lithostatic. This play presents technological problems in both drilling and hydrocarbon production. Although the first discoveries were made in the early 1980s, it is only recently that production from the Embla field in Norway (Figure 2) has commenced. This field produces gas and oil from an overpressured pre-Jurassic sandstone which underlies Upper Jurassic claystones.

This paper reviews the distribution and origin of the overpressures within the Central Graben. The author supports the conclusions of a number of recent papers (Buhrig, 1989; Gaarenstroom et al., 1993) that hydrocarbon generation is the principal cause of the overpressures in the pre-Cretaceous formations. When pore pressures exceed the fracture gradient (minimum confining stress), there will be a loss of fluid volume and a reduction of pore pressure. This interdependency between overpressure generation and leakage through

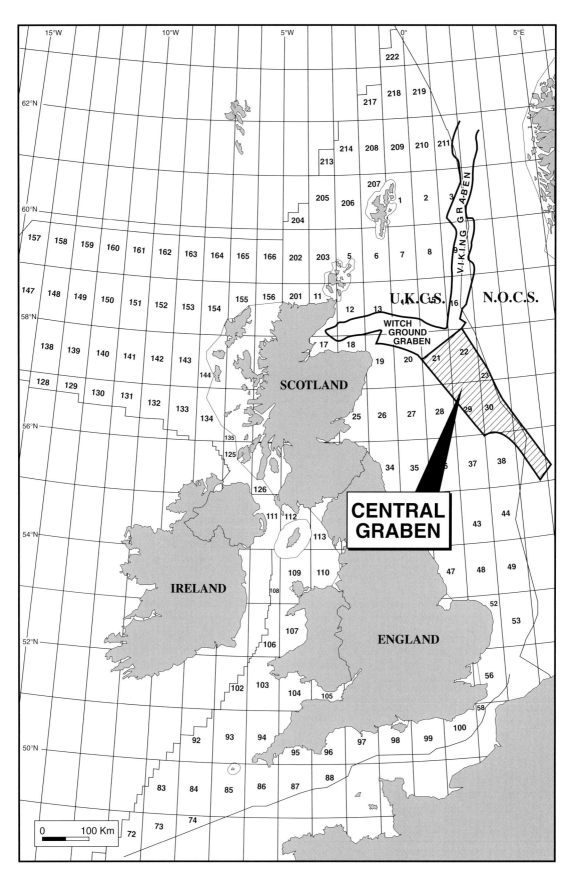

Figure 1. Location Map of the Central Graben and the quadrant licensing structure in the United Kingdom Continental Shelf Area (UKCS).

pressure-induced fractures has been previously recognized (Momper, 1978; Chen et al., 1990; Capuano, 1993; Cartwright, 1994; Roberts and Nunn, 1995; Bredehoeft et al., 1995). This process is similar to the stress cycling effects observed by Sibson (1994) in relation to earthquake activity and a similar mechanism has been recognized in the Central Graben by Holm (1996). Understanding this type of system is important, because it recognizes high overpressures as a key component of hydrocarbon migration.

The Viking Graben (Figure 1) has a similar pattern of overpressure as the Central Graben. The maximum pre-Cretaceous overpressures are present in the deepest part of the graben and similar in magnitude to the Central Graben. The areal distribution of the overpressure forms a north-south trending axial pattern with decreasing pressures towards the graben margins. Overpressure is not normally encountered within the Witch Ground Graben (Figure 1). This may be due to the shallower depth of burial of the pre-Cretaceous

sediments in the Witch Ground Graben compared to the deeper burial in the Viking and Central Grabens.

PREVIOUS WORK

Overpressured rocks were recognized during early exploration of the North Sea. Initially, comparisons were made to overpressured systems in the Gulf of Mexico and the overpressure was considered to be the result of compaction disequilibrium. With additional data, and in particular, the discovery of extreme overpressures in the Central Graben, it became obvious that the origin of overpressures was considerably more complex than initially considered. Currently, hydrocarbon generation is considered to be an important mechanism for generating overpressures.

Carstens (1978) discussed the Tertiary of the Norwegian Central Graben, while other early work on overpressure in the North Sea focused on the Viking

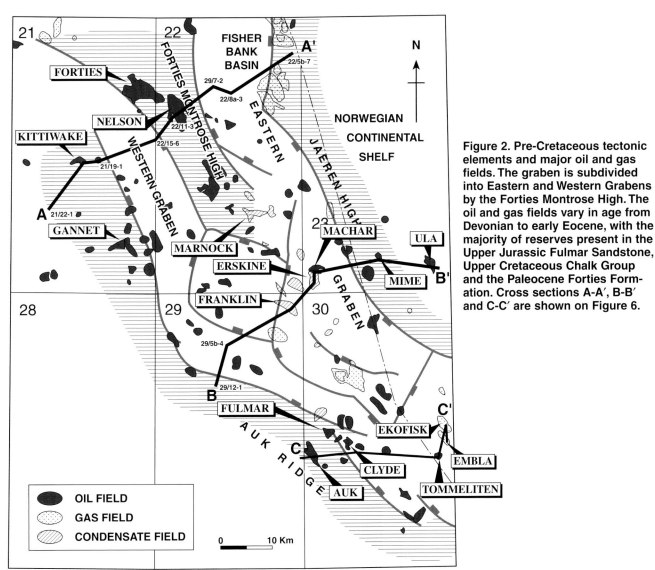

Figure 2. Pre-Cretaceous tectonic elements and major oil and gas fields. The graben is subdivided into Eastern and Western Grabens by the Forties Montrose High. The oil and gas fields vary in age from Devonian to early Eocene, with the majority of reserves present in the Upper Jurassic Fulmar Sandstone, Upper Cretaceous Chalk Group and the Paleocene Forties Formation. Cross sections A-A', B-B' and C-C' are shown on Figure 6.

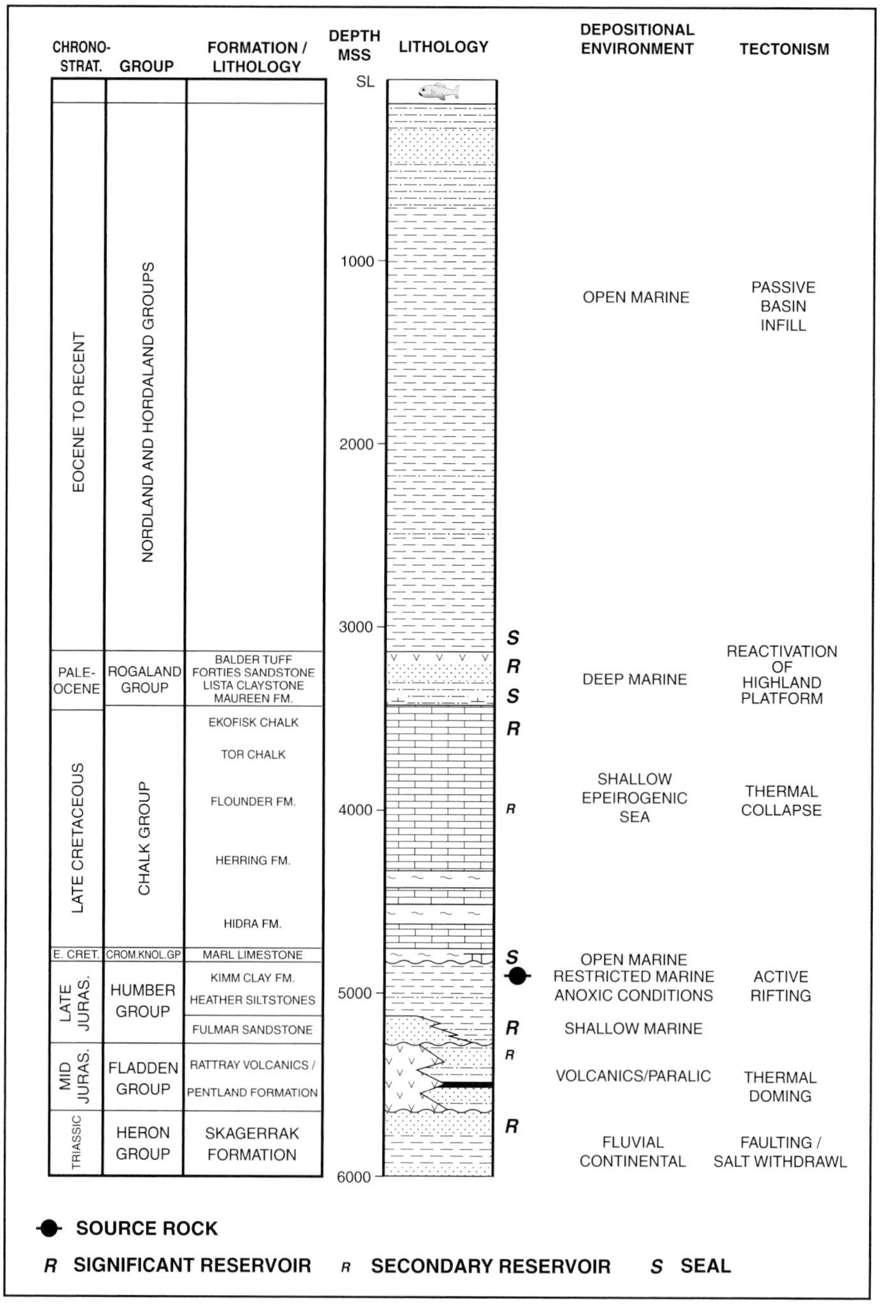

Figure 3. Generalized geological column for the Central Graben. This is typical of the geology that may be penetrated in a well drilled within the deep graben areas. On the graben margins, there are subcrops of formations dating from the Devonian to the Permian.

Graben, (Chiarelli and Duffraud, 1980; Carstens and Dypvik, 1981). These authors all considered that overpressures were related to compaction disequilibrium. Goff (1983), studied overpressure, associated with the expulsion and subsequent migration of oil from the Kimmeridge Claystones of the East Shetland Basin and Viking Graben. He believed the origin of the overpressure was compaction disequilibrium.

An important paper on overpressure in the Central Graben was published by Cayley (1987), who mapped the overpressure at two horizons, the top of the Jurassic and the base of the Paleocene. He noted that lateral and vertical seals must exist within the Jurassic, to prevent dissipation of overpressures at the graben mar-

gins. Cayley (1987) considered that the overpressuring was a result of both claystone undercompaction and aquathermal expansion.

Buhrig (1989) studied the Norwegian sectors of the Viking and Central Grabens and introduced a number of concepts critical to understanding the origin and retention of overpressure in the North Sea. He proposed two models to explain the association between rapid increases in pore pressure gradients and gas generation. First, he considered areas of moderate overpressure (up to 0.015 MPa/m (0.65 psi/ft)) to result from a combination of rapid loading, aquathermal expansion and oil generation. Secondly, he assumed that overpressure gradients greater than 0.02 MPa/m

(0.78 psi/ft) in the Jurassic sections of the deep graben, resulted from gas generation. This was the first publication to discuss the concept of hydrocarbon generation as an overpressuring mechanism in the North Sea. Buhrig (1989) also noted the presence of gas chimneys overlying areas of extreme overpressure and considered that the these might indicate vertical hydrocarbon migration following local seal failure.

Leonard (1993), in a study of the Norwegian Central Graben, subdivided the observed pressure regime into three vertically stacked compartments, the Tertiary, the Chalk Group and the pre-Cretaceous. The uppermost compartment is normally pressured, the second is overpressured by 15-20 MPa (2,500-3,000 psi), and the lowermost compartment is overpressured by 38-52 MPa (5,500-7,500 psi). Leonard (1993) believes these high overpressures in the lowermost compartment are the result of thermal cracking from oil to gas. Hunt (1990) and Leonard (1993) considered that the boundaries to these pressure compartments cut across lithological boundaries. Gaarensstroom et al. (1993) studied the relationship between the generation of overpressure and the sealing mechanisms in the Central Graben of the U.K.. They proposed that the evolution of overpressure cells in the Central Graben is the result of three processes. In their work, overpressure was initiated by rapid sedimentation and burial of low permeability claystones. This early pressure history was followed by an increase in the magnitude of the overpressure as a consequence of the onset of hydrocarbon generation (source rock catagenesis). Finally, as the pore pressure continued to increase, the fracture gradient (minimum confining stress) was exceeded with subsequent loss of fluid and pressure. Gaarensstroom et al. (1993) assumed that these intermittent losses occurred over several million years and were related to Tertiary tectonic events. Holm (1996) has expanded on the association of overpressure and intermittent seal failure and termed this a dynamic overpressure system.

In summary, hypotheses of overpressure development and maintenance in the Central Graben have evolved from models of compaction disequilibrium to more complex models. These more recent pressure models account for compaction, aquathermal expansion and hydrocarbon generation, and include dynamic models where a variety of pressure-generating processes are associated with episodic breaching and healing of seals.

GEOLOGICAL SETTING OF THE CENTRAL GRABEN

For a detailed review of the geology of the North Sea, the reader is referred to Glennie (1990). The following is a short description of the geological history of the Central Graben and its relevance to overpressure generation and maintenance.

The Central Graben of the North Sea is a failed rift that trends southeast, separating the basement areas of Norway from the Caledonides of Scotland (Figure 2). There is some uncertainty regarding the timing of rifting. Fisher and Mudge (1990) believed that rifting commenced in the Triassic, while Cornford (1994) considered that the principal phase of rifting occurred during the Late Jurassic. Cornford (1994) argues rifting followed thermal upwelling in the Middle Jurassic (Figure 3). A major marine incursion occurred in the Late Jurassic, coeval with normal faulting of the rifting phase. Following the rift development, subsidence, possibly associated with thermal collapse, continued through the Cretaceous and Tertiary to the present day. The geological history of the Central Graben can be neatly divided into pre-, syn- and post-rift.

Pre-Rift Sediments

Pre-rift sediments, varying in age from Devonian to Middle Jurassic, are found in the Central Graben. The Devonian consists of sandstones deposited in an eolian environment. The Carboniferous rocks consist of interbedded sandstones, claystones and coals that were deposited in a paralic environment. Lower Permian rocks are composed of sandstones deposited in an eolian environment. A major evaporite basin formed in the Late Permian and thick halite layers were deposited. Halokinesis, commencing in the Triassic, affected the deposition pattern of many younger sediments. Regional desiccation continued through the Triassic, with deposition of red, continental claystones with occasional fluvial sandstones.

Lower Jurassic rocks are missing throughout most of the Graben, because of non-deposition or deposition and subsequent erosion, associated with Middle Jurassic thermal doming. The Middle Jurassic Fladden Group consists of the volcanic Rattray Formation, and a sequence of paralic sediments termed the Pentland Formation (Figure 3).

Reservoirs are present throughout the pre-rift sequences. Hydrocarbons are found within rocks of Devonian to Permian age in the Auk and Argyll fields, the Skagerrak Formation of the Triassic in the Marnock and Josephine oil fields, and the Middle Jurassic Pentland Formation (Figure 2). The significance of these discoveries is put in context by Cornford (1994) who estimates that only 3% of the in-place resources of the Central Graben are present in pre-rift formations.

Syn-Rift Deposits

The Late Jurassic was a period of eustatic sea level rises. Deposition during this time commenced with the Fulmar Sandstone, a series of progradational sandstone packages. Regional deposition of organic-rich siltstones of the Heather Formation and the Kimmeridge Claystone Formation (Figure 3) followed. The Kimmeridge Claystone Formation is the source rock for all the hydrocarbons discovered in the Central Graben (Cornford, 1994).

The Fulmar Sandstone was deposited from the Callovian to the Volgian, coeval with active fault block rotation and salt withdrawal. The lithology consists of a medium- to fine-grained sandstone with extensive bioturbation, which has destroyed the original sedimentary structures. Where the sandstone has been preserved, it varies in thickness from a thin basal lag, to over 200 meters. Donovan et al. (1993) believe the Fulmar Sandstone was deposited as a series of wave-dominated bars with subsequent truncation by intra-Kimmeridgian unconformities. An alternative model assumes that the sands were deposited below wave base, by intermittent storm events, but remained within an aerobic environment, where extensive bioturbation occurred. The Fulmar Sandstone is a good reservoir with average porosity varying from 15-25% (Stevens and Wallis, 1991). The discontinuity of the Fulmar Sandstone packages results in poor pressure and fluid communication. Thus, within pre-Cretaceous

reservoir rocks the overpressure is present in discrete pressure cells (Gaarensstroom et al., 1993).

The Kimmeridge Claystone Formation (KCF), which is equivalent to the Mandal Formation in Norway, is the major source rock in the North Sea. The depth of the KCF ranges from less than 2,000 m (6,600 ft) to more than 5,000 m (16,400 ft) and reaches a maximum thickness of more than 1,200 m (3,900 ft).

The KCF consists mainly of Type II kerogen and has an average total organic carbon (TOC) of 5.5% with a maximum of 15% (Cornford, 1994). The KCF thermal maturity, based on vitrinite reflectance (Figure 4), indicates that at the graben margin, where the KCF is shallow, it is immature to early-mature for oil. In the deep graben areas, the KCF is gas generative to post-mature. Cornford (1994) calculated that for a deep basinal position, oil generation would have commenced at approximately 100 Ma, peaking at 30 Ma, and is currently in the late stages of gas generation. He estimates the transition from late-mature for oil, to wet gas generation should occur at approximately 3,800 m (12,500 ft). For

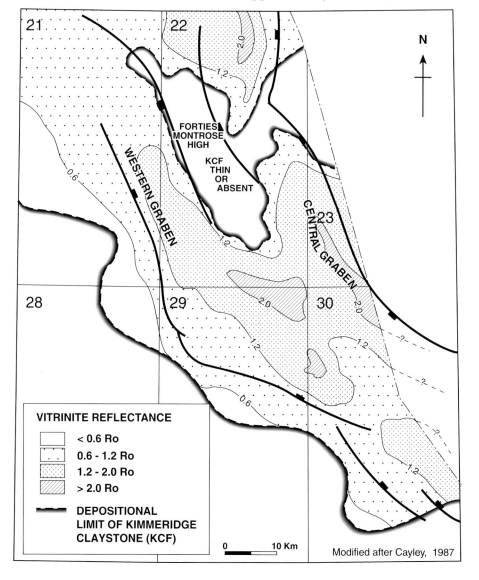

Figure 4. The distribution of the Late Jurassic Kimmeridge Clay Formation and current level of thermal maturity. Modified from Cayley (1987).

VITRINITE REFLECTANCE

< 0.6 Ro

0.6 - 1.2 Ro

1.2 - 2.0 Ro

> 2.0 Ro

DEPOSITIONAL LIMIT OF KIMMERIDGE CLAYSTONE (KCF)

0   10 Km

Modified after Cayley, 1987

a comprehensive discussion of the "Mandal-Ekofisk" petroleum system, refer to Cornford (1994).

Within the Upper Jurassic rocks, there are occasional, thin isolated sandstones found above the Fulmar Sandstones, known as Ribble and Freshney Sandstones (Richards et al., 1993). These sandstones are very clean, with sharp upper and basal contacts, and are considered to have been deposited by turbidity currents. The Ribble and Freshney Sandstones are typically completely encapsulated by the claystones and pressures measured within them are considered to give a good approximation of the pore pressure within the KCF.

Post-Rift Sediments

Following the active rifting phase of the Central Graben, a thermal subsidence phase began that has persisted through to the present day. This subsidence phase resulted in deposition of more than 5,000 m (16,400 ft) of sediments in the depocenter of the basin.

At the start of the Cretaceous Period a pattern of more open water circulation was established and marls and claystones of the Comer Knoll Group were deposited. These claystones and marls are considered by most workers to be seals for the underlying pre-Cretaceous overpressured rocks.

The Upper Cretaceous Chalk Group was deposited during a period of shallow warm sea conditions with little influx of clastic material from surrounding land masses. The chalk represents a pelagic ooze deposit formed from the skeletal remains of microscopic plates (coccoliths) of planktonic marine algae (Hancock, 1990).

Within the Chalk Group there are local deposits of marls and reworked chalks, the latter resulted from the slumping of the graben sides. The principal fields producing oil from the Chalk Group are in the Ekofisk complex of Norway (Figure 2), where porosities in the chalk reach 30-40% (Hancock, 1990). These porosities are considerably higher than could be expected at these depths under normal conditions of compaction and diagenetic alteration. It is considered likely that a combination of slumping, overpressure, and early hydrocarbon charge have been significant in porosity preservation (Hancock, 1990; Cornford, 1994). Despite the high porosity, the permeabilities are low, less than ten millidarcys (md). Drill stem test (DST) permeability calculations, of up to 150 md, are believed to be the result of fracture enhancement (Cornford, 1994).

The Paleocene Rogaland Group marks a change in post-rift sedimentation. Reactivation of the Caledonides to the west in the Scottish Highlands resulted in erosion and related deposition, through turbidity currents, into the basin (Lovell, 1990). Sandstones within the Paleocene Montrose and the Rogaland Groups were derived from the north and north-west and flowed southwards into the Central Graben. These sandstones form a normally pressured regional aquifer which acts as a pressure sink. At the southern end of the Graben, the Rogaland Group consists of claystones,

which form the seal to the overpressured chalk in the Ekofisk fields.

The Hordaland and Nordland Groups consist of claystones, siltstones, and sandstones that are over 3,300 m (10,800 ft) thick (Figure 3). This loading of the basin has been significant in increasing the maturity of the KCF and the volume of source rock actively generating hydrocarbons.

Halokinesis of the Permian Zechstein salt commenced during the Triassic and continues to the present day. The principal periods of halokinetic activity, determined by the shapes of diapirs and turtle-backs, were Triassic, Late Jurassic, and Eocene to Present.

METHODS OF STUDY

Formation Pressure Data

Data were acquired from released wells in the United Kingdom Continental Shelf (UKCS) records and supplemented with available published data. Data for the Norwegian Continental Shelf are from published information principally related to fields and discoveries.

Pressure data can be obtained from repeat formation testers (RFT), drill stem tests (DST), mud data, and "kicks" (influxes of pore fluids into the wellbore). The most accurate data are pressure readings made with the RFT or its equivalent. The RFT takes a pressure reading within a permeable formation, by setting a seal at a precise depth, determined by a gamma ray correlation curve. Numerous readings can be taken, thus allowing determination of in situ fluid density and, by analogy, gas-oil contacts (GOC) and oil-water contacts (OWC). The nature of the pressure build-up is observed and this allows discrimination between good formation pressure measurements, and those that have not built up fully, or where (RFT) seal failure has occurred. The majority of the deep wells within the UKCS Central Graben have extensive suites of RFT pressure measurements.

Drill stem tests (DST) are taken by perforating a section of cased-off formation and then flowing the fluid under controlled conditions back to the surface. The pressures are measured using gauges set at the surface and downhole. Surface pressures should never be used if downhole data are available. The pressures obtained during a test are the initial flowing pressure (IFP), when the well is flowed for a short period, usually 5 minutes, and the initial shut-in pressure (ISIP), after the pressure has built up for a period of approximately an hour. The ISIP is a good approximation of the formation pressure. During the IFP only a small amount of fluid is displaced, therefore the formation pressure rapidly stabilizes to the undisturbed formation pressure. The main flow period usually extends for 12-24 hours with a build-up period at least twice this time. The final shut-in pressure (FSIP), although often used, is not a true approximation to the original formation pressure, as it is unlikely that pressures will have resta-bilized over the monitoring period. Therefore it is nec-essary to calculate the formation pressure-P* (P star) by use of a Horner Plot.

Even when good downhole pressure measurements have been obtained, there are still a number of prob-lems in calculating formation pressures. The perfora-tions may cover a significant interval of the formation, often more than 30 m (100 ft), and the pressure is esti-mated for the mid-point of the perforations. This intro-duces an error, as it assumes that the pressure is even-ly distributed throughout the perforated interval. In fact, this is rarely the case as some perforations con-tribute significantly more than others. Inaccuracies are also introduced by the position of the gauges, located above the perforated interval. To compensate for the height difference between the perforations and the gauges, the fluid density must be known. This can be estimated but the compressibility of hydrocarbon flu-ids can vary, particularly if free gas is flowing with hydrocarbon liquids. Therefore, DST results must be

reviewed very carefully in order to obtain the best esti-mate of the formation pressure.

Mud weights must be greater than the formation pressure or a kick will occur. Therefore, if there are no pressure measurements, the pressures derived from the mud weight will supply an upper limit of the for-mation pressure within a permeable formation. The accuracy of this method can be improved when "con-nection and trip gases" (increases in the percentage of measured gas following a drill pipe connection or a trip out of the hole) are observed. These increases in gas percentage result from the pressure reduction when the drill pipe is pulled out of the hole. Thus, the pressure exerted by the mud weight is locally reduced and small amounts of gas may be swabbed into the wellbore. This is normally a good indication that the mud weight is close to the formation pressure.

Kicks, an influx of formation fluids, can occur when the formation pressure of a permeable zone is greater than the pressure exerted by the drilling mud. This is a dangerous situation and should be avoided. When the

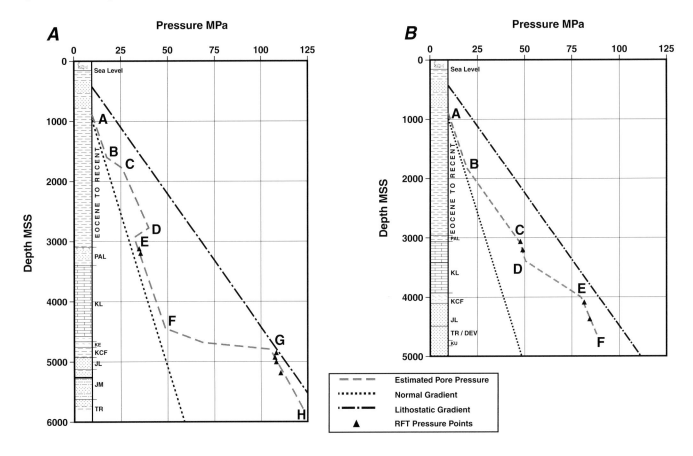

Figure 5. Examples of Pressure-Depth Plots. These have been created using pressure data from permeable form-ations, and estimated within impermeable rocks using indirect information (mud weights, D-exponents, connection gases, etc.). (A) An example of a pressure-depth plot from the deep part of the Central Graben. The significant fea-tures of this plot are the presence of the normally pressured Paleocene sandstones underlain by the normally pres-sured Chalk Group. The transition zone in the Lower Cretaceous (F-G) indicates a rapid build-up of pressure into the pre-Cretaceous. (B) This profile is typical of the Ekofisk Field. The top of the Chalk Group (C-D) is moderately over-pressured as the overlying Paleocene Rogaland Group, consisting of claystones, is an effective pressure seal.

kick is shut-in, pressures are measured at both the drill pipe (SIDP) and casing (SICP). The SIDP is a measure of the pressure at the bottom of the wellbore and the hydrostatic pressure of the vertical column of fluid within the drill pipe. Therefore, the formation pressure can be calculated using these data. SIDP calculations of the formation pressure can be inaccurate, as assumptions are made regarding the depth of the influx and the density of the kick fluid. The fluid density usually varies vertically up through the drill pipe, particularly if the pressure has fallen below the bubble point and gas has separated from oil.

The most accurate pressure measurements are obtained from RFT data, followed by DST data, particularly the initial shut-in pressure or the P* calculated from the Horner Plot. A kick can provide useful data, but the inaccuracies of estimating the density of the fluid column must be taken into account. If no other data is available then mud weights can be used, but these are only likely to approach the formation pressures when connection and trip gases have been observed.

Data from over 100 released wells were examined for this study and wherever possible, RFT data were used for reservoir intervals. When comparing pressures between wells, the pressures were calibrated to remove the effects of hydrocarbon buoyancy. In the construction of the pore pressure profiles, RFT data were used for the permeable formations. The estimation of pore pressure in impermeable horizons was derived from mud weights and drilling monitoring techniques. Particular importance was given to the presence of background and connection gas.

Leak-Off Test Data

The leak-off test (LOT) is conducted at each casing point prior to drilling ahead. The LOT is a measure of the formation strength at the casing shoe. This is the shallowest open hole section, and it is considered, often erroneously, to be the weakest place within the open-hole. The resulting pressure obtained is considered to be an approximation to the fracture gradient of the rock unit.

Following setting and cementing of the casing, the casing shoe and approximately 3 m (10 ft) of new formation is drilled. The well is then shut in at the annulus and excess fluid pumped down the drill pipe. The pressure increases directly in proportion to the amount of fluid pumped. The pressure eventually fractures the formation and some fluid enters the rocks. At this point there is a deviation from the straight line relationship between fluid pumped and pressure increase. This is the point of leak-off. The downhole pressure can be calculated using information from 1) the vertical height of the fluid column, 2) the mud fluid density, and 3) the surface measurement of excess pressure at the point of leak-off.

Measurements from a number of wells allow an approximation of the fracture gradient to be obtained.

The mud weight must always remain below the LOT strength in order to ensure that the borehole rock is not fractured with resulting fluid loss into the formation. It must be remembered that leak-off tests are typically taken in claystones or hard limestones (in order to obtain the maximum shoe strength) whereas fluid pressures are typically measured in sandstones. Sandstones, and other permeable formations, have different mechanical properties from the claystones in which the LOT was conducted. Therefore, in order to ensure that the fracture gradient will not be exceeded, the mechanical properties of the differing formations must also be taken into account. The leak-off test data used in this study were obtained from drilling completion reports or from mud logs.

VERTICAL DISTRIBUTION OF OVERPRESSURE WITHIN THE CENTRAL GRABEN

Two overpressure profiles (pressure vs. depth plots) are shown in Figures 5A and 5B. These pressure profiles represent two examples of a variety of profiles that may be encountered when drilling into the deep Central Graben. The significant points to note are:

- Pressures in pre-Cretaceous rocks can be >100 MPa (14,500 psi), and approach lithostatic pressure.
- The presence of a regional aquifer within the Paleocene rocks serves as a pressure sink.
- The top of the Chalk Group is only overpressured when the overlying Paleocene rocks are composed of claystone, providing a seal for the underlying overpressured rocks.

The first example (Figure 5A) is typical of the deep Central Graben, e.g., the Erskine field (Figure 2). The upper part of the Tertiary (A-B) is normally pressured. In Eocene claystones (C-D) there is evidence of overpressure, interpreted from borehole instability observed during drilling. The abnormally high pressure in these rocks has been interpreted to be due to high rates of sedimentation or compaction disequilibrium (Buchan, 1993).

The Paleocene sandstones form a normally pressured regional aquifer that acts as a pressure sink for high pore pressures that may have been developed in the underlying rocks. Normal pressures continue down through the Chalk Group (E-F). The abrupt transition to the highly overpressured sequence occurs within the lower Upper Cretaceous and the Lower Cretaceous rocks (F-G). The pressure transition is in the Lower Cretaceous marls and the abrupt transition is a function of the efficiency of the seal. The overpressured zone (G-H, Figure 5A) usually commences at the top of the Jurassic rocks and continues into underlying rocks. At the top of the overpressured section, the pore pressure gradient is closest to the fracture gradient. Diffi-

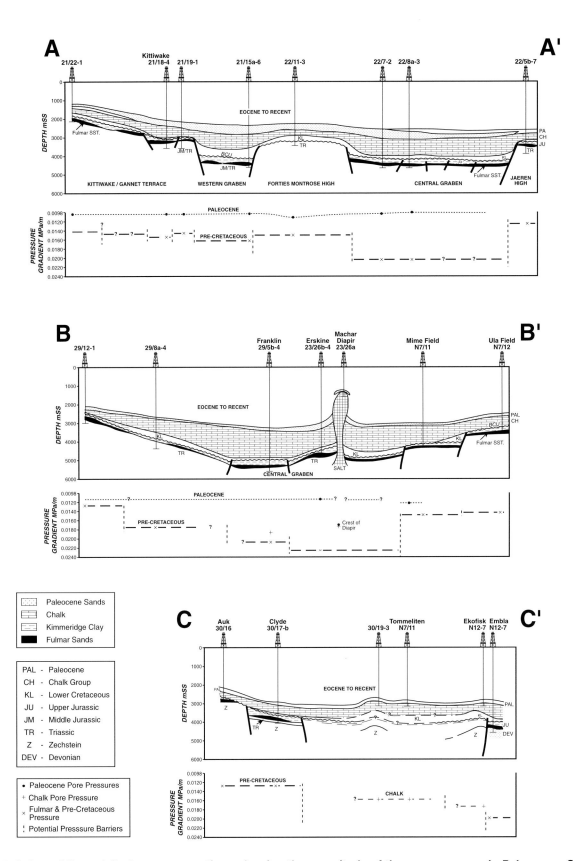

Figure 6. Selected Central Graben cross-sections showing the magnitude of the pore pressure in Paleocene, Cretaceous and pre-Cretaceous rocks. Note that many of the faults appear to be lateral pressure barriers. The locations of the cross-sections are shown in Figure 2.

cult pressure control problems can occur while drilling at this level.

Figure 5B is a pressure profile typical of the Norwegian Ekofisk fields (Figure 2) with normal pressures down to a depth of about 1,800 m (A-B). The pressure then gradually increases into the Chalk Group (B-C) where a pore pressure gradient of 0.016 MPa/m (0.69 psi/ft) is encountered (Leonard, 1993). The pressure transition (D-E) indicates a gradual increase between the moderately pressured Chalk Group into highly overpressured pre-Cretaceous rocks (E-F). The seal is usually marls in the Lower Cretaceous Cromer Knoll Group, and the Upper Cretaceous Chalk Group.

The significant difference between these two example pressure profiles is the absence of Paleocene sandstones in the southern part of the Central Graben. These sandstones act as a pressure sink, and overpressure in the Ekofisk and Tor Formations of the Chalk Group is only present when these overlying Paleocene sandstones are absent and replaced by a shale facies.

Figure 6 shows cross-sections traversing the graben (see Figure 2 for the location of the sections). The upper part of each cross-section shows the geology and lower part of each cross-section shows the corresponding pore pressure. These cross-sections illustrate the difference between the pore pressure in the shallower horizons and also indicates the probability of lateral seals forming pressure cells within pre-Cretaceous rocks.

DISTRIBUTION OF OVERPRESSURE

Maps were created of the overpressure (the difference between the measured pressure and the pressure from a normal hydrostatic gradient [0.01 Mpa/m, 0.45 psi/ft] at the same depth) for the pre-Cretaceous Formations (Figure 7) and the Chalk Group (Figure 8). The magnitude of the overpressure appears to be independent of the fluid phase in the reservoir, and within a single pressure cell, the pressures will lie on the same gradient after compensating for the effects of hydrocarbon buoyancy.

The Pre-Cretaceous Formations

The distribution and magnitude of overpressure within pre-Cretaceous rocks closely follows the structural morphology of the graben. The maximum overpressures are located close to the areas of maximum burial of pre-Cretaceous sediments. The pressures decline laterally toward the margins of the graben and also over the Forties Montrose High (Figures 6 and 7). There are two northwest to southeast trending maxima where overpressure exceeds 40 MPa (5,800 psi). These pressure maxima overlie the western and eastern portions of the Central Graben with the latter continuing south, into the Norwegian Sector.

The "pressure cells" appear to have boundaries which act as barriers to pressure communication (Gaarenstroom et al., 1993). A pressure cell can be considered as a three-dimensionally isolated volume of rock that contains fluid pressures of similar magnitudes. The boundaries to the cell are pressure seals (the transition zone in drilling terminology) where the pressure drop across the seal can be considerable. The seal is unlikely to be perfect, but is more likely a restriction to flow. Intermittent or slow diffusive fluid leakage will occur across the seal, as the higher pressured cell slowly equilibrates to the lower pressures in the adjacent cell. The boundaries separating pressure cells are probably the result of a combination of factors. The most important of these factors are lateral discontinuity of the sandstones and effectively sealed faults, although diagenesis may also contribute in the development of an effective seal.

The overpressure diminishes on either side of the graben. There is a relationship between the backward-stepping fault pattern of the graben and the pressure cells. Pore pressures are reduced toward the flanks of the graben (Figure 6) as there is an apparent tendency for abnormally high pressures to equilibrate to normal pressure conditions. Therefore, the pressure cells probably act as intermediate steps, between the highly pressured axial region and the graben margins. In some cases, faults probably act as pathways for fluid migration out of the deeper parts of the graben, with the high pressure serving as the driving force for hydrocarbon migration from the Kimmeridge (KCF) source rock into shallower, lower pressured reservoirs (Goff, 1983).

The Chalk Group

The Upper Cretaceous Chalk Group can be subdivided into 3 pressure regimes: 1) a normally pressured Chalk regime, 2) an overpressured intra-Chalk regime, and 3) an overpressured Upper Chalk regime.

The normally pressured chalk regime covers a large area of the Central Graben. The chalk in this region is overlain by, and in pressure communication with, normally pressured Paleocene sandstones (Figure 8).

Within the middle of the Chalk Group there are alternations of more permeable chalk and marls and overpressure is sometimes present within these rock units. In the Chalk Group there can be considerable overpressure in basal to middle sections, but, when the chalk is overlain by Paleocene sandstone, pressures at the top of the Chalk Group are usually normal. The intra-Chalk Group overpressure is typically observed in more porous lenses that are encased by impermeable chalk or marl. The compartments are usually very small, and rarely flow significantly on tests. The intra-Chalk Group overpressure appears to occur in areas where the Chalk Group rests unconformably on the Kimmeridge Claystone Formation (KCF). This superposition relationship facilitates the transmission of pressure upwards from the KCF source rocks into the Chalk Group.

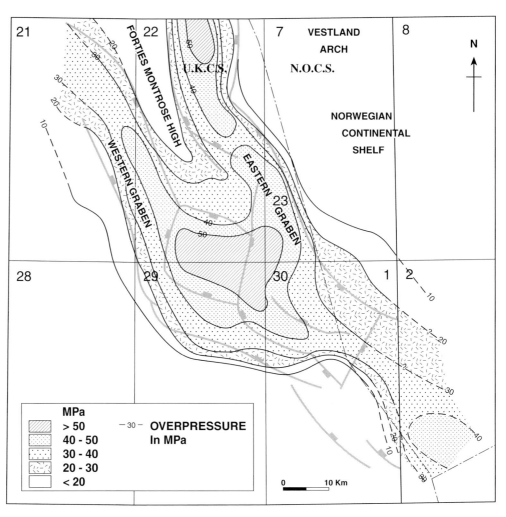

Figure 7. Distribution of overpressure within pre-Cretaceous formations, principally the Upper Jurassic Fulmar sandstone, the Middle Jurassic Pentland Formation and Triassic sandstones. Overpressure is calculated as the difference between the measured pressure and the pressure of a normal gradient (0.01MPa/m), at the same depth. Note the similarities between the pressure contours and the structural features shown in Figure 2.

At the Southern end of the Central Graben, the Ekofisk fields (Figure 2) are a major oil and gas-producing area, and form a distinct province of overpressured chalk. The overpressure is present in high-porosity chalk of the Danian Ekofisk and Maastrichtian Tor Formations of the Chalk Group (Figure 3). Overpressure measurements within rocks of this province are approximately 16 MPa (2,320 psi). Preservation of overpressure in this area is a result of the lack of overlying Paleocene sandstones (Figure 8). In the Ekofisk area, the Paleocene Rogaland Group consists of claystones that serve as an effective pressure seal (Figure 6, Section C-C′).

The Paleocene Rogaland Group

Sandstones of the Paleocene Rogaland Group are typically normally pressured. Exceptions can be found at the limits of sandstone deposition and overlying diapirs. At the margins of the sandstone distribution, some of the sandstones are isolated from the main sandstone body. These isolated sandstones are not in pressure communication with the main aquifer, therefore, they may contain fluid pressures elevated above the normal hydrostatic gradient. Locally, rafts of Pale-

ocene sandstone may overlie diapirs (Figure 6, Section B-B′). In these cases, the overpressure is a result of halokinetic uplift and isolation of the raft from the main sandstone aquifer.

There is a minor increase in the pore pressure overlying the Forties Montrose High (Figure 6, A-A′). This is an area of excellent sand development with good connectivity to the adjacent normally pressured Paleocene aquifer. This local anomaly may possibly be the result of active fluid flow out of the graben onto the structurally high area.

ORIGIN OF THE OVERPRESSURE IN THE CENTRAL GRABEN

Pre-Cretaceous Overpressures

Compaction disequilibrium and hydrocarbon generation are the principal processes responsible for generation of overpressure throughout the Central Graben. The author considers that the current overpressure regime within the pre-Cretaceous section is due to hydrocarbon generation within the Kimmeridge Claystone Formation. The overpressure is not generated directly within the pre-Cretaceous reservoirs, but is probably a result of hydraulic connectivity between the claystones and the sandstones. The pressure transfer most likely occurs during active hydrocarbon generation, as excess fluids are expelled from the claystone into the adjacent reservoirs.

The following evidence in favor of hydrocarbon generation as the primary mechanism for overpressure generation in the pre-Cretaceous, although circumstantial, is considered significant.

The overpressure maxima are coincident with areas of thermally mature Kimmeridge Claystone and, in particular, the extremes of overpressure are associated with areas that are currently gas generative (Figures 4 and 7). (The evidence of gas generation from the KCF in the deep graben is vitrinite reflectances greater than 1.2%, present day temperatures greater than 150°C, and thermal maturity modeling [Cornford, 1994]).

A plot of pre-Cretaceous pressures measured from different wells indicates that there are two overpressure regimes (Figure 9). The shallow overpressure regime occurs at depths between 2,500–3,500 m (8,200–11,500 ft) with a pressure gradient only slightly higher than the hydrostatic gradient. The second pressure regime occurs below 3,500m (11,500 ft), where pressures increase with depth until they are close to the lithostatic gradient. It is probable that these two overpressure populations have different origins.

The Late Jurassic and Early Cretaceous were periods of slow deposition with deposition rates estimated at between 2 and 25 meters per million years. Therefore, compaction disequilibrium, typically associated with high rates of deposition, is an unlikely cause of the pre-Cretaceous overpressure.

The D-exponent, shale factor, and sonic logs in the overlying Lower Cretaceous claystone intervals, do not contain any data that would indicate a deviation from a normal porosity curve.

Therefore, in consideration of these observations, abnormally high pressures in pre-Cretaceous rocks are most likely the result of hydrocarbon generation in the Jurassic Kimmeridge Claystone Formation.

Chalk Group Overpressures

The Chalk Group in the Ekofisk fields of the Central Graben is overlain by thick Tertiary claystones. Overpressures within the Chalk Group are believed to have been derived from compaction disequilibrium (Carstens, 1978; Hancock, 1990), although there is some evi-

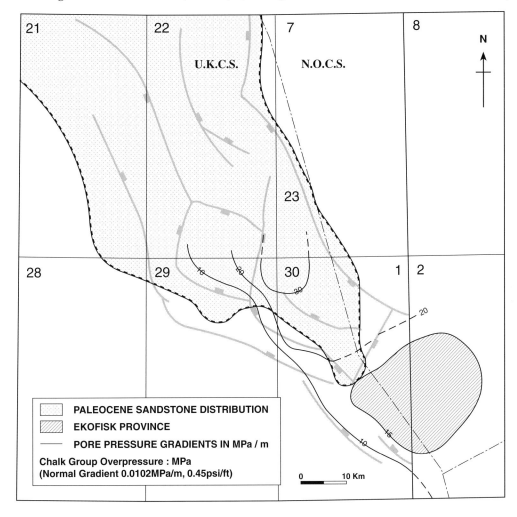

Figure 8. Distribution of Chalk Group Overpressure. Overpressure in the Ekofisk Complex occurs at the top of the Chalk Group, while the pressure measurements in United Kingdom Quadrants 29 and 30 are from measurements within intra-Chalk Group permeable horizons. The distribution of the Paleocene sandstone is outlined. Where the sandstone is present, it acts as a pressure sink, dissipating overpressure from the top of the Chalk Group.

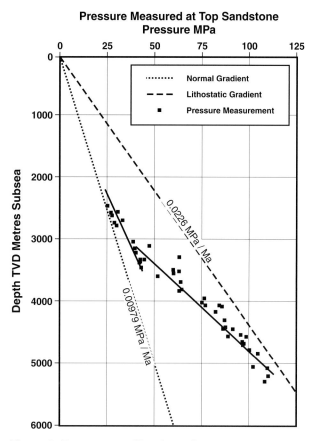

Figure 9. Pressure profile of pre-Cretaceous rocks in the Central Graben. The pressure measurements are from a large number of wells and individual pressure measurements in each well are from within the water leg or corrected for hydrocarbon buoyancy.

dence that hydrocarbon generation may also be a contributing mechanism. The evidence for supporting compaction disequilibrium as the principal mechanism of overpressure in the Chalk Group in the Ekofisk area is as follows:

Carstens (1978) reviewed wireline sonic data for the Tertiary claystones overlying the Cretaceous Chalk Group. Anomalous velocities are present in claystones, overlying the Chalk Group. He assumed that these claystones are in hydraulic connectivity with the underlying Chalk Group. Carstens (1978) concluded that a sonic anomaly, indicative of compaction disequilibrium, was present in the Tertiary claystones, and this was the cause of the overpressure within the Chalk Group.

Hancock (1990) observed that porosities in the Chalk Group of the Ekofisk field are between 30–40% at 2,900 m (9,500 ft). These high porosity values are considerably greater than porosity values of 2-25% that he quotes as being typical for normal compaction curves of chalk at these depths. Hancock (1990) considers that the porosity preservation is the result of early onset of overpressure and the generation of biogenic gas. Cornford (1994) argues that the excess porosity was a result of slumping which delayed water expulsion from the chalk. He believes that overpres-

sure commenced prior to emplacement of the oil. Thus, both of these authors consider that overpressure in the Chalk Group commenced prior to oil generation and emplacement, and that compaction disequilibrium is the dominant overpressuring mechanism.

The chalk is overlain by a thick claystone sequence with continuous sedimentation from the Paleocene to the present. The average sedimentation rate over that period has been calculated at approximately 50 m/million years, based on the present day compaction state of the rocks (if the sedimentation rate had been calculated for a decompacted lithologic column, the values would have been greater). High rates of sedimentation are a feature commonly associated with overpressure generation from compaction disequilibrium. These observations support the notion of an overpressure origin from compaction disequilibrium.

Alternatively, there is evidence that hydrocarbon generation may also contribute to the overpressure system in the Chalk Group. The Kimmeridge Claystone Formation (KCF) has a vitrinite reflectance of approximately 1.2% R_o underlying the Ekofisk fields (Figure 4) and modeling by Cornford (1994) indicates that the KCF is actively generating oil and gas. Furthermore, gas chimneys have been observed overlying the Ekofisk field (Buhrig, 1989). These may indicate that there is a vertical migration route for hydrocarbons generated from the underlying Kimmeridge Clay Formation. Therefore, evidence for the origin of overpressure within the Chalk Group is not conclusive. The author considers that the overpressure was initially generated by compaction disequilibrium, although there may be a contribution from hydrocarbon generation.

Paleocene and Tertiary Overpressures

The Paleocene sandstones are almost always normally pressured except for outliers in distal positions of sheet-sand deposition, and where rafts have been uplifted by salt diapirs (Figure 6, Section B-B').

In the cases of salt rafting, the overpressure is the result of active halokinetic uplift. The sandstones were originally normally pressured but, following uplift and isolation, the pressure was preserved, but at shallower depths. Diapirism has occurred from the Eocene to the present. These diapirs typically pierce through the Cretaceous chalk and Paleocene sandstone into the Tertiary claystones. Paleocene sandstone rafts that have undergone uplift are encased in claystones that act as a seal for pressure maintenance (Figure 6, Section B-B'). There is evidence that these claystones are not perfect seals, as gas chimneys have been observed overlying these diapirs.

The evidence for overpressure within the lower Tertiary claystones is derived from the weight of mud required to prevent shales from spalling into the borehole during drilling. There are few permeable horizons within this section and the author is not aware of direct pressure measurements that indicate overpressure. The apparent overpressures within the lower Tertiary claystones (Figure 5A) may be the result of compactional disequilibrium at the base of this series of claystones.

SEALING MECHANISMS

The seal is critical for the preservation of overpressure, and the significance of the seal has been discussed by a number of authors (Chapman, 1980; Watts 1987; Hunt, 1990). In an active pressure generative situation, overpressure can only be retained, if the rate of pressure generation is greater than the rate of pressure loss through the seal.

It is usually assumed that the top seal is the most important, but in the case of an overpressured cell there must also be bottom and lateral seals. Examples of bottom seals can be found in the Laramide Basins of the Western U.S.A. (Surdam et al., 1994) and the Anadarko Basin (Al-Shaieb et al., 1994). The presence of a bottom seal has not been recognized in the North Sea, probably because most wells in the Central Graben, have not drilled deep enough to penetrate the entire overpressured rock sequence.

Seals in the Central Graben

Evidence of the effectiveness of the seals is obtained from leak-off tests (LOT), which are conducted during drilling. The leak-off pressure is considered to be a good approximation to the minimum horizontal stress (σ_3) and the fracture gradient. A compilation of leak-off tests from numerous wells in the Central Graben indicates the regional strength of the formations at different depths (Figure 10). The leak-off data is concentrated in three groups, related to the setting depths of the casing shoes. The 20″ (50.8 cm) casing shoe is set at approximately 1,000m (3,300 ft), the 13³/₈″ (34 cm) between 2,000 and 3,500 m (6,600–11,500 ft) and the 9⁵/₈″ (24.4cm) shoe set below 3,500 m (11,500 ft). Leak-off tests at the 13³/₈″ (34 cm) shoe are either within the Tertiary claystones or in the top of the Chalk Group. The leak-off tests for the 9⁵/₈″ (24.4 cm) shoes are principally from the Lower Cretaceous Cromer Knoll Group, although they have also been conducted within the Kimmeridge Claystone.

Figure 10 indicates that below 4,000 m (13,100 ft) the fracture gradient is close to the lithostatic gradient and at these depths the minimum horizontal stress (σ_3) and the vertical stress (σ_1) converge. Therefore, at these depths both vertical and horizontal fractures may be generated, given sufficiently high pressures.

The principal top seal for pre-Cretaceous overpressured rocks is the Lower Cretaceous Cromer Knoll Group. This unit consists of claystone and marls with some minor limestones. It varies in thickness from 1-2 m (3–6 ft), on structural highs, to over 1,000 m (3,300 ft) in structurally low positions (Johnson and Lott, 1993). Below 3,000 m (9,800 ft), the claystones and marls tend

to be highly lithified and serve as hydraulic seals (Watts, 1987). Fluid (and pressure) loss through this interval is principally through pressure-induced fractures at the crests of structures.

The Upper Cretaceous Chalk Group may act as both seal and migration pathway. There are alternations of high and low permeability zones within the Chalk Group, and as the underlying pressures equilibrate to a normal pressure gradient, leakage occurs across the low permeability layers. It is uncertain whether the pressure reduction occurs through a network of fractures (micro-fractures) or by diffusion through clay-rich lithologies. Hunt (1990) considered the Ekofisk Province to be a sealed system with lateral and basal seals formed by diagenesis. Evidence that fluids can migrate vertically through the Chalk Group is provided by the presence of oil and gas in chalk of the Ekofisk fields and the Paleocene sandstones. Geochemical analyzes indicate that the oil and gas in the Ekofisk fields

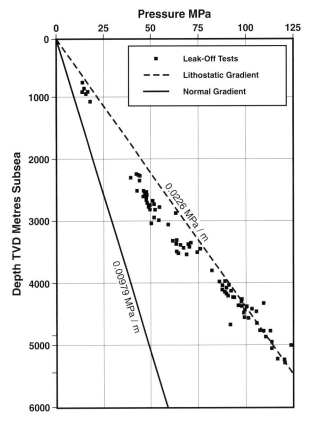

Figure 10. Cross-plot showing Leak-Off Test (LOT) values from wells drilled in the Central Graben. The concentrations of readings indicate the setting positions of the varying casing sizes. The 13³/₈″ (34 cm) casing is typically set between 2,000 and 3,500 m (6,600–11,500 ft) within the Paleocene Rogaland Group or the Chalk Group. The 9⁵/₈″ (24.4 cm) casing and 7″ (17.8 cm) liner are set between 3,500 and 5,500 m (11,500–18,000 ft) principally within Lower Cretaceous and the Upper Jurassic rocks. Below 4,000 m (13,100 ft), the fracture gradient is approximately equal to the lithostatic gradient.

A

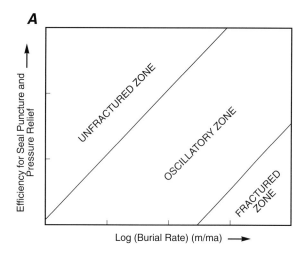

B

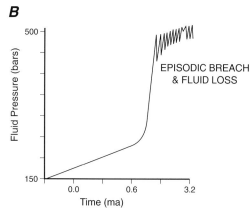

Figure 11. (A) Schematic cross-plot showing that during burial, the efficiency for seal puncture and pressure relief will go through an oscillatory zone, where pressure build-up and intermittent seal rupture can occur (Chen at al., 1990). (B) Cross-plot showing pressure changes during basin development. This plot demonstrates that during basin subsidence, overpressure will build up to a maximum, and then episodic breach and fluid loss will occur (Ortoleva, 1994).

were derived from the underlying Kimmeridge Claystone Formation (Cornford, 1994).

The upper seal of the Chalk Group in the southern part of the Central Graben is provided by claystones in the overlying Paleocene and Eocene rocks. These claystones have not been strongly lithified and most likely act as membrane seals, with slow continuous leakage. The presence of gas chimneys, locally overlying the Chalk Group (Pekot and Gersib, 1987; Leonard, 1993), provides evidence that fluid leakage, possibly through fractures, is occurring across the Tertiary claystones.

The seal for the Paleocene Forties Sandstone is the overlying Balder Tuff and younger claystones (Figure 3). The Paleocene sandstones are usually normally pressured, therefore, the seal does not have to be particularly effective. The Balder Tuff probably acts as membrane seal, although Cartwright (1994) considers that there is evidence of hydrofracturing within the overlying Tertiary claystones.

RELATIONSHIP BETWEEN PORE AND FRACTURE PRESSURE

The difference between the pore pressure and the lithostatic gradient is the effective stress. Thus, with the rapid increase in pore pressure observed near the bottom of the Cretaceous rocks, the effective stress approaches zero (Ward and Broussard, 1994) . It is at this point that hydraulic fractures can be generated. These pressure-induced fractures facilitate fluid leakage and, in the case of an overpressured compartment, the loss of fluids would result in a reduction of the pore pressure followed by the closure of fractures.

This process of episodic leaking through the seal has been studied by a number of authors. Du Rouchet (1981) indicated that the effective stress could not be negative; when it tended to zero, fractures opened, but the fractures would rapidly close following a reduction in fluid pressure. Sibson (1994) discussed a model of stress cycling in relation to faults and earthquakes and noted that there was evidence of seismic pulsing associated with entry of overpressured fluids into a fault system. Chen et al. (1990), described this process as "Oscillatory Fluid Ejection through Fracturing/Healing Cycles." Chen et al. (1990) modeled this process and noted that it was possible to go from a state of oscillatory expulsion to either an unfractured zone, where the pressuring process was less than the fracture gradient, or a totally fractured zone where the fractures remained open (Figure 11A). Fractures are only likely to be held open following partial cementation, with the sides of the fracture held apart. This process is similar to the use of a propant during a hydraulic fracturing operation. However, since fracture generation by excessive pore pressure is geologically fast, it is unlikely that the fractures would remain open long enough for cementation and crystallization to occur.

Figure 11B shows basin and pressure modeling by Ortoleva (1994). This modeling indicates that during basin subsidence, there will be a continual increase of fluid pressure until "episodic breach and fluid loss" occurs. At this point the pressure cannot increase, further hydrofracturing of the seals permits fluid release.

The author believes that in a pressure driven system, fluid (and pressure) loss will be episodic. This process of episodic leakage through pressure induced fractures in the seal will continue as long as the overpressure-generating mechanism continues to be operational. When this process is complete, the pore pressure will decline below the value of the fracture gradient, preventing further loss of fluids.

DYNAMIC OVERPRESSURE SYSTEM

The concept of fracturing rocks as a consequence of the development of abnormally high pore pressures was discussed by Momper (1978) in relation to oil expulsion from source rocks. He credits A. N. Snarskii with the original concept which was published in a series of papers (written in Russian) between 1961 and 1970. More recently, this phenomenon has been modeled by Chen et al. (1990) and Ortoleva (1994). Capuano (1993) discussed micro-fracture generation in a source rock and Cartwright (1994) proposed this as a method for fracture generation within Tertiary claystones in the North Sea. Bredehoeft et al. (1995) discussed the dynamic approach to overpressure generation and concluded that active hydrocarbon generation in the Uinta Basin (U.S.A.) resulted in overpressures that exceed the fracture gradient, with resultant leakage. These workers concluded that the high pressures would decline within a few hundred thousand years. Roberts and Nunn (1995) modeled similar times for their work on Eugene Island in the U.S. Gulf Coast.

Holm (1996) proposed that in the Central Graben episodic leakage was associated with overpressure and outlined the principal characteristics of such a system.

In the Central Graben, examination of the pore pressure profile from the deep graben indicates that at the top of pre-Cretaceous rocks, the pore pressure gradient is close to the fracture gradient (G in Figure 5A). The author considers that it would only require a small increase in pore pressure, most likely the result of active hydrocarbon generation, for the pore pressure to exceed the fracture gradient.

The model for the dynamic overpressure system predicts the following steps (Figure 12):

1. An initial hydrocarbon charge, primarily oil with some associated gas, migrates from the source

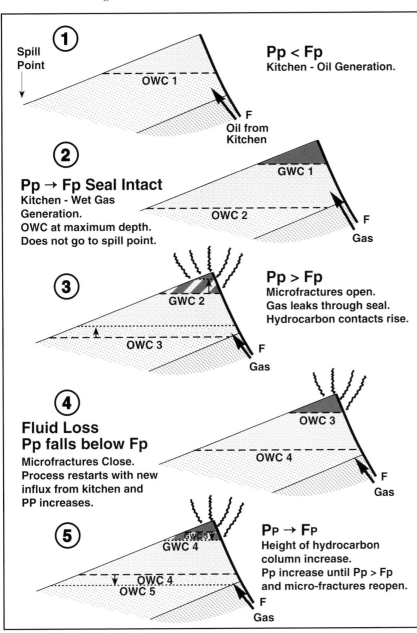

Figure 12. Hydrocarbon trap filling history in a dynamic overpressure system. This scheme demonstrates stages of hydrocarbon fill and leakage predicted by the dynamic overpressure system model.

rock into the reservoir. At this time the pore pressure is less than the fracture gradient. Charging of the reservoir continues and the pore pressure approaches the fracture gradient, while, with the increased volume of fluids entering the structure, the hydrocarbon contacts become deeper.

2. The pore pressure builds up until it exceeds the fracture pressure. Fractures open with loss of gas through the seal; consequently the gas/oil and oil/water contacts rise until the pore pressure falls below the fracture pressure. At this point, the fractures close.

3. Gas continues to migrate into the trap and the pore pressure rises again.

4. The process is autocyclic and will continue with episodic fluid loss and recharge as long as the source rock continues to supply hydrocarbons, or as long as the trap is in communication with an active hydrocarbon migration route. The autocyclicity of the proposed model will continue until the source is late mature, or tectonic activity diverts the migration route.

IMPLICATIONS OF THE DYNAMIC OVERPRESSURE SYSTEM

The dynamic overpressure system has a number of important implications for exploration that are listed below and shown schematically in Figure 13:

1. Oil and gas can leak intermittently through the seal as well as at the spill point.

2. Hydrocarbon contacts are not static but may rise and fall as a function of the state of discharge and recharge within a reservoir controlled by this type of system.

3. Gas can bypass oil, via intermittent gas loss through the top seal. Following an early oil charge into the structure, the source rock subsequently becomes gas generative, and in the Gussow Model (1954), the gas would be expected to displace oil with leakage at the spill-point. In the proposed model, the gas preferentially leaks from the crest, bypassing oil. The oil remains in the structure, despite the fact that gas is now the principal hydrocarbon migrating into the structure.

4. Gas chimneys, when overlying structures, are not necessarily indicative of fully breached seals; they may reflect the presence of an underlying, active hydrocarbon generating system.

In the Gussow model (1954) of oil and gas entrapment, the reservoirs would initially fill with oil and then, when the spill-point is exceeded, oil migrates to the next structure along the migration route. If the initial structure remained in communication with the source rock when the source rock began to generate gas, the gas would enter the structure and displace the oil. In the model proposed here, both oil and gas may leak through the seal. The level of the hydrocarbon contacts within the structure are dependent on the cycle of charge, displacement and leakage through the seal. Thus, in addition to examining the migration routes predicted by the Gussow Model (1954), the explorationist should investigate hydrocarbon migration routes directly overlying the overpressure system (Figure 13).

Roberts and Nunn (1995) modeled episodic discharge through a seal. They noted that the expulsive cycle was extremely rapid, while the pressure build-up cycle typically lasts 50-2,500 times the length of the expulsive event. Therefore, during the expulsive event, there is a very rapid rise in the hydrocarbon contacts, followed by a slow period of trap replenishment.

A dynamic model similar to the system proposed by the author has been described by Hodgson et al. (1992) in the Mungo field (UKCS Block 23/16a) of the North Sea. The Mungo field contains oil in a Paleocene reservoir with a large hydrocarbon column associated with a salt diapir. The source rock is currently supplying gas to the trap, but the gas has not displaced the oil. The oil has a high gas to oil ratio and is close to gas saturation. Hodgson et al. (1992) proposed that there was early emplacement of oil in the structure, followed by a gas charge. The added pressure from the buoyancy effect of the gas was sufficient to allow seal leakage by fracturing the crest of the structure. In this case, there was preferential leakage of gas over oil. Gas bypass occurred without displacement of oil from the structure. This process of gas bypass appears to be active in this structure and a gas chimney is visible on a seismic section over the structure.

Gas chimneys observed on seismic sections have traditionally been taken to indicate the presence of a structure with a breached seal, and unlikely to be prospective for hydrocarbon exploration. Gas chimneys have been observed overlying the oil fields of the Ekofisk complex (Leonard, 1993; Pekot and Gersib, 1987), overlying diapir oil fields (Hodgson et al., 1992), and are known to occur in other areas within the Central and Viking Grabens. Therefore, gas chimneys can be indicative of an active migration system, with leakage through a seal overlying a hydrocarbon field.

The Kimmeridge Claystone is currently gas generative below 4,000 m (12,700 ft, Cornford, 1994), and large volumes of "drilled gas" are released into the well bore during drilling. In the Central Graben, the KCF source rock has not been an exploration target to date. Given the large volumes of gas that have been generated in the KCF, the KCF may have fracture-enhanced permeability, and may therefore be a potential reservoir. Production from self-sourcing reservoirs is currently exploited from the Miocene Monterey Formation source rock in the Point Arguello field in offshore California (Mero et al., 1990) and from the Paleozoic Bakken Shale in the Antelope field of North Dakota (Meissner, 1978).

DRILLING PROBLEMS

Drilling for deep reservoirs in the Central Graben has proven to be difficult and dangerous. Many wells have experienced kicks (fluid influxes) and two wells have had major blowouts. Well 22/30b-3 blew out while drilling within Upper Jurassic rocks, with subsequent loss of the rig when the gas ignited.

Norwegian well N2/4-14 encountered well control problems (losses and gains) when drilling at 4,734 m (15,531 ft) in Upper Jurassic rocks. These problems eventually resulted in a release of condensate at the surface. Surface control was regained and the well was later re-entered, while a relief well (2/4-15) was simultaneously drilled (Olberg et al., 1990; Aadnoy et al., 1992). During re-entry, it was discovered that the original shut-in pressure had declined. The pressure decline was the result of casing failure and an underground blow-out from Upper Jurassic rocks into a Lower Pliocene sand at 828 m (2,717 ft). The control operation took 295 days with resultant costs making this the most expensive well drilled to date in the North Sea.

Buchan (1993) proposed an optimum casing program for high pressure wells in the Central Graben with the intermediate string (13⅜", 34 cm) set within the Upper Cretaceous Chalk Group, and 9⅝" (24.4 cm) casing set immediately above the high pressure zone in the Upper Jurassic (Figure 14). The advantages of set-

ting the 13⅜" (34 cm) string in the Chalk Group are: 1) a good leak-off is obtained and 2) the weak sandstone beds in the Paleocene are sealed off, reducing the possibility of an underground blow-out. The disadvantages of this program are that it is difficult to set 13⅜" (34 cm) casing at depths greater than 3,500 m (11,500 ft). In addition, the mud has to be weighted up to 0.014 MPa/m (0.6 psi/ft), to avoid borehole stability problems within the Eocene. These high mud weights can result in differential sticking of the drill pipe in the normally pressured Paleocene sandstones.

The overpressure protection string (9⅝", 24.4 cm) must be set as deep as possible within the pressure transition zone, to obtain the maximum leak-off, while being careful to avoid penetrating any overpressured permeable horizons. Detection of this zone, in real time, has proved to be very difficult and claystone porosity methods have not been satisfactory. A good seismic reflector present at the top of the overpressured Upper Jurassic rocks, has encouraged some operators to use a "Look-Ahead" Vertical Seismic Profile (VSP) to assist in determining the casing setting depth. "Drilling," "connection," and "trip gases" appear to give qualitative indications of the approach of the top of the overpressured zone.

Within the overpressured pre-Cretaceous section, the difference between the pore pressure and the fracture gradients is often less than 0.002 MPa/m (0.1 psi/ft). There are major drilling problems due to high

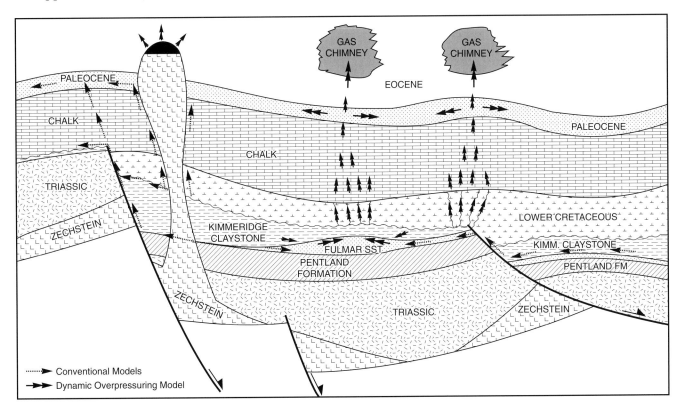

Figure 13. Schematic diagram showing hydrocarbon migration routes that may occur as a consequence of hydrocarbon generation, overpressure, and fracturing, compared to more conventional hydrocarbon migration routes.

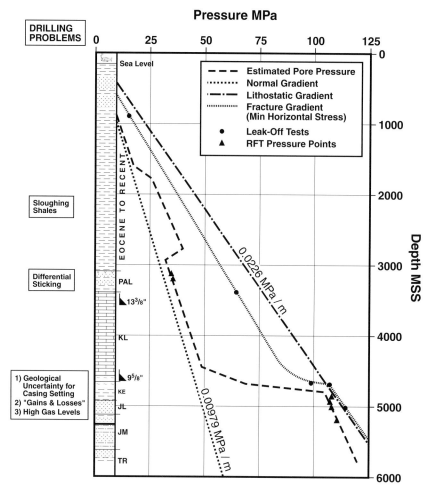

Figure 14. Pressure-depth profile for a well drilled in the center of the Central Graben. The optimum casing depths and potential drilling problems are highlighted.

levels of drilled gas entering the mud system and also the phenomenon of "gains and losses." This drilling problem is encountered when there are mud losses during mud circulation, and mud gains, when circulation stops. Buchan (1993) believes that gains and losses are the result of high mud weights supercharging formations, while Gill (1987) considers that this phenomenon is indicative of "well ballooning". In Gill's (1987) model, the high mud weight causes the borehole walls to expand slightly during circulation and therefore extra mud is required to fill the well. On stopping circulation, the extra mud returns at the surface and this apparent mud gain is similar in appearance to a kick.

Due to the fact that the pore pressures are close to the fracture pressures, only a small range of mud weights can be used to drill this section. The mud weight required to contain the pore pressure during static conditions (when mud circulation has stopped), may exceed the fracture gradient when the mud is circulating (effective circulating density). This could result in gains and losses.

Drilling of these wells must be planned with great care. The choice of casing setting depths is crucial to the success of any well drilled in the Central Graben. It must be understood also that high drilled gas levels and gains and losses while drilling within the over-

pressured section, are common phenomenon which can occur within this region.

CONCLUSIONS

Sedimentary rocks in the Central Graben of the North Sea contain complex pressure regimes consisting of: 1) a deep, highly overpressured system in the pre-Cretaceous rocks, 2) a variable overpressured regime in the Chalk Group, and 3) a normal pressured interval in the Paleocene sandstones.

The Upper Cretaceous Chalk Group has moderate overpressures in the Ekofisk complex and the preservation of this overpressure is related to the presence of a claystone seal in the overlying Paleocene rocks. The origin of overpressure in this interval is interpreted to be principally due to compaction disequilibrium.

The areal distribution of pre-Cretaceous overpressured rocks in the graben is characterized by a number of pressure cells. The seals separating these cells are formed by lateral facies changes in the permeable horizons and by faults.

Compaction disequilibrium and hydrocarbon generation are important processes in the development of overpressure within rocks of the Chalk Group and pre-

Cretaceous of the Central Graben, with hydrocarbon generation considered the dominant cause of overpressure in rocks below depths of 3,500 m (11,500 ft).

An episodic cycle of fracturing, caused by hydrocarbon generation and the development of high pore pressures, followed by diminished pore pressures and fracture closure, is proposed for the Central Graben. This process will continue episodically, with rapid rates of fluid loss followed by a long period of reservoir recharge. The process of episodic discharge and recharge is indicative of a dynamic overpressure system and will continue as long as the system is charged from an actively generating source rock.

The dynamic overpressure system has a number of important implications for exploration that include: 1) the existence of unusual hydrocarbon migration routes, 2) the preferential loss of gas from an oil and gas reservoir, and 3) gas chimneys overlying a structure are not necessarily indicative of a completely breached trap; they may indicate the presence of a petroleum accumulation directly below the gas chimney.

As a consequence of the presence of overpressured sequences in the Central Graben, drilling operations are very difficult, particularly when entering the top of the pre-Cretaceous overpressured zone. It is important that drilling programs are designed to encounter abrupt pore pressure changes. The more commonly used methods of overpressure detection cannot be used with confidence. Within the overpressured zone, large amounts of gas may be encountered and difficult control problems are associated with the gains and losses intervals.

ACKNOWLEDGEMENTS There are numerous people who have assisted either directly or indirectly in the production of this paper. I would like to particularly thank Bob McQuillin, Dave Darby, Phil Newman, and Ken Glennie, who advised on an early draft, and the drafting office of Elf Caledonia for production of the figures. I also appreciated the critical review provided by Chris Shaw (Exxon) whose comments vastly improved this paper, and Anne McGregor, who as a non-geologist, took the time to read it and make helpful comments.

REFERENCES CITED

Aadnoy, B.S. and P. Bakoy, 1992, Relief well breakthrough at problem well 2/4-14 in the North Sea: SPE 20915, pp. 329–338.

Al-Shaieb, Z., A. Puckette, A. Abdalla, and P. B. Ely, 1994, Mega-compartment complex in the Anadarko Basin: A completely sealed overpressure phenomenon, *in* P. J. Ortoleva, ed., Basin Compartments and Seals: AAPG Memoir 61, AAPG, Tulsa, pp. 55–68.

Bredehoeft, J.D., J.B. Wesley, and T.D. Fouch, 1995, Simulations of the origin of fluid pressure, fracture generation and the movement of fluid in the Uinta Basin, Utah: Bull. AAPG, v. 78, pp. 1729–1747.

Buchan, R., 1993, High-pressure, high-temperature drilling: Data management and interpretation: SPE/IADC 25764, pp. 831–840.

Buhrig, C., 1989, Geopressured Jurassic reservoirs in the Viking Graben: modeling and geological significance: Marine and Petroleum Geology, v. 6, pp. 31–48.

Capuano, R.M., 1993, Evidence of fluid flow in microfractures in geopressured shales: Bull. AAPG, v. 77, pp. 1303–1314.

Carstens, H., 1978, Origin of abnormal formation pressures in Central North Sea lower Tertiary clastics: The Log Analyst, March–April, pp. 24–28.

Carstens, H. and H. Dypvik, 1981, Abnormal formation pressure and shale porosity: Bull. AAPG, v. 65, pp. 344–350.

Cartwright, J.A., 1994, Episodic basin-wide hydrofracturing of overpressured Early Cenozoic mudrock sequences in the North Sea Basin: Marine and Petroleum Geology, v. 11, pp. 587–607.

Cayley, G.T., 1987, Hydrocarbon migration in the Central North Sea, *in* J. Brooks and K. W. Glennie, eds., The petroleum geology of north west Europe, Graham and Trotman, pp. 1305–1313.

Chapman, R. E., 1980, Mechanical versus thermal cause of abnormally high pore pressures in shales: Bull. AAPG, v. 64, pp. 2179–2183.

Chen, W., A. Ghaith, A. Park and P. Ortoleva, 1990, Diagenesis through coupled processes: Modeling approach, self-organization and implication for exploration, *in* I.D. Meshri and P.J. Ortoleva, eds., Prediction of Reservoir Quality Through Chemical Modeling, AAPG Memoir 49: AAPG, Tulsa, pp. 103–130.

Chiarelli, A. and F. Duffraud, 1980, Pressure origin and distribution in Jurassic of Viking Basin (United Kingdom - Norway): Bull. AAPG, v. 64, pp. 1245–1266.

Cornford, C., 1994, Mandal-Ekofisk(!) petroleum system in the Central Graben of the North Sea, *in* L.B. Magoon and W.G. Dow, eds., The Petroleum System-from Source to Trap: AAPG Memoir 60, pp. 537–571.

Donovan, A.D., A.W. Djakic, N.S. Ionnides, T.R. Garfield and C.R. Jones, 1993, Sequence stratigraphic control on Middle and Upper Jurassic reservoir distribution within the U.K. Central North Sea, *in* J.R. Parker, ed., Petroleum geology of the northwest Europe: Proceedings of the 4th Conference, Geological Society of London, pp. 251–269.

Du Rouchet, J., 1981, Stress fields, a key to oil migration: Bull. AAPG, v. 65, P 74–85.

Fisher, M.J. and D.C. Mudge, 1990, The Triassic, *in* K.W. Glennie, ed., Introduction to the Petroleum Geology of the North Sea, Blackwell Scientific Publications, pp. 191–218.

Gaarenstroom, L., R.A.J. Tromp, M.C. De Jong and A.M. Brandenburg, 1993, Overpressures in the North Sea: implications for trap integrity, *in* J.R. Parker, ed., Petroleum Geology of Northwest

Europe: Proceedings of the 4th Conference, Geological Society of London, pp. 1305–1314.

Gill, J.A., 1987, Well logs reveal true pressures where drilling responses fail: Oil and Gas Journal, March 16, pp. 41–45.

Glennie, K.W., 1990, Introduction to the petroleum geology of the North Sea, Blackwell Scientific Publications, 402 pp.

Goff, J.C., 1983, Hydrocarbon generation and migration from Jurassic source rocks in the E. Shetland Basin and Viking Graben of the northern North Sea: Jour. Geol. Soc., v. 140, p. 445–474.

Gussow, W. C., 1954, Differential entrapment of oil and gas: a fundamental principle: Bull. AAPG, v. 38, pp. 816–853.

Hancock, J.M., 1990, The Cretaceous, *in* K.W. Glennie, ed., Introduction to the petroleum geology of the North Sea, Blackwell Scientific Publications, pp. 255–272.

Hodgson, N.A., J. Farnsworth, and A.J. Fraser, 1992, Salt-related tectonics, sedimentation and hydrocarbon plays in the Central Graben, North Sea, UKCS, *in* R.F.P. Hardman, ed., Exploration Britain: Geological Society of London, Special Publication No. 67, pp. 31–63.

Holm, G.M., 1996, The Central Graben - a dynamic overpressure system, *in* A.M. Hurst, ed., AD 1995: NW Europe's hydrocarbon industry: Geological Society of London, Special Publication.

Hunt, J.M., 1990, Generation and migration of petroleum from abnormally pressured fluid compartments: Bull. AAPG, v. 74, pp. 1–12.

Johnson, H. and G.K. Lott, 1993, Lithostratigraphic nomenclature of the U.K. North Sea, 2 Cretaceous of the Central and Northern North Sea, R. W. O'B. Knox and W. G. Cordey, eds.: British Geology Survey, 169 pp.

Law, B.E., and W.W. Dickinson, 1985, Conceptual model for origin of overpressured gas accumulations in low permeability reservoirs: Bull. AAPG, v. 69, pp. 1295–1304.

Leonard, R.C., 1993, Distribution of sub-surface pressure in the Norwegian Central Graben and applications for exploration. *in* J.R. Parker, ed., Petroleum geology of the northwest Europe: Proceedings of the 4th Conference: Geological Society of London, pp. 1295–1303.

Lovell, J.P.B., 1990, The Cenozoic, *in* K.W. Glennie, ed., Introduction to the petroleum geology of the North Sea, Blackwell Scientific Publications, pp. 255–272.

Meissner, F.F., 1978, Petroleum geology of the Bakken Formation, Williston Basin, North Dakota, *in* D. Rehrig, ed., 24th Annual Conference, Williston Basin Symposium: Montana Geological Society, pp. 207–227.

Mero, W.E., S.P. Thurston, and R.E. Kropschot, 1990, The Point Arguello field - giant reserves in a fractured reservoir, *in* M.T. Halbouty, ed., Giant oil and gas fields of the decade 1978–1988: AAPG Memoir 54, pp. 3–25.

Momper, J.A., 1978, Oil migration limitations suggested by geological and geochemical considerations: Notes for AAPG short course on physical and chemical constraints on petroleum migration, Continuing Education Course Notes No. 8, AAPG, Tulsa.

Olberg, T., T. Gilhuus, F. Leraand and J Haga, 1990, Re-entry and relief well drilling to kill an underground blowout in a subsea well: A case history of well 2/4-14: SPE/IADC 21991, pp. 775–783.

Ortoleva, P., 1994, Basin compartmentation: Definitions and mechanisms, *in* P.J. Ortoleva, ed., Basin Compartments and Seals: AAPG Memoir 61, AAPG, Tulsa, pp. 97–118.

Pekot, L.J. and G.A. Gersib, 1987, Ekofisk, *in* A.M. Spencer et al., eds., Geology of the Norwegian oil and gas fields, London, Graham and Trotman, pp. 73–87.

Richards, P.C., G.K. Lott, H. Johnson, R.W.O'B. Knox and J.B. Riding, 1993, Lithostratigraphic nomenclature of the U.K. North Sea, 3. Jurassic of the central and northern North Sea, R.W.O'B. Knox, and W.G. Cordey, eds., British Geology Survey, 219 pp.

Roberts, S.J. and J.A. Nunn, 1995, Episodic fluid expulsion from geopressured sediments: Marine and Petroleum Geology, v. 12, pp. 195–204.

Sibson, R.H., 1994, Crustal stress, faulting and fluid flow, *in* J. Parnell, ed., Geofluids: origin, migration and evolution of fluids in sedimentary basins, Geological Society of London, Special Publication No 78, pp. 69–84.

Stevens, D.A. and Wallis, R.J., 1991, The Clyde field, block 30/17b, *in* I.L. Abbots, ed., The United Kingdom oil and gas fields 25 years commemorative volume, Geological Society of London, pp. 279–286.

Surdam, R.C., S.J. Zun, J. Liu, and H.Q. Zhao, 1994, Thermal maturation, diagenesis and abnormal pressures in Cretaceous Shales of the Laramide Basins of Wyoming, *in* Abstracts from AAPG Hedberg Research Conference, Abnormal Pressures in Hydrocarbon Environments, Golden, Colorado, June 8–10 1994.

Ward, C.D., K., and M.D. Broussard, 1994, The application of petrophysical data to improve pore and fracture pressure determination in North Sea Central Graben HPHT wells: SPE 28297, pp. 53–68.

Watts, N.L., 1987, Theoretical aspects of cap-rock and fault seals for single and two phase hydrocarbon columns: Marine and Petroleum Geology, v. 4, pp. 274–307.

Carlin, S., and J. Dainelli, 1998, Pressure regimes and pressure systems in the Adriatic foredeep (Italy), *in* Law, B.E., G.F. Ulmishek, and V.I. Slavin eds., Abnormal pressures in hydrocarbon environments: AAPG Memoir 70, p. 145–160.

Chapter 9

Pressure Regimes and Pressure Systems in the Adriatic Foredeep (Italy)

Sandro Carlin
Jacopo Dainelli
ENI-Agip E & P Division
Milan, Italy

Abstract

The exploration for oil and gas in the Adriatic Basin of Italy has resulted in a large quantity of pressure data. Abnormally high pressures in the area are mainly caused by compaction disequilibrium resulting from the high sedimentation rate of the Pliocene to Quaternary strata. The comparison of pressure profiles in the northern and central Adriatic basins has shown the presence of five pressure regions. Three regions are present in the post-Messinian siliciclastic succession that infills the Adriatic foredeep, and two pressure regions have been identified in the Cretaceous to Miocene carbonates of the Apulian continental margin. The boundaries of these regions are coincident with the main structural features of the Apenninic belt, indicating that the major thrusts act as pressure barriers. In the post-Messinian strata, the innermost region (with respect to Apenninic vergence) includes the inner buried thrusts in front of the Apenninic chain. This region is characterized by moderate to low overpressures, hydrostatic gradients, and good lateral hydraulic continuity. In the second region, in proximity of the outermost thrusts, overpressures are high and compartmentalization is pronounced. In the third pressure region, the undeformed foredeep of the Apennines, lateral hydraulic continuity prevails and high overpressures are present in the two principal Pliocene depocenters (the Romagna foredeep and the Pescara Basin). The fourth region, in the Cretaceous to Miocene carbonates, includes strata involved in Apeninnic thrusting; it is characterized by moderate overpressures. The fifth pressure region, in the foreland of the Apulian margin, has normal pressure conditions. Gas pools in the lower Pliocene interbedded sandstone-shale sequence in the first pressure region are mostly found below regional mudrock seals; whereas in the second and third pressure regions the overpressured shale beds in the lower Pliocene provide excellent seals for the interbedded gas-bearing sandstone reservoirs.

INTRODUCTION

The Adriatic Basin, located between the Apennine and Dynarid thrust belts, covers an area that roughly coincides with the Adriatic Sea between the Italian and Croatian shorelines (Figure 1). The Italian side of the Adriatic Basin is a maturely explored area in which exploration has been conducted since the early 1950s. More than 250 exploration and development wells have been drilled, resulting in significant oil and gas discoveries. Most of the exploited traps occur in conventional closed domes or in faulted anticlines where lateral closure is enhanced by normal or reverse faults.

Many studies have been carried out in the Adriatic Basin and some of the results relevant to the geologic framework of the oil and gas resources have been published by Pieri and Mattavelli (1986), Mattavelli and Novelli (1988), and Mattavelli et al. (1991).

The assessment of further potential, however, requires better knowledge of all the factors that modify or stop the (commonly upward) migration of hydrocarbons. In this regard, regional analysis of pore pressures can be used to identify pressure barriers that affect fluid movement in the subsurface and that may also act on hydrocarbon migration and entrapment. We have studied the distribution of pressure regimes in an area of approximately 20,000 km², that covers the Italian side of the northern and central Adriatic Basin (Figure 1). A comparison of pressure profiles from many wells has revealed the hydrodynamic setting of the

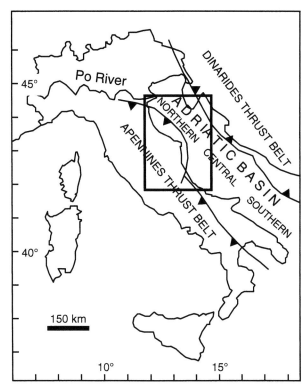

Figure 1. Location of study area within the Adriatic Basin, Italy.

Basin and has facilitated the identification of pressure systems and vertical and lateral seals that can influence the migration and trapping of hydrocarbons. Dainelli and Vignolo (1993) have analyzed the pressure distribution in a small area in the northern part of the study zone to determine the distribution of sealing faults. These authors have also discussed the difficulties of determining hydraulic continuity in the strata that characterize the Adriatic foredeep, and they have developed a method for pressure comparison that was also utilized in this study.

GEOLOGIC SETTING

The complex structural setting of the Italian Adriatic area is the result of two important tectonic events. The first event consists of extensional tectonics, from Jurassic to Late Cretaceous time, and corresponds to the segmentation of the Apulian passive continental margin (Mattavelli et al., 1991). The second event consists of compressional tectonics, from Late Cretaceous to present time, that gave rise to the Apennine Belt (Klingfield, 1979).

The Adriatic Basin (Figure 1) is generally considered a mildly deformed remnant of the African Promontory (Apulia) which acted as a foreland for both the Apennine and Dinaride thrust systems (Dewey et al., 1973; Channel and Horvath, 1976). Thrusting in the Apennines resulted in the generation of a NW-SE oriented foredeep which progressively migrated eastward as

thrusting involved its western margin. This compression produced tectonic features of regional importance such as the Adriatic folds and the Coastal Structure (Pieri and Groppi, 1981; Pieri and Mattavelli, 1986; Ori et al., 1991) (Figure 2). The structural development of the Adriatic folds and Coastal Structure started in Miocene time, reached maximum development in mid to late Pliocene time, culminating in the generation of two principal depocenters. These depocenters, the Romagna foredeep and the Pescara Basin (Figure 2), are separated by a structurally high zone that is coincident with the Pop-up zone (Figure 2), and are infilled by Miocene to Pleistocene siliciclastic sediments of mainly turbidite origin.

The stratigraphic succession includes the Cretaceous to Miocene predominately carbonate succession of the Apulian passive continental margin, Miocene mudstones of the foreland ramp, and the Pliocene to Pleistocene siliciclastic infill of the Adriatic foredeep. The Cretaceous to Miocene succession reflects the presence of a platform margin running sub-parallel to the Italian/Croatian territorial waters boundary. Toward the Italian peninsula, the typical succession includes the Cretaceous to Eocene basinal limestones of the Scaglia Formation and the overlying Oligocene and Miocene Scaglia Cinerea, Bisciaro and Schlier Forma-

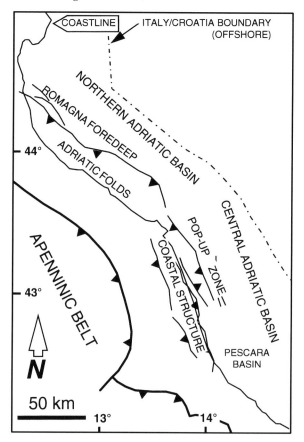

Figure 2. Simplified tectonic map showing the location of the northern and central Adriatic Basin, and the major structural features of the Apennines, Italy.

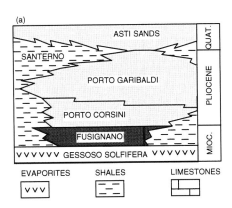

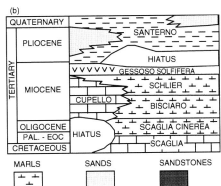

Figure 3. Generalized lithostratigraphy of the Apulian foreland and Apenninic foredeep succession, Italy: (a) lithostratigraphy in the northern Adriatic Basin, Quat. = Quaternary, Mioc. = Miocene; (b) lithostratigraphy in the southern Adriatic Basin, Pal.-Eoc. = Paleocene to Eocene. Redrawn from Mattavelli et al. (1991).

tions. In the offshore, toward the territorial waters boundary, the topmost unit of the Apulian margin is the platform carbonates of the Cupello Formation. Figure 3a shows the lithostratigraphic model for the northern Adriatic Basin. The elongated shape of the Romagna foredeep and the abundance of sediment supply generated by the prograding Po River Delta, favored the development of turbidite systems that fit the model of "highly efficient" fans of Mutti (1979), where turbidite layers can be correlated for more than 100 km (Mattavelli et al., 1991). In this area, the base of Pliocene rocks is deep, and only a few wells have reached the underlying foreland ramp mudstones or the Cretaceous to Eocene Scaglia Formation.

The lithostratigraphy of the siliciclastic infill of the Adriatic foredeep in the northern Adriatic Basin has been reconstructed from data interpreted from many wells. The lower Pliocene Porto Corsini Formation consists of interbedded sandstones and shales and is characterized by a high degree of overpressure. The transition to the overlying Porto Garibaldi Formation is marked by an increase in sandstone to shale ratio. This formation is characterized by moderate values of overpressure. The Santerno Formation consists of shale with rare sandstone beds, and is a top seal for the underlying overpressured rocks. The uppermost formation is the Pleistocene age Asti Sands that consists of sand and shaly sand. Towards the east, the lower and middle Pliocene strata gradually thin out and onlap the foreland ramp. In fact, the outermost wells, near the offshore Italian-Croatian boundary, have encountered a well developed Pleistocene sequence above a thin Pliocene one. In this part of the Adriatic Basin no direct pressure measurements are available in the Cretaceous to Miocene carbonates, however, mud weight data from the few wells that have reached these strata indicate near normal pressure conditions.

In the central Adriatic Basin, several wells have been drilled through the Pleistocene to Pliocene succession and reached the underlying platform carbonates of the Cupello Formation in the more external areas, or the Scaglia, Bisciaro, and Schlier Formations in the more internal areas (Figure 3b). The principal Pliocene depocenter is located in the Pescara Basin (Figure 2). The lithological evolution of the central Adriatic Basin

is similar to the northern Adriatic, but turbidite sand bodies are less developed and exhibit relatively poor lateral continuity (Casnedi, 1983). Sediment input from lateral sources tied to the emerging Apennines (Ori et al., 1991) and the poorly efficient character of the turbidites (in the sense of Mutti, 1979), resulted in the deposition of deep-sea fan deposits of limited extent (Casnedi, 1983).

In this area few wells have pressure measurements in the lower Pliocene rocks. However, data derived from acoustic logs indicate that overpressures are high. In the Scaglia and Cupello Formations many tests have been carried out and pressures are mostly normal except in the Coastal Structure and Pop-up zone (Figure 2) where moderate to high overpressures are present in the Scaglia Formation.

The main gas discoveries are located in the Romagna foredeep, in the Adriatic folds (northern Adriatic Basin), and in the Pescara Basin (central Adriatic Basin), whereas the main oil discoveries occur in the southern part of the area within the Pop-up zone and to the north of the Pescara Basin (Figure 4). For a more detailed review of oil and gas field in this area the reader is referred to papers by Pieri and Mattavelli (1986), Mattavelli and Novelli (1988), Mattavelli et al. (1991).

DATA ANALYSIS AND METHOD

Pressure data

We have examined most of the available pressure data in the northern and central Adriatic Basin (Figure 5). These data mainly consist of wireline formation tests (WFT), although pressure data from drill stem tests (DST) and production tests (PT) are also present, especially in older wells. Bottom hole pressure data from DST and PT were considered only if the build-up pressure was stabilized or if extrapolation to static reservoir pressure was carried out with Horner plots (Horner, 1951) or similar methods. In DST and PT-derived data that involve a considerable difference in elevation between the location of the pressure gauge and the top of the tested interval, the pressure value was corrected to the top of the interval by adding the

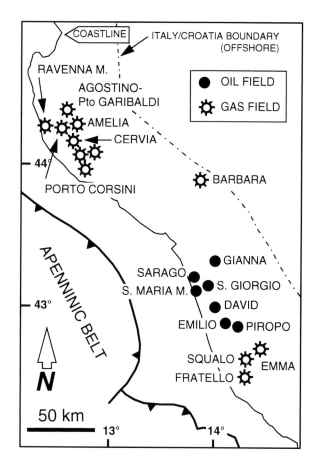

Figure 4. Main gas and oil fields in the northern and central Adriatic Basin, Italy.

pressure exerted by the fluid column that lies between the gauge and the top of the interval. Pressure values from reservoirs where depletion was evident were also disregarded. In order to compare pressure values independently of the fluid present in the reservoir, pressure measurements carried out in thick gas bearing layers were compensated for the overpressure generated by the hydrocarbon column. In wells where only PT or DST data are present, the reconstruction of the pressure profile is difficult because of data clustered around the pay zone. In contrast, WFTs are usually recorded through thicker intervals outside the pay zone, thus a better reconstruction of the pressure profile of the well can be obtained. Once a reliable pressure data base was compiled, pressure profiles of wells were reconstructed on pressure-depth (PD) plots. In the area we have studied, several developed hydrocarbon fields are present, thus pressure values from adjacent development wells have been merged into a single pressure profile representative of the field. All pressure values are referred to subsea depth.

Method for pressure data analysis

On a PD plot, normal pressures plot along a line whose slope is tied to formation water density (points

R through T, Figure 6). The reference gradient used in this study for normal pressures is 10.14 kPa/m (0.43 psi/ft), which corresponds, at temperature and pressure conditions of the Adriatic Basin, to a formation water salinity of 32 g/l (NaCl). Pressure measurements from overpressured zones will plot to the right of this line (points U through Y, Figure 6) and they may reach values approaching lithostatic pressure, depending on the fracture pressure of the rock. The overpressure value is defined as the distance of the point from the normal pressures line (points U and X, Figure 6).

On PD plots, lines parallel to the normal pressures line are termed hydrostatic lines (Figure 6) (Dainelli and Vignolo, 1993); all points falling on any one such line have the same overpressure value (points U through W, Figure 6). The overpressure difference between two points on the graph can be determined from the pressure difference between the hydrostatic lines passing through the points (e.g., points X and Y, Figure 6). The use of hydrostatic lines allows a more immediate graphical comparison of pressure data to determine hydraulic continuity between wells because, in the absence of hydrodynamic conditions and variations in the density of formation water, all the points belonging to the same aquifer must fall on one such line. An overpressure gradient corresponds in the PD

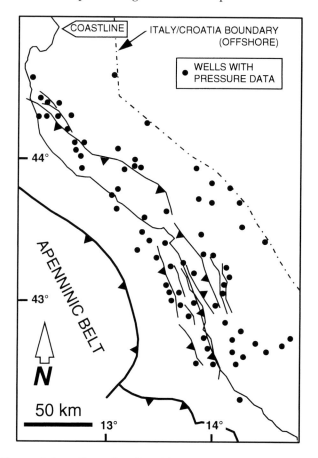

Figure 5. Location of wells with pressure data available in the northern and central Adriatic Basin, Italy.

plot to a line with a slope steeper than the hydrostatic line (interval C, Figure 6). The slope may even be greater than the lithostatic gradient (Dainelli and Vignolo, 1993; Leftwich and Engelder, 1994).

Hydrostatic gradients characterize strata with vertical hydraulic continuity and may be present either in normally pressured or in overpressured intervals (Powley, 1990). Overpressure gradients are common in mud-rich strata or in interbedded strata consisting of a cyclic repetition of permeable and impermeable layers in which each permeable layer is a confined aquifer (Dainelli and Vignolo, 1993).

With reference to the above terminology, pressure trends observed in the subsurface can be described and the profile of a well can be analyzed in terms of pressure regimes. Pressure regimes consist of intervals of constant or near-constant pressure gradient separated by pressure breaks or by gradient breaks (horizontal broken lines, Figure 6). Pressure regimes present in the Adriatic area studied can be subdivided into three main categories:

1. Normal pressure regimes (interval A of Figure 6),
2. Overpressure regimes with a hydrostatic gradient (interval B of Figure 6) and,
3. Overpressure regimes with an overpressure gradient (interval C of Figure 6).

Comparison of pressure profiles

The comparison of pressure profiles to determine compartmentalization is quite straightforward if strata are characterized by vertical hydraulic continuity. In this case, a single potentiometric map is enough to determine the presence of pressure barriers or the direction of fluid flow (Al-Shaieb, 1991). But in the Pliocene of the Adriatic Basin, where strata consist of a cyclic repetition of sand and shale layers characterized by overpressure gradients and in which each stacked sand is an aquifer isolated by mudrock pressure seals—a potentiometric map of a single layer is not enough to determine the presence of pressure barriers, nor is it representative of the pressure conditions of the area (Dainelli and Vignolo, 1993). For comparing pressure values, we have used the rules developed by Dainelli and Vignolo (1993) which use graphical relations between pressure profiles on PD plots to determine the presence of pressure barriers between wells. *The rules are valid only if the thickness of correlatable intervals remains relatively constant* (Dainelli and Vignolo, 1993). They are as follows:

Rule 1 If the pressure gradients in equivalent rock intervals of two wells are equal (interval between rock units E and G in Wells 1 and 2, Figure 7a), and overpressure values in equivalent rock units are the same (e.g., rock unit E, F, or G in Wells 1 and 2, Figure 7a) then, the

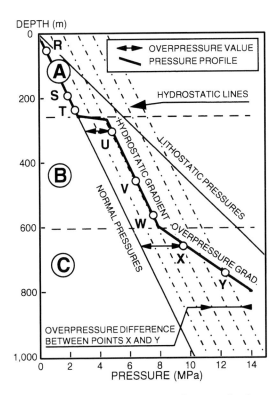

Figure 6. Pressure-depth plot showing terminology used to describe pressure regimes. The figure shows a hypothetical pressures profile of a well (defined by points R through Y). The three main types of pressure regimes that can be identified in strata of the northern and central Adriatic Basin are shown on this figure. A normal-pressure regime is indicated by A, an overpressured regime with a hydrostatic gradient is indicated by B, and an overpressured regime with overpressure gradient is indicated by C. See text for further explanations.

wells are in hydraulic continuity (Wells 1 and 2, Figure 7a).

Rule 2 If the pressure gradients in equivalent rock intervals of two wells are not the same (interval between rock units E and G of Wells 1 and 4, Figure 7b) then, a pressure barrier is present between the wells (Wells 1 and 4, Figure 7b).

Rule 3. If the pressure gradients in equivalent rock intervals of two wells are equal (interval between rock units E and G of Wells 1 and 2, Figure 7a) but overpressure values in equivalent rock units are different (e.g., rock unit E, F, or G in Wells 1 and 3, Figure 7a) then, the two wells are in hydraulic continuity and a fault has interrupted the geometrical continuity of the rock units (Wells 1 and 3, Figure 7a).

Rule 3 also allows the distinction between sealing and non-sealing faults. In the northern and central Adriatic Basin, sealing faults are the most common pressure barrier observed, although in the central Adriatic Basin facies variations may be responsible for

Figure 7. Graphical guide to the comparison of pressure profiles of wells. Wells 1, 2 and 3 are three hypothetical wells and A through G are hydraulically independent rock units. Case (a) shows hydraulic and layer continuity between Wells 1 and 2. Well 3 is separated by a non-sealing fault from the first two wells. In this case the geometrical continuity of layers is interrupted, but aquifers are continuous although they do not correspond to the same layer on both sides of the fault. Case (b) indicates the presence of a pressure barrier between wells 1 and 4.

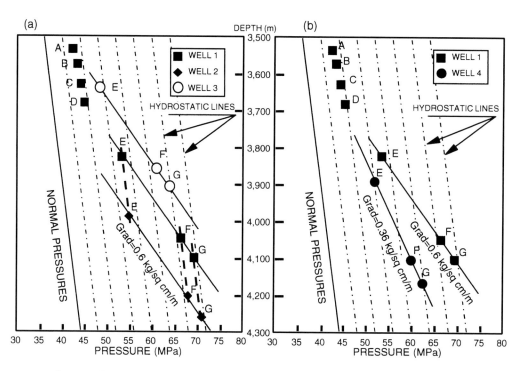

slight overpressure differences observed between equivalent units.

Sealing faults will interrupt the hydraulic continuity of layers thus generating a confined aquifer. In a multi-layered succession, each confined aquifer will correspond to a pressure cell. In this study we use the term "pressure system" to define a body of rock composed of a single cell, or a stack of cells, laterally bounded by the same pressure barriers and characterized by a constant or near constant pressure gradient. Where the sequence has vertical hydraulic continuity, cell and system will coincide.

DISCUSSION

Pressure Regimes and Pressure Systems

Our analysis of pressure data, following the criteria discussed above, has shown that overpressures are present throughout the area. They reach their highest values in the lower Pliocene strata where near lithostatic pressures are reached.

The pressure distribution in the northern and central Adriatic Basin is complex. Several pressure systems can be identified by comparing pressure profiles. Pressure systems that present similarities (pressure profiles that have the same gradient types and similar gradient values) have been grouped into pressure regions. The boundaries between these regions follow the main structural/stratigraphic features of the Apenninic belt, indicating that the major thrusts act as pressure barriers (Figures 8 and 9). Three pressure regions are present in the post-Messinian siliciclastic succession. These are the inner thrusts pressure region, the deformation front

pressure region, and the undeformed foredeep pressure region (Figure 9). The remaining two regions occur in Cretaceous to Miocene carbonate rocks, and divide the area into a thrust belt pressure region and a foreland pressure region (Figure 10).

Pressure regions in the post-Messinian siliciclastic succession

The inner thrusts pressure region (Figures 11 and 12) includes the Adriatic folds, in the northern Adriatic Basin, and the innermost part of the Coastal Structure in the central Adriatic Basin. This pressure region is characterized, in the Pliocene strata, by the absence of pronounced overpressures and by the presence of hydrostatic gradients, although restricted intervals with overpressure gradients are present in the northern part of the region (Figure 11). Overpressures are slightly higher in the northern part (Figure 11) than in the southern part (Figure 12).

Wells 1 and 3 (Figure 11) illustrate the typical aspect of pressure profiles in the northern part of the region. The pressure profiles consist of two regimes with hydrostatic gradient (the shallower with an overpressure of 2.5 MPa and the deeper with 8 MPa overpressure) separated by an interval with overpressure gradient (observable in Well 3, Figure 11). The full succession of these regimes is only documented in Well 1. Wells 4, 5 and 6 show only the upper 2.5 MPa regime. The equivalence between the gradients of Wells 1 and 3 and between Wells 1, 4, and 5 in Figure 11 suggests hydraulic continuity of the order of several tens of kilometers in this part of the region, and implies the absence of major sealing faults.

In the southern part of the area, pressure data from most wells indicate the presence of hydrostatic gradients (Figure 12). The fact that some wells have the same overpressure values (e.g., Well 11 with the composite of Wells 7, 8, 9, & 10 and also the composite of Well 14 & 15) suggests the presence of hydraulic continuity of the order of a few tens of kilometers. However, overpressure differences between some of the wells (Well composite 14 & 15, and Well 16, and Well 17; Figure 12), indicate the presence of some pressure barriers.

The deformation front (Figures 13 and 14) is a pressure region that includes the outermost thrusts of the Apenninic belt in the northern Adriatic Basin, and the Coastal Structure in the southern Adriatic Basin. This region is characterized by elevated overpressures, that often approach the lithostatic gradient, and by overpressure gradients that have high values (locally >50 kPa/m) especially in the lower Pliocene strata.

The comparison between pressure profiles of the wells indicates the presence of several pressure barriers, because on the PD plots (Figures 13 and 14) pressure profiles of wells, although similar, are not equivalent. Figure 15 (an enlargement of a portion of Figure 13) illustrates gradient differences in lower Pliocene strata, of Wells 19, 20, 21, and 22 in the northern Adriatic Basin. These gradient differences imply the presence of barriers that compartmentalize the region into

pressure systems. Note on Figure 15 that wells 22 and 26 have the same gradient and thus belong to the same pressure system.

In the Coastal Structure, near lithostatic pore pressures are also observed in shallow reservoirs (Well 31, Figure 14). These anomalous conditions may be the result of hydraulic communication with highly overpressured deeply buried reservoirs (e.g., Wells 27 and 33, Figure 14). Although Wells 28, 29, 30, and 33, do not have enough data to reconstruct pressure gradients, the relatively high pressure values of these wells indicate that they probably belong to the deformation front pressure region.

The undeformed foredeep pressure region extends from west of the outermost thrusts to the offshore Italy/Croatia boundary and it includes the two main depocenters of the Pliocene to Pleistocene strata. North of the region (Figure 16), in the Romagna foredeep, slight overpressures (1.5 MPa) appear below 1,000 m at the base of the Pleistocene. Below the Pleistocene strata, the pressure regime in the top part of the Pliocene strata has a hydrostatic gradient and an overpressure of approximately 2.5 MPa, suggesting the presence of vertical hydraulic continuity in these strata. Below this, overpressures gradually increase and reach high val-

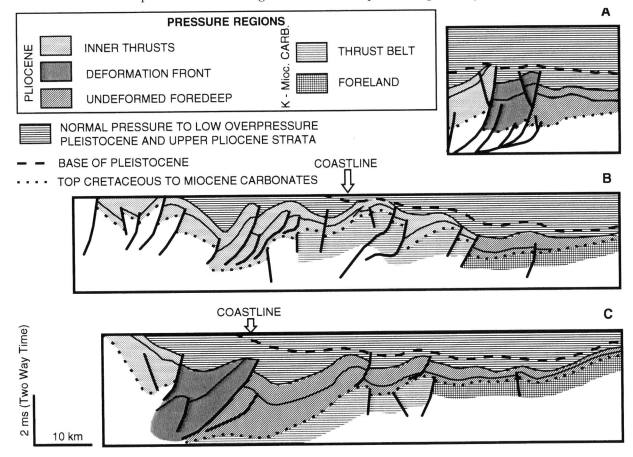

Figure 8. East-west cross-sections showing the relations between pressure regions and structural setting. Northern and central Adriatic Basins, Italy. Redrawn from Ori et al. (1991) and Dainelli and Vignolo (1993). Location of cross-sections is shown on Figures 9 and 10. K = Cretaceous, Mioc. = Miocene, CARB. = Carbonates.

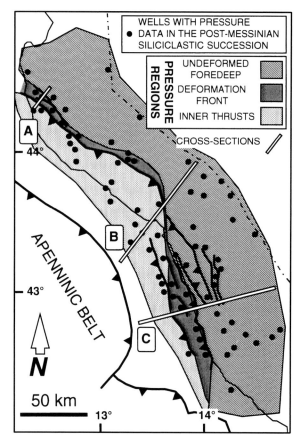

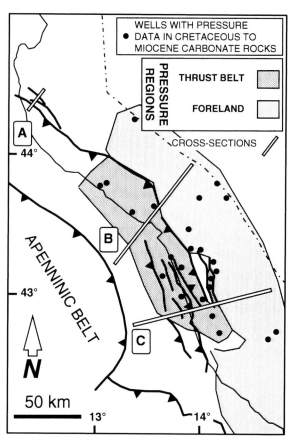

Figure 9. Pressure regions in the post-Messinian silici-clastic succession in the northern and central Adriatic Basin, Italy. The cross-sections A, B, and C are shown on Figure 8.

Figure 10. Pressure regions in Cretaceous to Miocene carbonate rocks of the northern and central Adriatic Basin, Italy. The cross-sections A, B, and C are shown on Figure 8.

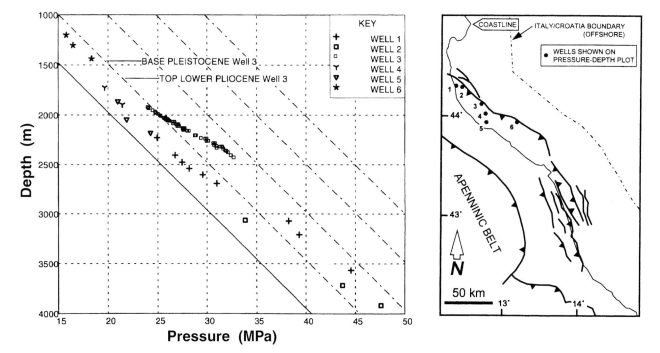

Figure 11. Inner thrusts pressure region in the northern Adriatic Basin showing pressure well profiles and locations of those wells.

ues in the lower Pliocene (Porto Corsini Formation) where the gradient can exceed 55 kPa/m (2.4 psi/ft). The equivalence between pressure gradients of Wells 34, 35, 36, and 37 (Figure 16) suggests that important lateral pressure barriers are absent and that hydraulic continuity of rock units exceeds 50 kilometers.

In the Pescara Basin (Figure 17), pressures are normal down to the bottom of the Pleistocene strata. In the upper Pliocene strata and part of the middle Pliocene strata, a low degree of overpressure (no more than 2.0 MPa) prevails. The step-like variations in overpressures that can be observed in these strata (Figure 18)

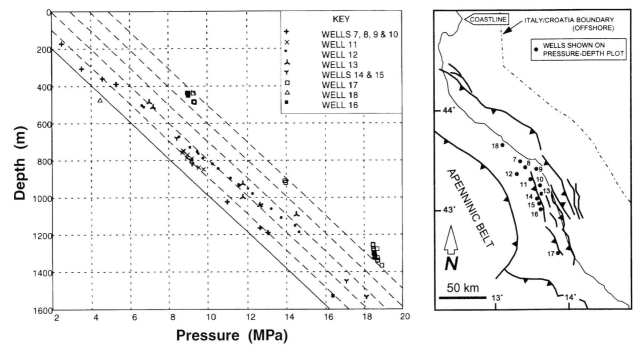

Figure 12. Inner thrusts pressure region in the central Adriatic Basin showing pressure well profiles and locations of those wells.

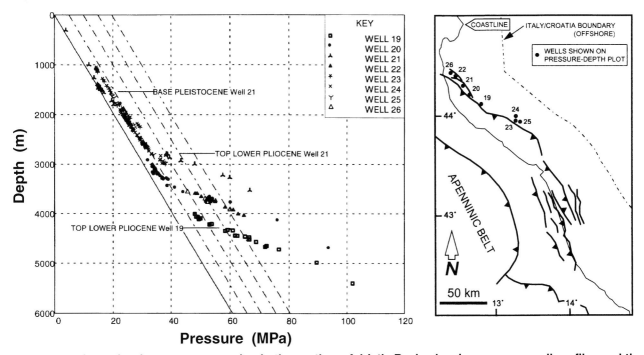

Figure 13. Deformation front pressure region in the northern Adriatic Basin showing pressure well profiles and the locations of those wells.

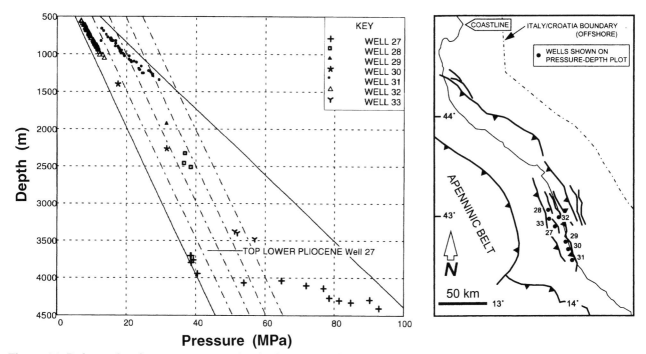

Figure 14. Deformation front pressure region in the central Adriatic Basin showing pressure well profiles and the locations of those wells.

can be linked to the presence of horizontal pressure barriers that isolate pressure regimes with slight over-pressure differences. These barriers probably correspond to low permeability mudrock seals that are correlatable at a basin-wide scale and have been tied to periods of inactivity between turbiditic events (Ori et al., 1991). Throughout the Pescara Basin, these pressure regimes are easily correlatable (as the good coincidence between pressure profiles on Figure 17 illustrates) indicating that important lateral pressure barriers are absent. Locally, slight differences between pressure regimes can be present (e.g., between Wells 40 and 44, Figure 17). These differences can be tied to the characteristics of the sedimentary sequence that consist of turbidite sandstones of limited lateral extent (Casnedi, 1983; Ori et al., 1991). Under these circumstances, each sedimentary unit would have a slightly different pressure regime. In the Pliocene strata overpressures start to increase in response to the presence of increasingly mud-rich rocks (lowermost pressure values of Wells 42 and 44, Figure 17). Although overpressures as high as those in the Romagna foredeep are not directly documented in the Pescara Basin, indirect data from acoustic logs (Figure 19) testify to the presence of abnormally slow transit time values that can be correlated with a high degree of overpressures in the lower Pliocene strata.

The outermost part of the undeformed foredeep (Figure 20) shows the low overpressures and near-hydrostatic gradients that are characteristic of the sand-rich Pleistocene strata. The normal pressure conditions of Pleistocene strata are a result of good permeability which has allowed the dissipation of the excess fluids

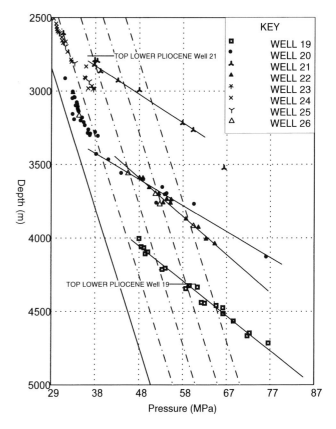

Figure 15. Enlargement of a portion of Fig. 13 showing pressure gradients in the deformation front pressure region in the northern Adriatic Basin.

expelled during compaction. The highly overpressured lower Pliocene strata, present in the depocenter areas, onlap westward against the foreland ramp (where they are absent). The near equivalence between pressures gradients of Wells 45 through 53 (Figure 20) indicates the absence of important pressure barriers.

Pressures regions in Cretaceous to Miocene carbonate rocks

In the Cretaceous to Miocene carbonate rocks, pressure data are mainly derived from the Scaglia and Cupello Formations. Figures 10 and 21 shows the few

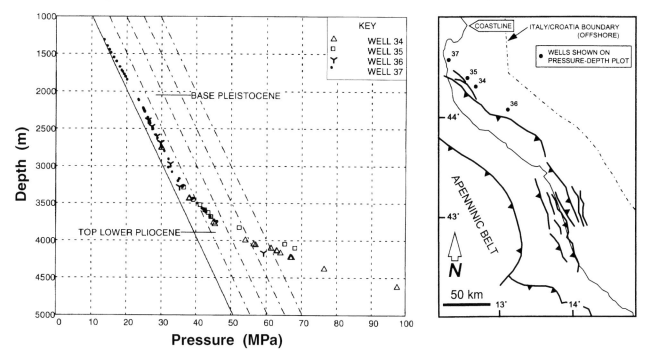

Figure 16. Undeformed foredeep pressure region in the northern Adriatic Basin showing pressure well profiles and the locations of those wells.

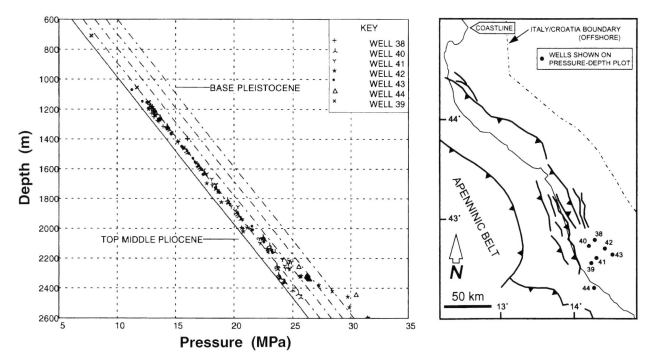

Figure 17. Undeformed foredeep pressure region in the central Adriatic Basin showing pressure well profiles and the locations of those wells.

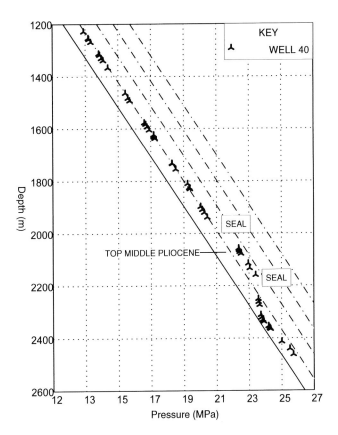

Figure 18. Pressure profile of Well 40, undeformed fore-deep pressure region, central Adriatic Basin. The shifts in the hydrostatic pressure gradient are the result of basin-wide horizontal seals. Note the thick gas column below the lowermost seal.

wells where pressure values are available. Note the poor resolution of pressure distribution compared to the overlying Plio–Pleistocene succession in Figure 9. Nonetheless, we have attempted to divide these strata into two pressure regions, a thrust belt pressure region and a foreland pressure region, on the basis of the presence or absence of overpressures.

The thrust belt pressure region (Figure 10) is characterized by moderate overpressures (Figure 21). Wells that plot along the same hydrostatic line have the same overpressure value and they are considered to be in hydraulic continuity (e.g., the composite of Wells 70 and 71, and Well 73, or the composite of Wells 55 and 56 and Wells 54, 57, 74, and 75, Figure 21). The difference in overpressure between wells indicates the presence of pressure barriers (e.g., between the composite of Wells 68 and 69, and the composite of Wells 70 and 71, Figure 21). These barriers probably correspond to the main thrusts and compartmentalize this pressure region into several pressure systems.

In the foreland pressure region (Figure 10), pressures are normal throughout the pre-Messinian strata (Figure 21). This region is characterized by large-scale lateral and vertical hydraulic continuity. In Figure 21, wells that have normal pressures in the Cretaceous to Miocene carbonate rocks have been plotted with the same symbol.

Causes of Overpressures

The analysis of acoustic logs has shown that the onset of overpressures can be correlated with the top of undercompacted strata (Figure 19 and 22), and that the build up of overpressures correspond to abnormally high porosities at depth (Figure 22). This suggests that

Figure 19. Acoustic log values (a) and pressure profile (b) for Well 32 in the Pescara Basin. Although no pressures data is available in lower Pliocene rocks, the abnormally high acoustic values in this interval suggest the presence of high overpressures.

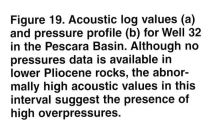

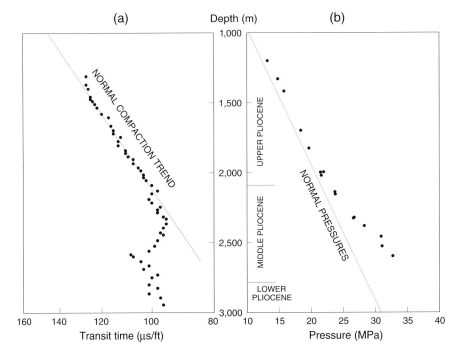

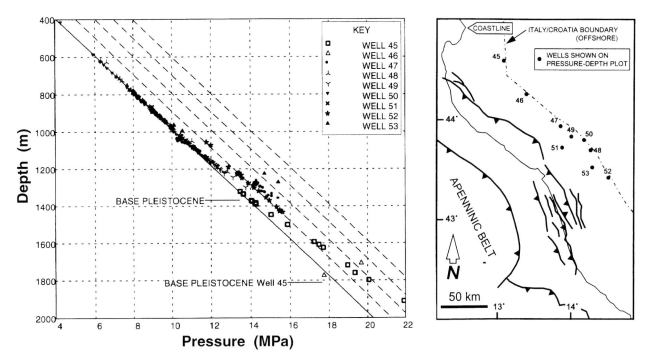

Figure 20. Outermost part of the undeformed foredeep pressure region in the northern and central Adriatic Basin showing pressure well profiles and the locations of those wells.

compaction disequilibrium may be the main cause of overpressuring. Furthermore, the high sedimentation rate of the Pliocene and Pleistocene succession (locally greater than 1,000 m/m.y.) and the presence of large amounts of shale in the lower Pliocene are favorable conditions for the development of abnormally high pressures (Mann and Mackenzie, 1990).

The extent of high overpressures is limited to the depocenters of the lower Pliocene strata. Within the deformation front pressure region, the distribution of overpressures is largely controlled by the presence of sealing faults that constrain the lateral transfer of pressures.

The presence of overpressure in carbonate reservoirs in the offshore central Adriatic oil province may be tied to the structural juxtaposition of these reservoirs to the highly overpressured lower Pliocene strata (see sections B and C, Figure 8). This geometrical relation may provide a pathway for the lateral transfer of overpressures from lower Pliocene strata to the upthrusted carbonate reservoirs.

Overpressures and Hydrocarbon Occurrence

In the northern and central Adriatic Basin, some of Italy's major gas accumulations are present. The gas is mainly produced from Pliocene turbidites of the Porto Garibaldi and Porto Corsini Formations. Traps usually consist of folds related to the outermost thrusts of the Apennines or in gentle folds in the foredeep. Isotope analyses indicate that the produced gases in this area are of biogenic origin (Mattavelli and Novelli, 1988) and occur in multiple pay-zone reservoirs at depth up

to 4,500 m. The cap-rock for each pay zone consists of shale that may be as thin as one meter.

The considerable amount of gas present in the northern Adriatic foredeep can be tied to factors such as the high sedimentation rates (up to 1,000 m/m.y.), syn-sedimentary tectonics, the cold present-day thermal regime (in the two Pliocene depocenters, the 70°C isotherm lies at approximately 4,000 m), and the widespread biogenic gas generation that, according to Mattavelli and Novelli (1988), occurs even at great depths (at least 4,000 m). Another factor that favors the presence of an extended gas province is the high content of organic matter in the turbiditic clays, preserved from oxidation by rapid burial. The regular interbedding of sands and shales with basin-wide extent, can also favor the migration of hydrocarbons over large distances (Mattavelli et al., 1991).

Gas fields of lesser importance are present in the Plio–Pleistocene sequences of the central Adriatic. According to Mattavelli et al. (1991), the poor characteristics of the reservoirs in this gas province and the lower reserves (in the central Adriatic Basin only 3 fields have reserves greater than 4 billion (standard) m³, compared to the 9 fields present in the northern Adriatic Basin) are mainly tied to the quality of the sand reservoirs that are largely formed by turbidite systems of limited lateral extent that fit the model of "poorly efficient" fans of Mutti (1979).

Besides biogenic gas offshore, gas of mainly thermogenic origin is present in the thrust belt. The accumulation of gas in this area is, in large part, due to tectonic loading and the consequent elevated levels of thermal maturity (Mattavelli et al., 1991).

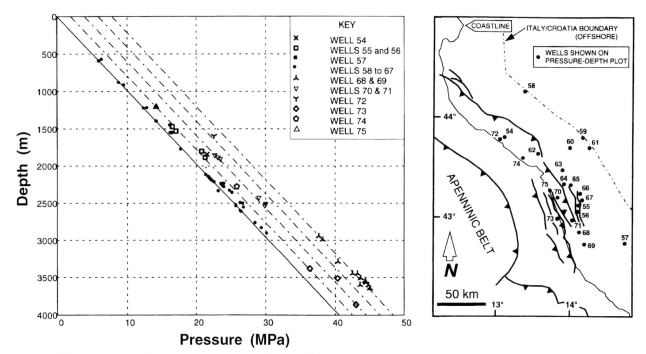

Figure 21. Thrust belt and foreland pressure regions of Cretaceous to Miocene carbonate rocks in the northern and central Adriatic Basins showing pressure well profiles and the locations of those wells.

There does not seem to be any noticeable relationship between the distribution of overpressure and the occurrence of gas fields, although the same factors that have made the area a prolific gas producing region (high subsidence rate, stacked sand-shale beds) also play an important role in the development of overpressure. There seems to be a relation, however, between the vertical distribution of gas pools and the pressure regime. In the deformation front and in the undeformed foredeep pressure regions, gas accumulations occur in multiple pay zone reservoirs where overpressure gradients are observed in the sand-shale interbeds (Figure 23a). Conversely, in the innermost thrusts region, gas accumulations tend to be localized in a single pay zone below regional mudrock seals, (although the reservoir consists of sand-shale interbeds, Figure 23b), or in limited intervals with an overpressure gradient.

The observation that thin shale layers serve as seals only where an overpressure gradient is present may explain the difference in the types of gas accumulations. The overpressure generated in the shale layers as a consequence of compaction disequilibrium creates an additional trapping factor. In this case, fluid flow directed towards the bed boundaries enhances the seal capacity of the shale layers. Conversely, in the presence of a hydrostatic gradient, the thin shales lose part of their sealing properties and vertical gas migration is possible. Where hydrostatic gradients are present gas will tend to accumulate below thick mudrock strata.

Although pressure distribution does not seem to have a basin-wide effect on the distribution of hydrocarbons, at a local scale pressure compartmentalization

indicates the presence of sealing faults. These faults may serve as seals for hydrocarbon traps, thus enhancing the hydrocarbon potential of pressure regions that show the most pronounced compartmentalization. In part of the northern Adriatic Basin, Dainelli and Vignolo (1993) have already shown that the deformation pressure region is a prospective area for sealing fault traps.

CONCLUSIONS

In the study area, the top of overpressures occurs at the base of Pleistocene strata. The highest overpressures are found in Pliocene depocenters in the Romagna foredeep and in the Pescara Basin, where a high sedimentation rate and a high shale to sand ratio have resulted in compaction disequilibrium and the development of abnormally high pressures.

The Cretaceous to Miocene carbonate rocks are generally characterized by normal pressures except in areas where they have been involved in Apenninic thrusting. In this case, overpressures may be due to transfer of pressures from overpressured Pliocene rocks. In the Pliocene to Pleistocene strata, overpressures are common only in the outermost thrust sheets of the of the Apenninic belt (where a pronounced compartmentalization is also present) and in the depocenters of the undeformed foredeep.

Although the pressure distribution is complex, five pressure regions have been recognized. Three pressure regions separated by sealing thrusts occur in the Pliocene to Pleistocene strata. The first region, which

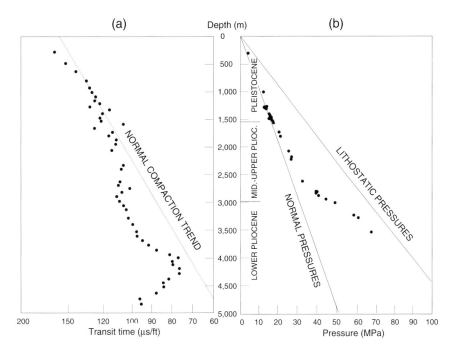

Figure 22. Acoustic log values (a) and pressure profile (b) for a well in the deformation front pressure region of the northern Adriatic Basin, Italy. Note the near coincidence of the top of overpressure at about 2,700 m (b) and the departure of acoustic values from the normal compaction line (a) at a depth of about 2,300 m.

includes the inner buried thrusts in front of the Apenninic chain, has low overpressures, frequent hydrostatic gradients and good lateral hydraulic continuity. The second region, in front of the thrust belt, has high overpressures and a pronounced compartmentalization (suggesting the presence of sealing faults that may also serve as lateral seals for oil and gas accumulations). The third region, in the undeformed Adriatic foredeep, has high overpressures where the lower Pliocene is present and also has excellent lateral hydraulic continuity.

Although there does not appear to be any obvious relationships between the distribution of overpressure and gas accumulations, there does seem to be a relationship between pressure regime and vertical gas distribution. Where overpressure gradients are present in the deformation front and undeformed foredeep pressure regions, thin overpressured shaly layers act as vertical seals for multi-layered reservoirs. In contrast, where hydrostatic gradients are present, gas accumulations tend to be located below thick regional mudrock seals.

ACKNOWLEDGMENTS *The authors wish to thank ENI-Agip E & P Division for permitting the publication of this paper. The authors would also like to thank Ben Law and two anonymous reviewers for their constructive comments and helpful suggestions.*

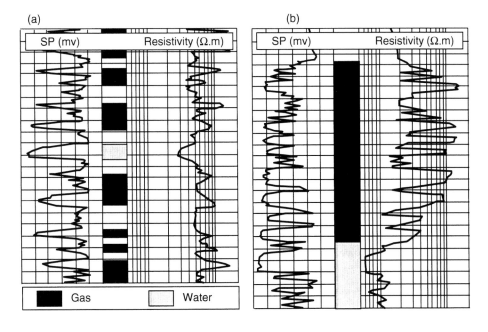

Figure 23. Schematic diagrams showing the vertical distribution of gas pools in the interbedded strata of the lower Pliocene of the Adriatic Basin, Italy: (a) multi-layered reservoir characteristic of intervals with overpressure gradient; (b) single pool reservoir characteristic of intervals with hydrostatic gradients.

REFERENCES CITED

Al-Shaieb, Z., 1991, Compartmentation, fluid pressure important in Anadarko exploration: Oil & Gas Journal, v. 89, p. 52–55.

Channel, J.E.T., and F. Horvath, 1976, The African/Adriatic Promontory as a paleogeographical premise for Alpine orogeny an plate movements in the Carpatho-Balkan Region: Tectonophysics, v. 35, p. 71–101.

Casnedi, R., 1983, Hydrocarbon-bearing submarine fan system of Cellino Formation, central Italy: American Association of Petroleum Geologists Bulletin, v. 67, p. 359–370.

Dainelli, J., and A. Vignolo, 1993, Regional assessment of sealing faults through pore-pressure analysis in the northern Adriatic: First Break, v. 11, p. 287–294.

Dewey, J.F., Pittman III, W.C., Ryan, W.B.F., and J. Bonnin, 1973, Plate tectonics and the evolution of the Alpine system: Geological Society of America Bulletin, v. 84, p. 3137–3180.

Horner, D.R., 1951, Pressure build-up in wells: Proceedings of the 3rd World Petroleum Congress, v. 2, p. 503–523

Klingfield R., 1979, The Northern Apennines as a collision orogen: American Journal of Science, v. 279, p. 679–691.

Leftwich, J.T., Jr., and T. Engelder, 1994, The characteristics of geopressure profiles in the Gulf of Mexico Basin, *in* P.J. Ortoleva, ed., Basin compartments and seals: AAPG Memoir 61, p. 119–129.

Mann, D.M., and A.S. Mackenzie,1990, Prediction of pore fluid pressures in sedimentary basins: Marine and Petroleum Geology, v. 11, p. 104–115.

Mattavelli, L., and L. Novelli, 1988, Geochemistry and habitat of natural gases in Italy, *in* L. Mattavelli and L. Novelli, eds., Advances in organic geochemistry 1987: Oxford, Pergammon Press, p. 1–13.

Mattavelli, L., Novelli, L., and L. Anelli, 1991, Occurrence of hydrocarbons in the Adriatic Basin, *in* A.M. Spencer, ed., Generation, Accumulation, and Production of Europe's Hydrocarbons: Oxford, Oxford University Press, p. 369–380.

Mutti, E., 1979, Turbidites et cones sous-marins profonds, *in* P. Homewood, ed., Sédimentation détritique (fluviatile, littorale et marine): Fribourg, Institut de Géologie Université de Fribourg, p. 353–419.

Ori, G.G., Serafini, G., Visentin, C., Ricci Lucchi, F., Casnedi, R., Colalongo, M.L. , and S. Mosna, 1991, The Pliocene–Pleistocene Adriatic foredeep (Marche and Abruzzo, Italy): an integrated approach to surface and subsurface geology, 3rd European Association of Petroleum Geoscientists & Engineers Conference, Adriatic foredeep field trip guide book, Milan, Agip S.p.A..

Pieri, M., and G. Groppi, 1981, Subsurface geological structure of the Po plain, Italy. Progetto finalizzato geodinamica, sottoprogetto 'modello strutturale': Consiglio Nazionale delle Ricerche pubblicazione 414, Rome, CNR.

Pieri, M., and L. Mattavelli, 1986, Geologic framework of the Italian petroleum resources: AAPG Bulletin, v. 70, p. 103–130.

Powley, D.E., 1990, Pressures and hydrogeology in petroleum basins: Earth-Science Reviews, v. 29, p. 215–226.

Nashaat, M., 1998, Abnormally high fluid pressure and seal impacts on hydro-carbon accumulations in the Nile Delta and North Sinai Basins, Egypt, *in* Law, B.E., G.F. Ulmishek, and V.I. Slavin eds., Abnormal pressures in hydrocarbon environments: AAPG Memoir 70, p. 161–180.

Chapter 10

Abnormally High Formation Pressure and Seal Impacts on Hydrocarbon Accumulations in the Nile Delta and North Sinai Basins, Egypt

M. Nashaat
Egyptian General Petroleum Corporation
Cairo, Egypt

Abstract

The Nile Delta and North Sinai Basins are active geodynamic (high subsidence rate) basins with a thick, clay-dominated Oligocene to Recent sedimentary section. Abnormally high formation pressures have developed in this section and in the underlying pre-Tertiary section primarily due to rapid sedi-mentation. Secondary mechanisms may be locally superimposed where the Messinian evaporite super seal is present. The abrupt development of pore pressure in the southern part of the Nile Delta is believed to be due to changes in the volume of pore fluids or rock matrix as a consequence of either aquathermal expansion, hydrocarbon generation, or thermal cracking of oil to gas in the lower Miocene-upper Oligocene compartment. Fluid flow in the Nile Delta and North Sinai Basins is mainly due to compaction-and thermal-driven forces.

The sedimentary sequence in the study area is divided into eight pressure compartments, separated by seals, some of which are associated with major unconformities. Four seals are clearly demonstrated in the North Sinai, Early Cretaceous basin and are referred to as: 1) Upper Jurassic-Lower Cretaceous; 2) Aptian; 3) Albian; and 4) Upper Cretaceous-Eocene carbonates. A total of four seals are also recognized in the Nile Delta and are referred to as: 5) Aquitanian-Burdgalian; 6) Langian; 7) Serravalian-Tortonian; and 8) Messinian.

INTRODUCTION

The Nile Delta and North Sinai Basins cover an area of 76,600 km² (29,540 sq.mi.). These basins are further subdivided into three basins that include the middle to late Miocene deltaic basin, the late Oligocene to early Miocene basin, and the Early Cretaceous basin (Figure 1). These basins are bounded on the south by the hinge fault zone on the stable carbonate platform. The north-ern boundary extends offshore, to the deeper parts of the Mediterranean Sea (Figure 1). The middle to late Miocene deltaic basin is bounded on the west by the upper Cretaceous platform, west of Abu Qir field (Fig-ure 1). The east and west boundaries of the early Miocene-late Oligocene basin are defined by shear zones (Figure 1) and the eastern boundary of the Early Cretaceous basin is believed to be bounded by a north-east-southwest trending transcontinental megashear (Jenkins, 1990).

A total of 172 wells have been drilled in this region. The first exploratory well in the Nile Delta was drilled in 1966. Since that time, 6 fields have been put onstream. However, more than 15 new gas (and con-densate) discoveries in the Nile Delta, and one oil dis-covery in the North Sinai have been defined. Seven of these are awaiting development operations in the very near future. Cumulative gas production from these fields through 1995 is 235.5 billion ft³ (6.7 x10⁹ m³) of gas from late Miocene sandstone reservoirs. The deep-est well (Mango-1) in the region was drilled to a total depth of 15,188 ft (4629 m) in Upper Jurassic rocks. The oldest rocks penetrated in the North Sinai Basin are those of the Jurassic Masajid Formation, while the old-est penetrated in the Nile Delta are in the Oligocene Tineh Formation.

Overpressured zones have been reported while drilling in the Nile Delta and North Sinai Basins. The evaluation of overpressure in these basins is becoming very important due to the association of abnormally high pressure with the recent offshore hydrocarbon discoveries (Maghara, 1978). Drilling operations in this region are confronted with complications associated with abrupt increases of pore pressure which are difficult to anticipate. As a consequence of the association of hydrocarbon accumulations with abnormally high pressure zones and the problem of drilling in abnormally pressured rocks, this study was initiated in order to predict the magnitude and occurrence of abnormally high pressure zones in both the Nile Delta and North Sinai Basins.

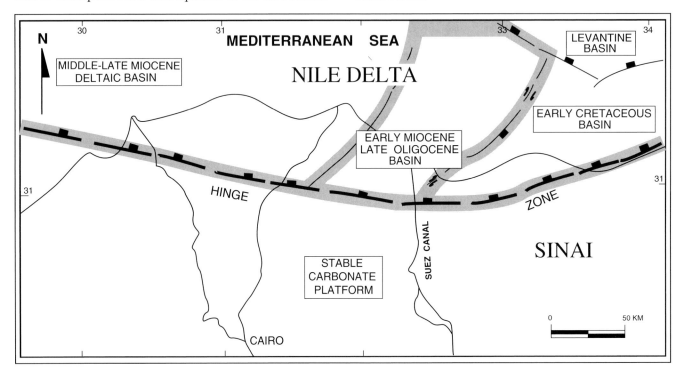

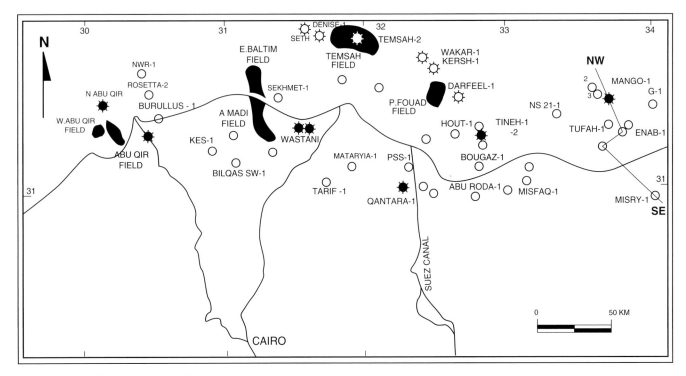

Figure 1. Location map of the Nile Delta and North Sinai Basins showing the principal structural areas, oil and gas fields, and selected wells.

GEOLOGIC SETTING

The Nile Delta and North Sinai regions are located in the northern continental to marine transitional margin of the African and Sinai Plates (Figure 1). The Nile Delta and North Sinai Basins are divided into three distinct structural regions (Figure 1). From west to east they are: 1) the middle to late Miocene Deltaic basin, 2) the early Miocene to late Oligocene basin in the eastern part of the Nile Delta, and 3) the Early Cretaceous basin in the northern offshore part of Sinai Peninsula. A hinge line separates the southern carbonate platform from the Nile Delta and Sinai Basins (Figure 1).

The break-up (rifting) of this area started in Late Triassic time and probably extended through Early Cretaceous time (Tethys opening). By the end of the Early Cretaceous, an extensional phase took place causing the northern part of Egypt, including the study area, to sink (Paul, 1991).

The second tectonic phase was dominated by northwest to southeast oriented oblique compression in Late Cretaceous (Laramide) time, ending in the Eocene. This phase reflects the closing of the Tethys which produced a series of en echelon northeast-southwest trending, doubly plunging anticlines (Syrian Arc structures) in the northern Sinai.

The last significant tectonic phase in the area occurred during late Oligocene to early Miocene time, with uplift and separation of the Arabian Platform from the Levantine Basin (Figure 1), culminating in middle Miocene time. Finally, during early Pliocene time, a northwest-southeast trending listric (wrench) fault pattern developed in the northern offshore area.

Sedimentation of the Nile Delta was initiated during late Oligocene-early Miocene time, west of the present-day location (Said, 1981). During late Oligocene to early Miocene time, coincident with the rifting of the Gulf of Suez, the delta abruptly shifted to a more northeast position, in the vicinity of the Port Said South-1 (PSS-1) well location (Figure 1). The present-day location of the Nile Delta was established during middle to late Miocene time.

A generalized stratigraphic column for the region is shown on Figure 2. The Lower Cretaceous rocks in the North Sinai Basin were deposited in fluvial dominated environments and are mainly composed of shale and sandstone with minor, thin beds of dolomitic limestone (Paul, 1991). Maximum thickness of the Lower Cretaceous sediments penetrated in the basin is 5,965 ft (1,818 m). During Late Cretaceous to Eocene time, chalky and marly limestones were uniformly deposited through the area in an open marine environment. The upper Oligocene to lower Miocene sequence (Tineh Formation) is represented by a marine shale section with some sandstone interbeds (possibly turbidite sand influx). In the Nile Delta, lower to upper Miocene rocks include the Qantara, Sidi Salim, Qawasim, Abu Madi, and Rosetta Formations. The Qantara Formation consists mainly of shale, limestone (dolomitic in some cases) and marl with some interbeds of sandstone. The Sidi Salim Formation is composed of shales with a few interbeds of limestone and sandstones. The Qawasim Formation is mainly sandstone and conglomerate in the southern parts of the region changing to shale northward. The Abu Madi Formation is represented by a series of sandstones, conglomeratic sandstones, and shales. The overlying Rosetta Formation consists of evaporites, interbedded with thin claystone beds. From Pliocene to Recent time, the area has experienced very high rates of deposition of claystone and siltstone and a few thin interbeds of sandstone.

METHODS OF STUDY

Data from 45 wells were utilized in the study. The pressure data were derived from wireline formation tests (WFT), drillstem tests (DST), mud weights, well kicks, leak-off tests (LOT), and wireline acoustic logs. The wireline formation tests, drillstem tests, and well kick data are the most reliable for pressure evaluation.

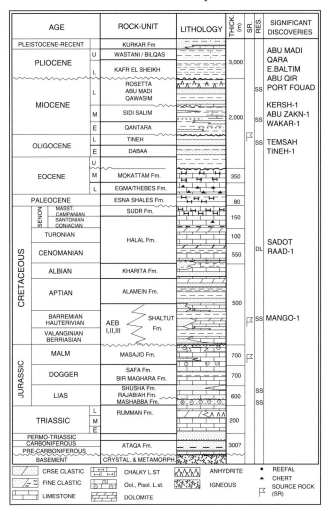

Figure 2. Generalized lithostratigraphic column of the Nile Delta and North Sinai Basins (after EGPC, 1994).

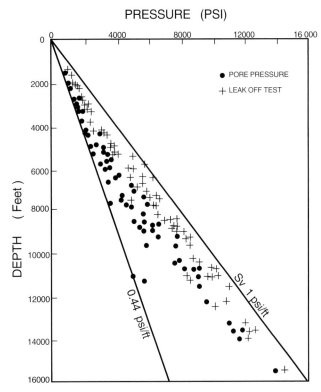

Figure 3. Composite pore pressure vs. depth profile in the Nile Delta and North Sinai Basins. Note the abnormal increase in the pore pressure and the leak-off test (fracture pressure) magnitudes at approximately 6,000 ft.

From these data, pressure versus depth plots (PD plots) were constructed in order to identify the pressure regimes in each basin. The pressure regimes were differentiated as hydrostatic, near hydrostatic, and overpressured regimes.

In order to evaluate the pressure systems in the study area, pressure gradient maps were constructed for upper Miocene and middle to upper Miocene rocks. The efficiency of seals was evaluated by calculating pressure changes across suspected seals and then constructing excess pore pressure maps for selected seals.

The assessment of excess fluid pressure development was facilitated by using a fluid flow compaction model (BasinMod 1D, version 4.0, Platte River Associates, Denver, Colorado, U.S.A.). This model identifies the hydrocarbon generation window based on the burial history of the basin and also identifies the most appropriate compaction model which, in turn, can be used to quantify pore pressure development.

The term absolute excess pressure (AEP) is defined as the magnitude of pressure at a given depth point above normal hydrostatic pressure for the same depth. It is used for the purposes of quantification of the fluid pressure compartments. Thus, it is believed that reporting pressure in each compartment as AEP rather than pressure gradient is preferable. In seals, the terms differential excess pressure across seals, seal gradient, and seal efficiency factor were used.

PRESSURE COMPARTMENTS

The concept of pressure compartments was first introduced by Bradley (1975), Powley (1987) and Hunt (1990). Pressure compartments must have vertical and lateral seals. Each pressure compartment has similar pressure conditions. Abnormally high pressure was defined in the Nile Delta and North Sinai to depths down to nearly 16,000 ft (4,900 m) (Figure 3). The pressure compartments were found to occur in several Jurassic to Pliocene intervals. These pressure compartments generally have well defined vertical and lateral seals. Lateral seals may occur as faults, salt or mud diapirs, or along facies change boundaries.

For reference, a normal hydrostatic pressure gradient in the region is 0.44 psi/ft (9.95 kPa/m), equivalent to 30,000 ppm formation water salinity. Maximum pressure gradients in the study area are as high as 0.89 psi/ft (20.2 kPa/m). The depth to the top of abnormally high pressured rocks is quite variable, ranging from about 1,700 ft (520 m) to 12,140 ft (3,700 m) (Figure 3 and 4a). The shallower depths to the top of overpressured zones in Pliocene rocks occur in areas of relatively high sedimentation rates (Figures 4a and 4b). Sedimentation rates in these areas are as high as 70 cm/1,000 yr (Figure 4b), which according to Mann and Mackenzie (1990) and Osborne and Swarbrick (1995) are sufficient to cause compaction disequilibrium and abnormally high pressures. Because of the variability of the magnitude and occurrence of overpressure, the pressure compartments within the study area are discussed with reference to the principal structural domains. Figures 5 and 6 show the pressure compartments and their stratigraphic positions in the Nile Delta and Sinai Basins, respectively.

Stable Platform and Hinge Zone

The stable carbonate platform, located south of the Hinge zone (Figure 1), has a normal hydrostatic pressure regime. The Tertiary rocks are thin and sandy compared to the thicker and more shale-rich equivalent rocks in the more distal, offshore areas of the delta.

Nile Delta Basin

The oldest rock unit penetrated in the western side of the Nile Delta Basin is the middle Miocene sandstone, at a depth of 14,997 ft (4,571 m) in the Bilqas SW-1 well (Figure 1). There are only a few wells in this side of the delta that have penetrated rock units older than the middle Miocene Sidi Salim Formation and consequently, there are no reliable pressure data for these older rocks in this area. The oldest and deepest rocks penetrated in the northeastern part of the Nile Delta area are Oligocene strata at a depth of 15,529 ft (4,733 m) in the Wakar-1 well (Figure 1).

Pliocene rocks in the Nile Delta have different pressure profiles and consequently, are subdivided into dif-

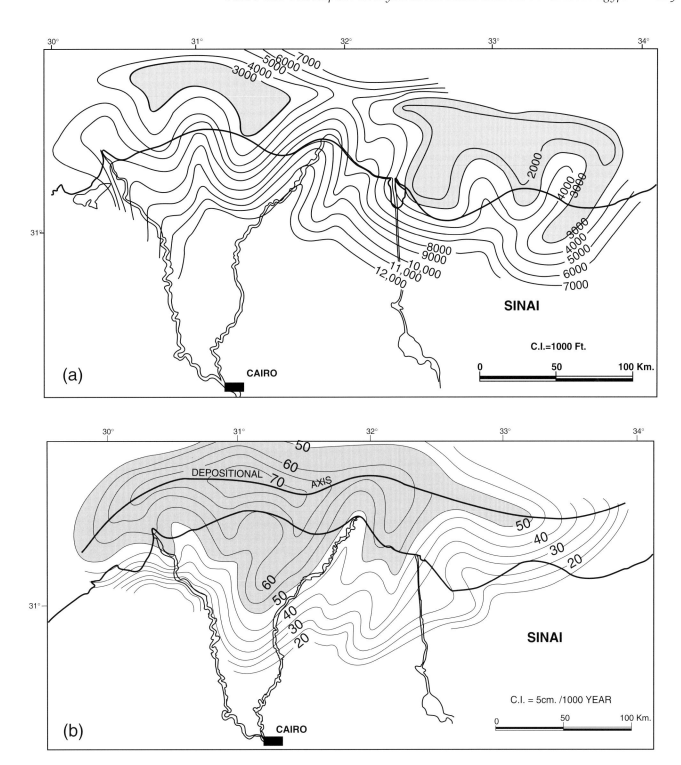

Figure 4. Maps of the Nile Delta and North Sinai Basins showing: (a) depth to the top of overpressure zones (in ft); (b) rates of sedimentation for Pliocene and Pleistocene sediments (reprinted from Nashaat, 1992).

ferent pressure regimes. These regimes may be either normally pressured (near Hinge Zone) or overpressured in the offshore area. In the overpressured regimes, a gradual increase (step-like in some cases) of pore pressure with depth usually occurs in shale intervals within the Pliocene sequence.

In the offshore, eastern part of the Nile Delta, four pressure compartments are recognized (Figure 5). These pressure compartments, in ascending order, include the following: (1) the upper Oligocene to lower Miocene Tineh compartment (I), (2) the lower Miocene Qantara compartment (II), (3) the middle Miocene Sidi

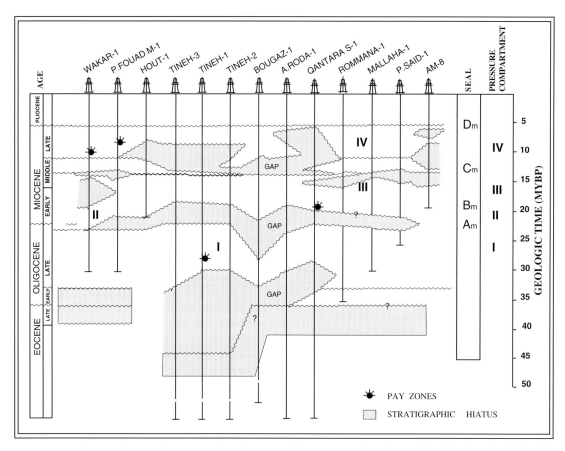

Figure 5. Chronostratigraphic chart of the Northeastern part of the Nile Delta illustrating the main stratigraphic gaps, the main pressure seals, and pressure compartments (from Nashaat, 1994).

Salim compartment (III), and (4) the upper Miocene Qawasim and Abu Madi compartment (IV).

Compartment I

The upper Oligocene-lower Miocene compartment (I) (Figure 5) contains the Tineh Formation. The pressure gradient is 0.83 psi/ft (18.77 kPa/m) in Tineh-1 and Temsah-1 wells and 0.84 psi/ft (19 kPa/m) in the Wakar-1 well (Figures 1, 7, and 8). The absolute excess pressure (AEP) ranges from 5,000 to 5,400 psi (34.5 to 37.2 MPa) above normal hydrostatic in the offshore area. In the onshore area, the pressure gradient is 0.87 psi/ft (19.68 kPa/m) in the Tarif-1 well (Figures 1 and 9) and the AEP is 5,600 psi (38.6 MPa). Structural blocks adjacent to the hinge fault complex have low pressure gradients because of pressure dissipation along fault planes. This compartment contains the best Tertiary hydrocarbon source rocks in the region. The average total organic carbon content (TOC) of upper Oligocene to lower Miocene shales is 2.5% and the thermal maturity of these rocks ranges from 0.55 to 0.7% R_o (Nashaat, 1994). The gas and condensate in the Tineh-1 and Qantara-1 wells (Figure 1) occur in this compartment.

Compartment II

Pressure compartment II contains the lower Miocene Qantara Formation (Figure 5). The pressure gradi-

ent ranges from hydrostatic near the hinge line in the southern part of the area, to 0.8 psi/ft (18.1 kPa/m) in the Port Said South-1 and Matariya-1 wells (Figure 1). The AEP in the Port Said South-1 well (Figure 1) is 3,870 psi (26.7 MPa). In the Tarif-1 well (Figure 9), the pore pressure build-up in this compartment is abrupt in a step-like manner, implying some sort of disequilibrium in vertical fluid flow across minor seals. To the north, in the Temsah field (Figure 1), the pressure gradient is 0.78 psi/ft (17.74 kPa/m) and the AEP is 4,500 psi (31 MPa). The Qantara Formation may have some source rock potential, but it is less thermally mature compared to the upper Oligocene-lower Miocene source rock in compartment I. Vitrinite reflectance (R_o) is less than 0.6 % in compartment II (Nashaat, 1992).

Compartment III

Compartment III contains the middle Miocene Sidi Salim Formation (Figure 5). The pressure gradient map of the middle Miocene Sidi Salim Formation (Figure 10b) indicates a lateral change of pressure gradient from 0.7 psi/ft (15.83 kPa/m) and an AEP of 3,450 psi (23.8 MPa) in the Temsah field (Figure 1) to a normal, hydrostatic gradient near the hinge line. Pressure reversals are also recognized in this compartment in the northeastern part of the Nile Delta at Wakar (Figures 1 and 8), Temsah, and Port Fouad condensate and gas fields (Figure 1). The pore pressure drops in the Tortonian sandstone reservoirs from 0.75 to 0.70 psi/ft (16.96 to 15.83 kPa/m) in the Temsah Field, and from 0.76 to 0.65 psi/ft (17.19 to 14.7 kPa/m) in the Wakar-1 well (Figure 8).

Compartment IV

Compartment IV (Figure 5) consists of the upper Miocene Qawasim and Abu Madi Formations. The upper boundary of the compartment is composed of evaporites in the upper Miocene Rosetta Formation. The Qawasim and Abu Madi Formations are composed dominantly of shale and sandstone. The maximum thickness of this compartment is 2,300 ft (700 m).

Reservoirs in Abu Madi gas field are characterized by a pressure reversal back to normal hydrostatic. Condensate and gas are produced from overpressured reservoirs (Tortonian sands) at the Port Fouad Marine and Wakar fields in the northeastern part of the Nile Delta region (Figure 1).

In Miocene rocks below the Messinian evaporite seal, pressure always increases, either very abruptly or gradually. The pore pressure gradient in this compart-

ment varies from 0.75 psi/ft (16.96 kPa/m) in the northern offshore area to normal hydrostatic in the onshore area (Figure 10a). The AEP at Temsah field (Tortonian sands) is 3,600 psi (24.8 MPa).

North Sinai Offshore Basin (Eastern Mediterranean)

The Early Cretaceous basin in the North Sinai offshore (Figure 1) has a thin Tertiary section compared to the Nile Delta. In this area, the maximum pressure gradient in Lower Cretaceous rocks is 0.85 psi/ft (19.2 kPa/m) in the NS 21-1 well. The western extent of the Lower Cretaceous overpressured zone is not defined by drilling to date. The eastern extent, however, has recently been defined by the G-1 well (Figure 1). In this well, the pressure gradient is abruptly reduced from 0.73 psi/ft (16.5 kPa/m) in Mango-1 well to 0.59 psi/ft (13.3 kPa/m) in the Lower Cretaceous equivalent rocks (Compartment IIIc).

In the Misfaq-1 onshore well (Figure 1), the pore pressure undergoes a significant pressure reduction from 0.62 psi/ft (14 kPa/m) in the Upper Cretaceous-

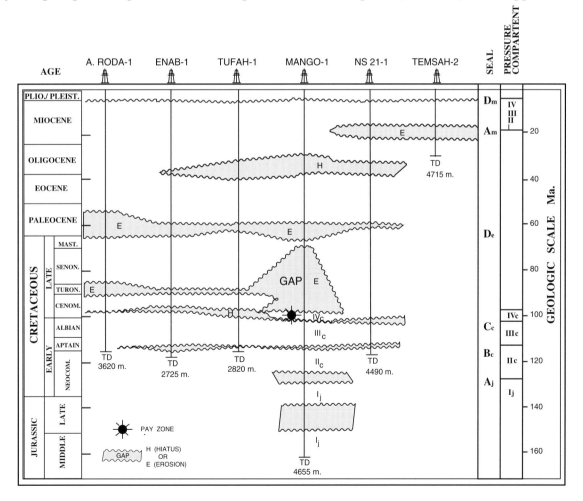

Figure 6. Chronostratigraphic chart of the North Sinai Early Cretaceous Basin illustrating the main stratigraphic gaps, the main pressure seals, and compartments.

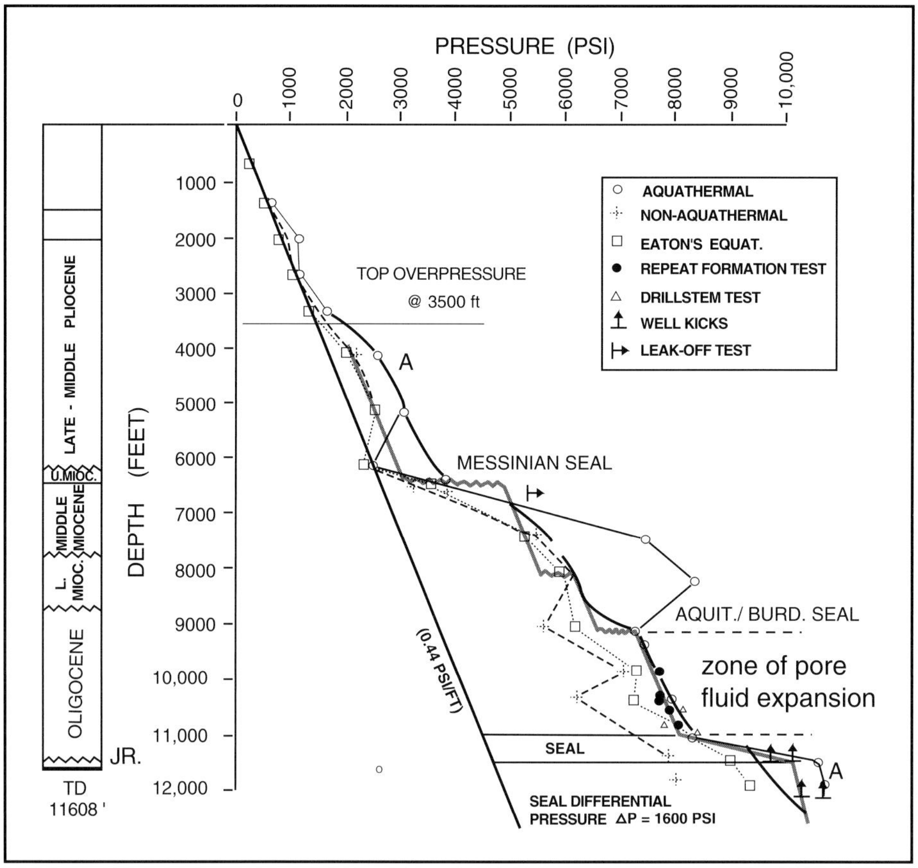

Figure 7. Pressure-depth profile of Tineh-1 well, showing coincidence of direct pressure values with those calculated from the aquathermal equation in the Oligocene to lower Miocene rocks (from Nashaat, 1991).

PRESSURE (PSI)

TOP OVERPRESSURE @ 3500 ft

○ AQUATHERMAL
⋮ NON-AQUATHERMAL
□ EATON'S EQUAT.
● REPEAT FORMATION TEST
△ DRILLSTEM TEST
⊥ WELL KICKS
⊢ LEAK-OFF TEST

A

MESSINIAN SEAL

AQUIT./ BURD. SEAL

zone of pore fluid expansion

SEAL

(0.44 PSI/FT)

SEAL DIFFERENTIAL
PRESSURE ΔP = 1600 PSI

A

DEPTH (FEET)

LATE - MIDDLE PLIOCENE
U.MIOC.
MIDDLE MIOCENE
L. MIOC.
OLIGOCENE
JR.
TD 11608'

Eocene carbonates to normal hydrostatic pressure, 0.44 psi/ft (9.9 kPa/m), in Lower Cretaceous clastic rocks over a vertical distance of 410 ft (125 m).

Internally, the Lower Cretaceous rocks are subdivided into three nested compartments in addition to the Upper Jurassic compartment. These compartments are shown in Figures 6 and 11. The rocks are mostly composed of clastics with minor stringers of carbonate. In ascending order the pressure compartments are as follows: (1) the Upper Jurassic compartment (Ij), (2) the Aptian-Neocomian compartment (IIc), (3) the Aptian-Albian compartment (IIIc), and (4) the Albian compartment (IVc).

Compartment Ij

The Upper Jurassic compartment (Ij on Figure 6) has been penetrated in only the Mango-1 well (Figures 1, 11, and 12). This compartment contains the principal mature source rock in the North Sinai region. The top of the compartment is bounded by the Upper Jurassic-Lower Cretaceous Aj seal (Figures 6, 11, and 12). The bottom of the compartment is not known because of insufficient drilling depths. The pressure gradient is 0.85 psi/ft (19.2 kPa/m) and the AEP is approximately 5,400 psi (37.2 MPa).

Compartment IIc

The Aptian-Neocomian compartment (IIc on Figures 6, 11, and 12) is dominantly composed of shale interbedded with sandstone and limestone (dolomitic in parts). The upper boundary of this compartment is the thick Aptian seal. The thickness of this compartment is 1,975 ft (602 m) in the Mango-1 well (Figure 12). Maximum pressure in this compartment occurs in the NS 21-1 well with a pressure gradient of 0.85 psi/ft (19.2 kPa/m) and an AEP of 5,775 psi (39.8 MPa). The lowest pressure occurs in the Mango-3 well with a pressure gradient of 0.71 psi/ft (16 kPa/m) and an AEP of 2,400 psi (16.6 MPa). The shallow Enab-1 well (Figures 1 and 11) has a high pressure gradient of 0.83 psi/ft (18.8 kPa/m) and the AEP is 3,200 psi (22.1 MPa).

Compartment IIIc

The Aptian-Albian compartment (IIIc, on Figures 6, 11, and 12) is composed of shale and sandstone with thicknesses ranging from 1,475 ft (450 m) in the Mango-1 well to 758 ft (231 m) in the Enab-1 well. The highest pressure gradient in compartment IIIc occurs in the NS 21-1 well (Figure 1) where the gradient is 0.81 psi/ft (18.3 kPa/m) and the AEP is 4,987 psi (34.38 MPa). The lowest pressure gradient occurs in the G-1 well where the pressure gradient is 0.59 psi/ft (13.3 kPa/m) and the AEP is 1,440 psi (9.9 MPa). In the Mango-1 well the pressure gradient is 0.73 psi/ft (12.2 kPa/m) and the AEP is 2,800 psi (19.3 MPa) (Figures 1, 11, and 12). The pressure gradient in the nearby Mango-3 well is similar to the Mango-1 well, while the AEP is considerably different at 2,316 psi (15.96 MPa). In the Enab-1 well

(Figures 1 and 11), the pressure gradient is 0.72 psi/ft (16.3 kPa/m) and the AEP is 2,000 psi (13.8 MPa).

Compartment IVc

The Albian pressure compartment (Figure 6) as exemplified by the Mango-1 well (Figure 12) is bounded above by the Upper Cretaceous-Eocene carbonate seal. The maximum thickness of this compartment is 790 ft (240 m). In the Mango-1 well (Figure 12), the pressure gradient is 0.735 psi/ft (16.6 kPa/m) and the AEP is 2,488 psi (17.15 MPa). The lowest pressure in this compartment is less than 0.55 psi/ft (12.4 kPa/m) in G-1 well (Figure 1). A relatively low pressure gradient also occurs in the Enab-1 well (Figures 1 and 11), where the pressure gradient is 0.63 psi/ft (14.2 kPa/m) and the AEP is 1,400 psi (9.6 MPa). The highest pressure gradient in this compartment is in the NS-21 well (Figure 1) where the pressure gradient is 0.76 psi/ft (17.2 kPa/m) and the AEP is 3,900 psi (26.9 MPa).

Water Quality

There are two genetic water types in two separate water chemistry zones recognized in the North Sinai Basin (Nashaat, 1990a). The first represents the Lower Cretaceous aquifer, and is dominated by $NaOHCO_3$

water type I (10,000–30,000 ppm). The second represents the Upper Cretaceous-Eocene aquifer and is dominated by $CaCl_2$ water type II (>35,000 ppm). The Lower Cretaceous aquifer has been subjected to mixing and dilution conditions caused by shale ion-filtration during compaction history (Nashaat, 1990a).

The definitions of the these pressure compartments based on pressure conditions alone are, to some extent, reinforced by the salinity of formation water in each of the compartments. Three to four different hydrochemical systems are coincident with the fluid pressure compartments (Figures 6 and 11). The deepest compartment, (Ij), which exhibits 22,000 ppm water salinity, is significantly different from the salinity of compartment IIc, with a salinity of 15,000 ppm. The range of water salinity in compartments IIc and IIIc is 13,500 to 18,000 ppm. In this case, there is not enough difference to distinguish the compartments on the basis of water salinity alone. However, compartment IVc has a water salinity of 30,000 ppm, considerably different from the underlying compartments in most wells. The only exception observed is in the Enab-1 well (Figure 1), where compartments IIIc and IVc have essentially the same formation water salinity of approximately 30,000 ppm. In this case, perhaps there

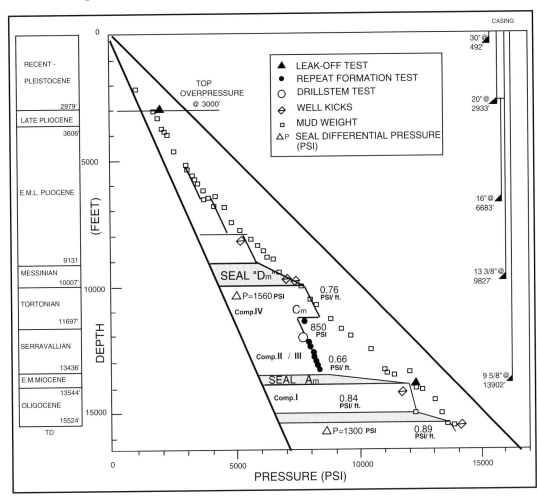

Figure 8. Pressure-depth profile of the Wakar-1 well which shows seals and reversed pressure interval (shaded). (from Nashaat and Campel, 1993). Location of well shown on Figure 1.

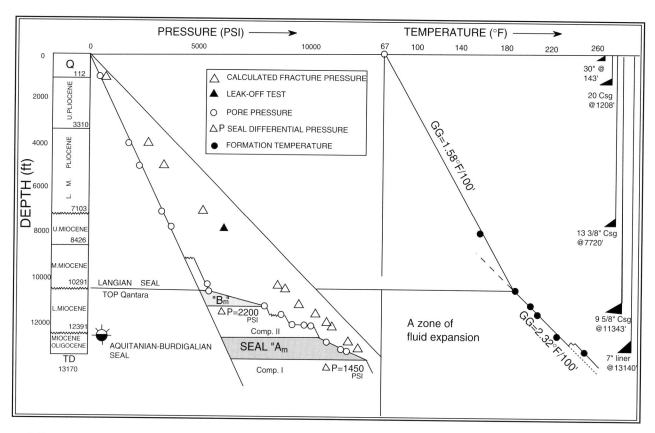

Figure 9. Pressure and temperature profiles of Tarif-1 well showing the zone of fluid expansion (aquathermal pressuring and/or hydrocarbon generation) in the early Miocene-Oligocene rocks (from Nashaat, 1994). Location of well is shown on Figure 1.

has been some communication between compartments along fault planes.

PRESSURE SEALS

The development of effective pressure seals is of critical importance in the definition of the pressure compartments. Effective pressure seals may be formed by faults, salt and mud diapirs, and vertical or lateral facies changes. In the Nile Delta, there is a coincidence between the occurrence of unconformities and pressure seals (Figure 5). This is due to the presence of impermeable lithologies such as evaporites or limestones adjacent to the unconformity. The development of diagenetic seals may also play a significant role in seal efficiency at these unconformable surfaces. Lateral, seal-bounding overpressure domains in the southern areas of the Nile Delta and North Sinai Basins are probably associated with the hinge fault zone (Figure 1).

Seals in the Nile Delta

Four regional seals are recognized in the eastern Nile Delta Basin (Figure 5). In ascending order they are referred to as: (1) the Aquitanian-Burdgalian seal (Am), (2) the Langian seal (Bm), (3) the Serravalian-Tortonian

seal (Cm), and (4) the Messinian seal (Dm). In the northern offshore areas, an additional pressure seal is recognized in lower Pliocene rocks. For example, in the Temsah-2 well (Figure 1), this seal is recognized in lower Pliocene rocks in the depth interval 8,300 ft to 8,800 ft (2,530 m to 2,680 m). The differential pressure across this Pliocene seal is 1,000 psi (6.9 MPa). In the same well, the Langian seal (Bm), equivalent to the top of the Qantara Formation, is not recognized. However, the seal Bm is well defined in the southeastern part of the Nile Delta.

Seal Am

The Aquitanian-Burdgalian Seal, Am (Figure 5), is well recognized throughout the eastern and northeastern part of the Nile Delta. It forms the upper seal for pressure compartment I. It may include the uppermost part of Oligocene rocks. The seal is mostly composed of shale interbeds with a few thin limestone beds. The thickness of the seal ranges from 250 ft (75 m) in the Temsah-2 well (Temsah field on Figure 1) to 300 ft (90 m) in the Tarif-1 well (Figure 1). The seal has an important role in regulating hydrocarbon migration from underlying source rocks.

The seal Am is the most efficient seal on the eastern side of the Nile Delta, where the depocenter of the basin was located during deposition of the seal (Figure

13b). The maximum difference in pressure across the seal is approximately 3,350 psi (20.7 MPa). In the northeastern offshore part of the Nile Delta in Temsah-2 well (Figure 1), the differential pressure across the seal is 1,100 psi (7.6 MPa). However, in the southeastern part of the Nile Delta (onshore) in the Tarif-1 well (Figures 1 and 9), the differential pressure across the seal is found to be slightly greater (1,450 psi or 10 MPa).

Figure 10. Pressure gradient maps for (a) the upper Miocene Qawasim and Abu Madi Formation and (b) the middle to lower Miocene Sidi Salim Formation in the Nile Delta and North Sinai Basins (from Nashaat, 1991).

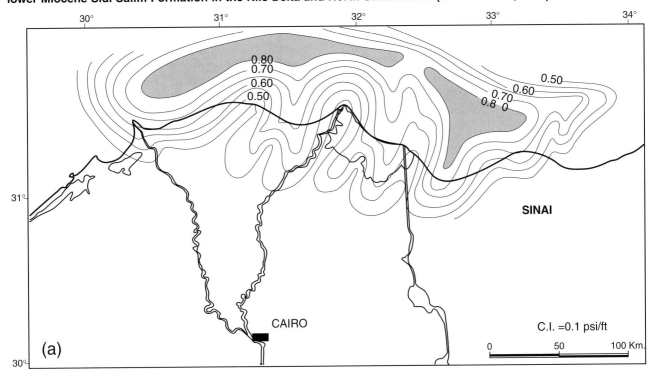

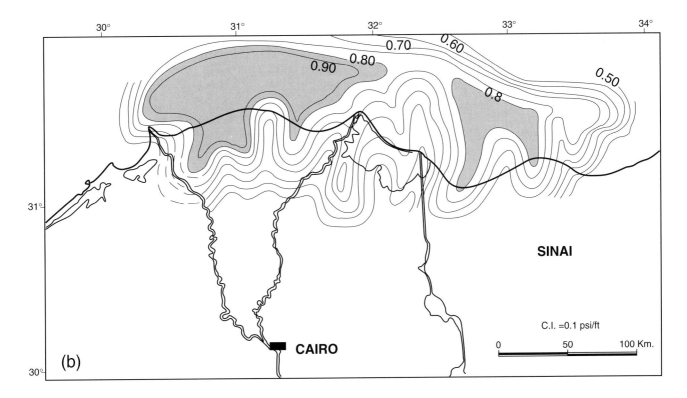

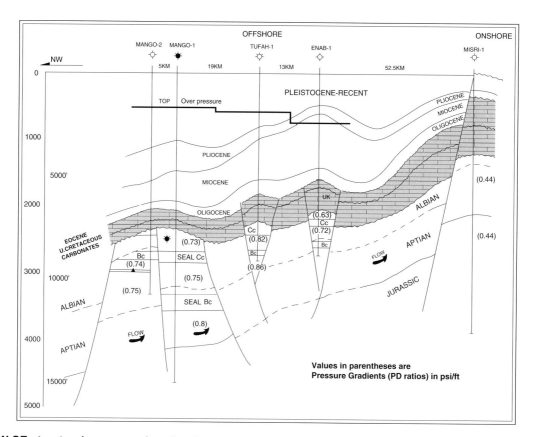

Figure 11. NW-SE structural cross section showing relationships among different pressure compartments in the North Sinai Basin. The overpressure regime in the offshore changes to normal hydrostatic pressure in the southern onshore block across the fault bounded Misri-1 block (from Nashaat, 1990b; reprinted by permission).

Seal Bm

The Langian seal (Bm on Figure 5) is of early to middle Miocene age and is easily traced throughout most of the southeastern part of the Nile Delta. The seal forms the upper boundary of the early Miocene Qantara Formation pressure compartment II and is mostly composed of shale and limestone beds with an average thickness of 650 ft (200 m). The maximum differential pressure of this seal occurs in the Tarif-1 (Figure 9) and PSS-1 wells in the southeastern part of the onshore Nile Delta (Figure 1). Differential pressures across the seal in these wells are 2,450 psi (17.1 MPa) and 2,200 psi (15.2 MPa), respectively. Near the hinge fault complex area (Figure 1), pressure differences across this seal have not been recognized, most likely because of pressure dissipation along fault planes.

Seal Cm

The middle to late Miocene Cm seal (Figure 5) forms the upper boundary of the middle Miocene (Sidi Salim Formation) pressure compartment III. The seal is primarily composed of limestone with shale and sandstone intercalations. The thickness of the seal ranges from 100 ft (30 m) in the PSS-1 well (Figure 1) to 1,850 ft (565 m) in the Matariya-1 well (Figure 1). The maximum seal efficiency occurs in the PSS-1 and Matariya-

1 wells (Figure 1). A differential pressure of at least 3,000 psi (20.7 MPa) was measured in the Matariya-1 well. The seal differential pressure in the Temsah-2 well (Figure 1) is 1,100 psi (7.6 MPa). It is generally accepted that the Cm seal is also the seal for hydrocarbon accumulations in the Temsah field.

Seal Dm

The Messinian seal (seal Dm in Figure 5) forms the upper boundary of the late Miocene pressure compartment IV. The Dm seal is composed of evaporites and is as thick as 200 m. In some areas the evaporite seal is absent. The maximum differential pressure across the seal is approximately 3,000 psi (20.7 MPa) in the Burrullus-1 well (Figures 1 and 13a). The late Miocene evaporites also form an efficient seal in the North Sinai Basin. For example, in the Mango-1 well (Figures 1 and 12), the pressure difference across this seal is 882 psi (6.1 MPa).

The high efficiency of the Messinian seal is mainly attributed to the presence of evaporites in the Rosetta Formation. Where the seal is absent, hydrocarbon-bearing fluids can migrate to younger reservoirs. For example, the recent discovery in Pliocene sandstone reservoirs in the western extension of the Temsah Field

(Figure 1) is due, in part, to the absence of Rosetta Formation evaporites.

Seals in the North Sinai Basin

In the North Sinai Early Cretaceous basin, four regional seals are identified (Figure 6). In ascending order they include: (1) the Upper Jurassic-Lower Cretaceous seal (Aj), (2) the Aptian seal (Bc), (3) the Albian seal (Cc), and (4) the Upper Cretaceous-Eocene seal (De).

In a regional sense, one of two possible scenarios could have resulted in lateral sealing at the western margin of the Early Cretaceous rocks in the North Sinai Basin. In the first scenario, the lateral seal was formed by the convergence of the Upper Cretaceous-Eocene carbonates (seal De) beneath the younger Tertiary rocks. Thus, the carbonate section (mentioned later) may serve as a top seal in the North Sinai offshore basin or as a lateral seal where it occurs beneath the very thick Tertiary sediments in the Nile Delta Basin. The second scenario consists a of single overpressure regime including both the Nile Delta and North Sinai Basins. The overpressure domains in the North Sinai Early Cretaceous basin may have extended westward into the Nile Delta Tertiary rocks. In such a case, seals

would have to cross stratigraphic boundaries at the basin margins. It is commonly believed that both scenarios have some merit, but which is more likely remains unresolved.

Seal Aj

The Upper Jurassic-Lower Cretaceous (Oxfordian-Neocomian) seal Aj (Figure 6) has only been identified in the Mango-1 well (Figure 12), where it is composed of limestone and shale. The seal overlies compartment Ij which has some source rock potential in this area. In the Mango-1 well, the top of the seal occurs at a depth of 13,125 ft (4,000 m) and is 575 ft (175 m) thick. The difference in pressure across the seal is 1,030 psi (7.1 MPa)

Seal Bc

The Aptian seal Bc (Figure 6) overlies the Early Cretaceous (Barremian/Neocomian) pressure compartment IIc and is composed mostly of shale, dolomite, and dolomitic limestone. The sealing efficiency of this seal may be enhanced by the occurrence of calcite mineralization in the lower most part of the seal. The thickness of the seal ranges from 1,000 to 3,000 ft (305 to 915 m) and the differential pressure across the seal ranges from 1,400 psi (9.6 MPa) in the Mango-1 well, to 80 psi (0.6 MPa) in the Mango-3 well.

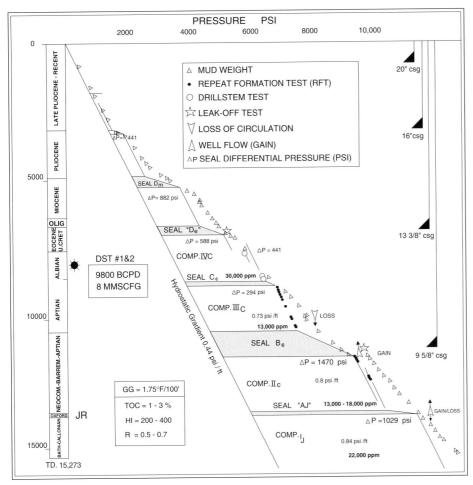

Figure 12. Pressure-depth profile in the Mango-1 well showing pressure compartments and seals (Nashaat, 1990b, 1994; reprinted by permission).

Seal Cc

The Albian seal Cc (Figure 6) forms the upper boundary of the Aptian pressure compartment IIIc and consists mainly of shale with minor thin beds of limestone. The average thickness of the seal is 300 ft (90 m). The difference in pressure across the seal ranges from 300 psi (2.1 MPa) in the Mango-1 well to 1,100 psi (7.6 MPa) in the NS 21-1 well (Figure 1).

Seal De

Seal De separates Tertiary from Mesozoic rocks (Figure 6). Seal De is a regional seal throughout both the Nile Delta and North Sinai Basins. The seal is composed of limestone and ranges in thickness from 300 to 900 ft (90 to 275 m).

The evaluation of seal De is inconclusive because of the unreliable pressure data collected from carbonate sequences. The pressure difference between Tertiary and Mesozoic rocks is at least 1,000 psi (6.9 MPa) as determined in the Mango-1 well (Figure 12). In the Tineh-1 (Figure 1) well, the difference in pressure between Tertiary and Mesozoic rocks is 1,600 psi (11.0 MPa on Figure 7).

Seal Efficiency

In order to compare the seal efficiency in this area, the following equation was utilized to quantify the seal efficiency factor (SEF):

$$SEF(psi/ft) = \Delta P \cdot H \cdot LHF$$

where ΔP is the difference in pressure across the seal (psi), H is the thickness of the seal (ft), and LHF is the lithological homogeneity factor which ranges from 1 to 0.9 for evaporite seals, 0.9 to 0.8 for overpressured shale, 0.6 to 0.4 for carbonates, and 0.4 to 0.2 for siltstone. Detailed sealing analysis was performed on the seals Bc and Cc taking into account the SEF, the pressure gradient in the seal rocks (SG), and the seal differential pressure (ΔP).

Application of these analyses to seal Bc shows that in the Mango-1, Ziv-1, and Enab-1 wells (Figure 14) seal Bc has a very good to excellent seal efficiency, while seal Cc is poor (leaky) in the same wells. In contrast, seal Cc in the NS 21-1 well (Figure 14) has an excellent efficiency but seal Bc is less efficient. In the Mango-3 and Tufah-1 wells, both seals Bc and Cc have fair to poor (leaky) efficiencies (Figure 14).

Pressure is released either in seismically active areas (fault re-activation) or when the magnitude of overpressure approaches the fracture gradient of the seal. In the case where the pore pressure equals or exceeds the fracture gradient of the seal, pressure-induced fractures develop and fluid loss will occur, causing an abrupt drop in the pore-fluid pressure (Caillet, 1993). The Mango-3, Dag-1, and Tufah-1 wells were drilled near the crest of structural highs, and were interpreted to have experienced seal failure, most likely due to the influence of tectonic deformation or pressure-induced fracturing. In contrast, wells drilled on the flanks of structures, such as the Mango-1, Ziv-1, and Enab-1 wells, showed very good to excellent sealing efficiency.

ORIGIN OF ABNORMAL PRESSURES

Of the several causes of overpressuring that have been proposed (see Swarbrick and Osborne, 1998 - this volume), the occurrence of abnormally high pressures in the Nile Delta and North Sinai is interpreted to be mainly due to high sedimentation rates (Figure 4b) of clay dominated sediments during middle Miocene to Pliocene time (Nashaat, 1992). The maximum rate of sedimentation during Pliocene and Pleistocene time was as high as 70 cm/1,000 yr., (Figure 4b). These rapid sedimentation rates caused rapid burial of sediments which, according to Mann and Mackenzie (1990), are sufficiently high to cause compaction disequilibrium and abnormally high pressures. Therefore, the top of overpressuring is mainly attributed to the rate of sedimentation during Pliocene to Pleistocene time. It has also been determined that the depths to the top of overpressuring are related to rates of sedimentation; higher rates of sedimentation are coincident with relatively shallow depth to the top of overpressuring. Hence, the top of overpressuring occurs at progressively shallower depths northward because of high rates of sedimentation and subsidence along the depositional axis during Pliocene and Pleistocene time (Figures 4a and 4b).

Evidence in support of compaction disequilibrium in older rocks (lower Miocene Qantara Formation) during Pliocene-Pleistocene deposition is provided by simulation modeling of pressure in the Abu Madi-8 well (Figure 15). The modeling shows a marked decrease in the pore pressure of the Qantara Formation during Messinian time (5.4 Ma), most likely due to uplift and erosion. Subsequently, excess fluid pressure rapidly developed during early to middle Pliocene time as a consequence of the high sedimentation rates of the lower Pliocene Kafr El Sheikh Formation. The increase of sand influx together with decreasing rates of sedimentation 3.5 Ma, during deposition of the Wastani/Bilqas Formations, also resulted in a dramatic decrease in pore pressure. Modeling indicates that the overpressuring observed in the Abu Madi-8 well was initiated 5 Ma, reaching maximum values about 4 Ma.

From the modeling, the question arises as to how high pressure could develop in the Pre-Messinian sediments after pressure dissipation. Lerche (1990) suggested that excess fluid pressure takes a considerably longer time than the porosity (by nearly a factor of 3) to reach equilibrium. According to Lerche (1990), excess fluid pressure is nonlinearly dependent on porosity and permeability, so that, even though the porosity has nearly completed its evolutionary approach to equilibrium, a very small porosity, in excess of the isostatic equilibrium value, corresponds to a relatively large

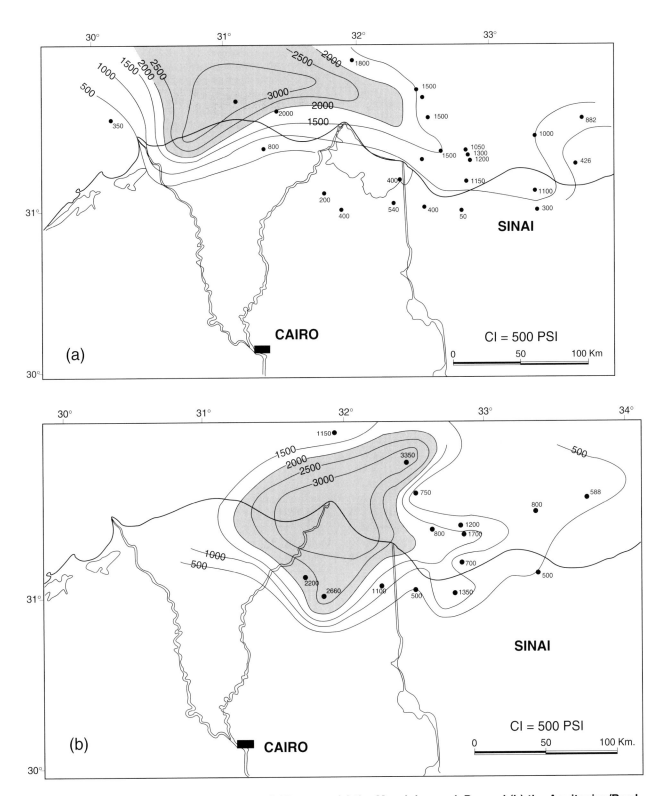

Figure 13. Maps showing differential pressure (ΔP) across (a) the Messinian seal, Dm and (b) the Aquitanian/Burdgalian seal, Am (from Nashaat, 1994).

excess fluid pressure. Since the permeability of the shale is now very low, it takes a very long time for the remaining amount of excess fluid pressure to re-equilibrate to hydrostatic (Lerche, 1990).

Accordingly, in the Nile Delta, the sharp increase of pore pressure below the Messinian evaporites is believed to be due to the presence of the Messinian seal that may have inhibited pressure dissipation during deposition of Pliocene sediments. The pressure modeling used in this example clearly demonstrates the role of compaction disequilibrium as an effective mechanism in generating overpressure in the study area.

Sand bodies interbedded with clay can be overpressured due to expulsion of fluids from the clay into the sandstone rocks (transference), and hence maintain pressure equilibrium in the entire rock system (Osborne and Swarbrick, 1995). Frequently, a combination of additional pressure processes are superimposed as well. Additional pressure-producing causal mechanisms which are believed to locally contribute to overpressuring in the Nile Delta are discussed below.

The role of aquathermal fluid expansion (Barker, 1972) as a pressure mechanism remains controversial, but overpressures caused by aquathermal expansion may be present in the Nile Delta. The coincidence of the direct pressure values from well testing and kicks with pressure values calculated from solving the aquathermal formula suggest that aquathermal expansion may have contributed to the development of overpressure. For example, in the eastern part of the Nile Delta, in the Tineh-1 well (Figure 1), there is good agreement between measured pressure values and calculated pressure values in the lower Miocene to upper Oligocene compartment (Figure 7). In addition, the

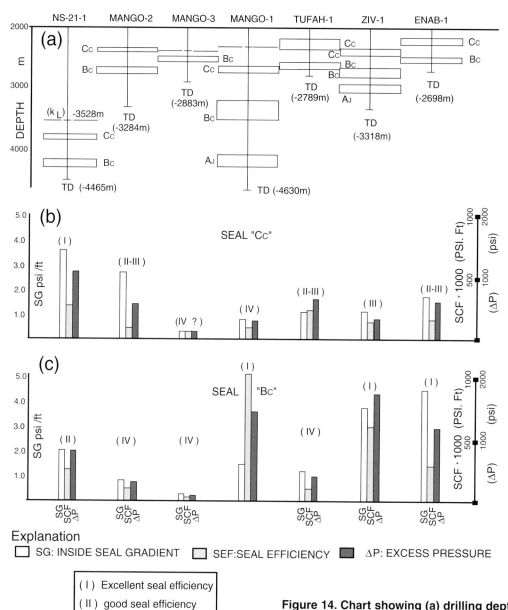

Figure 14. Chart showing (a) drilling depth to pressure seals in selected wells, and seal gradient, seal efficiency factor, and excess pressure for (b) seal Cc and (c) seal Bc in the North Sinai Basin (Nashaat, 1990b; reprinted by permission).

rapid development of pore pressure below seal Bm in conjunction with the temperature increase, as exhibited in the Tarif-1 well (Figures 1 and 9), in the southeastern part of the Nile Delta, is interpreted to be due to a combination of processes that include changes in the volume of pore fluids or rock matrix either by aqua-thermal expansion (Barker, 1972) or hydrocarbon generation and thermal cracking of oil to gas in the source rock. Geochemical modeling of lower Miocene to upper Oligocene source rocks indicates that the ratio of cumulative hydrocarbons expelled to cumulative hydrocarbons generated from source rock is very low (less than 20%) in some localities. This retained hydrocarbon (gas) may have contributed to the generation of overpressure in this compartment similar to that reported in the Scotian Shelf by Yassir and Bell (1994). In addition, the presence of listric growth faults in Pliocene rocks that are detached in Messinian evaporites probably has produced some tectonically induced pressure. This tectonically-induced pressure may have contributed to the development of overpressure in the Tineh-Bougaz area in the west-central part of the North Sinai Basin (Figure 1).

EFFECTS OF RESERVOIR PRESSURE AND TEMPERATURE ON HYDROCARBON ACCUMULATIONS

Formation temperature and pressure are important factors which significantly affect the physical state of hydrocarbons at depth (Modelevskiy and Parnov, 1967; Fertl, 1976).

In 1970, Timko and Fertl (1970) reported important pressure-temperature relationships in about 60 U.S. Gulf Coast wells. The most important findings were related to pressure gradient domains ranging from 0.5 psi/ft (11.31 kPa/m) to 0.94 psi/ft (21.26 kPa/m) and corresponding formation temperatures between 90° and 365°F (104° and 185°C). A temperature range of 215°–290°F (102°–143°C) coincides with the range of the highest pressure gradients in the hydrocarbon bearing zone. Timko and Fertl (1970) also concluded that with further increases of pressure and temperature, extremely high pressure gradients, encountered with high temperatures, are associated with zones of non-commercial oil and gas production.

By analogy to the U.S. Gulf Coast, a generalized correlation of temperature and pressure data between the Nile Delta-North Sinai Basins and the U.S. Gulf Coast data was established. Then, a pressure-temperature relationship was developed to determine favorable hydrocarbon accumulation environments in the study area. With such a correlation, the limits of each envelope or domain was modified after Modelevskiy and Parnov (1967) and Timko and Fertl (1970, 1972).

Thus, the pressure-temperature relationship in the study area is considered to be an important exploration tool in the identification of the occurrence and type of hydrocarbons (oil or gas) in traps. It also shows whether the hydrocarbon accumulation in the overpressured trap is economic or non-economic. This relationship can be studied by plotting static reservoir pressure against static reservoir temperature (Figure 16). Application of this technique indicates that Abu Madi gas field lies in the oil and gas envelope, while the Abu Qir field lies marginally in the oil and gas envelope. Both fields produce from upper Miocene (Messinian) sandstone reservoirs. The rest of the fields are consistent with the oil and gas limits (domain). The NS 21-1 well (Figures 1 and 16) which was drilled to a total depth of 14,732 ft (4,490 m) in Lower Cretaceous rocks (compartment II), having a pressure gradient greater than 0.85 psi/ft (19.23 kPa/m) and a temperature of 316°F (158°C), was found to be in the envelope of non-economic hydrocarbon accumulations. This example shows that using a P/T plot could save

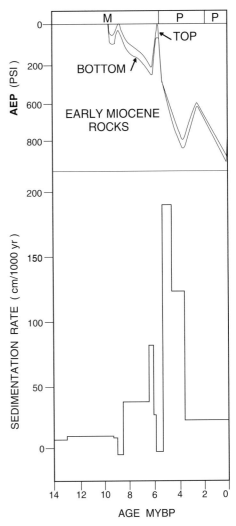

Figure 15. Chart showing the modeled pore pressure developments in the lower Miocene Qantara Formation relative to sedimentation rates during the last 5 million years in Abu Madi-8 well. Location of Abu Madi field shown on Figure 1.

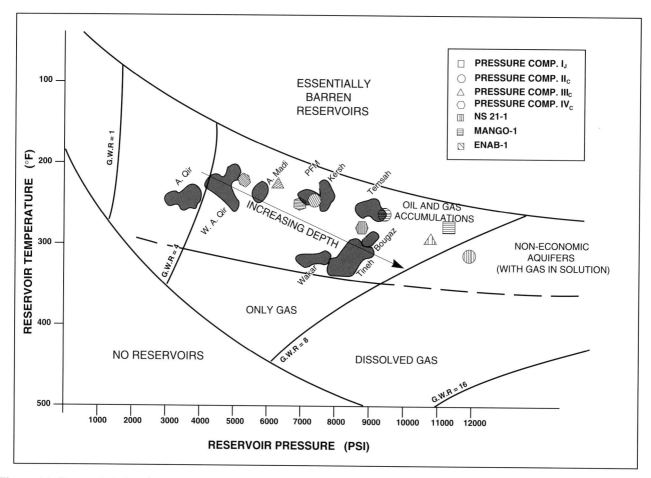

Figure 16. Cross-plot showing the relationship between reservoir pressure and temperature for oil and gas fields in the Nile Delta and different pressure compartments of three wells in the North Sinai Basin. Note the lower most pressure compartment (IIc) in the NS 21-1 well lies in the envelope of the non-economic hydrocarbon accumulation system. G.W.R. = gas-water ratio (from Nashaat, 1992).

money by avoiding unnecessary drilling. Conversely, P/T plots may show the hydrocarbon potential of deeper objectives. For example, this technique was used to drill a well on the western side of Abu Qir gas field (Figure 1) where oil was discovered in deeper stratigraphic horizons.

HYDROCARBON MIGRATION AND ENTRAPMENT

Highly efficient seals are necessary for the accumulation of hydrocarbons, however, in some cases they may hinder hydrocarbon migration. Compartmentalized rocks are, by definition, a relatively closed system with vertical and lateral seals of variable efficiency. Therefore, the temporal relationships among trap formation, sealing efficiency, and hydrocarbon generation, expulsion, and migration should be carefully evaluated.

In the northeastern part of the Nile Delta, growth faults sole-out into Oligocene and Miocene overpressured rocks. It seems likely that these faults and associ-

ated fractures, may experience cycles of periodic opening and closing as the pressure within each compartment builds up and then dissipates. During early stages of this cycle, fluids may flow along these pathways. Later, when pressure differences across the seal equilibrate, the fluid pathways probably close and fluid flow ceases until sufficiently high pore pressure develops to re-open or create new fluid flow pathways. This cyclic process is similar to that described by Momper (1978). The process could conceivably permit pulses of hydrocarbon-bearing fluids to migrate along fault planes into shallower compartments.

The anticipated short migration pathways in the study area favor the accumulation of hydrocarbons in close proximity to thermally mature source rocks in the basin depocenter. Structural features with four-way dip closure are believed to be the most prospective in the upper Oligocene-lower Miocene compartment (I), since faults may have been leaky because of episodic faulting due to the cyclic pressure buildup that may have occurred in this compartment.

Conversely, the overpressuring in the northeastern part of the Nile Delta may create some problems related to incomplete charging of reservoirs. In cases where the reservoir pore pressure is higher than the pressure of the source rock, migration of hydrocarbons to the reservoirs may be impeded. This may have occurred because of pressure communication between sandstone reservoirs and the more highly pressured systems.

In the Mango area in North Sinai (Figure 1), pressure differences between pressure compartments may be the dominant mechanism controlling fluid migration. For example, in a well such as the Mango-1, the pore pressure in the Jurassic compartment (a probable source rock) is about 5,400 psi (37.2 MPa) above normal pressure, and is higher than all of the overlying pressure compartments. Consequently, the pressure gradient between these compartments is favorable for fluid migration. There is a high vertical fluid flow potential for hydrocarbons in this area along fault planes. In addition, the updip southward migration pathways from the deeply buried thermally mature source rock of Jurassic age are enhanced as a result of vertical and lateral fluid potential changes.

The Abu Madi and Qawasim Formations in the Nile Delta are two potentially excellent reservoirs, and both are reversely pressured to a normal hydrostatic regime (underpressured in some cases). The Abu Madi is the major gas reservoir in the Nile Delta. No fields have been found in the underlying Qawasim. In an effort to explain this, sand to shale ratio maps of each formation were integrated with maps of potential for hydraulic communication. It was found that hydraulic communication between Qawasim and Abu Madi reservoirs was likely in the southern part of the Nile Delta, while in the northern, offshore areas, the Qawasim Formation becomes more prospective where the sand to shale ratio decreases.

CONCLUSIONS

The main cause of abnormally high pressure in both the Nile Delta and North Sinai regions is rapid sedimentation and burial of sediments during Pliocene and Pleistocene time. Secondary pressure generating mechanisms include possible volume changes of pore fluids or rock matrix by aquathermal expansion, hydrocarbon generation, or thermal cracking of oil to gas in the upper Oligocene-lower Miocene compartment in the eastern part of the Nile Delta. The presence of the Messinian evaporites also contributed to the development of overpressuring in pre-Messinian rocks by inhibiting pressure dissipation. The occurrence of the top of overpressure at shallow depths in the northern parts of the region compared to deeper depths in the southern part of the region, is attributed to the higher rates of basin subsidence and sedimentation in the northern regions.

In a regional sense, the sedimentary sequence in the study area can be divided into eight well defined pressure compartments that are separated by effective seals. Some of the seals are associated with major unconformities. Four seals are clearly demonstrated in the Early Cretaceous North Sinai basin. Four pressure seals are also recognized in the Nile Delta.

Pressure and temperature relationships of Tertiary reservoirs reveal that the Nile Delta and North Sinai Basins are favorable sites for oil and gas entrapment and deeper drilling objectives may have some potential. Pressure-temperature relationship analysis is highly recommended in the lower most abnormally pressured compartments to avoid non-economic hydrocarbon accumulations.

Growth faults in the Nile Delta sole-out in the overpressured upper Oligocene-lower Miocene compartment. It is likely that these growth faults periodically become conduits for hydrocarbon migration due to the cyclic development of pore pressures sufficiently high enough to forcibly expel hydrocarbons along fault planes into shallower compartments where the hydrocarbon-bearing fluids accumulate in structural and/or stratigraphic traps.

ACKNOWLEDGEMENTS *The author is most grateful to the EGPC management for authorization to publish this study, particularly to Magid A. Halim, vice chairman of exploration. I thank Joen Davis and Mohamed El Baz from ARCO for their helpful suggestions. I also wish to express sincere thanks to the AAPG reviewers Ben Law and R.E. Swarbrick for their constructive criticism. The paper has been significantly improved by their editorial assistance.*

REFERENCES CITED

Barker, C., 1972, Aquathermal pressuring - the role of temperature in the development of abnormal pressure zones: AAPG Bulletin, v. 56, p. 2068–2071.

Bradley, J.S., 1975, Abnormal formation pressure: AAPG Bulletin, v. 59, p. 957–973.

Caillet, G., 1993, The cap rock of the Snore field, Norway: a possible leakage by hydraulic fracturing: Marine and Petroleum Geology, v. 10, P 42–50.

Egyptian General Petroleum Corporation (EGPC), 1994, Nile Delta and North Sinai fields, discoveries and hydrocarbon potentials (a comprehensive overview): Cairo, 387p.

Fertl, W.H., 1976, Abnormal formation pressures, implication to exploration drilling and production of oil and gas resources: Elsevier, Amsterdam-Oxford-New York, 365 p.

Hunt J. M., 1990, Generation and migration of petroleum from abnormally pressured fluid compartments: AAPG Bulletin, v. 74 No. 1, p 1–12.

Jenkins, A.D., 1990, North and central Sinai, in R. Said, ed., The geology of Egypt: A.A. Balkema-Rotterdam-Brookfield, p. 361–389.

Lerche, I., 1990, Basin analysis: quantitative method: Academic Press, San Diego-New York, v.1, 561 p.

Maghara, K., 1978, Compaction and fluid migration-practical petroleum geology: New York, Elsevier, 319 p.

Mann, D.M., and A.S. Mackenzie, 1990, Prediction of pore pressures in sedimentary basins: Mar. Pet. Geol., v. 7, p. 55–65.

Modelevskiy, M.S., and Y.I. Parnov, 1967, Physical state of oil and gas at depth: Dokl. Akad. S.S.S.R., 175, p. 1372–1374.

Momper, J.A., 1978, Oil migration limitations suggested by geological and geochemical considerations, *in* Physical and chemical constraints on petroleum migration: v.1, Notes for AAPG short course, AAPG National Meeting, Oklahoma City, 60 p.

Nashaat, M., 1990a, Hydrochemistry of undercompacted sediments in North Sinai Basin, Egypt: EGPC, Tenth Exploration and Production Conference, 18 p.

Nashaat, M., 1990b, Hydrodynamic impacts on petroleum entrapment in northeastern part of North Sinai Basin, offshore: EGPC, Tenth Exploration and Production Conference, 32 p.

Nashaat, M., 1991, Detailed seal analysis, Lower Cretaceous rocks of the North Sinai Basin, Egypt, effects on hydrocarbon migration and accumulations in overpressured regimes: Seals, Abstract, 1991 AAPG annual convention, Dallas, U.S.A.

Nashaat, M., 1992, Geopressure and geothermal studies in the Nile Delta, Egypt, M.Sc. thesis, faculty of science, Al Azhar University, Cairo, 133 p.

Nashaat, M., 1994, Abnormal pressure origin and impacts on hydrocarbon prospectivity in the Nile Delta and North Sinai Basins, Egypt: AAPG Hedberg abnormal pressure conference, Golden, Colorado, U.S.A., June 8–10, 1994.

Nashaat, M., and D. Campel, 1993, Geopressure study in block A area, North Sinai, Egypt, EGPC and Phillips unpublished report.

Osborne, M.J., and R.E. Swarbrick, 1995, A review of mechanisms for generating overpressure in sedimentary basins: GeoPop - Geosciences into Overpressure, 2nd. six monthly report, p. 5–42.

Paul, R. M., 1991, The eastern Mediterranean Mesozoic basin: evolution and oil habitat: AAPG Bulletin, v.75, no.7, p. 1215–1232.

Powley, D.E., 1987, Subsurface fluid compartments: Gas Research Institute Workshop Unpublished Notes, 16 p.

Said, R., 1981, The geological evolution of the River Nile: Springer-Verlag, New York, 151 p.

Timko, D.J., and W.H. Fertl, 1970, Hydrocarbon accumulation and geopressure relationship and prediction of well economics with log-calculated geopressures: SPE 2990, 45th. AIME Fall Meeting, Houston, Texas, October, 1970.

Timko, D.J., and W.H. Fertl, 1972, How downhole temperature-pressure affect drilling, The shale resistivity ratio-a valuable tool for making economic drilling decisions: World oil, 175(7), p. 59–63.

Yassir, N.A., and J.S. Bell, 1994, Relationships between pore pressure, stresses and present day geodynamics in the Scotian Shelf, offshore eastern Canada: Geomechanics Group Publication, PMRI, University of Waterloo, Ontario, Canada, 44 p.

Polutranko, A.J., 1998, Causes of formation and distribution of abnormally high formation pressure in petroleum basins of Ukraine, *in* Law, B.E., G.F. Ulmishek, and V.I. Slavin eds., Abnormal pressures in hydrocarbon environments: AAPG Memoir 70, p. 181–194.

Chapter 11

Causes of Formation and Distribution of Abnormally High Formation Pressure in Petroleum Basins of Ukraine

A.J. Polutranko
UkrDGRI
Lviv, Ukraine

Abstract

The maintenance of abnormally high formation pressure (AHFP) over long periods of geologic time cannot be explained by compaction or structural deformation without the addition of fluids from underlying rocks. Abnormally high pressures develop in the deepest parts of basins that contain 8–10 km of sedimentary rocks. The deeper part of the sedimentary fill typically occurs in the zones of late catagenesis and incipient metamorphism with temperatures ranging from 200° to 300°C. The observed increase of formation pressure above normal hydrostatic pressure with increasing depth and temperature, in conjunction with other factors, indicates that extended zones of AHFP first appear at temperatures of 175°C and higher. At lower temperatures, abnormally high pressures occur only locally at the crests of anticlinal structures.

The main cause of abnormal pressure in the zone of late catagenesis is the generation of large volumes of methane, carbon dioxide, hydrogen, water, and other volatile components. The devolatilization process is commonly associated with elevated temperatures, loss of porosity and permeability, and fracturing of rocks. In rocks which have experienced a high level of catagenesis, such as those in the structurally inverted parts of the Donbas foldbelt and the western Lviv depression adjacent to the Carpathian foldbelt, abnormally high formation pressures are absent.

INTRODUCTION

Abnormally high pressures in Ukraine are known to occur in the following three petroleum provinces: (1) the Dnieper-Donets Basin; (2) the Carpathian Basin; and (3) the North Black Sea-Crimean Province (Figure 1). In these regions, AHFP is the main source of energy for hydrocarbon migration and accumulation. AHFP also poses significant problems for the drilling, completion, and production of hydrocarbons.

Previous work on aspects of the formation and distribution of AHFP in petroleum basins of Ukraine include those by Ali-Zade et al. (1984); Anikeev (1977); Dobrynin and Serebryakov (1978, 1989); Fertl (1980); Fertl and Timko (1972); Kucheruk and Shenderey (1975); Novosiletsky (1969, 1975, 1982, and 1984); Novosiletsky et al. (1987); Novosiletsky and Polutranko (1981, 1982, 1984); Ozerny (1984); Orlov (1980, 1981); and Shpak and Novosiletsky (1979). For this study a large amount of data collected with R.M. Novosiletsky (deceased) is included.

Historically, geologists have proposed a number of causes for the development of AHFP. For example, Dobrynin and Serebryakov (1989) suggest that the main factors leading to AHFP are rock compaction, tectonic stress, deformation of deep reservoirs, and the influx of highly pressured fluids. Kucheruk and Shenderey (1975) list 10 such factors that could lead to the development of AHFP and Fertl and Timko (1972) lists 16 factors. Swarbrick and Osborne (1998-this volume) have discussed several of the more commonly cited causes of AHFP.

Regardless of the causes responsible for the formation of AHFP, abnormally high pressures are found in rocks of various ages and compositions in basins of different types. Law and Spencer (1998-this volume) have discussed the wide variety of geographic and geologic conditions in which abnormal pressure occurs. The increase of the differential pressure (difference between formation pressure and hydrostatic pressure, ΔP_f) with increasing depth is observed through the sedimentary section to the top of basement rocks; in the basement,

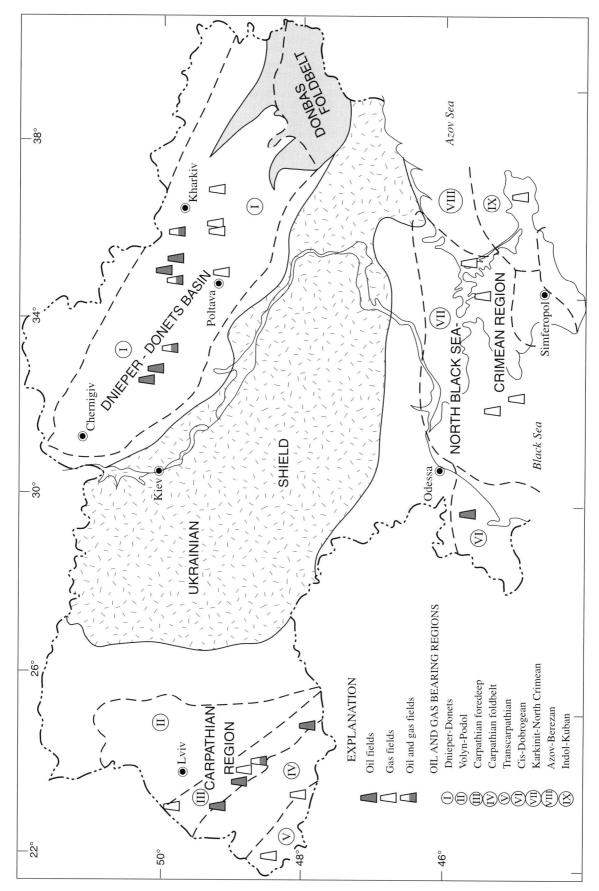

Figure 1. Map of Ukraine showing location of the Dnieper-Donets Basin, Carpathian Basin, and North Black Sea-Crimean region and the principle oil and gas fields in each region.

the magnitude of ΔP_f probably remains stable. AHFP is widespread in the deeper parts of sedimentary sections which have experienced high temperatures and changes in geochemical conditions. The existence of AHFP during long periods of geologic time requires a constant influx and generation of new volumes of fluids to compensate for the volume of fluids that are expelled or migrate out of the AHFP zone.

Investigations into the causes of abnormal pressure facilitate the prediction of the magnitude and distribution of AHFP in petroleum basins. The purposes of this study are to provide information on the stratigraphic and areal distribution of AHFP in petroleum basins of Ukraine and determine the relationships among AHFP, geochemistry of pore fluids, thermal history, and hydrocarbon generation.

METHODS OF STUDY

In this study we used a large amount of information collected through the years in petroleum regions of Ukraine. The information included data on the geology of oil and gas fields, results of well testing, measurements of formation pressure (P_f) and temperature (T_f) during testing, vitrinite reflectance, and petrophysical studies of reservoir rocks. In particular, more than 5,000 measurements of T_f and about 4,000 measurements of P_f in 300 fields and prospects were made. In addition, measurements of thermal gradients and the estimation of pore pressure based on d-exponents were used for the study of the thermodynamic regime of the Dnieper-Donets Basin. In the Carpathian region, 1,200 T_f and 1,000 P_f measurements from 130 fields were

used and in the North Black Sea-Crimean region, 500 T_f and 300 P_f measurements were used.

For detailed investigations of the changes of the thermodynamic conditions with depth, data on measurements of T_f and P_f in each field were plotted against depth. A series of maps showing depths to the top of abnormally pressured zones and to the zones of overpressure (ΔP_f) of 5, 7.5, 10, and 20 MPa were compiled. Interpretation of the results involved consideration of the structural and geothermal peculiarities of local areas, location of major seals in the section, and variations of rock thicknesses between surfaces of different overpressure values.

GEOLOGIC SETTING

Dnieper-Donets Basin

The Dnieper-Donets Basin is an intracratonic depression, the central part of which (Dnieper graben) is a segment of the Pripyat-Dnieper-Donets aulacogen (Figures 1 and 2). The basin is bounded on the south by the Ukrainian shield. The north end of the basin is bounded by the Voronezh high on which basement rocks occur at very shallow depths. The Dnieper-Donets Basin is 650–700 km long and between 60 and 160 km wide.

Structurally, the basin consists of the central graben and northern and southern margins. In the graben, the basement is strongly faulted into a large number of blocks. Cross-sections I through V (Figures 3 and 4) exhibit the structural style of the basin. Intense salt tectonics related to deformation of Devonian salt further

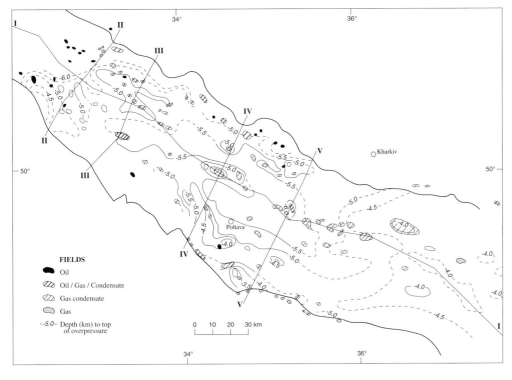

Figure 2. Map of Dnieper-Donets Basin showing lines of cross section I through V, oil and gas fields, pressure zones, and depth to top of overpressure. Cross sections I-V shown on Figures 3 and 4.

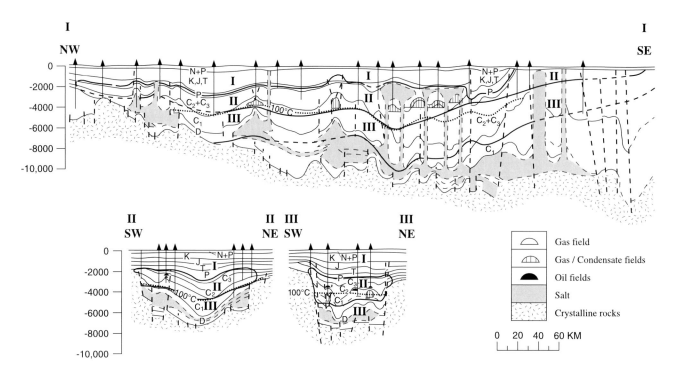

Figure 3. Structural cross sections I through III showing the structural style in the Dnieper-Donets Basin. Superimposed on the cross sections are the zones of catagenesis (I-III) and the 100°C isotherm. Location of cross sections are shown on Figure 2. (Symbols: I = oil and gas accumulation zone; II = oil generation and accumulation zone; III = gas generation and AHFP zone; D = Devonian; $C_{1,2,3}$ = lower, middle, and upper Carboniferous, respectively; P = Permian; T = Triassic; J = Jurassic; K = Cretaceous; N+P = Neogene and Paleogene).

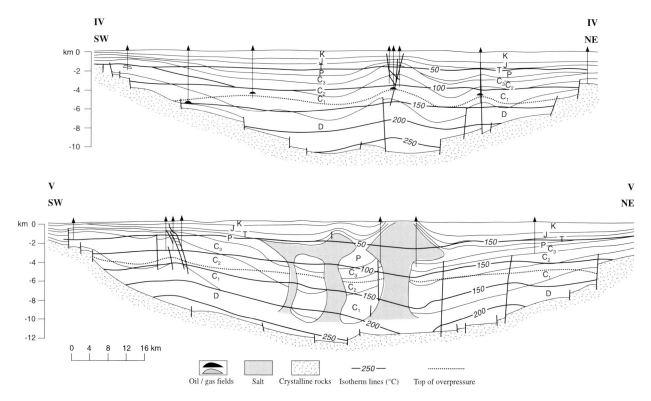

Figure 4. Structural cross sections IV and V through the Dnieper-Donets Basin showing structural style, isotherms, and top of overpressuring. Location of cross sections shown on Figure 2. (Symbols as per Figure 3).

complicates the structure. The margins are slightly faulted monoclines covered by as much as 4.5 km of Upper Visean and younger rocks on the northern margin and up to 3 km on the southern margin. The basin boundary with the Donbas foldbelt (Figure 1), which is a structurally inverted and deformed part of the aulacogen, is uncertain. Donbas folds gradually plunge into the basin and dip under younger rocks. The boundary is conditionally drawn along the pinch-out line of Triassic rocks.

According to Vitenko and Kabyshev (1977), the history of Dnieper-Donets Basin development includes three stages that include rift, syn-rift, and early sag. The Late Devonian rift stage was characterized by intense faulting and volcanism. The syn-rift sequence is terminated by a large pre-Carboniferous unconformity. The early sag stage embraces Carboniferous and Early Permian time. In the early part of the development of the basin, sedimentation was limited to the central graben, but in late Visean time sedimentation extended onto the margins. The area of sedimentation began to shrink in middle Carboniferous time and the stage was terminated by the major pre-Triassic unconformity. The following late sag stage resulted in formation of a gentle shallow depression.

The sedimentary section is composed of Devonian through Quaternary rocks which are 1–2 km thick in the northwestern part of the basin and 16–23 km thick in the southeastern part of the basin, near the boundary with the Donbas foldbelt. Devonian rocks are composed of carbonates, clastics, volcanics, and evaporites with a total thickness of 5–6 km. Devonian rocks are limited to the central graben, whereas Carboniferous rocks, which are dominantly composed of clastics, are 8 km thick, and extend far onto the basin margins. The overlying Permian rocks are composed of clastic, carbonate, and evaporites that are as thick as 2 km. Permian rocks occur mainly in the central and northern margin of the basin. Younger Mesozoic and Cenozoic rocks are mainly clastic deposits that are widespread and overlie various older sedimentary and basement rocks with pronounced unconformity.

More than 200 hydrocarbon fields, most of which are gas fields, have been discovered in the Dnieper-Donets Basin (Figure 2). Three of the fields are large, while others are of medium and small sizes. The productive reservoirs occur in 6 complexes which include the: Devonian; Tournaisian-lower Visean; upper Visean-Serpukhovian; middle Carboniferous; Upper Carboniferous-Lower Permian; and Mesozoic complexes (Figure 4).

Carpathian Region

The western part of Ukraine consists of four major structural units: (1) the Volyn-Podol plate of the Russian platform, (2) the Carpathian foredeep, (3) the Carpathian foldbelt, and (4) the Transcarpathian depression (Figures 1, 5, and 6).

The Volyn-Podol plate is a monocline dipping westward from the Ukrainian shield. The Riphean and Cambrian-Carboniferous sections of the monocline are overlain by Jurassic, Cretaceous, and Miocene rocks. The eastern part of the plate experienced uplift during late Paleozoic time and its western part continued to subside and formed the Paleozoic Lviv Basin.

The Carpathian foredeep is located southwest of the Volyn-Podol plate (Figure 1). The foredeep is divided into the Bilche-Vilitsa, Sambir, and Borislav-Pokutya zones (Figures 5 and 6) based on age of rocks and character of tectonic deformation. The Bilche-Volitsa zone overlies the Paleozoic basement in its northwestern part and the Precambrian Russian platform in the southeastern part. The sedimentary section of the zone consists of Jurassic, Cretaceous, Badenian, and Sarmatian (both Miocene) rocks. The thickness of Miocene rocks in most areas does not exceed 2.5 km and only locally increases to 3–5 km. The Sambir zone (Figures 5 and 6) is a tectonic nappe transported about 15–18 km to the northeast. It partly cuts and overlaps rocks of the Bilche-Volitsa zone. Orogenic clastics of the Sambir zone are thrusted and folded. Five thrust sheets deformed into steep, overturned to the northeast anticlinal and synclinal folds, are distinguished.

The Borislav-Pokutya zone is thrusted onto the Sambir zone in the northeast and is overthrusted by the Carpathian foldbelt on the southwest (Figures 5 and 6). The sedimentary section in the zone is composed of two structural stages; the Cretaceous-Paleogene stage and the Paleogene-Neogene stage. The Cretaceous-Paleogene stage is composed of flysch and the Paleogene-Neogene stage is formed by molasse, orogenic clastics. In post-Sarmatian time, both structural stages were thrust to the northeast and deformed into asymmetric, commonly overturned anticlinal folds (Figure 6). Thrusts with displacement of more than 5 km separate the zone into as many as four nappes. Each nappe consists of several anticlines also separated by thrusts and reverse faults with displacements of 2–3 km. Vertical thickness of the nappe system reaches 5–7 km and possibly more.

The Carpathian Foldbelt is a complex thrust-fold system which consists of several tectonic zones (Figures 1 and 5). The outer zone, or flysch Carpathians, is a system of nappes composed of Cretaceous-Paleogene flysch deposits. The nappes have a complicated folded and imbricated structure and are tectonically transported a significant distance to the northeast (Figure 6).

The Transcarpathian depression is filled with more than 3 km of Neogene orogenic clastics which overlie deformed sedimentary and magmatic rocks of Mesozoic age and Paleozoic crystalline schists. Salt domes are found in the inner part of the depression. The depression is actually a marginal structure of the Pannonian Basin.

Only one small oil-gas field has been discovered in Middle Devonian rocks of the Volyn-Podol plate (Figure 5). The Bilche-Volitsa zone of the Carpathian foredeep contains more than 30 gas and gas condensate fields and only 2 oil fields. Conversely, the Borislav-Pokut zone is mostly oil-prone. About 30 oil fields and only a few abnormally pressured gas condensate fields in the deep (up to 5.1 km) parts of the sedimentary cover have been found. In the Flysch Carpathians, several shallow, small oil fields have been discovered.

And, in the Transcarpathian depression, three small gas fields have been discovered at depths to 1.5 km.

North Black Sea-Crimean Region

Three areas of this region contain thick sedimentary sections: 1) the western part of the Indol-Kuban Basin,

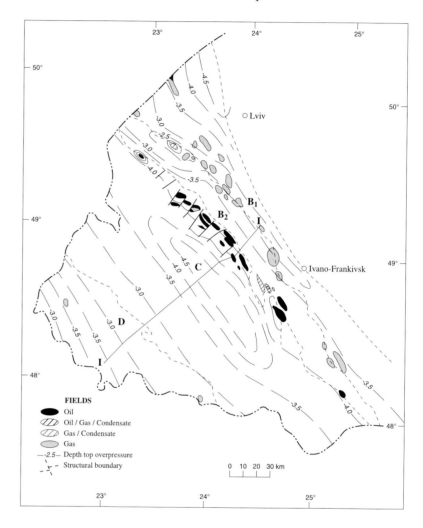

Figure 5. Map of Carpathian Basin showing structural zones, oil and gas fields, location of cross section I, and depth to top of overpressured rocks. (B1 and B2 = Bilche-Volitsa, Borislav-Pokutian, and Sambir zones of Pre-carpathian foredeep, C = Carpathian folded zone, D = Transcarpathian zone).

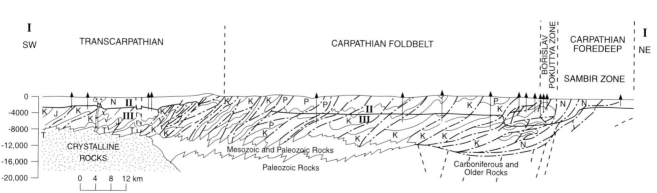

Figure 6. Structural cross section I showing structural style through Carpathian Basin. Location of cross section shown on Figure 5. (T = Triassic, J = Jurassic, K = Cretaceous, P = Paleogene, N = Neogene).

2) the Karkinit-North Crimean, and 3) the Cis-Dobrogean Basins (Figures 1, 7, and 8). The Cis-Dobrogean and Karkinit-North Crimean Basins overlie a junction zone of the Precambrian Russian platform and epi-Hercynian Scythian plate. The basement of the latter is composed of Riphean through Middle Jurassic deformed and partly metamorphosed rocks.

The Karkinit-North Crimean Basin is filled by Tertiary, Cretaceous, and in places, older rocks that are as thick as 7–8 km. On the west, it adjoins the Cis-Dobrogean Basin, in which the sedimentary section consists of Upper Proterozoic, Paleozoic, and Mesozoic rocks with a total thickness of 7–9 km. The western Indol-Kuban Basin is the largest structure in the region. It occupies the southern part of the Azov Sea and adjoining land areas and extends eastward into the North Caucasus region of Russia. The basin is filled mainly by shale of the Oligocene-lower Miocene Maykop series that is as thick as 4–5 km. Younger Cenozoic clastic rocks in the basin attain thicknesses of 4–4.5 km. On geophysical data, basement rocks occur at depths of 21 km in the deepest part of the basin, suggestive of the presence of 9–10 km of as yet unknown pre-Oligocene sedimentary rocks. A number of anticlinal folds related to shale diapirism and mud volcanoes are known in the central and southern areas of the basin (Figure 7). Southwest of the basin is the eastern slope of the Crimean high. The sedimentary cover of the slope is composed of thick Cretaceous and Paleocene-Eocene rocks and the relatively thin Maykop series. Younger Miocene and Pliocene rocks are absent. Shale diapirs and mud volcanoes are present in the Maykop series. Older rocks are characterized by intense block faulting.

About 25 small oil and gas fields have been discovered in the North Black Sea-Crimean region (Figure 7). Most of them are found in the Karkinit-North Crimean and Indol-Kuban Basins and on the slope of the Crimean high. The fields produce mainly gas and gas condensate from Cretaceous, Paleogene, and Miocene sandstone and clayey carbonate reservoirs at depths of up to 4.5 km in the Karkinit-North Crimean Basin. Oil and gas condensate fields in Paleogene-Miocene sandstones and limestones at depths of up to 3,680 m are known in onshore and offshore areas of the Indol-Kuban Basin and slope of the Crimean high. Two small oil fields have been discovered in Devonian rocks of the Cis-Dobrogean Basin.

SPATIAL DISTRIBUTION OF AHFP

Depths to the top of the abnormally pressured zone in the Dnieper-Donets Basin vary from 3.5 to 6.2 km (Figures 2, 9, and 10). The top of the zone is located in Lower Permian sandstones in the central zone of the basin. Toward the margins and in the northwestern parts of the basin, the top of the zone occurs in older stratigraphic units, in Lower Carboniferous clastic and Devonian carbonate, clastic, and evaporite rocks. The

ΔP_f increases with depth and reaches 55 MPa (Figure 11). The top of the abnormally pressured zone is located between ΔP_f surfaces of 10 and 20 MPa (Figure 10).

In the Carpathian region, AHFP was detected in the Carpathian foredeep, Transcarpathian depression, and Carpathian foldbelt. Depths to the AHFP zone vary from a few meters in a single outcrop to 5 km in the central part of the Carpathian foldbelt (Figure 5). Abnormally high pressures were detected in Neogene, Paleogene, and Cretaceous rocks and the ΔP_f increases from 10 MPa at a depth of about 1 km to 50 MPa at 7 km (Figure 12).

In the North Black Sea-Crimean region, abnormal pressures are found in the Karkinit-North Crimean and Indol-Kuban Basins (Figure 7). AHFP in the Indol-Kuban Basin is detected at depths from several meters in the area of active mud volcanism on the Kerch Peninsula, to 4 km in Neogene, Paleogene, Cretaceous, and Jurassic rocks of the central basin areas. In the Karkinit-North Crimean Basin, abnormally high pressures are encountered at depths of 2 to 4.5 km in Neogene, Paleogene, and Cretaceous rocks. The ΔP_f increases from 1.2 MPa at a depth of about 1 km to 42 MPa at 4.5 km (Figure 13).

MAIN CAUSES OF AHFP FORMATION

An analysis of data from the petroleum basins of Ukraine and other countries indicate that the development and long preservation of AHFP cannot be explained by compaction of clays or tectonic deformation without a constant influx of additional volumes of fluids from underlying rocks. The zone of overpressure is commonly accompanied by a change in water mineralization from a chlorine-calcium type above the AHFP zone to a hydrocarbonate-sodium type within the AHFP zone. Abnormal pressures have been detected in rocks of various lithologic composition and in basins of various age and different tectonic activity. The magnitude of AHFP increases with increasing thickness of sedimentary rocks and increasing sedimentation rates of relatively young deposits. For example, the area of maximum water head in the Karkinit-North Crimean Basin is coincident with the area of maximum thickness of rocks (Figure 14). In the deep basins with thicknesses of 17 to 23 km, AHFP is expressed on the surface as mud volcanoes, as exhibited on the Kerch Peninsula (Figure 7). These mud volcanoes are commonly associated with high concentrations of methane with an admixture of heavier hydrocarbon gases, hydrogen, nitrogen, and some other components.

Drilling into the deep horizons of the Dnieper-Donets Basin has revealed that AHFP is widespread in rocks with temperatures exceeding 110°C. Similar situations exists in the Karkinit-North Crimean Basin, Carpathian foredeep, and in some other regions of the world. In the zone of hydrostatic pressure of these

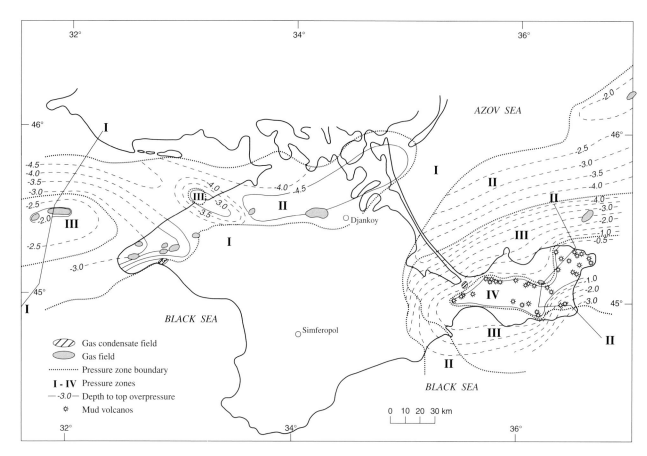

Figure 7. Map of the northern Black Sea-Crimean region showing oil and gas fields, depth to top of overpressure, and location of cross sections II and III. Cross sections shown on Figure 8. (I = hydrostatic pressure zone, II = sealed AHFP zone, III = partially breached AHFP zone, IV = completely breached AHFP zone).

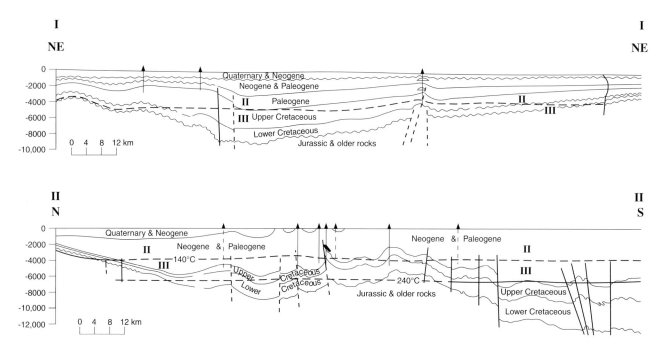

Figure 8. Structural cross sections I and II through the northern Black Sea-Crimean region with isotherm lines and catagenetic zones. Location of cross sections shown on Figure 7. (Catagenetic zones: II = early; III = late).

regions, the ΔP_f gradually increases to 5 MPa which occurs in rocks characterized by the transition from early to late catagenesis (i.e., brown coal to long-flame coal substages).

Depths to the boundaries of the catagenetic substages which correspond to vitrinite reflectances of 0.5, 0.67, and 0.9% R_o were mapped in the Dnieper-Donets Basin and changes of T_f and P_f at these boundaries were studied. An R_o of 0.9% corresponds to the boundary between the early and late catagenetic stages (Logvinenko, 1968). The lower boundary of the late catagenetic stage (which includes fat, coke, lean-baking, and lean coal substages) with the early metagenetic stage (semi-anthracite substage) is drawn at an R_o of 2.45%.

This study shows that at an R_o boundary of 0.67%, formation temperature varies from 15° to 110°C and ΔP_f varies from less than 0.1 to 5.5 MPa. Above the abnormally pressured zone, these values are 75°–110°C and 2.5 to 5.5 MPa, respectively. At the boundary of the early and late catagenetic zones ($R_o = 0.9\%$), T_f varies from 25° to more than 150°C and ΔP_f varies from 0.2 to more than 30 MPa. In the abnormal pressure zone, the temperature at this boundary increases from 120°–135°C to 155°C and ΔP_f is more than 10 MPa. In two areas where the temperature at this boundary is highest (above 150°C), ΔP_f is also at a maximum and exceeds 30 MPa. These data indicate that near the boundary between early and late catagenesis, when approaching the top of abnormally pressured rocks, ΔP_f abruptly increases from about 5 to 10–15 MPa. In overpressured rocks within the late catagenetic zone, ΔP_f rapidly increases to 40–55 MPa and more.

Regional data discussed above show that AHFP occurs in those parts of the basins where the thickness of the sedimentary section exceeds 8–10 km and part of the section is within the late catagenetic and metagenetic zones with temperatures of 200°–300°C. This suggests that fluids with AHFP are generated from rocks which occur in these zones. For rocks that are in the late catagenetic zone, the length of time necessary to generate hydrocarbon fluids depends upon the duration of the transformation processes and temperature. For Paleozoic rocks the required temperature is 110° ±15°C and for Mesozoic rocks the temperature is 140° ±15°C (Levenshtein, 1969). The passage of rocks from early to late catagenetic zones, accompanied by changes in the geothermal gradient and increasing thickness of overlying rocks during burial and subsidence, has the effect of impeding the discharge of volatile components, thereby diminishing the role of geologic time. At depths of 6 km and more, the role of geologic time is negligible.

Subsidence of beds into the late catagenetic zone coincides with the main qualitative leap in lithification of rocks and maturation of organic matter. The rocks undergo a number of processes that result in rapid structural and mineralogical changes. For example, kaolinite and montmorillonite are transformed into hydromica minerals and shales and mudstones lose the ability to swell in water. Recrystallization of carbonate rocks also takes place in the late catagenetic zone. Large amounts of adsorbed, crystallized, bound, and free pore water with low mineralization become available to the pore system of the rocks. Highly porous sandstones become tight, their porosity decreases from 10% and more to 3–5% and their permeability approaches zero. The rocks become a seal for underlying beds.

All organic matter, independent of composition and concentration, undergoes intensive transformation and generates methane and heavier hydrocarbon gases. For example, one gram of coal generates 70 to 120 cm³ of gas. The loss of plasticity by shales, decreasing porosity, and generation of large volumes of gas and water lead to an increase of pore pressure above geostatic pressure and the consequent fracturing of rocks. Partial closing of fractures at a later time results in transmission of part of the geostatic pressure to the pore fluids. Massive fracturing of rocks in the late catagenetic zone increases permeability, which explains significant large yields of gas from low-porosity overpressured zones and commonly observed stable values of ΔP_f down the section to the top of basement rocks. For example, within the depth interval 2,141–3,875 m, in the abnormally pressured zone of the Golitsin field in the Karkinit-North Crimean Basin ΔP_f remains stable at 14 MPa. The abnormality ratio (ratio of formation pressure to hydrostatic pressure) in the same interval decreases downward from 1.66 to 1.36. A number of geologists interpret such a decrease of the abnormality ratio as the gradual disappearance of overpressure with depth and confinement of overpressure to certain isolated stratigraphic intervals. In reality, however, fluids in the upper part of the abnormally pressured zone and in the underlying fractured rocks are hydrodynamically connected which results in a stable value of ΔP_f. Therefore, a decrease of the abnormality ratio with increasing depth does not indicate a decrease in the magnitude of overpressure if the value of ΔP_f remains stable or increases.

A significant decrease of ΔP_f downward in the section has not been observed and should not be present in natural environments. Alternation of beds with normal and abnormally high formation pressures and a decrease of ΔP_f with depth occurs in a number of cases in limited stratigraphic intervals and is related to the discharge of fluids from the overpressured zone through more permeable beds. However, in deeper parts of the section affected by injection of overpressured fluids, the value of ΔP_f exceeds that in the shallower horizons.

Typically, abnormally high pressure is confined to rocks occurring in the late catagenetic zone, but in some cases the top of the overpressure rises in the lower part of the early catagenetic zone. The confinement of abnormally pressured fluids is facilitated by precipitation of less soluble components along the migration paths owing to the mixing of chemically different waters of the overpressured and normally pressured zones. Claystones in this zone of catagenesis still preserve some elasticity and, being reinforced by precipitating minerals (CaCO$_3$, SiO$_2$), form a reliable seal

for the AHFP zone (Novosiletsky et al., 1987). Therefore, two different hydrodynamic zones, each characterized by the dominance of lateral migration, exist in basins with a sedimentary section more than 7–8 km thick. The upper zone above the seal is characterized

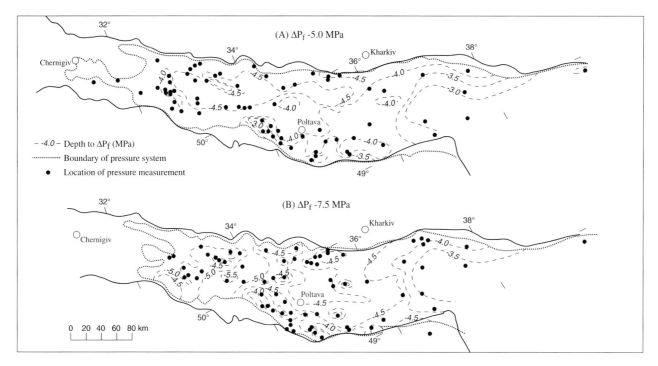

Figure 9. Maps of the depths, in kilometers, to the ΔP$_f$ of (A) 5.0 MPa and (B) 7.5 MPa in the Dnieper-Donets Basin.

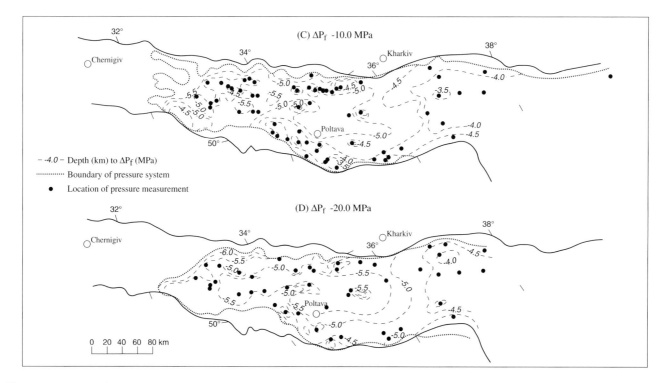

Figure 10. Maps of the depths, in kilometers, to the ΔP$_f$ of (C) 10.0 MPa and (D) 20.0 MPa in the Dnieper-Donets Basin.

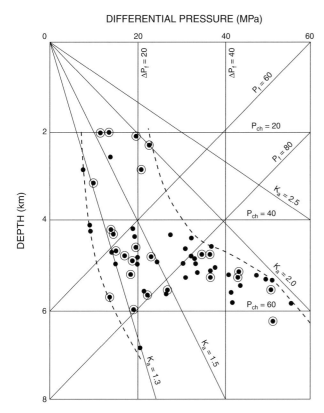

Figure 11. Cross plot of ΔP_f versus depth in the Dnieper-Donets Basin. Solid dots = pressure measurements in water-bearing rocks; circled dots = pressure measurements in oil and gas rocks; K_a = abnormality coefficient (the ratio of formation pressure to hydrostatic pressure); P_f = formation pressure; P_{ch} = hydrostatic pressure.

by mainly matrix permeability, whereas fracture permeability dominates in the lower zone. The two zones are separated by nearly impermeable seals and the hydrodynamic connection between them occurs only through faults or fractures as local break-throughs of overpressured fluids.

Free and water-dissolved thermogenic hydrocarbon gas in the overpressured zone partly migrates through faults into the overlying normally pressured section. The resistance to migration results in the drop of pressure by 30–70 MPa along the migration paths. This decrease of pressure leads to separation into the free phase of up to 75–90% of the dissolved gas. The following accumulation of gas in traps, its partial adsorption by organic particles and mineral grains, and dissolution in formation waters may result in the reduction of the amount of free gas phase and the decrease of fluid pressure in upper horizons of the early catagenetic zone to values lower than hydrostatic pressure. The abnormally low pressure in these horizons is observed in some gas fields of the Dnieper-Donets Basin and Carpathian foredeep.

The best conditions for the accumulation of hydrocarbons exist in rocks deposited at moderate sedimentation rates and where generation occurs in the late

catagenetic zone (Nesterov, 1969; Vassoevich, 1967). At higher sedimentation rates, clastic deposits are poorly sorted and commonly lack beds with high porosity and permeability. High rates of subsidence into zones of late catagenesis, results in relatively rapid rates of hydrocarbon generation and overpressuring in short periods of time which, for example, is characteristic of the Indol-Kuban Basin. At some critical values of ΔP_f, overpressured fluids break through into overlying rocks and form a semi-open AHFP zone. A totally open AHFP zone is formed where overpressured fluids break through to the surface as mud volcanoes.

Most commonly, intrusion of overpressured fluids into the early catagenetic zone occurs in deep basins in which temperatures at the base of a sedimentary section exceeds 300°C and a zone of metamorphism is present. This zone expels the mobile pore water, mineral transformation-derived water, and also carbon dioxide, methane, hydrogen, and water formed during the graphitization of organic matter. A large amount of carbon dioxide is also formed in the zone, owing to reduction of metal oxides by carbon (Sokolov, 1971). At high temperatures, water undergoes partial decomposition. Especially common is the reaction of water with FeO,

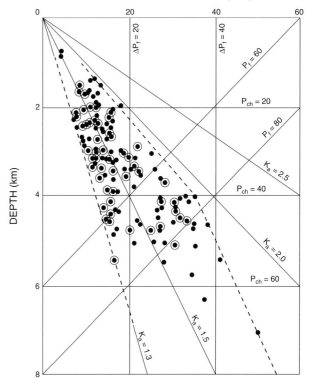

Figure 12. Cross plot of ΔP_f versus depth in the Carpathian Basin. Solid dots = pressure measurements in water-bearing rocks; circled dots = pressure measurements in oil and gas-bearing rocks; K_a = abnormality coefficient (the ratio of formation pressure to hydrostatic pressure); P_f = formation pressure; P_{ch} = hydrostatic pressure.

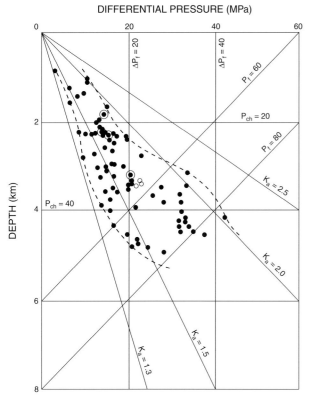

DIFFERENTIAL PRESSURE (MPa)

Figure 13. Cross plot of ΔP_f versus depth in the north Black Sea-Crimean region. Solid dots = pressure measurements in water-bearing rocks; circled dots = pressure measurements in oil and gas-bearing rocks; K_a = abnormality coefficient (the ratio of formation pressure to hydrostatic pressure); P_f = formation pressure; P_{ch} = hydrostatic pressure.

which produces free hydrogen. Other reactions of water with carbon, carbon monoxide, and other mineral components, also produce hydrogen and carbon dioxide (Sokolov, 1971). In addition, fluids of magmatic origin, dominantly carbon dioxide, can play a significant role in some areas. Therefore, carbon dioxide is the main product of the metamorphic zone; in many cases it is present as a free gas phase.

Earthquakes can facilitate break-through of overpressured fluids into the early catagenetic zone and to the surface. For example, the discharge of gas through fractures on the Starun mud volcano in the Carpathian foredeep increased by tens of times after earthquakes in Mexico and Romania in 1977 and 1978 (Bilous and Petruniak, written communication). Probably, this relates to the increase of permeability of the seal under the effect of the ultrasound and infrasound parts of the earthquake's oscillation spectrum. In lab experiments the permeability of marble samples increased by 1,000 to 100,000 times under the effect of ultrasound, depending on the frequency and size of capillaries, reaching 1–2 md.

In Ukraine, the most intense manifestation of AHFP is observed in the Indol-Kuban Basin, on the Kerch Peninsula (Figures 1 and 7). In this region, the thick-

ness of rocks occurring under conditions of late catagenesis and metamorphism is approximately 10–17 km and a large, open to the surface AHFP zone is present.

Significant discharge of fluids to the surface results in lower ΔP_f values in the early catagenetic zone than those values in the Dnieper-Donets and Carpathian regions. Conversely, ΔP_f in the late catagenetic zone on the Kerch Peninsula, at depths greater than 3,800 m, is higher than in the Dnieper-Donets and Carpathian regions. The Kerch Peninsula area is also characterized by higher subsurface temperatures owing to convective heat transfer during discharge of overpressured fluids. From the central area of the Indol-Kuban Basin toward its southern margin, the discharge of overpressured fluids becomes less intensive and ΔP_f and formation temperature at a depth of 4 km decrease from 42.5 to 20–24 MPa and from 172° to 139°C, respectively. Above the closed AHFP zone on the Kerch Peninsula, the temperature is 10°–20°C higher than in the zone of hydrostatic pressure and 10°–20°C lower than the semi-closed zone of AHFP.

CONCLUSIONS

The above discussion on processes occurring in the zones of late catagenesis and metamorphism suggest that the main cause of AHFP formation is the transformation of organic matter and other constituents of sedimentary rocks under temperatures exceeding 175°C. Other important factors are the hydrogenation of organic matter and products of its transformation by free hydrogen and interaction of hydrocarbon gases with organic matter and high-molecular compounds of oil. These processes lead to the decrease of formation water salinity, generation of large amounts of methane, hydrogen, water, and other volatile compounds. The accompanying reduction of porosity and permeability results in the formation of a pressure seal and subsequent fracturing of rocks.

Transformation processes are terminated in highly lithified rocks that have been structurally uplifted. For example, areas with low present-day temperatures at the boundary between the early and late catagenetic zones in the uplifted northwestern part of the Dnieper-Donets Basin and in its southeastern part and adjoining Donbas foldbelt coincide with areas of low ΔP_f. An analogous situation exists in the Cis-Dobrogean Basin and in the western Lviv Basin.

Generation of hydrocarbons and the expulsion and migration of formation fluids are affected by the pressure history. Expulsion of fluids occurs when formation pressure increases to values that significantly exceed hydrostatic pressure. AHFP plays an important role in fluid migration and strongly influences the resulting distribution of hydrocarbon accumulations. Discharge of AHFP through permeable beds of the early catagenetic zone leads to formation of pools with tilted oil-water and gas-water contacts. A shift of the

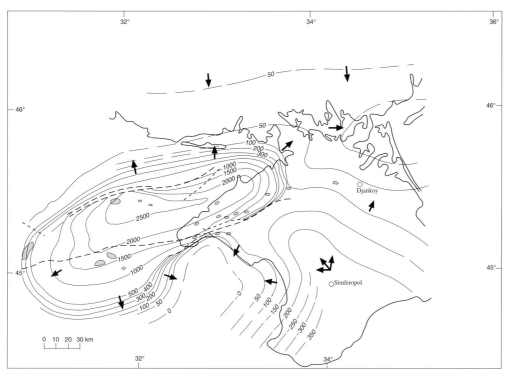

Figure 14. Map of north Black Sea-Crimean region showing potentiometric surfaces, in meters, and oil and gas fields. Vector arrows indicate direction of fluid flow, short dashed lines are faults.

pools from the crests to the flanks of anticlinal structures and formation of hydrodynamic traps is feasible. Exploration for hydrocarbons in such traps is underway in the Karkinit-North Crimean Basin.

In the Dnieper-Donets Basin, the most favorable conditions for discharge of overpressured fluids into the zone of early catagenesis are present on the northeastern basin margin. Maximum temperatures and ΔP_f at the boundary between early and late catagenetic zones are indicative of the presence of an interval of very active generation of overpressured fluids adjacent to the marginal fault. In this area an intensive discharge of fluids parallel to the bedding of sedimentary rocks and the fractured top of the basement, directed toward the northern shoulder of the rift graben, is proposed. Hydrocarbon pools in hydrodynamic and structural traps are expected.

AHFP hampers the catagenetic transformation of rocks and effectively extends the temperature boundary conditions for the preservation of hydrocarbon accumulations (Novosiletsky et al., 1987). The normal sequence of oil accumulations in the upper part of the section changes downward into gas condensate and gas accumulations. The thermal destruction of gas occurs at about 165°C and is observed only in the zone of hydrostatic pressures. By increasing ΔP_f to 25.9, 32.2, and 35.8 MPa in the geothermal gradient interval of 3.4° to 3.66°C/100 m, the maximum temperatures of oil pools discovered in Ukraine progress from 174°, 180°, and 186°C, respectively. Gas pools with temperatures of 187° and 196°C have been found in the overpressure zone where ΔP_f equals 49 and 85 MPa, respectively.

In general, as the ΔP_f increases and the geothermal gradient decreases, the thermal limit of hydrocarbon

pools increases. Theoretically, oil pools can exist at temperatures as high as 240°–260°C under conditions of ΔP_f of 85 MPa, which is the maximum ever measured in gas accumulations. A high content of organic matter and low geothermal gradient favor preservation of oil.

Almost all methane and heavier hydrocarbon gases are generated in the zones of late catagenesis and metamorphism under conditions of abnormally high formation pressure. Based on observations in Ukraine, significant hydrocarbon pools cannot be formed in rocks that do not contain abnormally pressured rocks in the late catagenetic zone. Normally pressured basins with thin sedimentary sections in the late catagenetic zone do not contain significant hydrocarbon reserves.

REFERENCES CITED

Ali-Zade, A.A., Shnyukov, E.F., and Grigiryants, B.V., 1984, Geotectonic conditions of world's mud volcanoes and their importance for prediction of gas productivity, *in* Neftyanye i gazovye mestorozhdeniya, XXVII MGK, Sektsiya 13 (The oil and gas fields, XXVIIth IGC, Sec. 13): Moscow, Nauka, p. 166–172.

Anikeev, K.A., 1977, O geologicheskikh osnovakh prognozirovaniya sverkhvysokikh plastovykh davleniy i problemy glubokogo bureniya (Geologic foundations for prediction of very high formation pressures and problems of deep drilling): Trudy VNIGRI, v. 397, Leningrad, p.14–54.

Dobrynin, V.M., and Serebryakov, V.A., 1978, Metody prognoza anomalno vysokikh plastovykh davleniy (Methods for prediction of abnormally high formation pressures): Moscow, Nedra, 232 p.

Dobrynin, V.M., and Serebryakov, V.A., 1989, Geologo-geofizicheskiye metody prognoza anomalnykh plastovykh davleniy (Methods of prediction of abnormal formation pressure based on geologic and geophysical data): Moscow, Nedra, 287 p.

Fertl, W.H., 1980, Anomalnye plastovye davleniya (Abnormal formation pressures): Moscow, Nedra, 395 p.

Fertl, W.H., and Timko, D.J., 1972, How downhole temperatures, pressures affect drilling, 2. Detecting and evaluating formation pressures: World Oil, v.175, no. 1, p. 45–50.

Kucheruk, E.V., and Shenderey, L.P., 1975, Sovremennye idei o prirode anomalno vysokikh plastovykh davleniy (Present concepts on abnormally high formation pressures): Moscow, VINITI, v.6, 186 p.

Law, B.E., and C.W. Spencer, 1998, Abnormal pressures in hydrocarbon environments, *in* Law, B.E., G.F. Ulmishek, and V.I. Slavin eds., Abnormal pressures in hydrocarbon environments: AAPG Memoir 70, p. 1–11.

Levenshtein, M.L., 1969, Main problems of regional metamorphism of coals, in Geologiya ugolnykh mestorozhdeniy, Materialy III Vsesoyuznoy konferentsii po tverdym poleznym iskopaemym (Geology of coal fields, Materials of III All-Union Conference on solid fossil fuels): Moscow, Nauka, p. 113–123.

Logvinenko, N.V., 1968, Postdiageneticheskie izmeneniya osadochnykh porod (Postdiagenetic changes of sedimentary rocks): Leningrad, Nauka, 94 p.

Nesterov, I.I., 1969, Kriterii prognoza neftegazonosnosti (Criteria of prediction of oil and gas productivity): Moscow, Nedra, 335 p.

Novosiletsky, R.M., 1969, Plastovye davleniya fluidov v nedrakh Ukrainy (Fluid formation pressures in the subsurface of Ukraine): Kiev, Tekhnika, 164 p.

Novosiletsky, R.M., 1975, Geologo-gidrodinamicheskiye i geokhimicheskiye usloviya formirovaniya neftyanykh i gazovykh zalezhey Ukrainy (Geologic, hydrodynamic, and geochemical conditions of formation of oil and gas fields of Ukraine): Moscow, Nedra, 228 p.

Novosiletsky, R.M., 1982, Zony anomalno vysokikh plastovykh davleniy i ikh neftegazonosnost (Zones of abnormally high formation pressure and their oil and gas productivity): Geologichesky Zhurnal, no. 3, p. 60–80.

Novosiletsky, R.M., 1984, Fiziko-khimicheskiye protsessy obrazovaniya uglevodorodnykh fluidov (Physico-chemical processes of hydrocarbon generation): Geologichesky Zhurnal, no. 5, p. 113–120.

Novosiletsky, R.M., and Polutranko, A.J., 1981, Connection of abnormally high formation pressures in the Dnieper-Donets depression with subsurface temperatures, *in* Geologiya neftegazonosnykh rezervuarov (Geology of oil and gas reservoirs): Moscow, Nauka, p. 154–158.

Novosiletsky, R.M., and Polutranko, A.J., 1982, Vliyaniye gidrodinamicheskikh protsessov na raspredeleniye zalezhey uglevodorodov v neftegazonosnykh regionakh Ukrainy (Effect of hydrodynamic processes on distribution of hydrocarbon pools in petroleum regions of Ukraine): Sovetskaya Geologiya, no. 11, p. 41–47.

Novosiletsky, R.M., and Polutranko, A.J., 1984, Geothermal characteristics of oil- and gas-productive formations. Temperature maps of Paleozoic formations, *in* Arsiry, J.A.,Vitenko, V.A., Paly, A.M., and Tsypko, A.K., eds., Atlas geologicheskikh struktur i neftegazonosnosti Dneprovo-Donetskoy vpadiny (Atlas of geologic structures and oil and gas productivity of the Dnieper-Donets depression): Kiev, Ministry of Geology of the USSR, 190 p.

Novosiletsky, R.M., Vitenko, V.A., and Polutranko, A.J., 1987, Zony neftegazonakopleniya Ukrainy (Zones of oil and gas accumulation of Ukraine), Trudy UkrDGRI, v. 35: Moscow, Nedra, 196 p.

Orlov, A.A., 1980, Anomalnye plastovye davleniya v neftegazonosnykh regionakh Ukrainy (Abnormal formation pressures in petroleum basins of Ukraine): Lviv, Ukraine, Vyshcha Shkola, 188 p.

Orlov, A.A., 1981, Analiz vliyaniya tektonicheskogo faktora na velichiny plastovykh davleniy v zalezhakh hefti i gaza Ukrainy (Analysis of effect of tectonics on formation pressures in oil and gas fields of Ukraine): Geologiya Nefti i Gaza, no. 2, p. 48–52.

Ozerny, O.M., 1984, Prognoz AVPD v Chernomorsko-Krymskoy neftegazonosnoy oblasti (Prediction of abnormally high formation pressure in the Black Sea-Crimean petroleum region): Geologiya Nefti i Gaza, no. 6, p. 35–41.

Shpak, P.F., and Novosiletsky, R.M., 1979, Proiskhozhdeniye i rasprostraneniye anomalnykh davleniy plastovykh fluidov v neftegazonosnykh basseynakh (Origin and distribution of abnormal formation pressures in petroleum basins): Geologichesky Zhurnal, v.39, no. 3, p. 1–11.

Sokolov, V.A., 1971, Geokhimiya prirodnykh gazov (Geochemistry of natural gases): Moscow, Nedra, 336 p.

Swarbrick, R.E. and M.J. Osborne, 1998, Mechanisms that generate abnormal pressures: an overview, *in* Law, B.E., G.F. Ulmishek, and V.I. Slavin eds., Abnormal pressures in hydrocarbon environments: AAPG Memoir 70, p. 13–34.

Vassoevich, N.B., 1967, Teoriya sedimentatsionno-migratsionnogo proiskhozhdeniya nefti (Theory of sedimentary-migrational genesis of oil): Doklady Akademii Nauk SSSR, ser. geol., no. 11, p. 37–72.

Vitenko, V.A., and Kabyshev, B.P., 1977, Geologicheskaya istoriya i neftegazonosnost lokalnykh struktur Dneprovsko-Donetskoy vpadiny (Geologic history and hydrocarbon productivity of local structures of the Dnieper-Donets depression): Moscow, Nedra, 192 p.

Wilson, M.S., B.G. Gunneson, K. Peterson, R. Honore, and M.M. Laughland, 1998, Abnormal pressures encountered in a deep wildcat well, southern Piceance Basin, Colorado, *in* Law, B.E., G.F. Ulmishek, and V.I. Slavin eds., Abnormal pressures in hydrocarbon environments: AAPG Memoir 70, p. 195–214.

Chapter 12

ABNORMAL PRESSURES ENCOUNTERED IN A DEEP WILDCAT WELL, SOUTHERN PICEANCE BASIN, COLORADO

Michael S. Wilson
Consultant
Evergreen, Colorado, U.S.A.

Bret G. Gunneson
Legacy Energy Corporation
Denver, Colorado, U.S.A.

Kristine Peterson
Consultant
Lakewood, Colorado, U.S.A.

Royale Honore
Matthew M. Laughland
Mobil Oil Corporation
Dallas, Texas, U.S.A.

Abstract

A deep wildcat well in the southern Piceance Basin of western Colorado (Mobil O'Connell F11X-34P) encountered three separate and distinct overpressured zones. The shallowest overpressured zone occurs within the Upper Cretaceous Mesaverde Group and coincides with gas-charged, thermally mature coal beds with vitrinite reflectance (R_o) ranging from 0.8 to 2.0%. This overpressured zone is located in the center of the basin where Mesaverde coal beds are thickest and where burial beneath Tertiary sediments is deepest. The overpressured zone is surrounded by subnormally pressured zones, and normally pressured, water-saturated strata occur along the basin margins.

A deeper overpressured zone occurs within marine shales of the Upper Cretaceous Niobrara and Frontier Formations. These gas-charged, depleted source rocks have R_o values ranging from 2.8 to 3.5% in the O'Connell well. The overpressured zone occupies the center of the basin. Subnormal pressures apparently occur in a ring around this overpressured zone. Low pressure gas has also been found below the Niobrara-Frontier section in fractured quartzitic sandstones of the Lower Cretaceous Dakota Group.

Highly overpressured salt water flows were encountered in the Pennsylvanian Minturn Formation and caused severe drilling and casing complications in the O'Connell well. This overpressure zone occurs in fractured quartzitic sandstones and is sealed by overlying argillaceous limestone beds. Salinity and isotope data indicate that the water is probably original formation water. The lateral extent of this pressure system is not known.

A reservoir with excellent porosity was discovered in dolomite beds of the Mississippian Leadville Formation. Measured borehole temperatures ranged from 441° to 464°F (239°C), indicating a present-day geothermal gradient of 2.2°F/100 ft (40°C/km). Measured R_o values above the reservoir are as high as 4.4 to 6.1%. During testing, the Leadville reservoir flowed carbon dioxide gas, traces of methane, nitrogen, hydrogen sulfide, and salt water to the surface at approximately normal pressure. Isotope data indicate that the carbon dioxide gas was derived from alteration of carbonate rocks, probably due to hydrothermal activity associated with Tertiary igneous intrusions along the southeastern margin of the basin.

Burial and thermal history reconstructions indicate that maximum paleogeothermal gradient was 3.0°F/100 ft (52°C/km). Maximum paleotemperature in the Leadville Formation was as high as 644°F (340°C), based on estimates from vitrinite and fluid inclusion data from core samples. Maximum paleotemperature occurred at approximately 27 Ma, based on argon thermochronology of cuttings from the Pennsylvanian Maroon Formation. Analyses of apatite fission tracks in cuttings from the Wasatch Formation and upper Mesaverde Group indicate that significant erosion and cooling have occurred since 5 Ma. Cooling, removal of overburden and gas leakage along faults and fractures have contributed to gradual pressure decline in the Niobrara and Mesaverde overpressured zones.

INTRODUCTION

Mobil Oil Corporation's O'Connell F11X-34P wildcat well (34-T7S-R92W) was drilled to a depth of 18,422 ft (5,620 m) during 1990–1992 to evaluate a deep gas prospect in the southern Piceance Basin of western Colorado (Figure 1). Although many wells have evaluated the Upper Cretaceous Mesaverde Group in the Piceance basin, very few Paleozoic tests have been drilled, and little is known about the pre-Cretaceous section in the central part of the basin. This paper presents the geologic setting, exploration objectives and case history of the O'Connell wildcat and our interpretation of the abnormally pressured zones penetrated by this and other deep wells in the area (Wilson et al., 1994). Vitrinite reflectance measurements, fluid inclusion studies, thermochronology, and apatite fission track analyses (Laughland, 1995) are integrated with regional geologic trends to define the burial and thermal history of the stratigraphic section penetrated by the O'Connell well.

TECTONIC AND STRATIGRAPHIC FRAMEWORK OF THE PICEANCE BASIN

The Piceance Basin is an intracratonic depocenter in the Rocky Mountain foreland province (Figure 1). It is surrounded by basement uplifts and contains more than 20,000 ft (6,100 m) of Cambrian to Miocene strata (Figure 2). The stratigraphy and tectonic evolution of the Piceance Basin have been described previously by Johnson and Nuccio, 1986; De Voto et al., 1986; Waechter and Johnson, 1986; Johnson, 1989; Lorenz and Finley, 1991; Grout et al., 1991; Grout and Verbeek, 1992; Johnson, 1992; Johnson and Nuccio, 1993; and Pearl, 1986.

During the early Paleozoic era, western Colorado was a shallow marine shelf with carbonate and clastic deposits above the Precambrian basement. The Cambrian through Mississippian section thins gradually to the east and contains numerous erosional unconformities. The Mississippian Leadville Limestone is an

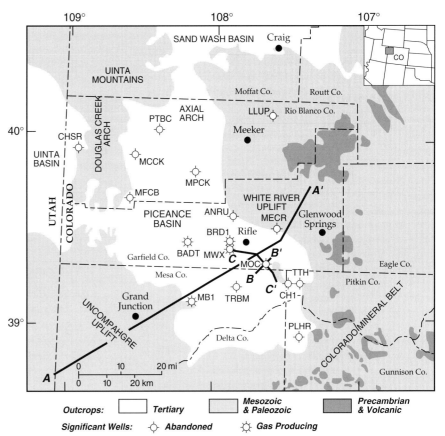

Figure 1. Map showing the location of the Piceance Basin and adjacent structural features in western Colorado. Cross sections A-A', B-B' and C-C' are shown in Figures 4, 5, and 9. Significant wells described in the text are shown.

Symbol	Company	Well Name
LLUP	Louisiana Land & Exploration	Uranium Peak #13-9
PTBC	Pacific Transmission	Barcus Creek #22-12
MCCK	Munson	Corrall Creek #36-1-100
MPCK	Mobil Oil Corp	Piceance Creek #T52-19G
MFCB	Mountain Fuels	Cathedral Bluff #1
CHSR	Chorney	SW Rangely #1-9
ANRU	ARCO	North Rifle Unit #1
MECR	Mobil Oil Corp	Elk Camp Ryden #F12X-22P
BADT	Barrett Resources	ARCO Deep Test #1-27
BRD1	Barrett Resources	Rulison #1 Deep
MWX	U.S. DOE & CER	Superior MWX #1
MOC	Mobil Oil Corp	O'Connell #F11X-34P
CH1	California	Hurd #1
TTH	Texaco	Thunderhawk #1
TRBM	Terra Resources	Brush Creek McDaniel #1-11
MB1	Martin	Blair #1
PLHR	Petro Lewis	Hotchkiss Ranch #1

Outcrops: ☐ Tertiary   ☐ Mesozoic & Paleozoic   ■ Precambrian & Volcanic

Significant Wells: ⬦ Abandoned   ☼ Gas Producing

important reservoir objective due to widespread karst-ification and porosity development in dolomite and limestone beds (DeVoto, 1985; Robertson et al., 1995).

During Pennsylvanian time, collision of North America (Laurasia) and South America (Gondwana) caused extensive basement faulting in Colorado (Kluth, 1986; Johnson, 1992) and several fault-bounded marine depocenters were created (Waechter and Johnson, 1986). The Uncompahgre Uplift in western Colorado and the Ancestral Front Range in central Colorado formed along deep reverse faults. These arches flanked the Colorado Trough, a restricted marine basin which filled with Pennsylvanian carbonates and black shales (Belden Formation), evaporites (Eagle Valley Formation), and fan delta deposits (Minturn Formation). The Colorado Trough was gradually covered by thousands of feet of arkosic conglomerates of the Maroon Formation, eolian sandstones of the Pennsylvanian-Permian Weber Formation, red Triassic sandstones and shales of the State Bridge Formation and eolian, fluvial, and lacustrine deposits of the Jurassic Entrada Sandstone and Morrison Formations (De Voto et al., 1986; Houck, 1991). During Cretaceous time, the region subsided as the Sevier and Cordilleran overthrust belts developed to the west. The Piceance Basin region was gradually inundated by the Western Interior Seaway. Fluvial (lower Dakota Sandstone), shallow marine (upper Dakota Sandstone), deep marine (Frontier Sandstone and Niobrara and Mancos Shales) and shallow marine, fluvial and deltaic sediments (Mesaverde Group) were deposited as the Seaway subsided and then filled (Johnson, 1992).

During Late Cretaceous and early Paleocene time (early Laramide), extensive basement deformation occurred throughout the Rocky Mountain region. In Colorado, basement-cored arches developed above deep reverse faults. The Front Range, Williams Fork, White River, Sawatch, Uinta, and Douglas Creek arches formed during this tectonic episode (Johnson and Nuccio, 1986; Johnson, 1992; Nuccio et al., 1989; Lorenz; and Finley, 1991). Several igneous intrusions were emplaced along the Colorado Mineral Belt (Bookstrom, 1990).

During Paleocene and Eocene time, the eastern and northwestern margins of the Piceance Basin continued to develop by progressive uplift of basement blocks (White River Uplift, Uinta, Douglas Creek and Axial arches) along deep reverse faults. Sediments eroded from these arches were deposited in fluvial (Wasatch Formation) and lacustrine (Green River Formation) environments in the center of the Piceance Basin, reaching a cumulative thickness of more than 10,000 ft (3,050 m) along the basin axis. Late Laramide deformation along the White River, Douglas Creek, and Axial arches (Figure 3) folded the Green River Formation, Wasatch Formation, and Mesaverde Group along the margins of the basin, and caused several fold structures to develop over basement faults in the central part of the basin. Coal ranks in the center of the basin

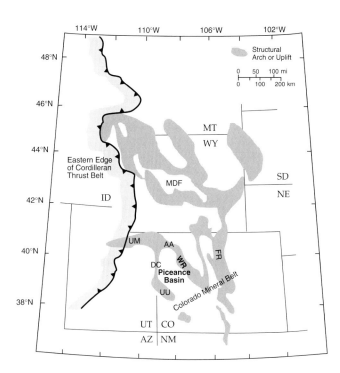

Figure 2. Map showing location of the Piceance Basin and major structural features in the Rocky Mountain region: UU–Uncompahgre Uplift; FR–Front Range Uplift; WR–White River Uplift; DC–Douglas Creek Arch AA–Axial Arch; UM–Uintah Mountains; and MDF–Madden Deep Field. Modified after Dyman et al., 1993.

reached their present-day levels by late Eocene time (Johnson and Nuccio, 1986; Johnson and Nuccio, 1993). During Oligocene time, the Sopris, Snowmass, Chair Mountain, and Coal Basin igneous intrusions (Figure 3) were emplaced along the southeastern margin of the basin (Wallace, 1993).

Figure 4 shows generalized cross section A-A' through the Piceance Basin and several deep wells. The White River Uplift forms the steeply dipping eastern margin of the basin. The Uncompahgre Uplift forms the gently dipping southwestern margin. Several basement fault blocks have been identified in the bottom of the basin with seismic and aeromagnetic data.

Since Miocene time, Tertiary sediments have been deeply eroded by the Colorado and White River drainage systems, resulting in the present-day topography of canyons, plateaus and mesas. Johnson and Nuccio (1986) estimate that up to 5,000 ft (1,525 m) of erosion has occurred throughout most of the basin. At the O'Connell well location, the Paleocene Wasatch Formation is exposed at the surface at +6,550 ft (2,000 m) elevation. West of the well site, Miocene basalt flows cover the Eocene Green River Formation on Battlement Mesa at +11,165 ft (3,405 m) elevation (Tweto et al., 1978). The pre-Miocene surface can be extrapolated eastward over the O'Connell well site to the summit of Mt. Sopris, an Oligocene igneous intrusion, at +12,959 ft (3,930 m) elevation (Figure 3). We estimate that at

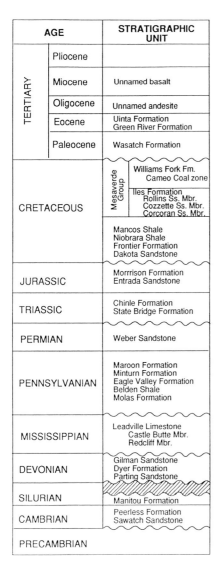

AGE		STRATIGRAPHIC UNIT
TERTIARY	Pliocene	
	Miocene	Unnamed basalt
	Oligocene	Unnamed andesite
	Eocene	Uinta Formation / Green River Formation
	Paleocene	Wasatch Formation
CRETACEOUS	Mesaverde Group	Williams Fork Fm. / Cameo Coal zone / Iles Formation / Rollins Ss. Mbr. / Cozzette Ss. Mbr. / Corcoran Ss. Mbr.
		Mancos Shale / Niobrara Shale / Frontier Formation / Dakota Sandstone
JURASSIC		Morrrison Formation / Entrada Sandstone
TRIASSIC		Chinle Formation / State Bridge Formation
PERMIAN		Weber Sandstone
PENNSYLVANIAN		Maroon Formation / Minturn Formation / Eagle Valley Formation / Belden Shale / Molas Formation
MISSISSIPPIAN		Leadville Limestone / Castle Butte Mbr. / Redcliff Mbr.
DEVONIAN		Gilman Sandstone / Dyer Formation / Parting Sandstone
SILURIAN		Manitou Formation
CAMBRIAN		Peerless Formation / Sawatch Sandstone
PRECAMBRIAN		

Figure 3. Stratigraphic units in the southern Piceance Basin, Colorado. Modified after Western GeoGraphics, 1985 and Pearl, 1986.

least 4,460 ft (1,360 m) of Tertiary strata have been removed from the O'Connell location.

DEEP GAS EXPLORATION IN THE PICEANCE BASIN

The Piceance Basin contains thick, thermally mature Cretaceous shales, coal beds and tight sandstone reservoirs (Johnson and Nuccio, 1986; Johnson and Rice, 1990). Numerous natural gas fields have been discovered in sandstones and coal beds of the Dakota Group, Mesaverde Group and Wasatch Formation (Johnson, 1989; Johnson and Rice, 1990; Johnson et al., 1994). However, as of 1990 only eight deep wells had been drilled to evaluate pre-Cretaceous reservoirs in the central part of the basin (Figure 3), and only one of these wells (Terra Resources McDaniel #1-11 in

11-T9S-R94W) reached the prospective Mississippian carbonate section.

During 1988–1992, Mobil Oil Corporation acquired regional seismic lines across the central Piceance Basin and identified several deep exploration targets on structural features interpreted from the seismic data. Figure 5 shows part of a seismic line across the tilted basement fault block evaluated by the O'Connell well. Figure 6 shows a generalized cross section B-B' based on the seismic and well data. The deep structure is formed by two northwest-trending reverse faults. The western fault is a Pennsylvanian age reverse fault which offsets Precambrian basement through Pennsylvanian age rocks, but terminates up-section in the Pennsylvanian and Permian Maroon Formation. The northeastern fault is a mid-Eocene age (Laramide II) fault which clearly offsets the Dakota Group but splays out in the Mancos Shale, forming a fault propagation fold which deformed the overlying Mesaverde Group and Wasatch Formation.

The primary reservoir objective on this structure was porous dolomite in the Redcliff Member of the Leadville Limestone. Studies of other ultradeep gas fields, especially Madden Deep Field in central Wyoming, indicated that some types of dolomite can maintain excellent porosity and permeability in spite of deep burial and extremely high reservoir temperature. At Madden Deep, dolomite beds in the Mississippian Madison Formation and Ordovician Bighorn Formation have excellent vuggy, coarse crystalline and sucrosic porosity ranging from 3 to 17% at 23,800 ft (7,260 m) (Brown and Shannon, 1989). The gas recovered from the Madison reservoir contains 68% methane, 20% carbon dioxide, and 12% hydrogen sulfide. Reservoir temperature has been measured at 425°F (216°C) and vitrinite reflectances of 3.92 and 4.14% Rm have been measured near the Madison reservoir (Pawlewicz, 1993).

Pre-drill thermal models for the O'Connell location predicted temperatures of 425° to 475°F (218° to 246°C) in the Leadville Limestone, based on thermal gradients from other wells in the area (Johnson and Nuccio, 1986). It was hoped that porous dolomite reservoirs similar to those at Madden Deep Field might be present. Source rock availability, timing of hydrocarbon expulsion, possible hydrogen sulfide and carbon dioxide gases, and hydrocarbon stability at high temperature were significant risks associated with the deep Leadville objective. Oil residue has been found in the Leadville, Minturn, Maroon and Weber Formations, probably derived from hydrocarbons generated from Pennsylvanian black shales (Waechter, 1989). Organic-rich black shales have been identified in outcrops of the Pennsylvanian Belden and Minturn Formations along the east side of the basin (Nuccio and Schenk, 1986; Nuccio et al., 1989), however, the presence and generative potential of these hydrocarbon source rocks in the deep basin near the O'Connell location was unknown prior to drilling. The structural interpretation (Figure 6) indicated sufficient reverse fault offset to juxtapose a Leadville reservoir in the hanging wall against possible Belden black shales in the foot wall.

The seismic data (Figure 5) indicated a possible wedge of evaporite beds of the Pennsylvanian Eagle Valley Formation along the east flank of the tilted fault block (Gunneson et al., 1995). Anhydrite deposits, if present, might have generated hydrogen sulfide gas due to thermochemical sulfate reduction at high temperature (Heydari and Moore, 1989). The possibility of sour gas in the Pennsylvanian and Mississippian section was considered to be a major risk. The possibility of finding carbon dioxide gas was also evaluated, due to the presence of igneous intrusions along the southeastern margin of the basin (Figure 3) which might have caused metamorphic alteration of adjacent Paleozoic carbonates. The Coal Basin laccolith near Redstone (Collins, 1977), Haystack Mountain volcanic neck south of Divide Creek Anticline (Tweto et al., 1978) and Yule Marble Quarry at Marble, Colorado, were graphic reminders of this risk. The stability of methane gas at great depth and high reservoir temperature was another uncertainty. Modeling by Hunt (1979) and Takach et al. (1987) indicate that methane can be stable at extremely high temperatures in the subsurface, however the effects of contaminant gasses are poorly understood. In the Arkoma Basin and Ouachita thrust belt, methane is produced from Spiro and Arbuckle reservoirs with measured vitrinite reflectance values ranging from 2.5 to about 5%R_o (Houseknecht and Spotl, 1993; Houseknecht, 1995).

The Cretaceous Dakota, Jurassic Entrada, and Pennsylvanian-Permian Weber sandstones were other deep exploration objectives at the O'Connell prospect. However, reduction of sandstone porosity and permeability due to deep burial and high thermal exposure was considered a significant risk (Dyman et al., 1993; Schmoker and Schenk, 1994). Core data and porosity logs from other deep wells in the Piceance Basin indicate generally poor reservoir quality in these formations.

DRILLING SUMMARY, TEMPERATURE AND VITRINITE DATA

The Mobil O'Connell F11X-34P wildcat was "spudded" in November, 1990 and reached total depth of 18,422 ft (5,620 m) in the Devonian Parting Formation during October, 1991 (Figure 7). Overpressured zones were encountered in the Cretaceous Mesaverde Group and Niobrara Shale. The Cretaceous Dakota Sandstone was cored and evaluated with a drill stem test. While drilling through the Pennsylvanian Minturn Formation, an unexpected overpressured zone was encountered which caused severe drilling and casing complications. The Mississippian Leadville Limestone was evaluated with cores, logs and an open hole test. During the test, carbon dioxide gas and salt water flowed to the surface. The well was abandoned in early 1992.

Figure 7 shows the formations penetrated, drilling mud weight, measured bottom hole temperature and vitrinite reflectance data for the O'Connell well. Mud weights between 8 and 9 lb/gal (±1 g/ml) generally indicate that the drilling mud was balancing "normal" formation pore pressures, equivalent to a column of salt water from surface with a pressure gradient of approximately 0.45 psi/ft (10.2 kPa/m). Higher mud weights indicate that formation fluids or gas were entering the wellbore at above-normal pressures and that higher mud densities were needed to maintain control of the well and prevent blowouts. Bottom hole temperatures were measured with well logging tools and drill stem tests, and are presented here without corrections. Temperature predictions and corrections for the O'Connell well have been described previously by Honore et al. (1993) and Laughland (1995). Vitrinite reflectance measurements (%R_o) were made on air-dried cuttings and core samples from the O'Connell well by Mobil Exploration and Producing Services, Inc. The vitrinite reflectance values range from 0.56% in the upper Mesaverde Group at 2,580 ft (790 m) to 6.1% in Pennsylvanian shales at 17,911 ft (5,460 m). The lack of vitrinite data between 11,950 ft and 15,800 ft (3,645–4,820 m) is due to the absence of organic-rich shales in the oxidized upper Pennsylvanian through Jurassic continental clastic section.

THE MESAVERDE OVERPRESSURED ZONE

The O'Connell well penetrated the Mesaverde Group at 2,590 ft (695 m) with 9.2 lb/gal (1.1 g/ml) drilling mud. Gas shows commenced below 2,750 ft (839 m). Mud weight was increased to 10 lb/gal at 3,400 ft (1,035 m) because of increasing background gas, and was gradually raised to 10.6 lb/gal (1.3 g/ml) in the Cameo Coal Zone due to strong and frequent gas shows (Figures 6 and 7).

The Mesaverde overpressure zone (Figures 8a and 8b) coincides with coals and carbonaceous shales with measured vitrinite reflectance ranging from 0.8 to 1.3%

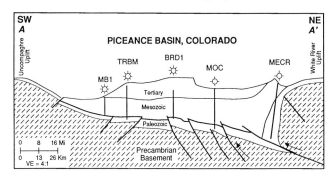

Figure 4. Diagrammatic cross section A-A' showing structural style in the southern Piceance Basin. See Figure 1 for names of the projected wells along the cross section.

in the O'Connell well (Figure 7). These levels of thermal maturity indicate that the coal beds have passed through the stage of water expulsion and have entered the gas generation and expulsion window (Law, 1983; Decker and Horner, 1987). Gas expelled from the Mesaverde coal beds and carbonaceous shales has migrated into nearby fine-grained, low-permeability Mesaverde and Wasatch sandstone reservoirs, forcing moveable water out of pore spaces and fractures (Spencer, 1987; Law and Dickinson, 1985). Many natural gas wells have been drilled to exploit these tight gas sandstones, especially at Grand Valley, Parachute, and Rulison Fields and at Hunter Mesa Unit (Johnson, 1989; Reinecke et al., 1991).

Figure 8a shows the lateral extent of the Mesaverde overpressured zone, which is located along the basin axis where the total thickness of Mesaverde coal deposits is greatest (Tyler and McMurray, 1995) and where maximum burial occurred beneath eastward-thickening upper Cretaceous and Tertiary sediments (Johnson and Nuccio, 1986). Figure 8b shows the position of the Mesaverde overpressured zone with respect to regional vitrinite reflectance trends, based on previous work by Nuccio and Johnson (1983), Tyler et al. (1995), and data contributed by Snyder Oil Corporation. The overpressured zone at the O'Connell well is interpreted to be continuous with the overpressured zone at the U.S. Department of Energy's Multi-Well Experiment (MWX) Site (Spencer, 1989; Johnson, 1989). At the MWX Site, 14 to 15 lb/gal mud (1.7 g/ml) was required to control high background gas (Spencer, 1987). Measured vitrinite reflectance ranges from 0.9 to 2.1% (Law et al., 1989; Johnson and Nuccio, 1993). Pressure data reported by Spencer (1987) indicate normal pressures down to approximately 5,500 ft (1,680 m) within the Mesaverde fluvial section. Pressure increases gradually below 5,500 ft in

the "inactive hydrocarbon generation zone" above the Cameo Coal Zone (Spencer, 1987). Law et al. (1989) describe a bend or "kink" in the vitrinite profile above the Cameo Coal zone and interpret the kink as the paleo-top of overpressure. Within the "active" hydrocarbon generation zone, overpressure increases from 0.65 psi/ft (14.7 kPa/m) at about 7,000 ft (2,135 m) to nearly 0.8 psi/ft (18 kPa/m) in the Cozzette Sandstone at 7,830 ft (2,390 m). Measured pore pressures reach the highest gradient (>0.8 psi/ft or >18 kPa/m) in the Corcoran Sandstone at 8,110 ft (2,470 m). It is curious that pore pressures continue to increase *below* the gas-generating Cameo Coal Zone. The pressure gradient should be highest within the primary source rock section where gas is being generated and expelled, and should decrease away from the source rocks. A hydrocarbon source deeper than the Cameo Coal may be forcing high-pressure gas into the Cozzette and Corcoran Sandstones from below. The MWX wells may be in vertical communication with the highly overpressured Mancos-Niobrara system encountered in Barrett Resources Corporation's Rulison Deep Unit No. 1 well.

The Mesaverde overpressured zone is surrounded by slightly overpressured to subnormally pressured zones, where previously overpressured areas have lost gas pressure due to cooling, removal of overburden by the Colorado River system, and gas leakage along natural fracture systems (Spencer, 1989). Water-saturated rocks occur along outcrops and uplifted areas, especially along the steeply dipping Grand Hogback and at Divide Creek Anticline on the eastern edge of the basin. Approximate positions of the overpressured, subnormally pressured and normally pressured areas in the Piceance Basin have been mapped and described previously by McFall et al. (1986), Decker and Horner (1987), Kaiser (1993), Tyler et al. (1995) and Tyler and McMurray (1995).

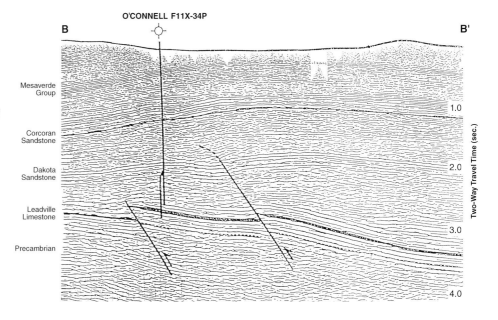

Figure 5. Seismic line B-B′ showing the tilted fault block tested by the Mobil O'Connell F11X-34P well. The location of cross section B-B′ is shown on Figure 1. Interpreted seismic reflectors and approximate position of O'Connell wellbore are labeled.

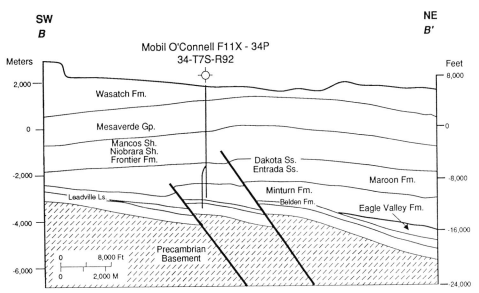

Figure 6. Interpretive cross section B-B′ along seismic line B-B′ (Figure 5) showing the tilted fault block tested by the Mobil O'Connell F11X-34P well. The cross section B-B′ location is shown on Figure 1.

Figure 9 shows cross section C-C′ extending southeast from Rulison Field through the O'Connell location to the west flank of Divide Creek Anticline. The section shows how the pressure and vertical extent of the gas-saturated zone decreases away from the basin center. Mud weight and thermal maturity are lower at the O'Connell well than at the MWX Site. The overpressured, gas-saturated section gradually loses pressure up-dip and interfingers with normally pressured, water-saturated rocks. Several wells drilled along the west flank of Divide Creek Anticline required 11 to 12.6 lb/gal (1.5 g/ml) drilling mud in the Mesaverde section, and produced varying mixtures of water and natural gas. These high mud weights and water recoveries are interpreted to indicate the downdip margin of artesian water flows migrating downward along fractures and permeable zones from high elevation recharge areas (10,000 ft or 3,050 m) at Divide Creek Anticline.

THE NIOBRARA OVERPRESSURED ZONE

After casing was set below the Mesaverde Group, the O'Connell wildcat was deepened through several thousand feet of Cretaceous Mancos Shale with 8.8 to 9.1 lb/gal (1.1 g/ml) mud. A strong gas kick occurred at 10,740 ft (3,280 m) in dark gray calcareous shales of the Niobrara Shale. The well was shut in and mud weight was increased to control the gas kick. A mud weight of 10.4 lb/gal (1.25 g/ml) was used to control high background gas while the main part of the Niobrara overpressured zone was penetrated.

The Niobrara overpressured zone coincides with source rocks with vitrinite reflectance ranging from 2.8 to 3.5% at the O'Connell well (Figure 7). Marine shales with vitrinite reflectance in this range have passed through the hydrocarbon generative window and generally have little to no remaining generative potential.

However, the original source potential of the Niobrara Shale was substantial. Mancos and Niobrara shale samples collected from outcrops along the basin margins contained a mixture of Type II and III kerogens and had total organic carbon values ranging from 0.18 to 3.36% (Johnson and Rice, 1990).

The Niobrara overpressured zone has been penetrated by several other deep wells in the central part of the Piceance Basin (Figure 10). Gas-charged overpressure was encountered in the Niobrara and Frontier shale section at the Mobil Piceance Creek T52-19-G well (19-T2S-R96W), where 13.1 lb/gal (1.57 g/ml) mud was needed to control high levels of background gas below casing at 14,527 ft (4430 m). Measured vitrinite reflectance ranges from 1.85 to 3.11% in the lower Mesaverde to Dakota section in this well (Nuccio and Johnson, 1986; Johnson and Nuccio, 1993). The northern extent of the Niobrara overpressured zone is defined by the Pacific Transmission Supply Company Barcus Creek #22-12 well (12-T1N-R99W) where 15.6 lb/gal mud (1.87 g/ml) was needed to control gas kicks and high background gas in the Niobrara and Frontier section. A cased hole log measured a bottom hole temperature of 320°F (160°C) at 15,642 ft (4,770 m) in this well. The southwestern extent of the overpressured zone (Figure 10) is defined by the Terra Resources Brush Creek McDaniel #1-11 well (11-T9S-R94W). At this location, 13.8 lb/gal mud (1.65 g/ml) was used to control the well as the Niobrara-Frontier section was penetrated.

At the Barrett Resources Corporation's Rulison #1 Deep well (27-T6S-R94W), highly overpressured, gas-charged fractures were encountered within the upper part of the Mancos Shale. This wildcat was drilled during late 1994 to test potential Dakota and Entrada Sandstone reservoirs on a structural closure (T. Barrett, 1995, pers. comm.). The shallow Mesaverde overpressured zone was encountered at a depth of 6,100 ft (1,860 m). High levels of background gas were

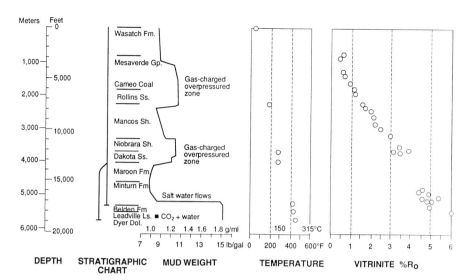

Figure 7. Diagram showing formations penetrated, mud weight, bottom hole temperatures and measured vitrinite reflectance data for the Mobil O'Connell F11X-34P well, Piceance Basin, Colorado.

controlled with 9.9 to 11.9 lb/gal (1.4 g/ml) drilling mud as the Cameo Coal Zone, Rollins, Cozzette and Corcoran sandstone intervals were penetrated (Figure 12). Casing was set in the upper Mancos Shale to cover the Mesaverde Group. The well was deepened with 9.5 to 9.8 lb/gal (1.1 g/ml) mud, however an unexpected gas kick occurred in the upper Mancos Shale at 9,248 to 9,259 ft (2,825 m) and 10.5 to 13 lb/gal (1.6 g/ml) mud was used to control the well. As the well was deepened, 15 to 17.9 lb/gal (2.1 g/ml) drilling mud was required to control frequent gas kicks and high background gas. The upper Mancos gas kick was eventually tested through casing perforations (9,227 to 9,542 ft or 2,810 to 2,910 m), and flowed gas-cut drilling mud and pieces of fractured shale. A pressure buildup test recorded shut-in tubing pressure of 4,180 psi (28.8 MPa) with pressure still increasing slowly as of August, 1995. This equates to a bottom hole pressure of at least 5,150 psi (35.5 MPa, 0.55 psi/ft or 12.4 kPa/m) near the top of the overpressured zone.

Hard, tight, well cemented Dakota Sandstone beds were penetrated below 13,423 ft (4,095 m) in the Barrett well. Tests before and after hydraulic fracture stimulation indicated tight, nonproductive reservoirs. Tests of the Jurassic Entrada Sandstone and Morrison Formation recovered formation water. A bottom hole temperature of 365°F (185°C) was measured at 13,600 ft (4,150 m) after casing had been cemented, indicating a present-day geothermal gradient of 2.3°F/100 ft (42°C/km). The thermal gradient at this location is slightly higher than at the Mobil O'Connell well.

SUBNORMAL PRESSURE IN THE NIOBRARA-FRONTIER-DAKOTA SECTION

Subnormal pressures are present in the Cretaceous shale section and Dakota Sandstone updip from the central overpressured zone (Figure 11). There is an extensive area near Grand Junction and along the

southern part of the Douglas Creek Arch where moderate mud weights are typically used in the Niobrara-Frontier section, many wells have been air drilled, and drill stem tests of gas-filled Dakota Sandstone indicate low pressures. The low pressure zone along the west side of the basin has not been mapped in detail by the authors, however approximate boundaries are shown on Figure 10, based on publicly available test data. At the Martin Blair #1 well (1-T10S-R97W), a pressure buildup test of combined Frontier and Dakota Sandstone perforations measured 2,175 psi (15 MPa) at 7,103 to 7,412 ft (2,165–2,260 m), indicating a gas pressure gradient of 0.3 psi/ft (6.8 kPa/m). At the Munson Corrall Creek #36-1-100 well (36-T1S-R100W), a pressure buildup test of the gas productive Dakota Group measured 3,072 psi (21.2 MPa) at 11,700 ft (3570 m), indicating a subnormal pressure gradient of 0.26 psi/ft (5.9 kPa/m). Fine-grained, thinly bedded Mancos B sandstones produce very low pressure gas along the Douglas Creek Arch (Johnson and Rice, 1990). Gas pressure gradients as low as 0.17 psi/ft (3.8 kPa/m) have been described in Mancos B reservoirs along the Douglas Creek Arch (Mathias, 1971). Many wells within this low pressure area have been drilled with gas or other underbalanced drilling methods to minimize formation damage.

On the southeast side of the basin (Figure 10), the California Hurd #1 wildcat well at Divide Creek Anticline (36-T8S-R91W) tested a highly fractured interval within the Niobrara Shale after a gas kick and blowout occurred. A final shut in pressure of only 3,735 psi (25.8 MPa) was measured at 10,973 to 11,020 ft (3,345–3,360 m) after the drill stem test flowed gas to surface at an estimated rate of 7 million ft³/d (198 x 10³ m³). This equates to a gas pressure gradient of 0.34 psi/ft (7.7 kPa/m), indicating a low pressure gas accumulation in the fractured Cretaceous shale deep within the anticline (Gunneson et al., 1995).

Transition to subnormal pressure may occur *below* the Niobrara overpressured zone as well as laterally. In the Mobil O'Connell well, two cores were cut to evalu-

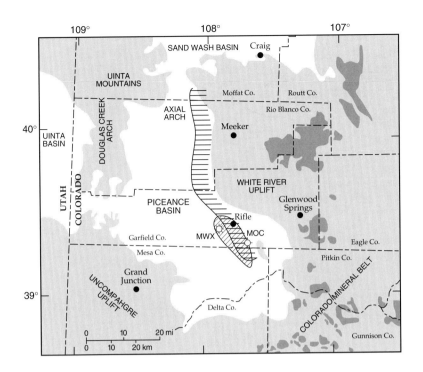

Figure 8a. Map showing the location of the Mesaverde overpressured zone and thickest Mesaverde coals in the Piceance Basin. Overpressure coincides with thickest coals and deepest burial beneath east-thickening Upper Cretaceous and Tertiary sediments. Modified from Johnson and Nuccio, 1986; Kaiser, 1993; and Tyler and McMurray, 1995.

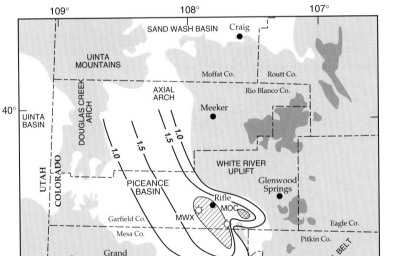

Figure 8b. Map showing the location of the Mesaverde overpressured zone and vitrinite reflectance trends in the Cameo Coal Zone. Overpressured zone coincides with high vitrinite reflectance. Gas pressure has depleted along the structurally deformed eastern edge of the basin. Modified from Nuccio and Johnson, 1983; Kaiser, 1993 and Tyler et al., 1995.

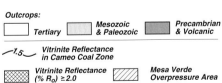

ate the reservoir quality of the Dakota Sandstone. These cores contained hard, well cemented quartzitic sandstone with porosity ranging from 2 to 5% and less than 0.01 md permeability. Open vertical fractures were observed in the cores with quartz crystals lining the fracture walls. A drill stem test was performed to evaluate the fracture system. No gas or fluids were recovered to surface, however a small amount of gas and drilling mud was recovered in the sample chamber. During the final shut in period of 245 minutes, the formation pressure built up slowly to 4,488 psi (30.9 MPa) at 12,018 ft (3,665 m), indicating a pressure gradient of approximately 0.37 psi/ft (8.4 kPa/m). While the pressure probably would have increased somewhat with a longer shut in period, the Dakota zone was evidently *not* overpressured. This pressure gradient and the small gas recovery indicate a subnormally pressured, low-permeability gas system which may be part of a low pressure ring or "halo" surrounding the Niobrara overpressured zone. A mud weight of only 7.2

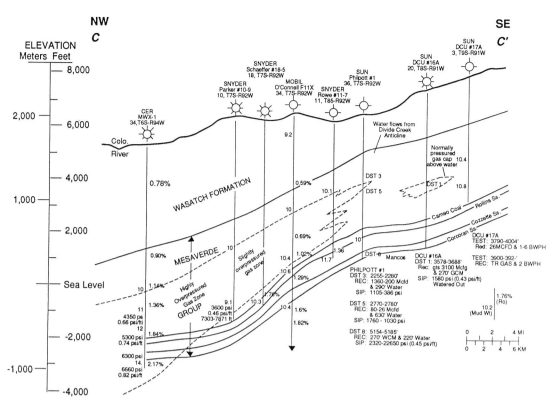

Figure 9. Cross section C-C′ showing the southeastern part of the Mesaverde overpressured zone in the Piceance Basin, Colorado. Location of cross section C-C′ is shown on Figure 1. Pressure and %R$_o$ are greatest at the MWX site and decrease toward the southeast, where artesian water flows have been encountered down-dip from the Divide Creek Anticline.

lb/gal (0.86 g/ml) would be needed to balance the gas pressure measured in the Dakota. It is likely that the natural fracture system was invaded by the 10.4 lb/gal (1.24 g/ml) drilling mud.

Pore pressures apparently revert back to normal in the Jurassic Morrison Formation and Entrada Sandstone at the O'Connell location. Mud weight was reduced to 10.0 lb/gal (1.2 g/ml) as the Entrada Sandstone was penetrated, and no significant hydrocarbon shows were observed. As noted previously, the tilted fault block tested by the O'Connell well was formed by a Pennsylvanian-age reverse fault which dies out in the Maroon Formation. The Entrada Sandstone was not thrust over the Cretaceous (Figure 6), so it was probably not charged with hydrocarbons expelled from Niobrara and Frontier source rocks.

NORMAL PRESSURE IN THE NIOBRARA-FRONTIER-DAKOTA SECTION

Normal pressures are found in the Dakota Sandstone (Figure 10) at the northwest margin of the basin near Rangely Field. In the Chorney Southwest Rangely #1 well (9-T1S-R103W), a shut in pressure of 3,138 psi (21.6 MPa) was recorded in the Dakota Sandstone at 7,425 to 7,522 ft (2,265–2,295 m), indicating an almost

normal pressure gradient of 0.42 psi/ft (9.5 kPa/m). Localized oil and gas caps above an extensive water system are found in this structurally complex area. This normally pressured system has not been mapped in detail by the authors.

DISCUSSION OF THE NIOBRARA OVERPRESSURED ZONE

We interpret the Niobrara overpressured zone as part of a basin-center gas system (Masters, 1979) with a "gas kitchen" of highly mature source rocks surrounded by an outer zone of less thermally mature, subnormally pressured, gas saturated strata and an updip zone of thermally immature, water-saturated rocks. Gas expelled from mature source rocks has migrated upward through fracture systems in the Mancos Shale, especially on fault-fold structures such as Rulison Field, and has probably escaped downward and then laterally through sandstones of the Dakota Sandstone. Overpressure in the Niobrara section has dissipated along the uplifted margins of the basin and has decreased where Tertiary sediments have been removed by erosion. With the notable exception of the Barrett Rulison No. 1 Deep well, mud weights needed to control background gas from the Niobrara overpressured zone are highest in wells drilled where mesas or plateaus of

Eocene age Green River Formation are still present (e.g., Piceance Creek field and Battlement Mesa), and are lower where the Green River Formation has been eroded away (e.g., the O'Connell location). The highest temperatures, mud weights and gas pressures encountered in this overpressured system are at Rulison Field (Figure 12), which has an unusually high geothermal gradient and well developed natural fracture systems. High pressure gas from this system may be in vertical communication with shallower Mesaverde Corcoran and Cozzette sandstones tested at the MWX Site.

The low porosity of deeply buried Dakota, Morrison, and Entrada sandstones, high bottom hole temperatures, and lack of other commercially viable reservoirs have discouraged additional drilling in the deep parts of the basin. The Dakota Sandstone may actually be subnormally pressured in many areas, and its natural fracture systems may have been plugged by overbalanced drilling mud. Fractured Niobrara limestone beds, thin Frontier sandy siltstones and fractured siliceous black shales immediately above the Dakota Sandstone are possible unconventional reservoir objectives in this extensive but lightly explored gas system.

MINTURN PRESSURE COMPARTMENTS

After casing was set at the top of the Maroon Formation to cover the Niobrara overpressured zone in the O'Connell well, the drilling mud was changed to an oil-based system in anticipation of high bottom hole temperatures. The well was deepened with 8.1 lb/gal

(0.97 g/ml) mud through arkosic sandstones and conglomerates of the Maroon Formation, and reached the Pennsylvanian Minturn Formation (Figure 7) at 14,950 ft (4,560 m). No previous wells had penetrated the Minturn Formation in the center of the basin. Due to westward thinning of the Pennsylvanian section along the east flank of the Uncompahgre Uplift (Figure 3), the Minturn and Belden Formations are absent along the west side of the basin.

The Minturn Formation consists of thinly bedded shallow marine sandstones, shales, limestones, and small algal mounds where exposed in outcrops along the White River Uplift, and has been interpreted as a series of fan deltas deposited along shorelines of the Colorado Trough (Bartleson, 1972; Leighton, 1987; Levorsen, 1987). Evaporites of the Eagle Valley Formation were deposited within several restricted subbasins and interfinger with the Minturn Formation. The O'Connell well did not encounter any salt or anhydrite beds in either the Minturn or Belden Formations. The western edge of the Eagle Valley evaporite beds can be interpreted on seismic (Figure 5) where strong reflectors pinch out on the east side of the tilted fault block. This edge is shown on cross section B-B' (Figure 6).

As the well penetrated the Minturn Formation (Figure 7), unexpected flows of salt water were encountered below 16,300 ft (4,970 m). The flows increased in pressure and intensity to a depth of 17,000 ft (5,185 m). The mud weight was increased to 10.9 lb/gal (1.3 g/ml) to control the water flows, however whenever the mud weight was raised to more than 10.6 lb/gal (1.27 g/ml), lost circulation occurred. The borehole

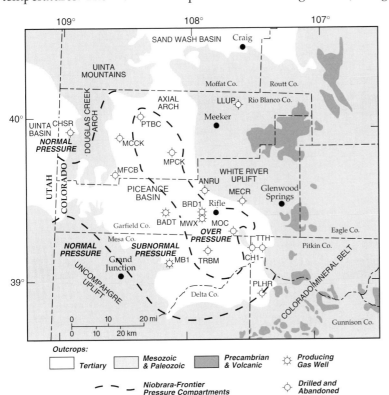

Figure 10. Map showing the approximate location of overpressured, subnormally pressured and normally pressured zones in the Niobrara Shale and Frontier Formation in the Piceance Basin, Colorado.

Symbol	Company	Well Name
LLUP	Louisiana Land & Exploration	Uranium Peak #13-9
PTBC	Pacific Transmission	Barcus Creek #22-12
MCCK	Munson	Corrall Creek #36-1-100
MPCK	Mobil Oil Corp	Piceance Creek #T52-19G
MFCB	Mountain Fuels	Cathedral Bluff #1
CHSR	Chorney	SW Rangely #1-9
ANRU	ARCO	North Rifle Unit #1
MECR	Mobil Oil Corp	Elk Camp Ryden #F12X-22P
BADT	Barrett Resources	ARCO Deep Test #1-27
BRD1	Barrett Resources	Rulison #1 Deep
MWX	U.S. DOE & CER	Superior MWX #1
MOC	Mobil Oil Corp	O'Connell #F11X-34P
CH1	California	Hurd #1
TTH	Texaco	Thunderhawk #1
TRBM	Terra Resources	Brush Creek McDaniel #1-11
MB1	Martin	Blair #1
PLHR	Petro Lewis	Hotchkiss Ranch #1

alternately flowed and lost fluid. Blade shaped, partially polished pieces of red arkosic sandstone were retrieved from the bottom of the hole with a junk basket bit. These rock samples indicated that the borehole wall was fractured and caving in somewhere uphole in the Maroon Formation.

Inflows of high temperature salt water with variable amounts of methane, hydrogen sulfide and carbon dioxide gas filled the lower part of the hole. The water flows were circulated out, only to occur again. It became evident that the salt water filled reservoir or fracture system was extensive and was not depleting rapidly. A bottom hole temperature of 420°F (215°C) was measured during a logging attempt. It became apparent that if control of the well was lost, a hot water and steam blowout (geyser) might develop. Hydrothermal activity along the basin margin was reviewed, including Glenwood, South Canyon, and Penny Hot Springs (Pearl, 1980).

A short core was collected at 16,861 ft (5,140 m) where a water flow occurred. The core consisted of hard, tight quartzitic sandstone with thin black shale laminations and a calcite filled vertical fracture. Vitrinite reflectance measured in a shale lamination was 5.05%. Salt water circulated from the wellbore contained 42,250 mg/l chlorides. Isotope analysis of the water sample showed the $^{87}Sr/^{86}Sr$ ratio to be 0.72. Based on the core, salinity and isotope data, we interpret the salt water to be original formation water which was trapped within fracture networks in sandstone beds of the Minturn Formation. The moderate salinity indicates that the salt water was *not* connected to salt beds of the Eagle Valley Formation.

The salt water flows gradually contaminated the oil-based mud system and turned it into an ineffective, frothy oil-water-mud emulsion. The oil-based mud

was replaced with a new water-based mud system so that the salt water flows could be treated. Repeated attempts at controlling the water flows with high mud weights resulted in lost circulation; treatments with lost circulation materials and cement squeezes were ineffective. The drill pipe became stuck, probably due to borehole damage and fluid flow from the overpressured water zone into lost circulation zones uphole in the Maroon Formation. Attempts to free the stuck drill pipe were unsuccessful, and the wellbore was cemented off and sidetracked.

The sidetrack hole started at 13,149 ft (4,010 m) in the Maroon Formation (Figure 7) and was drilled with 10.2 lb/gal (1.22 g/ml) mud down to a dark gray shale bed at 16,612 ft (5,065 m) in the upper part of the Minturn water flow zone. A bottom hole temperature of 397°F (203°C) was measured at this depth during a logging run (Figures 7 and 13). A short core cut at 16,612 ft (5,065 m) contained massive, dark gray argillaceous lime mudstone without fractures. A thin section from the core contained Atokan foraminifera, micrite and clay. The forams were intact and undeformed, indicating that metamorphic recrystallization to marble had not occurred. The calcareous shale was very fine grained, massive, and ductile looking, and lacked visible microfractures. This calcareous shale layer is one of several thin but effective pressure seals within the Minturn Formation which form the transitional top of the overpressured salt water zone. As noted by Wyman (1993), some rock types which are brittle at low temperature can become ductile at high temperature and pressure. Handin and Hager (1958) have demonstrated the ductile behavior of limestones at high temperature in laboratory experiments.

After casing was set at 16,612 ft (5,065 m) in the sidetrack hole, the well was deepened with 10.4 lb/gal (1.25 g/ml) mud to a depth of 16,690 ft (5,090 m), where a high pressure salt water kick occurred in a quartzitic sandstone bed. The well was shut in and mud weight was increased to 15.3 lb/gal (1.83 g/ml) to control the water flow. The 5 lb/gal mud weight transition indicated a vertical pore pressure change of at least 4,000 psi (27.6 MPa) across less than 20 ft (6 m) of rock. Drill cuttings immediately above and below this water entry zone contained gray limestone and calcareous shale similar to the core samples recovered from 16,612 ft (5,065 m).

DISCUSSION OF THE MINTURN OVERPRESSURED ZONE

We interpret the Minturn overpressured salt water zone to be a series of brittle, extensively fractured quartzitic sandstone beds containing high temperature, high pressure formation water and dissolved gasses. These fractured reservoirs are sealed by overlying and underlying ductile calcareous shales and argillaceous limestones. Due to the lack of other deep

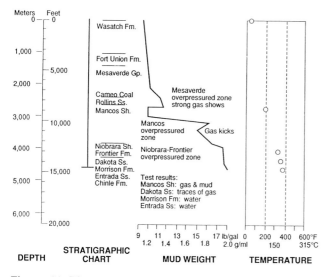

Figure 11. Diagram showing the formations penetrated, mud weight, bottom hole temperature, and test results for Barrett Resources Corporation Rulison No. 1 Deep. The location of this well is shown on Figures 1 and 4.

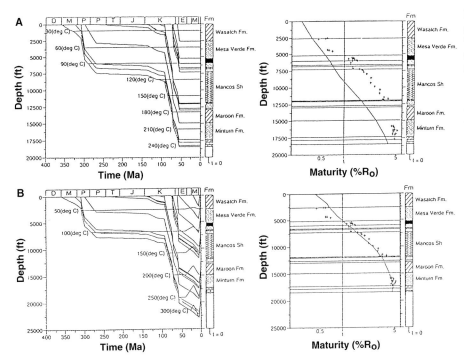

Figure 12(a) Burial history diagram for the O'Connell F11X-34P well, based on present-day geothermal gradient of 2.2°F/100 ft (40°C/km). The burial history is undefined since 55 Ma because Paleocene age rocks outcrop at surface. The predicted %R_o values (solid line) do not match the actual measured values (crosses). (b) Burial history diagram for the O'Connell F11X-34P well using revised inputs from thermal analyses. Deposition continued until uplift at 5 Ma. Approximately 4,460 ft (1,360 m) of section have been removed since 5 Ma. The maximum paleogeothermal gradient of 3°F/100 ft (52°C/km) occurred at 27 Ma. The empirically constrained model yields a good fit between the predicted %R_o (solid line) and the actual measured vitrinite reflectance values (crosses).

wells in the area and the highly variable seismic character of this zone, we have not mapped the lateral extent of the Minturn salt water system. It may be an isolated pressure compartment which occurs only on this fault block. The moderate water salinity indicates lack of fluid communication with laterally adjacent salt deposits of the Eagle Valley Formation. However, the steep dip of the fault block and possible ductile salt movement or differential compaction downdip on the east flank of the structure may be factors in causing the abnormally high water pressure.

LEADVILLE PRESSURE COMPARTMENT

With casing set just above the Minturn overpressured zone and 15.3 lb/gal (1.83 g/ml) mud in the hole to control the water flows, the well was deepened through the Minturn and Belden Formations into the thin red shale of Pennsylvanian Molas Formation at 17,750 ft (5,415 m). The Molas Formation was deposited on the karstified, weathered erosional surface at the top of the Mississippian Leadville Limestone. The Leadville Limestone often contains normally to subnormally pressured water and carbon dioxide gas in the central Rocky Mountain Region (Robertson et al., 1995). In order to avoid drilling into a potentially permeable reservoir with highly overbalanced mud, it was decided to set casing (again) at the top of the Leadville Formation in order to cover and isolate the Minturn overpressured salt water zone. Well logs were run and a bottom hole temperature of 426°F (219°C) was recorded. A liner was cemented at 17,772 ft (5,420 m), however during the cementing procedure a cement plug apparently failed and allowed cement slurry to

flow back inside the liner, perhaps assisted by overpressured salt water from the Minturn Formation. The new liner twisted and ruptured. The hole was nearly lost; however, persistent drilling with milling bits eventually reopened the hole to the top of the Leadville Limestone. Overpressured salt water entered the borehole through punctures in the casing whenever the mud weight was lowered. As the well was deepened into the Leadville Formation, 15.4 lb/gal (1.84 g/ml) mud was used to maintain control of the Minturn water flows.

A reservoir with excellent porosity and permeability was discovered in recrystallized dolomite grainstone beds of the Redcliff Member of the Leadville Limestone at 17,900 ft (5,490 m). In spite of the overbalanced 15.4 lb/gal (1.84 g/ml) drilling mud, there were no major problems with lost circulation as the reservoir was penetrated. Minor hydrocarbon gas shows and traces of carbon dioxide gas were noted by the mudlogger. Six cores were cut to evaluate porosity zones. Core plugs tested with 9,000 psi (62 MPa) overburden stress had 3 to 13% porosity and 0.02 to 81 md permeability. Thin sections from the cores showed vuggy and coarse crystalline dolomite with good pore interconnectivity.

Due to high bottom hole temperature, the drilling mud dehydrated and thickened during each trip out of the hole. Congealed mud cake had to be drilled and circulated out of the lower part of the hole before returning to bottom with a new drill bit. Several high temperature well logs were run. Bottom hole temperatures of 442 to 464°F (240°C) were recorded at 18,378 ft (5,605 m) in the Devonian Parting Formation, indicating a present-day geothermal gradient of 2.2°F/100 ft (40°C/km). Because of the hydrocarbon gas shows, overbalanced mud, and temperature related difficul-

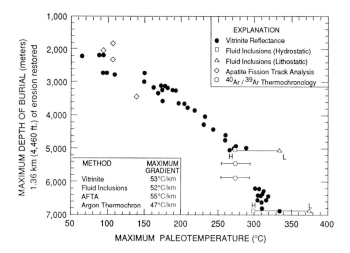

Figure 13. Chart showing maximum paleotemperature versus maximum depth of burial, with estimates derived from four independent analytical methods. Maximum paleogeothermal gradients derived from the different methods range from 2.59°F/100 ft (47°C/km) to 3.03°F /100 ft (55°C/km). Argon thermochronology indicates that the maximum temperature occurred during a brief (2 to 4 m.y.) thermal event at 27 Ma ±4 m.y., probably caused by Oligocene igneous intrusions along the Colorado Mineral Belt. Measured vitrinite reflectance values were converted to estimates of paleotemperature using the empirical correlation of Barker (1988): T(°C) = 104 (ln %R_o) = 146.

ties with the well logging tools, the fluid and/or gas content of the reservoir content was uncertain.

An open hole test of the Leadville reservoir was completed by Benson Mineral Group during early 1992. A 3.5 in. (8.9 cm) liner was cemented at 17,820 ft (5,435 m) inside the previously ruptured 5.5 in. (14 cm) liner to cover the Minturn overpressure zone. Solidified drilling mud was cleaned out of the hole down to 18,159 ft (5,540 m), tubing was installed, and the hole was filled with 9.8 lb/gal (1.17 g/ml) potassium chloride water. No kicks or flows were noted. This fluid was unloaded with coiled tubing and nitrogen, and the reservoir was tested without stimulation or acid treatments. During the test, the reservoir flowed nonflammable gas (97% carbon dioxide, 2% methane and traces of nitrogen and hydrogen sulfide) to surface at rates of approximately 500 mcfd (14.2 x 10³ m³/d, plus salt water at 60 to 100 barrels per day (9.5–15.9 m³/d). The salt water contained 64,000 mg/l chlorides, which is typical for Leadville water throughout the region. At the end of the open hole test, the water flow was "killed" with 8.6 lb/gal (1.03 g/ml) potassium chloride water, indicating that the reservoir was approximately normally pressured (0.45 psi/ft, 10.2 kPa/m). This normally pressured Leadville reservoir is separated from the overlying Minturn overpressured salt water zone by approximately 1,000 ft (300 m) of limestone and calcareous shale beds. These two deep salt water systems were evidently *not* in pressure communication.

Analysis of the methane gas from the Leadville Formation showed a $\delta^{13}C$ isotope ratio of –27.5, indicating "overmature" methane. Traces of pyrobitumen observed in the cores indicate that minor amounts of hydrocarbons may have migrated through the Leadville at some time, however the reservoir had never been filled with hydrocarbons. The Belden section penetrated by the O'Connell well contained mostly gray limestone beds with very few black or dark gray shale layers, indicating deposition in shallow, open marine environments with poor hydrocarbon source rock potential, rather than in a deep, anoxic marine basin. Though the source rock potential of the Belden Formation in the footwall of the structure was still unknown, it was apparent that the Leadville reservoir in the hanging wall had not been charged with hydrocarbons.

Isotope analysis of the carbon dioxide gas from the Leadville reservoir showed a $^{12}C/^{13}C$ ratio of –4.6, indicating that the carbon dioxide in the reservoir was probably derived from metamorphic alteration of carbonate rocks (Hunt, 1979; Houseknecht and Spotl, 1993). Contact metamorphism and hydrothermal alteration probably occurred near the Tertiary igneous intrusions along the southeastern margin of the basin, or near as yet undiscovered intrusions which may be present in the center of the basin. Carbon dioxide and "nonflammable" gas have been previously reported in pre-Cretaceous reservoirs in several other deep tests in the region, including the Louisiana Land and Exploration Company's Uranium Peak No. 13-9 well (9-T2N-R92W) (Robertson et al., 1995), the Munson Corrall Creek No. 36 (36-T1S-R100W), and the Texaco Thunderhawk Unit No. 1 (4-T9S-R91W) (Gunneson et al., 1995).

POST-DRILLING ANALYSIS OF THERMAL AND BURIAL HISTORY

Core and cuttings samples from the O'Connell well were analyzed by four independent methods to refine the burial, uplift, and thermal history of the southern Piceance Basin (Laughland, 1995). Modeling was performed using BasinMod software (available from Platte River Associates), the present-day geothermal gradient of 2.2°F/100 ft (40°C/km), and the Easy %R_o kinetic model (Burnham and Sweeney, 1989). The initial model generated predictions of %R_o which did not match the measured vitrinite reflectance values from the well (Figure 12a). The mismatch indicated that the stratigraphic section is not presently at conditions of maximum temperature or maximum burial. Similar mismatches between present-day and paleogeothermal gradients in the Piceance Basin were found by Johnson and Nuccio (1993).

Apatite fission track analyses (AFTA) were performed on cuttings from the Wasatch Formation (1,395 and 2,190 ft, 425 and 670 m) and Mesaverde Group

(3,015, 3,810, 4,605, and 6,175 ft or 920, 1,160, 1,405, and 1,880 m) from the O'Connell well. Maximum paleotemperatures for the Wasatch samples ranged from 203° to 225°F (95° to 107°C) based on the AFTA analyses. The shallowest Mesaverde sample showed a maximum paleotemperature of 225°F (107°C). The three deeper Mesaverde samples showed maximum temperatures exceeding the annealing temperature for apatite fission tracks (>230°F or >110°C). Note that paleotemperatures in the Wasatch and upper Mesaverde exceed the 180° to 200°F (82°to 93°C) thresholds for the typical top of overpressuring in Rocky Mountain basins noted by Law and Dickinson (1985) and Spencer (1987).

Modeling of the AFTA data indicates that uplift began at approximately 5 Ma and that approximately 4,460 ft (1,360 m) of overburden have been removed. The timing of uplift at the O'Connell location is synchronous with the timing of regional uplift of the Colorado Plateau (Johnson, 1992). The amount of overburden removed is consistent with previous estimates for erosion resulting from the down-cutting of the Colorado River (Bostick and Freeman, 1984; Johnson and Nuccio, 1986; Johnson and Nuccio, 1993). These AFTA data fit with previous interpretations of fission tracks from Mesaverde cores collected at the MWX Site at Rulison Field (Kelley and Blackwell, 1990).

Two samples of potassium feldspars from arkosic sandstones of the Maroon Formation (13,350 and 14,750 ft or 4,070 and 4,500 m) were analyzed using $^{40}Ar/^{39}Ar$ thermochronology to determine the age of maximum paleotemperature. Both samples produced reliable age spectra during laboratory heating experiments due to the pristine quality of the feldspar grains. These samples showed identical timing: cooling at about 600 Ma during late Precambrian uplift and erosion; deposition at about 275 Ma (Permian age), and a thermal spike with maximum temperature of approximately 527°F (275°C) at 27 Ma ±4 m.y. (Oligocene age). This age coincides with Oligocene igneous activity along the southeastern margin of the basin (Bookstrom, 1990; Wallace, 1993). Kinetic modeling of the laboratory-derived age spectra indicated that the maximum temperature event lasted only 2 to 4 m.y.

The brief duration of the maximum temperature event indicates that peak temperature might have been caused by hydrothermal activity. Fluid inclusions from euhedral quartz crystals found in vugs and veins in both the Leadville and Dakota cores were analyzed to determine homogenization temperatures. Laboratory-derived homogenization temperatures were pressure corrected to true trapping temperatures by restoring the sedimentary section to maximum burial thickness (i.e., adding 4,460 ft [1,360 m] to the measured sample depths). The pressure gradient at the time of fluid inclusion entrapment was assumed to be between hydrostatic (0.433 psi/ft or 9.8 kPa/m) and lithostatic (1.0 psi/ft or 23.5 kPa/m). Maximum trapping temperatures for Dakota fluid inclusions ranged from 536° to 662°F (280° to 350°C). Maximum trapping

temperatures for Leadville fluid inclusions ranged from 572° to 707°F (300° to 375°C).

These independent estimates of paleotemperature from the vitrinite, fluid inclusion, argon, and AFTA analyses converge fairly closely (Figure 13). We interpret the maximum paleogeothermal gradient to have been 3°F/100 ft (55°C/km) at approximately 27 Ma, with maximum paleotemperature in the Leadville Formation of approximately 644°F (340°C). At least 4,640 ft (1,400 m) of Tertiary sediments have been eroded at the O'Connell location during the past 5 million years. The refined estimates for the magnitude and timing of paleogeothermal gradient, uplift and erosion were entered into the BasinMod program. The resulting model shows a close match between the kinetic model predictions of $\%R_0$ and the actual, measured vitrinite values (Figure 12b). The good fit gives a strong degree of confidence to the burial, temperature, and uplift reconstruction for the section penetrated by the O'Connell wildcat. Timing of maximum temperature and uplift fit with known geologic events in western Colorado.

DISCUSSION OF PICEANCE BASIN PRESSURE SYSTEMS

Figure 14 shows a pressure versus depth plot of the stratigraphic section penetrated by the Mobil O'Connell well, based on drilling mud weights and test data. Three separate and distinct overpressured systems were penetrated. The Cretaceous section in the central part of the basin contains gas-charged overpressured zones in the Mesaverde Group and Niobrara Shale. These evidently reached peak gas generation several million years ago and are now contracting.

Underlying the gas-charged Cretaceous systems is a water-saturated Jurassic, Triassic, and Paleozoic continental and marine section which apparently lacks extensive hydrocarbon source rocks. A highly overpressured salt water system was found within the Minturn Formation at the O'Connell well, but has not been encountered in any other deep wells drilled in the basin. Porosity development in dolomite beds of the Mississippian Leadville Formation may be laterally extensive, but the porosity is likely to be filled with water and carbon dioxide unless the reservoirs were charged with hydrocarbons from generating source rocks, and the traps preserved intact through several episodes of tectonic deformation and igneous activity.

At the O'Connell location, the geothermal gradient has decreased from a maximum of 3.0°F/100 ft (55°C/km) during Oligocene time to a present-day gradient of 2.2°F/100 ft (40°C/km). The formation temperature in the Dakota Sandstone has cooled at least 240°F (115°C), from a maximum of 536°F to 662°F (280° to 350°C) at 27 Ma, to a present-day temperature of about 300°F (150°C). Similar temperature decline has occurred in the Leadville Limestone. Four to five thousand feet (1.2 to 1.5 km) of Tertiary rocks have been

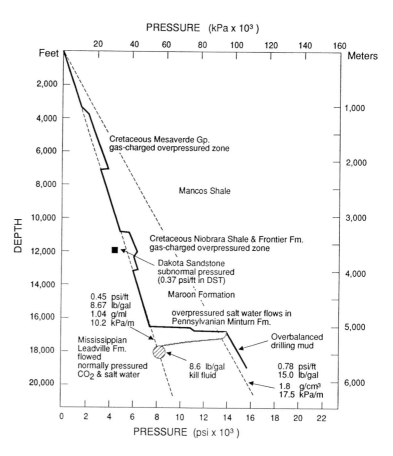

Figure 14. Pressure versus depth profile for the Mobil O'Connell F11X-34P well. The three distinct overpressured zones described in the text are labeled. Solid line shows actual pressure from drilling mud. Light dashed lines show reference gradients. Black square shows apparent subnormal pressure in Dakota Ss. measured by drill stem test. Diagonally ruled circle shows the approximately normal pressure of the Leadville Ls. reservoir based on the 8.6 lb/gal fluid used to "kill" the well after the flow test.

removed during the past 5 to 10 million years. Cooling, erosion, and gas leakage provided ample cause for significant loss of pressure from the gas-charged over-pressured compartments in the center of the basin. Contraction of the overpressured zones has created zones of low pressure gas which surround the over-pressured zones, not only laterally, as mapped by Kaiser (1993), but also vertically. Careful pressure testing of the stratigraphic sections immediately above and below present-day overpressured zones may reveal bypassed reservoirs of low pressure gas, especially if early exploratory wells were drilled with over-balanced drilling mud. Underbalanced drilling methods (Koch, 1994) may be preferred for exploiting these low pressure gas reservoirs.

CONCLUSIONS

The Mobil O'Connell well penetrated three vertically distinct overpressured zones. The shallowest overpressured zone occurs within the Cretaceous Mesaverde Group and is a gas-charged system associated with coal beds and carbonaceous shales with vitrinite reflectance ranging from 0.8 to 2%. These source rocks have generated and expelled gas into nearby tight sandstone reservoirs. The center of this overpressure zone is located near Rulison Field where the thickest coalbeds and deepest burial occurred. This over-

pressured zone grades laterally into less thermally mature, lower pressure, gas saturated rocks and then into an outer zone along the basin margins where rocks are water saturated and normally pressured.

A separate and deeper overpressured zone occurs within the Cretaceous Niobrara Shale in the center of the basin. Vitrinite reflectance in this source rock section ranges from 2.8 to 3.5% in the O'Connell well and up to 3.11% in the Mobil Piceance Creek T52-19G well, indicating overmature conditions and depleted hydrocarbon generation capacity. The greatest pressures encountered in this gas-charged system have been found at Rulison Field, where Barrett Resources Corporation's Rulison Deep Unit No. 1 well used 13 to 17.9 lb/gal (1.6 to 2.1 g/ml) drilling mud to control gas kicks in fractured Mancos, Niobrara and Frontier shale. The Niobrara overpressured zone is surrounded by a lower pressure, gas-saturated zone and an outer ring of normally pressured rocks along the basin margins. Limited test data from the O'Connell well indicate that low pressure gas may occur within the fractured quartzitic Dakota Sandstone below the Niobrara and Frontier Shales.

A highly overpressured, high temperature (420°F or 215°C) salt water system occurs within the Pennsylvanian Minturn Formation. The salt water is interpreted as original formation water trapped in permeable fracture systems within quartzitic sandstone beds. The system is apparently sealed by calcareous shale and argillaceous limestone beds. This unexpected over-pressured compartment caused substantial drilling complications at the O'Connell well. The lateral extent of the system is not known.

A deep reservoir with excellent porosity and permeability was found in dolomite beds of the Redcliff Member of the Mississippian Leadville Limestone at 17,900 ft (5,460 m), with bottom hole temperature of at least 464°F (240°C). During an open hole test, the reservoir flowed carbon dioxide and salt water to surface, with traces of methane, nitrogen, and hydrogen sulfide gas. The Leadville reservoir was evidently *not* filled by Belden-sourced hydrocarbons. Isotope analysis indicates that the carbon dioxide was derived from metamorphic alteration of carbonate rocks, probably due to hydrothermal activity associated with Tertiary igneous intrusions along the southeastern edge of the basin.

Apatite fission track data, fluid inclusion analyses, $^{40}Ar/^{39}Ar$ thermochronology and vitrinite reflectance data from core and cuttings samples have been used to refine the burial and thermal history of the southern part of the basin. Maximum thermal gradient of 3°F/100 ft (55°C/km) and maximum paleotemperature of 644° F (340°C) in the Leadville Limestone occurred at approximately 27 Ma, coinciding with Oligocene igneous intrusions along the Colorado Mineral Belt. Formation temperatures have decreased by at least 200°F (100°C) in the Leadville Limestone and by at least 240°F (110°C) in the Dakota Sandstone since the Oligocene thermal event. During the past 5 million years, at least 4,460 ft (1,360 m) of Tertiary sediments have been removed by the Colorado River and its tributaries. Cooling, erosion, and loss of gas through fractures and faults have caused significant pressure loss as well as vertical and lateral contraction of the Mesaverde and Niobrara overpressured zones.

ACKNOWLEDGEMENTS Mobil Oil Corporation and Benson Mineral Group gave permission to publish this paper. Mobil contributed well data, seismic data, and technical support. Barrett Resources Corporation contributed data for the Rulison No. 1 Deep well. Snyder Oil Corporation contributed data for Mesaverde wells in the Hunter Mesa Unit and technical assistance for a cross section. Roy Enrico measured and interpreted vitrinite reflectance values. Tim Dennison and Melody Rooney analysed and interpreted isotope data. Kirby Rogers analysed foraminifera in the Minturn core. Mary Barrett made petrographic analyses of the Dakota and Minturn cores. Dave Eby made petrographic analyses of the Leadville cores. John Divine assisted with interpretations and advice on wellsite geology and formation testing. Peter Groth made vitrinite measurements on samples from several of Snyder Oil Corporation's wells. Jerry Treybig assisted with completion engineering and formation testing. Jim Reynolds assisted with analysis and interpretation of fluid inclusions. Mark Harrison and his coworkers at the Dept. of Earth Sciences, University of California-Los Angeles assisted with $^{40}Ar/^{39}Ar$ thermochronology. Geotrack International (Australia) performed apatite fission track analyses.

AUTHORS' NOTE

The interpretations presented in this paper are based on a combination of public and proprietary data. Proprietary data have been shown only if permission was given by the owner(s) of the data. Well logs and historical data which have been released to the public are available from the Colorado Oil and Gas Conservation Commission, the Denver Earth Resources Library, or commercial data services such as Petroleum Information and Dwight's Energydata, Inc.

REFERENCES CITED

Bartleson, B., 1972, Permo-Pennsylvanian stratigraphy and history of the Crested Butte-Aspen region: Colorado School of Mines Quarterly, v. 67, p. 187–248.

Bookstrom, A.A., 1990, Igneous rocks and carbonate-hosted ore deposits of the Central Colorado Mineral Belt: Economic Geology, Monograph 7, p. 45–65.

Bostick, N.H. and V.L. Freeman, 1984, Tests of vitrinite reflectance and paleotemperature models at the Multiwell Experiment Site, Piceance Creek Basin, Colorado: USGS Open-File Report 84-757, p. 110–120.

Brown, R.G., and L.T. Shannon, 1989, The #2-3 Bighorn –an ultradeep confirmation well on the Madden Anticline: Wyoming Geological Association 40th Field Conference Guidebook, p. 181–187.

Burnham, A.K. and J.J. Sweeney, 1989, A chemical kinetic model of vitrinite maturation and reflectance: Geochima et Cosmochimica Acta, v. 53, p. 2649–2657.

Collins, B.A., 1977, Geology of the coal basin area, Pitkin County, Colorado, in Exploration frontiers of the Central and Southern Rockies: Rocky Mountain Association of Geologists Symposium, p.363–377.

De Voto, R.H., 1985, Sedimentology, dolomitization, karstification, and mineralization of the Leadville Limestone (Mississippian), central Colorado, in 1985 SEPM Midyear Meeting Field Guide, Field Trip No. 6, SEPM, 180 p.

De Voto, R.H., B.L. Bartleson, C.J. Schenk, and N.B. Waechter, 1986, Late Paleozoic stratigraphy and syn-depositional tectonism, northwestern Colorado, in D.S. Stone, ed., New interpretations of northwest Colorado geology: Rocky Mountain Association of Geologists Guidebook, p. 37–49.

Decker, A.D. and D.M. Horner, 1987, Source rock evaluation: a method of predicting dominant reservoir mechanisms of deeply buried, low-permeability coal reservoirs: SPE/DOE 16419, in Proceedings, SPE/DOE Joint symposium on low-permeability reservoirs, May 18–19, 1987, p. 297–306.

Dyman, T.S., D.D. Rice, J. W. Schmoker, C.J. Wandrey, R.C. Burrus, R.A. Crovelli, G.L. Dolton, T.C. Hester, C. W. Keighin, J.G. Palacas, W.J. Perry, Jr., L.C. Price, C.W. Spencer, and D.K. Vaughan, 1993, Geologic studies of deep natural-gas resources in the United

States, *in* D.G. Howell, ed., The Future of Energy Gases: USGS Professional Paper 1570, p. 171–203.

Grout, M.A., and E.A. Verbeek, 1992, Fracture history of the Divide Creek and Wolf Creek anticlines and its relation to Laramide basin-margin tectonism, southern Piceance Basin, northwestern Colorado: USGS Bulletin 1787-Z, 32 p.

Gunneson, B.G., M.S. Wilson and J.A. Labo, 1995, Divide Creek anticline—a decapitated pop-up structure with two detachment zones, *in* R.R. Ray, ed., High definition seismic: 2-D, 2-D swath & 3-D case histories: Rocky Mountain Association of Geologists Guidebook.

Handin, J., and R.V. Hager, Jr., 1958, Experimental deformation of sedimentary rocks under confining temperature: tests at high temperature: AAPG Bulletin, v. 42, no. 12, p. 2892–2934.

Heydari, E. and C.H. Moore, 1989, Burial diagenesis and thermochemical sulfate reduction, Smackover Formation, southeastern Mississippi salt basin: Geology, v. 17, p. 1080–1084.

Honore, R.S. Jr., B.A. Tarr, J.A. Howard and N.K. Lang, 1993, Cementing temperature predictions based on both downhole measurements and computer predictions: a case history: SPE 25436, *in* SPE Production Operations Symposium, Oklahoma City, OK, March 21–23, 1993, p. 31–38.

Houck, K.J., 1991, Structural control on distribution of sedimentary facies in the Pennsylvanian Minturn Formation of north-central Colorado: USGS Bulletin 1787-Y, 33 p.

Houseknecht, D.W. and C. Spotl, 1993, Empirical observations regarding methane deadlines in deep basins and thrust belts, *in* D.G. Howell, ed., The Future of Energy Gases: USGS Professional Paper 1570, p. 217–229.

Houseknecht, D.W., 1995, Assessing an exploration barrier—is there a "methane deadline" in the Arkoma basin and Ouachita thrust belt: Van Tuyl Lecture Series, Colorado School of Mines, Sept. 28, 1995.

Hunt, J.M., 1979, Petroleum geochemistry and geology: Freeman: San Francisco, 617 p.

Johnson, R.C., and V.F. Nuccio, 1986, Structural and thermal history of the Piceance Creek Basin, Western Colorado, in relation to hydrocarbon occurrence in the Mesaverde Group, *in* C.W. Spencer and R.F. Mast, eds., Geology of tight gas reservoirs: AAPG Studies in Geology 24, p. 165–206.

Johnson, R.C., 1989, Geologic history and hydrocarbon potential of Late Cretaceous-Age, low-permeability reservoirs, Piceance Basin, Western Colorado: USGS Bulletin 1787-E, 51 p.

Johnson, R.C., D.D Rice and T.D. Fouch, 1994, Evidence for gas migration from Cretaceous basin-centered accumulations into Tertiary reservoirs in Rocky Mountain basins, *in* Extended Abstracts, Rocky Mountain Association of Geologists First Biennial Conference on Natural Gas in the Western United States, October 17 & 18, 1994.

Johnson, R.C. and V.F. Nuccio, 1986, Structural and thermal history of the Piceance Creek Basin, western Colorado, in relation to hydrocarbon occurrence in the Mesaverde Group, *in* C.W. Spencer and R.F. Mast, eds., Geology of Tight Gas Reservoirs: AAPG Studies in Geology No. 24, p. 165–205.

Johnson, R.C. and V.F. Nuccio, 1993, Surface vitrinite reflectance study of the Unita and Piceance Basins and adjacent areas, eastern Utah and western Colorado—implications for the development of Laramide basins and uplifts: USGS Bulletin 1787-DD, 38 p.

Johnson, R.C., and D.D. Rice, 1990, Occurrence and geochemistry of natural gases, Piceance Basin, northwest Colorado: AAPG Bulletin, v. 74, p. 805–829.

Johnson, S.Y., 1992, Phanerozoic evolution of sedimentary basins in the Uinta-Piceance Basin region, northwestern Colorado and northeastern Utah: USGS Bulletin 1787-FF, 38 p.

Kaiser, W.R., 1993, Abnormal pressure in coal basins of the western United States, *in* Proceedings of the 1993 International Coalbed Methane Symposium, The University of Alabama, School of Mines and Energy Development, v. 1, p. 174–186.

Kelley, S.A., and D.D. Blackwell, 1990, Thermal history of the Multi-Well Experiment (MWX) Site, Piceance Creek Basin, northwestern Colorado, derived from fission-track analysis: Nucl. Tracks Radiat. Meas., vol. 17, No. 3, p. 331–337.

Koch, G., 1994, Primed for takeoff-underbalanced drilling is unlocking significant incremental production and reserves for more and more producers: Oilweek, v. 45, no. 50, p. 26–28.

Kluth, C.F., 1986, Plate tectonics of the ancestral Rocky Mountains, *in* Peterson, J.A., ed., Paleotectonics and sedimentation in the Rocky Mountain region, United States: AAPG Memoir 41, p. 353–370.

Laughland, M.M., 1995, A practical approach to reconstructing burial and temperature histories: an example from the Mobil O'Connell well, southern Piceance Basin, Colorado: SEPM Rocky Mountain Section Newsletter, vol. 20, no. 1.

Law, B.E., J. R. Hatch, G.C. Kukal and C.W. Keighin, 1983, Geologic implication of coal dewatering: AAPG Bulletin v. 67, p. 2255–2260.

Law, B.E., and W.W. Dickinson, 1985, Conceptual model for origin of abnormally pressured accumulations in low-permeability reservoirs: AAPG Bulletin, v. 69, no. 8, p. 1295–1304.

Law, B.E., V.F. Nuccio and C.E. Barker, 1989, Kinky vitrinite reflectance well profiles: evidence of paleo-pore pressure in low-permeability, gas-bearing sequences in Rocky Mountain foreland basins: AAPG Bulletin, v. 73, no. 8, p. 999–1010.

Leighton, C.D., 1987, Stratigraphy and sedimentology of the Pennsylvanian Gothic Formation in the Crested Butte area, Colorado: M.Sc. Thesis no. T-3214, Colorado School of Mines, 195 p.

Levorson, M., 1987, Stratigraphic analysis of the Gothic Formation, Pitkin and Gunnison Counties, Colorado: M.Sc. Thesis no. T-3274, Colorado School of Mines, 135 p.

Lorenz, J.C. and S.J. Finley, 1991, Regional fractures II: fracturing of Mesaverde reservoirs in the Piceance Basin, Colorado: AAPG Bulletin, v. 75, no. 11, p. 1738–1757.

Mathias, J.P., 1971, Successful stimulation of a thick, low pressure, water-sensitive gas reservoir by pseudolimited entry: Journal of Petroleum Technology, February, 1971, p. 185–190; SPE 2903.

Masters, J.A., 1979, Deep basin gas trap, Western Canada: AAPG Bulletin, v. 63, no. 2, p. 152–181.

McFall, K.S., D.E. Wicks, V.A. Kuuskraa, K.B. Sedwick, 1986, A geologic assessment of natural gas from coal seams in the Piceance Basin, Colorado: Gas Research Institute, Topical Report no. 87/0060, 76 p.

Nuccio, V.F. and R.C. Johnson, 1983, Thermal maturity map of the Cameo-Fairfield or equivalent coal zone through the Piceance Creek Basin, Colorado—a preliminary report: USGS Miscellaneous Field Investigations Map, MF-1575.

Nuccio, V.F., and and C.J. Schenk, 1986, Thermal maturity and hydrocarbon source-rock potential of the Eagle basin, northwestern Colorado, *in* D.S. Stone, ed., New interpretations of northwest Colorado geology: Rocky Mountain Association of Geologists Guidebook, p. 259–264.

Nuccio, V.F., S.Y. Johnson, and C.J. Schenk, 1989, Paleogeothermal gradients and timing of oil generation in the Belden Formation, Eagle Basin, northwestern Colorado: The Mountain Geologist, v. 26, no. 1, p. 31–41.

Pawlewicz, M., 1993, Vitrinite reflectance and geothermal gradients in the Wind River Basin, central Wyoming, *in* Keefer, W.R., Metzger, W.J. and Godwin, L.H., eds., Wyoming Geological Association Special Symposium on Oil and Gas and Other Resources of the Wind River Basin, p. 295–305.

Pearl, R.H., 1980, Geothermal resources of Colorado: Colorado Geological Survey, Map Series 14.

Pearl, R.H., 1986, colorado stratigraphic nomenclature chart, *in* D.S. Stone, ed., New interpretations of northwest Colorado geology: Rocky Mountain Association of Geologists Guidebook.

Reinecke, K.M., D.D. Rice and R.C. Johnson, 1991, Characteristics and development of fluvial and coalbed reservoirs of upper Cretaceous Mesaverde Group, Grand Valley Field, Colorado, *in* S.D. Schwochow, ed., Coalbed methane of western North America: Rocky Mountain Association of Geologists Guidebook, p. 209–225.

Robertson, G.C., D.W. McDermott, and R.L. Tang, 1995, Model based, 2-D seismic definition of subthrust structure in the Uranium Peak area, northwest Colorado, *in* R.R. Ray, ed., High definition seismic—2-D, 2-D swath, & 3-D case histories: Rocky Mountain Association of Geologists Guidebook.

Schmoker, J.W. and C.J. Schenk, 1994, Regional porosity trends of the upper Jurassic Norphlet Formation in southwestern Alabama and vicinity, with comparisons to formations of other basins: AAPG Bulletin, v. 78, p. 166–180.

Spencer, C.W., 1987, Hydrocarbon generation as a mechanism for overpressuring in the Rocky Mountain region: AAPG Bulletin, v. 71, no. 4, p. 368–388.

Spencer, C.W., 1989, Comparison of overpressuring at the Pinedale anticline area, Wyoming and at the Multiwell Experiment site, Colorado, *in* Law, B.E., and C.W. Spencer, ed., 1989, Geology of tight gas reservoirs in the Pinedale anticline area, Wyoming, and at the Multiwell Experiment site, Colorado: USGS Bulletin 1886, p. C1–C16.

Takach, N.E., C. Barker, and M.K. Kemp, 1987, Stability of natural gas in the deep subsurface-thermodynamic calculations of equilibrium compositions: AAPG Bulletin, v. 71, p. 322–333.

Tyler, R. and R.G. McMurray, 1995, Preliminary assessment of the regional genetic stratigraphy and coal occurrence of the upper Cretaceous Mesaverde Group, Williams Fork Formation, Piceance Basin, Colorado, *in* Tyler, R., W.R. Kaiser, R.G. McMurry, H. Seay Nance, A.R. Scott, and N. Zhou, 1995, Geologic characterization and coalbed methane occurrance: Williams Fork Formation, Piceance Basin, northwest Colorado: Gas Research Institute Annual Report no. GRI-94/0456, p. 107–132.

Tyler, R., W.R. Kaiser and A.R. Scott, 1995, Evaluation of the coalbed methane potential in the Greater Green River, Piceance, Powder River, and Raton Basins, Rocky Mountain Foreland, western United States: controls critical to coalbed methane producibility, *in* Kaiser, W.R., A.R. Scott and R. Tyler, 1995, Geology and hydrology of coalbed methane producibility in the United States: analogs for the world: Intergas '95 Short Course, May 1995, Tuscaloosa, Alabama.

Tweto, O., R.H. Moench, and J.C. Reed, 1978, Geologic map of the Leadville 1°×2° Quadrangle, northwestern Colorado: USGS Map I-999.

Waechter, N.B. and E. Johnson, 1986, Pennsylvanian-Permian paleostructure and stratigraphy as interpreted from seismic data in the Piceance Basin, NW Colorado, *in* D.S. Stone, ed., New interpretations of NW Colorado geology: Rocky Mountain Association of Geologists Guidebook, p. 51–64.

Waechter, N.B., 1989, Oil-impregnated outcrops and their relationship to petroleum generation in the late Paleozoic Eagle Basin, northwestern Colorado: AAPG Bulletin v. 73, no. 9, p. 1177 (abs).

Wallace, A.R., 1993, Summary of isotopic geochronology for the Leadville 1° x 2° Quadrangle, central Colorado: USGS Open-File Report No. 93-615.

Western GeoGraphics, 1985, Colorado geologic highway map.

Wilson, M.S., B.G. Gunneson, R. Honore, and R. Enrico, 1994, Characteristics of abnormally pressured compartments in the southeastern Piceance Basin, Colorado, *in* Extended Abstracts, Rocky Mountain Association of Geologists First Biennial Conference on Natural Gas in the Western United States, October 17–18, 1994.

Wyman, R.E., 1993, Challenges of ultradeep drilling, *in* D.G. Howell, ed., The future of energy gases: USGS Professional Paper 1570, p. 205–216.

Heppard, P.D., H.S. Cander, and E.B. Eggertson, 1998, Abnormal pressure and
the occurrence of hydrocarbons in offshore eastern Trinidad, West Indies, *in*
Law, B.E., G.F. Ulmishek, and V.I. Slavin eds., Abnormal pressures in
hydrocarbon environments: AAPG Memoir 70, p. 215–246.

Chapter 13

Abnormal Pressure and the Occurrence of Hydrocarbons in Offshore Eastern Trinidad, West Indies

P.D. Heppard
H.S. Cander
E.B. Eggertson
Amoco Corporation
Houston, Texas, U.S.A.

Abstract

Abnormal pore pressure is widespread in the Tertiary through upper Mesozoic, clastic-dominated section of the Eastern Venezuelan Basin and the eastern extension of the basin into Trinidad. Some of the largest oil and gas columns are found within abnormally pressured sandstones which account for 43 million bbl (6.8 million m³) of oil in Poui field and 882 billion ft³ (24.98 billion m³) of gas in Cassia field. Abnormal pressure within the Tertiary to Upper Cretaceous rocks resulted from the transfer of overburden stress to the pore system during the rapid subsidence and infilling of the foredeep basin during the Miocene and Pliocene. Primary migration from thick, Upper Cretaceous source rocks and secondary migration through the thick Tertiary clastics occurred principally through hydraulically induced fractures within a highly overpressured section. Final migration out of the overpressured section and charging of present-day reservoirs off the east coast of Trinidad occurred during the late Pliocene to Pleistocene uplift and associated complex normal faulting. The multiple pressure compartments within the six fields studied are separated by relatively thin, abnormally pressured shale. The shale seals are most effective in trapping hydrocarbons when the pressure difference across the shale is less than 4 psi/ft (90 kPa/m) regardless of the shale thickness. Normal faults form effective pressure seals throughout the basin, separating porous sandstone pressure compartments with pressure differences as great as 1,856 psi (12.8 MPa). The oil and gas fields of offshore Trinidad reveal a widely varying depth to the top of abnormal pressure, large pressure differences across faults, pressure reversals, and a narrow zone of transition from mild abnormal pressure (<11 PPG [lb/gal] equivalent) to highly overpressured conditions (>14 PPG equivalent).

INTRODUCTION

An extensive section of abnormal pore pressure exists in the Tertiary through late Mesozoic sedimentary section on the island of Trinidad and extends eastward offshore and westward into Venezuela (Figure 1). Trinidad and the adjacent, offshore areas are part of the Eastern Venezuelan Basin, an east-to-west-trending basin on the northern edge of the South American Plate. The basin is bounded on the north by highly deformed and metamorphosed rocks along the Caribbean/South American Plate Boundary Zone, a zone of right lateral strike-slip. The southern margin of the basin is defined by the Guyana Shield. Within this extensive sedimentary section abnormal pressure is estimated to form a continuous zone extending approximately 380 miles (611 km) in an east-west direction and about 100 miles (161 km) in a north-south direction. The principal area of study is off the southeast coast of Trinidad (Figure 1).

Exploration for oil and gas has occurred onshore and offshore of the island of Trinidad since 1866. The first commercial production was established in 1902 (Tiratsoo, 1986). Some of the earliest exploration took place near gas and oil seeps—some of which were associated with mud volcanos, features of high abnormal

Figure 1. A tectonic map of Trinidad and eastern Venezuela. The area of principal study is offshore of the southeast coast of Trinidad—an area of regional normal expansion faults within a thick Pliocene clastic section. Cross sections A-A′ and B-B′ are shown on Figure 5. The Eastern Venezuelan Basin extends from the deformed Mesozoic metamorphic rocks along the plate boundary zone of the Caribbean Plate and the South American Plate to the Guyana Shield. The extent of abnormal pressure in the basin is shown.

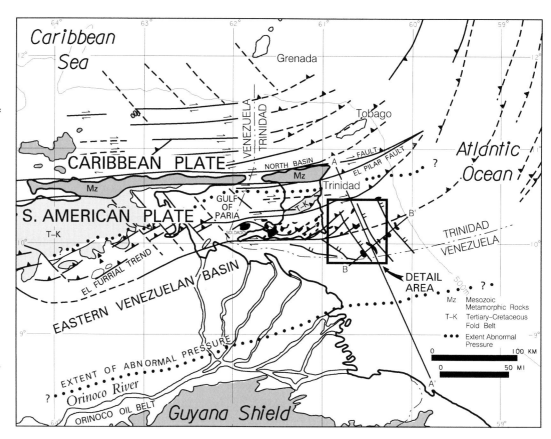

pressure in the subsurface. More than 2.6 billion bbl of oil (413 million m³) have been produced from Trinidad (Rodrigues, 1995), mostly from normally pressured reservoirs. However, significant volumes of oil and gas have been produced from many of the Trinidad's mildly to highly overpressured reservoirs.

Throughout the post-World War II period, abnormal pressure has played an important role in exploration, drilling, and development onshore and offshore Trinidad. However, no detailed study of the distribution of abnormal pressure in the region has been published. The most common problems for exploration and development in the abnormally pressured section in this region are establishing the geologic and pressure conditions in which hydrocarbons occur, and evaluating shale and fault seals within the abnormal pressure zones. The drilling community is most concerned with safety and efficiency related to drilling into the abnormally pressured rocks, a task which requires accurate pressure prediction. For these reasons, most of the over fifty exploration wells and many development wells have been evaluated for abnormal pressure conditions. Pressure predictions for exploratory wells are based on pressure interpretations of previous wells, geologic considerations, and pressure calculations from interval velocities derived from seismic lines. Predictions from seismic velocities are dependent on the quality of the reflection data.

Abnormal pressure is encountered in all the developed fields in the principal study area off the southeast coast of Trinidad. These oil and gas fields have produced over 724 million bbl (115 million m³) of oil and 3.6 trillion ft³ (102 billion m³) of gas from multiple, high porosity, marine sandstone reservoirs of Pliocene age. The fields are excellent examples of abnormally high pressures within hydrocarbon accumulations. They also indicate some of the conditions within the overpressured section necessary for successful exploration and development of hydrocarbons and provide some guideline for the evaluation of shale and fault pressure seals.

In this study, we documented the regional occurrences of abnormal pressure, magnitude of pressure changes both laterally and vertically (including absolute decreases in pore pressure with increasing depth [pressure reversals]), and the effectiveness of hydrocarbon seals within abnormally high pressures. We also studied the distribution of oil and gas relative to abnormal pore pressure to understand hydrocarbon migration within the abnormally pressured zones.

METHODS

Abnormal pressures occurring in the Trinidad area were evaluated by an integrated study of direct pressure measurements from drillstem tests (DSTs) and repeat formation tests (RFTs), conditions during drilling, drilling mud weights, and pressure calculations from well logs and seismic-derived interval velocities. The use of all these types of data and techniques was necessary to create an accurate interpretation of the pore pressure in both the sandstone and shale units. Almost all the producing reservoirs in separate fault blocks within the fields had virgin pressure measurements which were incorporated in the pressure profiles shown in the accompanying figures. Direct pressure measurements, such as RFTs and DSTs, are most accurate in determining pore pressure, but drilling records reporting kicks, flows, gas cut mud, tight hole, and other indications of underbalanced conditions are also valuable. Direct pressure measurements are not practical in very low-permeability shales which often make up the majority of the stratigraphic section. In shale, pore pressure must be estimated by inference from drilling responses such as sloughing, tight hole, or high gas content (which suggest underbalanced conditions), or from techniques using shale properties measured directly by well logs or indirectly by surface seismic. Shale velocity and resistivity from well logs and seismic, in conjunction with empirical equations, provided an estimate of shale pore pressure. Further, unmeasured pressures in adjacent sandstone or limestone units were inferred to be similar. Under some geologic conditions this inference could have been incorrect, for example, where lateral leaks have occurred, pressure may have decreased in sandstone beds independent of the bounding shale.

Well log resistivity, acoustic transit time, and density data were evaluated from a large number of wells for indications of abnormal pressure using proprietary Amoco variations of published techniques included in the PRESGRAF personal computer program. PRESGRAF was recently developed for the evaluation of pressure data by Martin Traugott of Amoco's Formation Evaluation Section. All of the plots of log and pressure data versus depth in this report were created using this program. The plots are shown in true vertical depth (TVD) from the drilling rig datum, which range from 50 to 100 ft (15 to 30 m), unless otherwise noted. Seismic interval velocities used to calculate pore pressure for this study were derived from stacking velocities and generated using depth migration before stack processing techniques.

The calculation of abnormal pressure from log data is based on the empirical relationship between observed resistivity or acoustic values of shale and a trend line reflecting normal compaction. Most of these techniques were developed in the Gulf Coast of the United States in the 1960s, when it was observed that at the onset of significant overpressure, shale resistivity and density decreased while acoustic travel time increased (Hottmann and Johnson, 1965). Early techniques used empirical equations relating observed resistivity or travel time to a normal or expected value for that depth (Hottmann and Johnson, 1965; MacGregor, 1965). An alternative 'equivalent depth' approach uses a vertical projection of an observed shale property onto the normal trend to establish the effective stress

at the depth of the observation (Foster and Whalen, 1966). Recently, most workers have attempted to refine the normal trend line and the empirical equations, which often had to be altered, as new basins or areas were evaluated. Numerous efforts have been made to apply these techniques to interval velocities derived from surface seismic data (Pennebaker, 1968; Reynolds, 1970; Scott and Thomsen, 1993; Davis and Jones, 1994).

A geologic model was used in this study to establish the normal trend line. A normal trend line expressed in travel time or resistivity should reflect normal compaction in terms of porosity reduction (in response to increasing effective stress with burial). While previous workers often had to define separate, empirical trend lines for different age rocks or use complex routines and equations to adjust for different clay rock types, the normal trend line used in this report has considerable worldwide application when local geologic conditions are taken into account.

The equation that defines the normal pressure and compaction trend was developed by Scott and Thomsen (1993). The form of the equation is similar to that developed by Hubbert and Rubey (1959). The normal pressure and compaction trend line is derived from the simultaneous solution of several equations relating compaction, effective stress, and overburden:

$$\phi = K + \phi_0 e^{-\frac{\sigma_e}{c}} \qquad (1)$$

Where ϕ is shale porosity, K is a constant, ϕ_0 is the initial shale porosity, σ_e is the effective stress (overburden pressure minus pore pressure, Terzaghi and Peck [1948]), and c is a constant in the equation which ranges from 4,000 to 7,000 for most areas.

Since the normal trend is defined in terms of porosity, equations must be used to relate porosity to well log measurements such as travel time, resistivity, and density. The work of Eberhardt-Phillips et al. (1989) was used to relate sonic log travel time to porosity. They demonstrated a relationship between compressional velocity, porosity, rock type (volume of clay), and effective stress. The normal, compaction trend line was converted from terms of porosity to resistivity in order to evaluate shale resistivity data for indications of abnormal pressure. This was accomplished by a series of equations that take into consideration shale porosity, water saturation, temperature, resistivity of the water, and the cation-exchange-capacity (CEC) of the clay minerals. The CEC of clay minerals varies greatly from the high CEC and conductive smectite clays to the low CEC and resistive kaolinite clays. The basic equation was taken from Dewan (1983, p. 242) who related conductivity and the above parameters in a Dual-Water model.

This series of equations relating velocity (travel time), resistivity, and density (in terms of porosity) constitute geologic models for how these measurements should vary for normally compacting clay rocks.

Figure 2. The normal compaction trend of shale is shown in terms of sonic travel time. The curved line is the reduction in porosity due to increased effective stress (Equation 1) converted to travel time. The symbols are measured values from clay rocks. The top of abnormal pressure is indicated by values that are higher than the normal trend. a.) A well near the center of the offshore basin reflects a normal compaction sequence relatively unaffected by uplift and erosion. b.) A well from the Galeota Ridge (see Figure 4) has much lower travel time, or higher velocity, within the normally pressured section, which was buried much deeper, uplifted, and had overlying sediments removed by erosion. The trend was adjusted only by reducing the surface porosity parameter in Equation 1.

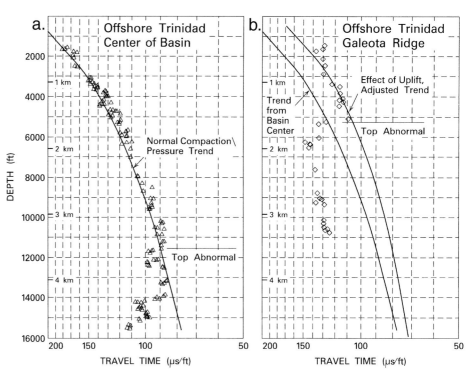

It has been found that shale porosity at the surface, effective stress, volume of clay, and CEC are the primary controls for defining a compaction curve. These variables account for the compaction state, rate of compaction, and reasonable lithologic variations between areas; the other parameters in the equations may be approximated in most cases.

In the study area the normal compaction trend was established from a number of deep, mostly normally pressured wells in the central portion of the offshore basin. Figure 2 illustrates the sonic travel time of clay rocks versus depth for a well in this area. This general compaction trend is curved on a semi-log plot of depth versus travel time. The scatter of values on Figure 2 reflects variation in clay content, cementation, and the relative accuracy of the tool (which may be affected by hole conditions). For the calculation of abnormal pressure, the trend line was adjusted to match the slowest material using the volume of clay parameter. This value should be chosen from the most clay-rich rocks within the section which have undergone normal compaction; this helps compensate for the natural rock variations. The decrease in velocity, or increase in travel time, below 11,500 ft (3,505 m) on Figure 2 marks the top of abnormal pressure in the well and was confirmed by DSTs, fluid kicks, and flows. The slower than expected velocity indicates a state of undercompaction at this depth.

Also shown in Figure 2 is a plot of travel time of shale from a well on the Galeota Ridge—an area that has undergone considerable tectonic uplift and erosion. The travel time is faster in the section above 5,000 ft (1,500 m) than in the well from the center of the basin. This is consistent with a greater compaction

state for rocks that have been more deeply buried, and later uplifted and eroded. The normal compaction trend was only adjusted for the increased compaction state at the surface; all the other variables remained the same. The shift in travel time to values greater than the adjusted normal compaction trend indicates a much shallower top of abnormal pressure than observed in the basinal well.

The normal resistivity trend is shown in Figure 3 for the same wells as shown in Figure 2. The scatter of resistivity values is similar to the scatter of the travel time data. For reasons similar to those used for sonic log data, the trend was shifted to match the lowest resistivity values using the CEC parameter in the most clay-rich rocks. The presence of abnormal pressure is indicated by a lower than expected resistivity (at the same depth as the sonic log data) below 11,500 ft (3,505 m). The PRESGRAF compaction trend derived from resistivity data closely fits the Hottmann and Johnson (1965) normal trend line for the U.S. Gulf Coast when using default values for the compaction parameters and a high CEC value.

The higher resistivity of shale in the well from the Galeota Ridge (Figure 3), relative to the well from the center of the basin, above 5,000 ft (1,500 m) reflects the greater compaction state of these rocks. The normal trend was adjusted only by the surface porosity variable. The surface porosity value affects the beginning compaction state, which led to the prediction of increased resistivity. The shift to lower resistivity below 5,000 ft (1,524 m) indicates the top of abnormal pressure, which was confirmed by DSTs and a rapid increase in mud weights to over 18 PPG (lb/gal) (0.935 psi/ft, 21.2 kPa/m).

The methods used in this report for the calculation of pore pressure from resistivity and acoustic log data for shale were based on the work of Eaton (1975) and modified by Martin Traugott of Amoco (personal communication, 1995). For acoustic travel time data, Eaton's (1975) acoustic method was modified with a substitution of a porosity volume ratio instead of Eaton's ratio of normal versus observed travel time:

$$P_{sh} = OB - (OB - P_n)\left(\frac{1-\phi_v}{1-\phi_n}\right)^x \qquad (2)$$

where P_{sh} is shale pressure, OB is the pressure of the overburden, P_n is normal pressure, ϕ_v is porosity from shale travel time, ϕ_n is the porosity at that depth on the normal trend line, and x is an exponent.

For resistivity data, the equation to calculate pore pressure for shale employed in this study was also from Eaton (1975) and modified by Martin Traugott (personal communication, 1995) to solve directly for pore pressure:

$$P_{sh} = D\left[\frac{OB}{D} - \left(\frac{OB}{D} - \frac{P_n}{D}\right)\left(\frac{R_o}{R_n}\right)^{1.2}\right] \qquad (3)$$

where P_{sh} is shale pressure, D is depth, OB is the pressure of the overburden, P_n is the normal pressure, R_o is the observed shale resistivity, and R_n is the resistivity on the normal trend.

The above techniques, based on well logs and seismic interval velocity, provided an estimate of the pore pressure in shale and were also used to infer a similar pore pressure in adjacent porous and permeable sandstone beds when their pressures were not measured. The assumption of a similar pore pressure was used for the development of these empirical techniques in the U. S. Gulf Coast. In most cases, the methods work well enough to accurately estimate both sandstone and shale pore pressures. However, in some areas in Trinidad, such as within the deepest section at Samaan Field, the calculated shale pressure was markedly different than the measured pressure in the adjacent sandstone. Since the techniques worked well most of the time, we inferred that there was probably a large pressure difference between the units.

The hydraulic fracture gradient of shale is the pressure gradient at which open fractures will form in rocks. Miller (1995) discussed how these fractures may be induced. He proposed that extremely high pore pressure causes sufficient compression or shrinkage of the rock matrix to form fractures. He interpreted that once the fractures form, fluid flows out of, or through, the fractured rock until the pore pressure decreases to a state where the rock matrix expands and the hydraulically induced fractures close, resealing the rock. The volume of fluid released during a single, hydraulic fracturing event was deemed relatively small. While Miller (1995) only considered the case of fractures formed from internal pore pressure, and not intrusive fluids, both cases are considered to have occurred in the study area.

For this study, a method developed by Traugott (1982, 1984) was used to estimate the hydraulic fracture gradient of shale. The hydraulic fracture gradient of rock is dependent on an estimate of overburden stress, pore pressure, and the horizontal to vertical stress

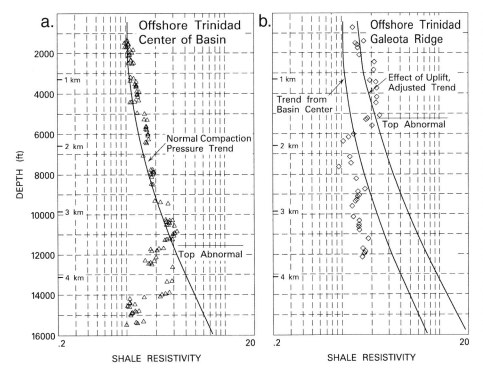

Figure 3. The normal compaction trend of shale is shown in terms of resistivity. The curved line is the reduction in porosity due to increased effective stress (Equation 1) converted to resistivity. The symbols are measured values from clay rocks. The top of abnormal pressure is indicated by values that are lower than the normal trend. a.) A well near the center of the offshore basin reflects a normal compaction sequence relatively unaffected by uplift and erosion. b.) A well from the Galeota Ridge (see Figure 4) has much higher resistivity within the normally pressured section which was buried much deeper, uplifted, and had the overlying sediments removed by erosion. The trend was adjusted only by reducing the surface porosity parameter in Equation 1.

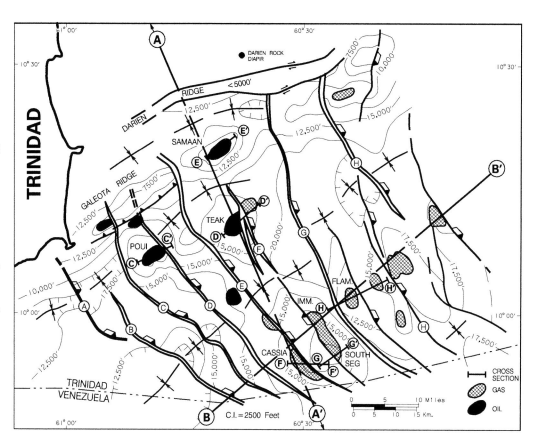

Figure 4. A map of the study area showing the depth to the top of the 0.73 psi/ft pressure gradient (14 PPG, 16.5 kPa/m), principal structural elements, and oil and gas fields. Regional cross sections A-A′ and B-B′ located here are shown in Figure 5. The location of cross sections C-C′ to H-H′, across the individual fields, are discussed in the text and shown in other figures. The top of the 0.73 psi/ft pressure gradient was chosen rather than the absolute top of abnormal pressure due to the relative ease of identification in wells and from seismic interval velocity.

ratio, K. In this study K for shale was based on Pennebaker's (1968) data from the Gulf Coast of the United States. For sandstone, K may be estimated using a relationship based on Poisson's ratio, which was assumed to be 0.27 for this study.

GENERAL GEOLOGY OF THE TRINIDAD AREA

The geologic history of Trinidad and surrounding areas is important for understanding the origins and present-day distribution of abnormal pore pressure and hydrocarbons.

The geology of the late Tertiary of Trinidad and eastern Venezuela was controlled by the tectonics associated with transpression along the plate boundary between the oceanic Caribbean Plate and the continental South America Plate (Figure 1) (Robertson and Burke, 1989; Ave Lallemant, 1991). Compressional strain throughout the island of Trinidad is evident as shown by the style of folding and faulting. The strain was partitioned by major east-west oriented, right lateral, strike-slip fault zones (Eggertson, 1996). Within these partitions are numerous, west-northwest trending anticlines and south-southeast verging thrusts which form the structural traps of the oil and gas fields within the Tertiary section (Figures 1 and 4). Transpressional deformation due to the relative plate movement progressed from west to east along the northern

boundary of the Eastern Venezuelan Basin beginning in the Oligocene (Parnaud et al., 1995) and continues eastward of Trinidad today. The degree of tectonic strain and the severity of uplift in Venezuela and Trinidad diminishes from north to south from the metamorphosed and highly deformed Mesozoic strata along the northern coast at the plate boundary to the homoclinal Pleistocene sediments far to the south along the Guyana Shield (cross section A-A′, Figure 5).

Off the southeast coast of Trinidad the dominant pattern of northwest to southeast trending, large normal faults is interpreted to be the result of combined structural and depositional controls (Figure 4 and cross section B-B′, Figure 5). These regional normal faults are syndepositional and have been active since the middle Pliocene. The east-to-west-trending plate boundary bends to the northeast east of Trinidad. The divergence of this zone of transpression to the north forms a zone of tensional stress. A set of large, syndepositional normal faults resulted from the tensional stress and when combined with rapid Plio-Pleistocene infilling of the area, loaded the poorly consolidated sediment pile which was unconstrained to the east.

Trinidad and the areas offshore of the island on the east, south, and west are considered part of the Eastern Venezuelan Basin (Figure 1). The southern flank of the Eastern Venezuelan Basin is formed by the homocline of the pre-Cretaceous section illustrated on Figure 5 (cross section A-A′) and dips to the north from its outcrop area on the Guyana Shield of South America

(Prieto Cedraro, 1987). The north flank of the basin is defined by metamorphosed Mesozoic rocks along the northern coast of eastern Venezuela and Trinidad.

Sedimentary infilling of the Eastern Venezuelan Basin began following the breakup of Pangea with Early Cretaceous age, and possibly older, sediments deposited in outer shelf and upper slope environments on the northern passive margin of the newly created South American continent (Ross and Scotese, 1988). During a Late Cretaceous flooding event a thick section composed of limestone, siliceous argillite, and organic-rich mudstone was deposited forming the Gautier and Naparima Hill Formations (Requejo et al., 1994) in Trinidad and the Querecual and San Antonio Formations (Figure 6) in Venezuela (Talukdar et al., 1988). Almost all of the oil and most of the condensate in the Eastern Venezuelan Basin, including Trinidad, correlate best with extracts from the organic-rich mudstone of this sequence (Rodrigues, 1988; Talukdar et al., 1988; Heppard et al., 1990).

During the Paleogene there was continued deposition of deep-water, condensed sequences of shale and marl on the passive margin. The earliest effects of the plate boundary collision occurred during the Oligocene in the northwestern part of the Eastern Venezuelan Basin. As the collision event moved eastward during the Miocene and Pliocene the passive margin was converted into a deep, foreland basin. During this time the basin was rapidly filled by pro-deltaic, deltaic and finally non-marine sediments provided by the proto-Orinoco River. This deltaic complex prograded from west to east as it filled the basin (Erlich and Barrett, 1992; Michelson, 1976; Rohr, 1990). In eastern Venezuela the maximum thickness of this sequence is about 20,000 ft (6,000 m). In the study area (off the southeast coast of Trinidad) this sequence is composed of clastic rocks that were as thick as 40,000 ft (12,000 m).

Almost all hydrocarbon production has been from Tertiary clastic rocks (Figure 6). Most reservoir sandstones are interbedded with shale in deltaic to marine shoreface sequences. They are usually thick, well-sorted, very fine-grained sandstones with excellent reservoir properties (15 to 30% porosity). The average initial production rate from oil and gas wells in the southeast offshore area of Trinidad is 2,000 bbl (318 m^3) of oil/day, and 25 million ft^3 (0.71 million m^3)/day,

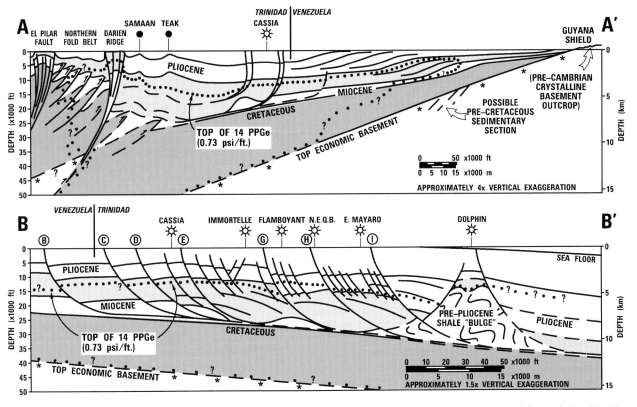

Figure 5. Representative cross-sections through the area of study showing regional structural style and the distribution of high abnormal pressure greater than 0.73 psi/ft (14 PPG, 16.5 kPa/m). Cross-section A-A′ illustrates the structural style along a NNW-SSE trending transect of the Eastern Venezuelan Basin off the east coast of Trinidad. The top of abnormal pressure in cross-section A-A′ occurs in the Pliocene clastic section and sub-parallels stratigraphic boundaries. The abnormally pressured sediments interfinger with coarse clastics at the basin boundaries where the sedimentary section also thins. Cross-section B-B′ illustrates a series of large normal faults oriented in a SW to NE direction in Neogene rocks. The top of abnormal pressure compartments along cross-section B-B′ "steps up" stratigraphy across some of the faults such as the E and G faults but generally follows stratigraphy within the larger fault blocks. The locations of the cross sections are shown on Figures 1 and 4.

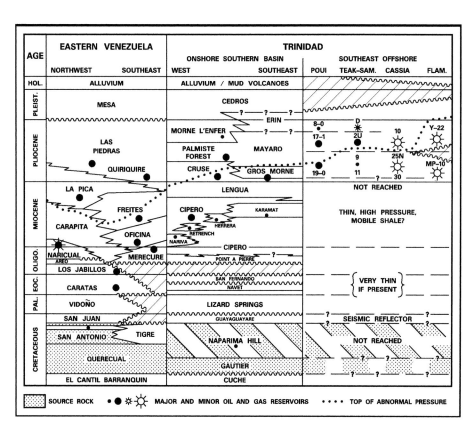

Figure 6. A simplified stratigraphic chart of the Eastern Venezuelan Basin. The dotted line indicates the stratigraphic intervals which have been identified to be, in part, abnormally pressured. The major source rocks in the basin have been identified as the Upper Cretaceous San Antonio and Querecual Formations in eastern Venezuela and the Naparima Hill and Gautier Formations in Trinidad. The units which have been important reservoirs are also indicated. Modified from Parnaud et al., (1995); Erlich et al., (1993); Rohr (1990); and Barr and Saunders (1965).

respectively. A few of the thick, sandstone reservoirs within the normally pressured section produce with a strong water-drive. However, most produce by a combination of fluid expansion and water-drive mechanisms, as do all of the reservoirs within the abnormally pressured section. Large volumes of water are produced from all reservoirs in the later stages of production. The basal sequence of the foredeep sediments was deposited in transitional to outer shelf environments and contains thin, well-sorted, sandstone reservoirs. Additional reservoirs occur as thinly laminated, low resistivity sandstones that were deposited in fan deposits of the shelf slope environment.

Within the study area (Figures 4, 5, and 6) the sedimentary section is predominately a thick, Pliocene to Pleistocene age clastic sequence of sand, clay, sandstone, and shale deposited in a generally shallowing upward trend of deep marine to very shallow marine and deltaic environments. This shallowing trend follows a general pattern progressing from clay-dominated deep water, upper slope or possibly bathyal sedimentation (greater than 600 ft [180 m] water depth), through outer shelf deposits (300 to 600 ft [90 to 180 m] water depth), to sand-rich, shallow marine shelf deposits in water depths generally less than 200 ft (60 m). The deep-water section is nearly always overpressured. Based on palynomorphs, benthic foraminifera assemblages, and the presence of a few meander channels observed on geophysical hazard surveys, the shallowest strata are interpreted as deltaic deposits in the western portion of the area. Through the study area

there is no formal subdivision of the strata. Each field or area is informally subdivided using an alphanumeric sequence (Figure 6). The top of the Pliocene is difficult to define in many areas but reasonable estimates are indicated on the cross sections for each of the fields discussed below.

Only a relatively thin, deep-water Miocene section is postulated to separate the Pliocene section from the Upper Cretaceous source rocks. Miocene rocks have possibly undergone extensive soft sediment deformation and, in places, may form ridge-like shale bulges. This interpretation is speculative since it is based solely on seismic expression. Very high pore pressure may have aided deformation of the Miocene shale.

During the Pleistocene, or possibly the latest Pliocene, further movement of the Caribbean Plate caused compressional deformation of the area southeast of Trinidad. During this time, east-northeast trending ridges including the Darien Ridge, Galeota Ridge, the Poui to Teak, and Cassia to Flamboyant anticlinal trends (Figures 4 and 5) were formed. The Galeota Ridge and Poui to Teak ridge are speculatively interpreted to overlie thrusts. Complex normal fault systems formed as a result of tension in the Pliocene to Pleistocene section along the crest of these ridges and anticlines. The ridges and anticlines apparently had surface expression and were deeply eroded during the Pleistocene. Pliocene to Pleistocene strata are truncated at the sea floor, such as at Poui, or lie unconformably below a thin and flat lying Pleistocene section (Carr-Brown, 1971) which increases in thickness from 25 ft

(7.6 m) at Samaan to over 200 ft (60 m) at Cassia and South SEG fields (Figure 4). At Samaan and Poui over 5,000 ft (1,520 m) of Pleistocene section is estimated to have been eroded. Only a thin, 5 to 10 ft (1.5 to 3.0 m) thick Holocene silt and mud layer now forms the sea floor where the Pleistocene does not subcrop.

DISTRIBUTION OF ABNORMAL PRESSURE

General

Abnormally pressured sedimentary rocks (>0.45 psi/ft, >10.2 kPa/m) in the Trinidad area occur in Upper Cretaceous to Tertiary rocks throughout much of the Eastern Venezuelan Basin (Figure 1). The base of abnormal pressure is unknown over much of the area. The deepest wells drilled in the area range from 15,000 to 17,000 ft (4,600 to 5,200 m) and, at these depths, are still within abnormally high pressured Cretaceous to Pliocene sedimentary rocks. Abnormal pressure is interpreted to extend from within the Pleistocene clastic wedge in the deep water east of Trinidad, northward to the thin Tertiary clastic section just south of the plate boundary along the northern coast of Trinidad and Venezuela. The southern limit of abnormal pressure extends onto the Guyana Shield where the clastic section becomes thin. Westward, abnormal pressure extends into eastern Venezuela. Abnormally high pressure extends into the underlying, regionally extensive Upper Cretaceous oil and gas source rocks. The depth to abnormal pressure ranges from less than 1,500 ft (460m), where structural deformation has uplifted previously deeply buried rocks, to below 12,000 ft (3,660 m), in areas of mild deformation. Abnormally high pressure conditions originated in the western part of the basin during the late Miocene when the sedimentary rocks attained thicknesses ranging from 10,000 to 20,000 ft (3,000 to 6,000 m). Development of abnormal pressure propagated eastward following the shifting depocenter of the proto-Orinoco deltaic complex.

Prior to the search for hydrocarbons in Trinidad, the presence of abnormal pressure was manifested by the surface expression of high pressure shale diapirs and mud volcanoes (Salvador and Stainforth, 1965). The mud volcanoes are interpreted as vents of high pressure water and minor amounts of hydrocarbons mixed with clay where abnormally pressured rocks are close to the surface (Barker and Roberts, 1965; Hedberg, 1974; Kugler, 1965; Woodside, 1981). Some of these features can be spectacular, such as the occasional explosions at the Devil's Woodyard mud volcano (Wharton and Hudson, 1995) and the transient appearance of mud islands off the southern Trinidad coast. In 1911, the Chatham Island mud volcano spontaneously ignited forming a 300 ft (90 m) gas flare (Arnold and Macready, 1956; Higgins and Saunders, 1967). The diapiric movement of mobile shale is probably initiated by structural deformation and partially aided by high pressure fluids.

Abnormal Pressure Onshore Trinidad

Abnormal pressure occurring within the oilfields of onshore Trinidad has been reported from Miocene to Pliocene rocks by a number of workers. The highest reported pressure gradients are 0.95 psi/ft (21.5 kPa/m) (Sealy and Ranlackhansingh, 1985). Oil-bearing sandstone pressure compartments with pressure gradients up to 0.74 psi/ft (16.8 kPa/m) have been reported from depths of 800 to 4,500 ft (250 to 1,375 m) (Jones, 1965). Evaluations of well logs and pressure data from a few available wells support these reported pressure gradients.

Abnormal Pressure in the Gulf of Paria Region

West of Trinidad, in the Gulf of Paria region (Figure 1), the oil bearing Mio-Pliocene Manzanilla and Cruse Formations in some fields are abnormally pressured at depths between 7,000 and 8,000 ft (2,100 to 2,400 m) to pressure gradients approaching 0.83 psi/ft (18.8 kPa/m)(McDougall, 1985; Radovsky and Iqbal, 1985). At Soldado Marine field (Figure 1), which has produced 400 million bbl of oil (63.6 million m^3)(Abelwhite and Higgins, 1968; Tiratsoo, 1986), abnormal pressure begins as shallow as 4,000 ft (1,200 m). Mud weights during logging in several wells in the Soldado Marine field were as high as 17 PPG (0.88 psi/ft, 19.9 kPa/m) at a depth of 9,000 ft (2743 m). The highly overpressured section includes a 3,500 ft (1,067 m) thick sequence of the Late Cretaceous age, Naparima Hill and Gautier Formations which contain the principal source rocks of the basin.

Abnormal Pressure in Eastern Venezuela

Abnormal pressure continues south and west of Trinidad within the upper Mesozoic to Tertiary sedimentary section into eastern Venezuela. In eastern Venezuela it appears that abnormally pressured strata extend from the deformed belt along the wrench fault system on the northern plate boundary zone to a poorly defined area southward and roughly coincident with the Orinoco oil belt where the sedimentary section thins and onlaps the Guyana Shield (Figure 1, Figure 5 cross-section A-A'). In these areas, the zone of abnormal pressure dissipates where the sedimentary section becomes less marine and contains greater amounts of coarse clastics. The limit of the abnormal pressure is apparently controlled by the general lack of shale seals and the thickness of sediments. Abnormal pressure extends westward within the Cretaceous through Miocene section at least to the El Furrial trend (Figure 1). Pressure data from a few deep wells and seismic interval velocities in the central portion of the

Eastern Venezuelan Basin (in Venezuela) show that abnormal pressure occurs as shallow as 3,000 ft (910 m). More typical depths to the top of overpressure vary between 7,000 and 9,000 ft (2,100 and 2,700 m) in the Miocene section. Well logs indicate that abnormal pressure gradients can be as high 0.88 psi/ft (20 kPa/m) at depths of 6,000 to 10,000 ft (1,800 to 3,000 m).

The recently discovered fields of the El Furrial trend (Figure 1) represent some of the largest, overpressured oil and gas fields in the world, and have estimated reserves of 3.9 billion bbl (620 million m^3) of oil and 34.1 trillion ft^3 (96.6 billion m^3) of gas (Carnevali, 1992; Prieto and Valdes, 1992). The main Oligocene sandstone reservoir, the Naricual Formation, is abnormally pressured and has a gradient as high as 0.846 psi/ft (19.1 kPa/m) near the crest of the trap at 13,000 ft (3,960 m) (Russo and Dumont, 1994). The greater than 2,000 ft (610 m) oil columns found within the fields of the El Furrial trend are likely due, in part, to the excellent top seal formed by the thick, deep marine shale of the Miocene Carapita Formation which is also abnormally pressured. The reservoir is estimated to be 3,900 psi (26.89 MPa) above normal hydrostatic pressure in water below the oil column.

ABNORMAL PRESSURE OFFSHORE SOUTHEAST TRINIDAD

Overview

Abnormally high pore pressure is present throughout the offshore area southeast of Trinidad (Figures 4 and 5). While most of the oil and gas has been produced from the normally pressured section, significant reserves have been found within abnormally pressured sandstone reservoirs at Poui, Cassia, and Flamboyant fields. Generally the top of abnormal pressure is parallel with stratigraphic boundaries, but in some areas the top of abnormal pressure is stratigraphically offset across the regionally extensive normal faults, such as from Cassia field to South SEG field and from the SEG/Immortelle area to Flamboyant field (cross section B-B′, Figure 5).

Abnormally pressured shale and sandstone pressure compartments closely follow stratigraphy within the petroleum fields and sub-regionally. A map of the depth to the 0.73 psi/ft (16.5 kPa/m) isopressure gradient line is shown on Figure 4. The equivalent pressure gradient of 14 PPG (0.73 psi/ft, 16.5 kPa/m) was chosen, rather than the top of abnormal pressure, because of the relative ease of defining this level of abnormal pressure and its importance in drilling operations. The pressure gradient mapped in Figure 4 was defined by pressure compartments or shales near the crest of structures since these were the areas of well control. In contrast, the absolute top of abnormal pressure, which can be only 100 to 300 psi (689 to 2,068 kPa) above normal pressure, is often subtle and unless

directly measured is masked during drilling by commonly used mud weights. Very low levels of overpressure are also difficult to discern from resistivity and sonic log response.

The top of abnormal pressure occurs at shallow depths in areas of greatest structural uplift in the area west of the E fault along the Galeota Ridge and along the crest of the Darien Ridge (Figure 4). The shallowest known occurrence of abnormal pressure is at the eastern end of the Darien Ridge (Figure 4). In this area, abnormal pressure begins above 2,400 ft (731 m) and 16.6 PPG (0.862 psi/ft, 22.6 kPa/m) mud was required for drilling to a depth of 3,660 ft (1,120 m) in a water-wet interval.

Abnormal pressure occurs within sandstone and shale sequences regardless of the depositional environment or relative abundance of shale or sandstone. However, there is a difference in the rate of pressure increase as a function of lithology. In the Poui, Teak, Samaan, and Flamboyant fields the average pressure increase ranges from 1.1 to 1.2 psi/ft (25 to 27 kPa/m) in shallow marine deposits that have >50% sandstone and siltstone. Pressure gradients are higher, 1.9 to 3.0 psi/ft (45 to 68 kPa/m), in the deep-water and shale-dominated deposits of the deepest drilled sections of Cassia, South SEG, Samaan, and Flamboyant fields. A difference in pressure sealing ability between shallow versus deep-water shales cannot be demonstrated (as will be discussed below in the section on shale seals). The difference in the rate of pressure increase of these sections may be related to the origin of abnormal pressure. Although they are not considered source rocks, deep-water shales likely contain enough dispersed organic matter to create high pressure by a net volume increase from the generation of small amounts of natural gas. However, the different pressure gradients may be a consequence of the greater amount of shale in the deep-water deposits and the higher pressure gradient associated with pressure sealing shales (versus a pressure gradient of 0.45 psi/ft [10.2 kPa/m] in water-filled sandstone). That is, the less sandstone in the section the more continuous the build-up of pressure at a higher rate in shale.

All the petroleum fields within the basin are complexly cut by normal faults. In most cases, the faults are pressure seals. In general, large hydrocarbon columns are found where the reservoir has lower pore pressure than the overlying shale and the rock section juxtaposed across a normal fault. Examples include the 30 Sand in Cassia field and the MP-50 Sand in Flamboyant field. Thin hydrocarbon columns are present where the reservoir is juxtaposed by normal faults to lower pressured rocks—for example, in the high pressure horst in Poui field.

Absolute decreases in pore pressure, or pressure reversals, below higher pressure compartments also exist in the offshore area. A significant pressure reversal at Flamboyant field may have resulted from lateral fluid leakage westward across the regional G fault to normally pressured sandstones in the Immortelle/SEG fields vicinity.

Poui Field

The Poui field is located 10 miles (16.1 km) off the east coast of Trinidad in 180 ft (55 m) of water (Figure 4). The field has produced over 197 million bbl of oil, 374 billion ft³ of gas, and 162 million bbl of water (30.8 million m³, 10.6 billion m³, and 25.7 million m³ respectively) from multiple, sandstone reservoirs of Pliocene age (Figure 7). The Poui field was discovered in 1972 and development began in 1974 with the first of two platforms. The trap is a northeast trending anticline and is cut by a complex set of northwest to southeast trending normal faults that have throws as large as 2,500 ft (762 m). The sandstone beds are remarkably continuous and contain shale markers as thin as 5 ft (1.5 m) which may be traced throughout the field.

While the majority of the oil in Poui field is from normally pressured sandstone reservoirs, 22% (43 million bbl, 6.8 million m³, of oil) has been produced from abnormally pressured sandstone reservoirs from the 18-0 to 20-0 sandstones—informally named units within the field (Figure 7). This is the largest accumulation of oil occurring within abnormally pressured reser-

voirs off the east coast of Trinidad. These reservoirs (between the Z and Y faults located on Figure 7) are 2,641 to 2,673 psi (18.2 to 18.4 MPa) above normal pressure and have combined gas and oil columns of 350 ft (107 m) and 643 ft (196 m), respectively.

Abnormal pressure begins as shallow as 7,000 ft (2,100 m) and increases in step-like fashion over a 5,000 ft (1500 m) interval of alternating sandstone and shale to a pore pressures equivalent of 0.78 psi/ft (17.6 kPa/m) (Figure 8). Correlative, sandstone reservoirs are at the same or very similar pressure across the field despite the complex normal faulting. That is, a correlative sandstone, such as the 18-2 sand, is at about the same pressure above normal even when offset by large normal faults. The pressure versus depth plot shown in Figure 8 illustrates three pressure profiles from the largest fault blocks within the field. The profiles demonstrate the large pressure differences across normal faults. For simplicity, only sandstone pressure compartments are shown.

Shale units average 200 ft (61 m) in thickness and separate pressure compartments such that the shale units have pressure gradients across them equivalent to 0.80 to 6.23 psi/ft (18.1 to 141 kPa/m) of shale. Normal faults in the field form hydrocarbon and pressure seals despite fault juxtaposition of high porosity sand-

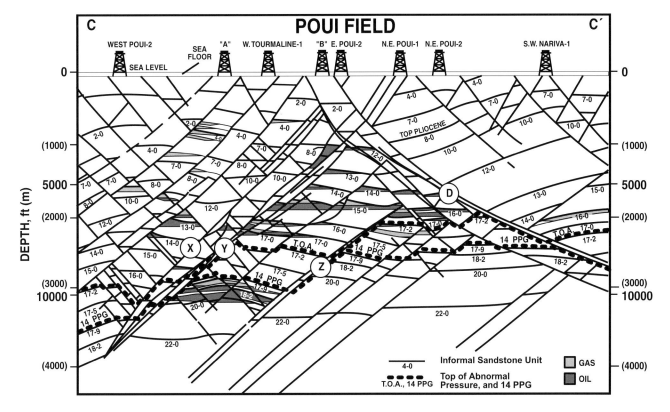

Figure 7. Geologic cross-section C-C′ through the Poui Field illustrating the complex normal faulting and stacked gas and oil zones in Pliocene to Pleistocene sandstones. The top of abnormal pressure (T.O.A.) consistently occurs in the shale above the 17-2 sand across the field. Large pressure differences are present across the larger normal faults. Location of cross-section shown on Figure 4. (after P. Heppard, F. Sobol, T. Tinl, E. Lewis and K. Sawyer; Amoco Trinidad Oil Company).

Figure 8. Three pressure profiles from the Poui Field, Trinidad. The three profiles included are 1) from the east side of the field, at the crest on the upthrown side of the Z fault, 2) at the crest of the central fault block between the Z and the Y faults, and 3) within the lowest fault block on the western side of the field downthrown by the X and Y faults. The profiles illustrate the large pressure differences across these pressure sealing, normal faults, and the lateral continuity of the pressure compartments despite the faulting. Only the pressure within the sandstone pressure compartments is shown.

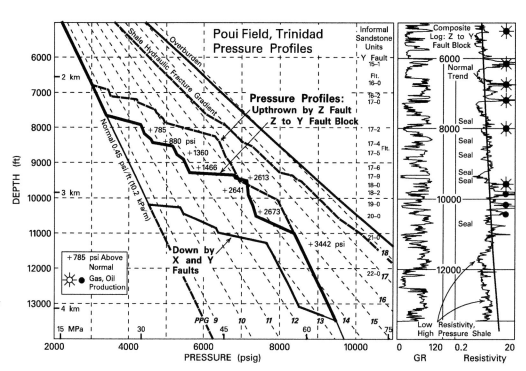

stones. The differences in pressure across some of the faults are as high as 1,856 psi (12.80 MPa) (Figure 9).

Large differences in pressure across normal faults are one of the significant features of Poui field and other Trinidadian fields. Detail of part of the cross section shown in Figure 7 through the Poui field illustrates these large cross-fault pressure differences and is shown in Figure 9. Along the Z fault, where the abnormally pressured section has been offset by about 1,500 ft (460 m) of throw, pressure differences range from 330 psi to 1,856 psi (2.28 to 12.80 MPa) where high porosity sandstones are juxtaposed across the fault. Only small hydrocarbon volumes are trapped on the high pressure side of the Z fault, mainly in the mildly pressured 17-1, 17-2 and 17-4 sands. But pressure in the 21-0 sandstone, on the upthrown side of the Z fault, forms a pressure barrier to flow and enhances the large oil column of the 18 through 20 series of reservoirs on the downthrown side of the Z fault.

Although the seal for oil and gas trapped in the 18-0 to 20-0 sands is enhanced by higher pressure to the east across the Z fault, on the west flank of Poui field there is a large pressure decrease from this accumulation across the Y fault complex and another large fault which together total 2,000 ft (610 m) of throw (Figure 7). The pressure difference is as high as 2,641 psi (18.21 MPa) from the abnormally pressured reservoirs to normal pressure in the 17-0 sand. Sand-on-sand juxtaposition through this complex fault zone provides potential paths for flow out of the trap between the Y and Z faults. Leakage to the west has likely occurred and is the most reasonable explanation for the absence of a large hydrocarbon column in the topmost sand, the 17-9, in the Y to Z fault block.

Teak Field

The giant Teak field is located 25 miles (40.2 km) off the east coast of Trinidad in 190 ft (57.9 m) of water (Figure 4). The field has produced over 308 million bbl (49.0 million m³) of oil, 1.26 trillion ft³ (35.7 billion m³) of gas, and 152 million bbl (24.2 million m³) of water from multiple Pliocene quartz sandstones (Figure 10). Teak field was discovered in 1969 and development began in 1972 with the first of five platforms (Lantz and Ali, 1990; Bane and Chanpong, 1980). The field extends over 4 mi (6.44 km) along an anticlinal crest and is about 2.5 miles wide (4.0 km).

The majority of the oil in Teak field is from normally pressured, shallow sandstone reservoirs. Cumulative production from mildly overpressured reservoirs (with pressure gradients of 0.51 to 0.55 psi/ft [11.5 to 12.4 kPa/m]) through 1995 totals 9.8 million bbl (1.6 million m³) of oil, 573 billion ft³ (91 billion m³) of gas, and 6.9 million bbl (1.1 million m³) of water. Low oil recovery from abnormally high-pressured reservoirs in Teak field is, in part, due to the difficulty of producing the paraffinic oil from a large accumulation in the informally named 9 sand between the F and F2 faults (Figure 10). Overpressure begins at a depth of about 10,000 ft (3,000 m) and increases in step-wise fashion over a 6,000 ft (1,800 m) interval of alternating sandstone and shale to pressure gradients above 0.88 psi/ft (20.0 kPa/m) at a depth of 14,000 ft (4,300 m) in the highest fault block upthrown by the 19 fault (Figures 10 and 11). Pore pressure gradients in the deeper sandstones are over 0.779 psi/ft (17.6 kPa/m) at a depth of 13,420 ft (4,090 m) and the deepest shale intervals are estimated to have pressure gradients greater than 0.88 psi/ft (20.0 kPa/m).

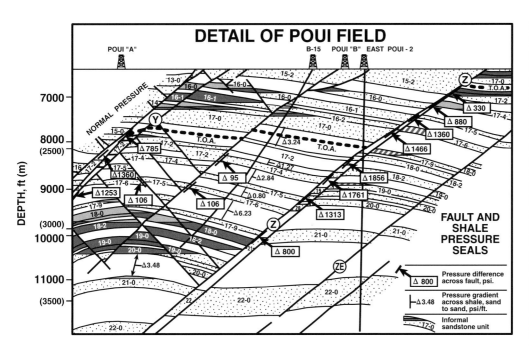

Figure 9. A detail of the Poui Field cross-section, shown in Figure 8, showing the pressure gradient or pressure difference across shale beds, and the pressure differences across normal faults. The Z fault forms a pressure seal from about 7,000 feet to 11,000 feet (2,100 to 3,350 m). The shale above the 17-9 sand has the largest pressure difference of 6.23 psi/ft (141 kPa/m). Only relatively minor oil and gas columns are trapped on the, up-thrown, high pressure side of the Z fault in the 17-2, 17-4, 17-6, 17-9, and 19-0 sands.

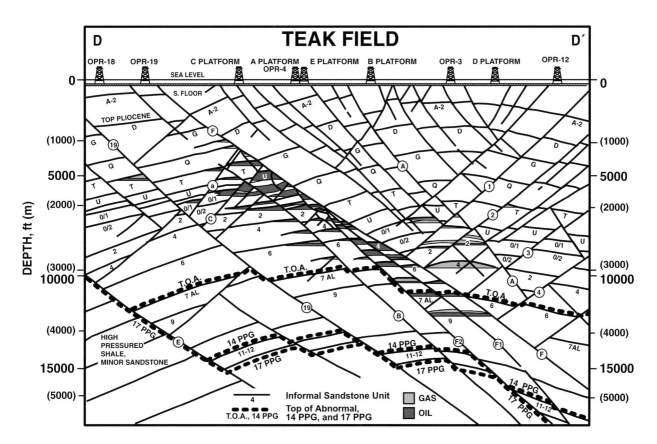

Figure 10. Geologic cross-section D-D′ through the Teak Field illustrating the complex normal faulting and stacked gas and oil zones in the Pliocene to Pleistocene, shallow marine sandstones. The F fault forms the major seal of a three-way, structural closure for the majority of the oil reservoirs. The top of very mild overpressure (T.O.A.) begins in the 7AL sand on the west and 6 Sand on the east. The overlying shale is normally pressured. Below these reservoirs both sandstones and shales are abnormally pressured. (after P. Heppard, H. Jackson., J. Lantz, and W. Wu; Amoco Trinidad Oil Company).

Figure 11. Three pressure profiles from the Teak Field representing 1) the east side of the field in the gas area downthrown by the F fault, 2) the major oil fault block between the F and the F2 faults, and 3) the western fault block upthrown by the 19 fault. The profiles illustrate the large pressure differences across the field due to the large throw of these pressure sealing, normal faults and the lateral continuity of the pressure compartments despite the faulting. The largest pressure difference across the 19 fault is most likely due to the predominance of impermeable deepwater shale in the high pressure block.

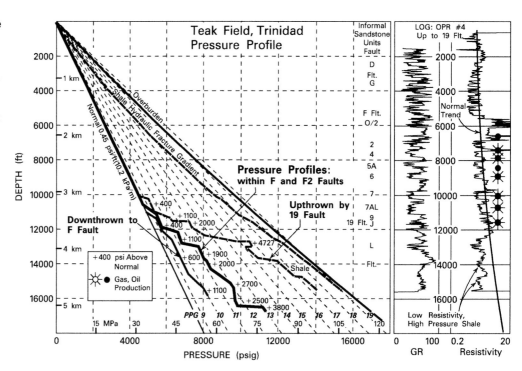

As in the Poui field and other offshore fields, shale beds and normal faults in the Teak field separate sandstone pressure compartments with large pressure differentials between them. Figure 11 shows three pressure profiles from the major fault blocks. Unlike the Poui field, the top of abnormal pressure in the Teak field occurs in different stratigraphic intervals, stepping up from the informally named 7AL sand to the 6 sand across the F fault. Also unlike the Poui field, the topmost abnormally pressured sands do not underlie abnormally pressured shale beds until (above) the 9 sand. These mildly pressured sands range from 360 to 600 psi (2.48 to 4.14 MPa) above normal. The absence of abnormally pressured shale above these sandstone units suggests that the origin of the abnormal pressure was not transference of overburden stress to the pore system since no abnormally pressured shale seal exists (see Origin of Abnormal Pressure, below). Rather, we suggest, that the abnormal pressure in these sandstones was caused by the migration of abnormally pressured fluids from below or laterally. The fact that the overlying shale is not abnormally pressured suggests that the shale is more permeable than the abnormally pressured shale below, and the abnormal pressure in these sandstones may be much more transient than in other, abnormally pressured sandstone units within these fields.

Within the Teak field, 100 to 500 ft (30 to 150 m) thick shale beds commonly have pressure differentials of 700 to 800 psi (4.83 to 5.52 MPa) and up to 2,730 psi (18.82 MPa) from underlying to overlying sandstone reservoirs. Shale seals separate sandstone pressure compartments such that pressure gradients between 1.5 to 5.56 psi/ft (33.9 to 125.8 kPa/m) exist across the shale units.

Large differences in pressure across the faults in the

Teak field are evident from the cross section and pressure profiles illustrated in Figures 10 and 11. In some cases, oil and gas are trapped in a high pressure block juxtaposed to a block with normal pressure or a lower abnormally high pressure. Within the F to F2 fault block, oil and gas are trapped in the 7AL and 9 sands which have pressures of 556 to 1,200 psi (3,833 to 8,274 kPa) higher than shales on the downthrown side.

Oil and gas migration into Teak field was mostly vertical from very highly pressured rocks below the current accumulations. To the west of the field, the very-high-pressure block upthrown by the regional E fault (Figure 10) is juxtaposed with normally pressured and lower abnormally pressured compartments. This suggests that some oil and gas may have migrated into the field laterally from the west. However, an oil sample from the highly overpressured section upthrown by the E fault has a significantly different composition from the oils from Teak field, indicating that migration into the field was not from the west. Further, it is unlikely that the principal hydrocarbon charge was from the east. The east side of the field, downthrown by the regional F fault, is mostly gas which, as suggested by Gibson (1994), most likely migrated across the F fault from accumulations trapped behind it. Gibson (1994) suggested that the buoyancy pressure of the oil and gas columns overcame the capillary properties of the F fault zone, and gas, at the top of the buoyant column and at the highest pressure, preferentially migrated to the east across the F fault zone.

Samaan Field

The Samaan field is a northeast-southwest trending anticlinal structure that lies along the Galeota Ridge about 22 miles (35.4 km) off the southeast coast of Trinidad (Figure 4). The large oil and gas reserves of Samaan field were discovered by Amoco Trinidad Oil Company in 1971 and development began with the first of three platforms, set in 185 ft (56.4 m) of water, in 1972. Cumulative production from 1972 to 1995 totals 207 million bbl (32.9 million m³) of oil, 636 billion ft³ (18.0 billion m³) of gas, and 280 million bbl (44.5 million m³) of water of which, 11 million bbl (1.75 million m³) of oil, 39.8 billion ft³ (1.1 billion m³) of gas, and 28.7 million bbl (4.56 million m³) of water have been produced from the abnormally pressured section. The field is cut by many normal faults that have throws between 200 and 900 ft (60 and 275 m) (Figure 12).

Samaan field has the highest pressure gradient in the offshore area, equivalent to 0.987 psi/ft (22.3 kPa/m). This pressure gradient occurs in a deep section which is interpreted to be very close to, or higher in pressure than, the hydraulic fracture gradient of shale. The lack of hydrocarbons in the deep section is likely due to failure of the shale seals through hydraulic fracturing. However, substantial oil columns are present in the shallower abnormally pressured section, including the 360 ft (110 m) oil column of the 11U sand which is 2,628 psi (18.12 MPa) above normal at

10,000 ft (3,353 m). Production is limited by poorer reservoir properties and more paraffinic oil compared to the overlying and normally pressured section (Farfan and Bally, 1990; Heppard et al., 1990; Ross and Ames, 1988; de Landro, 1985).

The top of abnormal pressure begins in the informally named 7A sand, which is 653 psi (4.50 MPa) above normal pressure, based on a single RFT test at about 8,500 ft (2,600 m) (Figure 12). Abnormal pressure increases in steps as illustrated in the pressure versus depth plot of Figure 13. The uppermost abnormally pressured sandstones are overlain by normally pressured shales with the exception of the shale above the 9 sand. As at Teak field, these sandstones are interpreted to be abnormally pressured due to the vertical and horizontal migration of high pressured fluids.

Below the relatively mildly overpressured zones, pressure increases dramatically in the sandstone units and is very much greater than in the bounding shales. The higher pressure in the sandstone units relative to the shales is partly explained by the effect of structural dip and good lateral continuity of these pressure compartments. We infer that high abnormal pressure deep in the basin to the east has been transmitted to these sandstone pressure compartments which are laterally

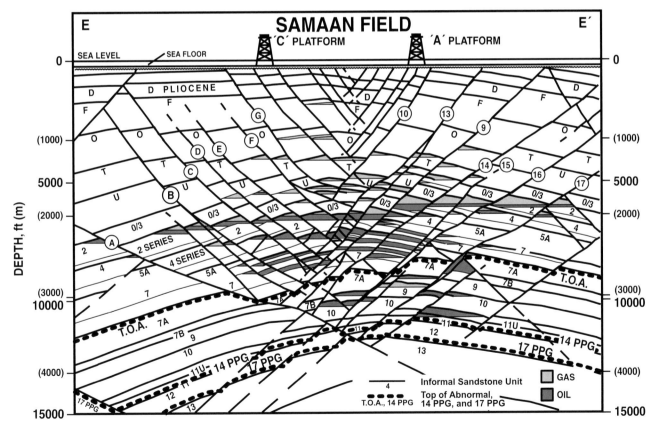

Figure 12. Geologic cross-section E-E′ through the Samaan Field illustrates the complex normal faulting and the stacked gas and oil pay zones in Plio-Pleistocene sandstones. The top of abnormal pressure (T.O.A.) is within the 7A sand at a depth of 8,500 to 11,000 feet (2,591 to 3,353 m). Pressure builds rapidly in the section below the 10 sand and reaches nearly 0.99 psi/ft (22.3 kPa/m) by the 13 sand. Location of cross-section shown on Figure 4. (after P. Heppard, M. Brew, and R. Cendrowski; Amoco Trinidad Oil Company).

Figure 13. Pressure profile for Samaan field illustrates the very high pressure of the lowermost 3,000 feet (910 m) of the drilled section. The lowermost section is interpreted to be "breached" and incapable of holding any significant hydrocarbon charge due to the extremely high pore pressure's proximity to the hydraulic fracture gradient of shale. Most of the sandstone pressure compartments at the crest are higher in pressure than the bounding shales, indicative of laterally and vertically continuous pressure compartments in a large structure.

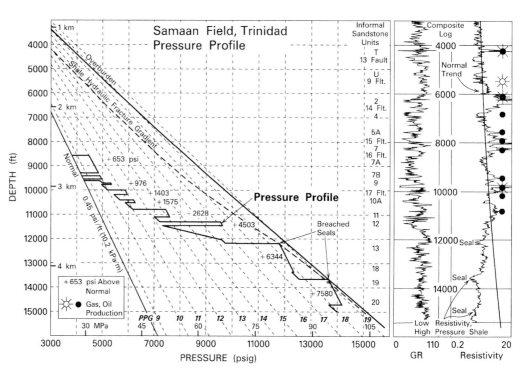

continuous to the crest of Samaan field. Conceptually, we assume that laterally continuous pressure compartments are affected by the vertical stress of the overburden over the entire distribution of the compartment. The pressure compartment and the enclosing shale only have a single point of equivalent pressure which is termed the "centroid" (Traugott and Heppard [1994]). Using the centroid concept, the vertical continuity of the 12, 13, and 19 sand pressure compartments at Samaan field is estimated to be 3,000 to 4,000 ft (910 to 1,200 m). This is compatible with interpretations from seismic lines and recent drilling.

The highest pressure gradient in the study area is equivalent to 19.17 PPG (0.996 psi/ft, 22.5 kPa/m) and occurs in the Samaan field at 13,600 ft (4,145 m) in the 19 to 20 sand pressure compartment (7,580 psi [52.26 MPa] above normal pressure) (Figure 13). The overlying 13 to 18 sand pressure compartment is 6,344 psi (43.74 MPa) above normal pressure at 12,000 ft (3,700 m). The seals above these compartments are considered to be ineffective hydrocarbon seals due to the extremely high pressure. The pressure at the crest of the 13 sand compartment is within 200 psi (1,379 kPa) of the overburden pressure gradient, and the pressure at the crest of the structure in the 19 sand is estimated to be higher than the overburden pressure.

Because of the expected high pressure in the deep section, a recent deep exploration well was located 300 ft (90 m) off the structural crest to allow a sufficient kick tolerance between high pressure in the sandstone beds and the hydraulic fracture pressure gradient of shale. Three previous attempts to drill this structure at the crest failed. The shale seal above the 13 sand may be only 130 ft thick (39 m) and therefore, would have a pressure gradient of 14.16 psi/ft (320.3 kPa/m) across

it—which is significantly higher than any known hydrocarbon seal in the offshore area. The shale seals for these very high pressure compartments were deposited in relatively deep water, marine environments during high stand events in contrast to the shallow-marine environments for sediments within the 13 to 18 and 19 to 20 sand pressure compartments. Shale pressure seals at shallower depths in the Samaan field separate pressure compartments with pressure gradients of 0.78 to 6.46 psi/ft of shale (17.6 to 146.1 kPa/m).

A study of the composition of the oils within Samaan field indicated that the majority of the oil migrated into the field vertically out of the underlying very highly overpressured section (Heppard et al., 1990; Ross and Ames, 1988). Only relatively small volumes of oil and gas migrated laterally from the east and west and accumulated behind the A, B and C faults on the west flank, and the 16 and 17 faults on the east flank of the field.

Cassia Field

The Cassia field is located 35 miles (56 km) southeast of Trinidad (Figure 4) and has produced from 1983 to 1995, 1.19 trillion ft[3] (33.7 billion m[3]) of gas, 18 million bbl (2.9 million m[3]) of condensate, and 0.8 million bbl (127,000 m[3]) of water from four sandstone reservoirs of Pliocene age. The field is on the western flank of the large, nearly domal, Southeast Galeota (SEG) anticlinal structure which is roughly 15 by 15 miles (24 by 24 km) in size. Within this structure are several large accumulations of gas in traps having three-way, structural closure formed by gently dipping strata against normal faults which have relatively large throws of 1,200 to 4,000 ft (370 to 1,200 m). The fields on the eastern flank are the Immortelle and South SEG fields. The Cassia field was discovered in 1973 by the SEG #9 well which was followed by a delineation well in 1979, the West SEG #1 well (Figure 14). The Cassia platform, which is operated by the Amoco Trinidad Oil Company, was set in 1982 in 220 ft (67 m) of water and production has continued to the present from nine development wells drilled between 1983 and 1987 (Alison and Farfan, 1989).

Of all the fields offshore of the southeast coast of Trinidad, Cassia field has produced the most gas from an abnormally highly pressured reservoir (882 billion ft^3, 24.98 billion m^3) (Figure 4 and Figure 14). All the gas has come from a single reservoir—the informally named 30 sand—which has a gas column of 1,250 ft (381 m) at a depth of 11,750 ft (3,581 m). It is the longest gas column in the offshore area. The large reserves and long hydrocarbon column are considered to be within a pressure enhanced trap. Both the overlying shale, and the shale and sandstone juxtaposed to the reservoir across the regional E fault to the east, are interpreted to be higher in pressure than the 30 sand creating nearly ideal hydrocarbon seals.

The top seal for the abnormally pressured 30 sand is an 840 ft (256 m) thick shale bed (Figures 14 and 15).

The shale has a pressure gradient across it of 1.79 psi/ft (40.5 kPa/m), from the 30 sand to the overlying 25 sand, and is 400 psi (2,758 kPa) higher in pressure at its base than the underlying 30 sand. The lateral seal is formed by the regional E fault and the high pressure section on the downthrown side (Figures 4 and 14). The pressure differential across the fault ranges from 970 to 2,820 psi (6.69 to 19.44 MPa) across the face of the fault juxtaposed to the 30 sand, and enhances the sealing capabilities of the E fault at the 30 sand. The normal E fault has about 4,000 ft (1,220 m) of throw at the 30 sand horizon. There is significant expansion of section across the E fault indicating continuous movement throughout the Pliocene. Higher pressure on the downthrown side of a normal fault is unusual in the Trinidad area. The top of abnormal pressure is stratigraphically offset across the E fault and is within younger rocks on the downthrown side.

The section below the 30 sand is composed of highly overpressured, deep water shales and thinly laminated sands (Figure 15). Pressure increases rapidly below 13,000 ft (3,962 m) at an average rate of 1.87 psi/ft (42.3 kPa/m) from 13,250 to 15,250 ft (4,040 to 4,650 m). Abnormally pressured shales in this section separate three sandstone pressure compartments.

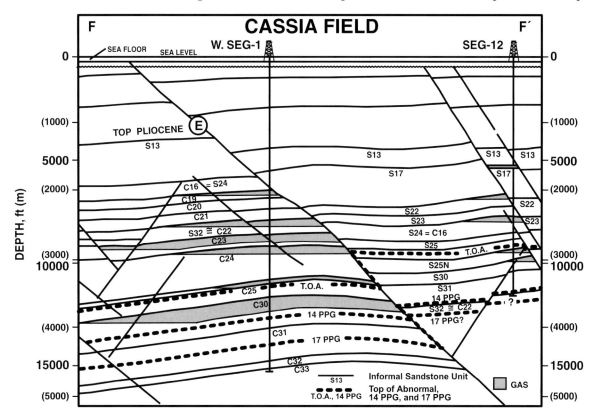

Figure 14. Geologic cross-section F-F′ through the Cassia field. The longest gas column from offshore Trinidad of 1,250 feet (380 m) is within the pressure enhanced trap of the 30 sand. Pressure "steps-up" stratigraphy to younger rocks on the expanded, downthrown side of the E fault from Cassia field. The top of abnormal pressure (T.O.A.) is within the shale below the C25 and S25 sands (C and S refer to naming conventions for the Cassia and SEG areas). Location of cross-section shown on Figure 4. (modified from J. Finneran, K. Ortmann, M. Staines, C. Sharpf, B. Alison, and W. Skelton; Amoco Trinidad Oil Company)

Figure 15. Cassia field pressure profile indicating the pressure enhanced trap for the 1,250 feet (381 m) gas column of the 30 sand. The overlying shale is estimated to be higher in pore pressure than the 30 sand. A pressure interpretation of the strata down thrown by the E fault (dashed line) and a fault plane map show that the shale and sandstone juxtaposed to the 30 sand is significantly higher in pressure than the 30 sand. Pore pressure approaches the hydraulic fracture gradient of shale at the depth of the deepest wells.

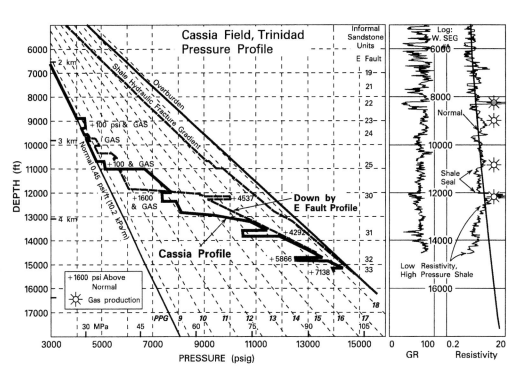

These shale seals have pressure gradients across them of 1.7 to over 4 psi/ft (38.4 to 90.5 kPa/m).

It is likely that near the total depth of the West SEG #1 well (Figure 14), shale has been repeatedly hydraulically fractured, allowing vertical migration of oil, gas, and water. Based on well log calculations and high drilling mud weights, the shale above the 33 sand is within 300 psi (2,069 kPa) of overburden pressure and the 33 sand is within 700 psi (4,826 kPa) of overburden at the W. SEG #1 location at a depth of 15,160 ft (4,621 m)(Figure 14). Current structure maps indicate that the crest of the 33 sand may be considerably shallower and therefore, the pressure much closer to lithostatic, possibly within 350 psi (2,413 kPa).

South SEG and Immortelle Fields

The South SEG field is located on the eastern flank of the large SEG structure (Figure 4). The South SEG field is at present being evaluated and has not yet been developed. The first well, the SEG #1 well, was drilled in 1968 but delineation drilling did not take place until 1993. The field is the southern portion of a three way, structural closure shared with Immortelle field to the north (Figure 4). A northwest trending, west-dipping normal fault forms the updip trap (Figures 4 and 16). This fault has about 1,200 ft (366 m) of vertical throw and has little to no expansion of section across it.

The majority of the gas reserves lie within the normally pressured section. There is evidence from pressure measurements, similar gas-water contacts, and well log correlation that the largest gas columns in the 22 and 24 sands are continuous over 9 miles (14.5 km) from the South SEG field to the Immortelle field to the northwest. The largest gas column is found in the 22

sand and is approximately 900 ft (270 m). The estimated gas reserves for the combined area are over 2.5 trillion ft³ (70.8 billion m³). The accumulations contain primarily gas and condensate of moderate to immature thermal maturity, but thin oil columns below gas caps are also present. The oil has been developed by horizontal wells at Immortelle field (Lanan et al., 1995). The only abnormally pressured compartment with any significant, proven hydrocarbons is the 25N sand, which is the topmost sandstone within the abnormally pressured section. It has a gas column slightly greater than 100 ft (30.48 m) (Figures 16 and 17).

The SEG #1 well penetrated a thick section of abnormally high pressure from 10,500 to 15,500 ft (3200 to 4720 m) (Figure 16). Pressure within this section increases at an average rate of 1.9 psi/ft (43.0 kPa/m) beginning in the shale above the 25N sand (Figure 17). The shale seals range in thickness from 160 to 390 ft (49 to 119 m) and have pressure gradients across them from 2.24 to 7.04 psi/ft (50.6 to 159.2 kPa/m). Pore pressure at the base of the existing wells is within 1,100 psi (7,584 kPa) of the overburden pressure. It is likely that below the total depth drilled, the pore pressure approaches the hydraulic fracture gradient of shale as observed in the other fields.

Short gas columns and water-bearing sandstones occur in the 25N to 30B sandstones which are juxtaposed across the trapping normal fault to the west with normal or less abnormally pressured reservoirs. Significant volumes of hydrocarbons were unlikely to have remained trapped under these pressure conditions.

A direct indication of hydrocarbon migration in the field is provided by an asphalt zone immediately below the 30B sand. This thin zone is associated with a rapid drop in pore pressure (Figure 17). It is thought

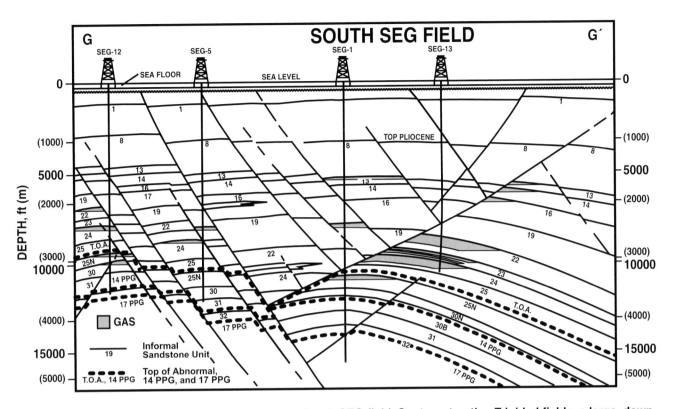

Figure 16. Geologic cross-section G-G′ through the South SEG field. Contrary to other Trinidad fields, a large, down-to-the-west normal fault forms the seal in a three-way closure. Within the overpressured section only a relatively thin gas column is known in the 25N sand. The top of abnormal pressure (T.O.A.) is within the shale above the 25N sand. The limited reserves within the abnormally pressured section may be due to leakage across the main fault into normally pressured sandstones and trapped only within structural closure. Location of cross-section shown on Figure 4. (modified from J. Finneran, K. Ortmann, M. Staines, F. Sobol, B. Golob, and W. Skelton; Amoco Trinidad Oil Company)

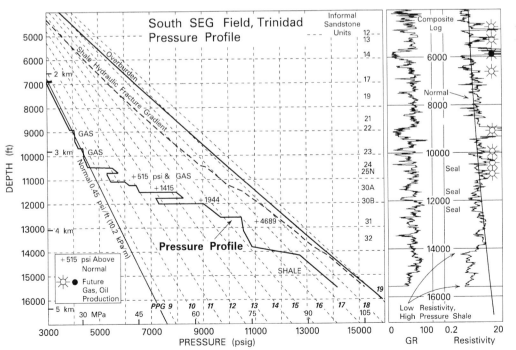

Figure 17. Pressure profile of the South SEG field illustrating the rapid increase in pressure below 10,500 feet (3,200 m). The shale appears to be higher in pore pressure than the intervening sandstone until 12,000 feet (3,660 m), at the 30B sand, possibly due to fluid leakage across the trapping fault to normal pressures to the west. An asphalt zone lies below the 30B Sand which is interpreted to be residual oil dropped out during migration as a result of a rapid, 2,700 psi (18.62 MPa) decrease in pressure. The low resistivity of the deepest section clearly reflects abnormally high pressure conditions.

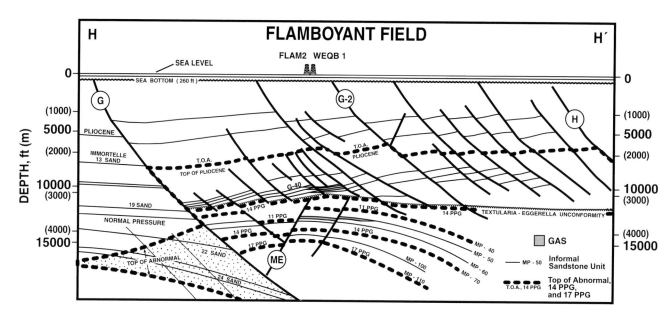

Figure 18. Geologic cross-section H-H′ through the Flamboyant field. Gas reservoirs are restricted to the overpressured section. The top of abnormal pressure (T.O.A.) "steps-up" stratigraphy and is dramatically shallower east of the regional G fault at Flamboyant than in the South SEG / Immortelle area to the west. A pressure reversal at Flamboyant field within the MP-30 to possibly the MP-100 sands is interpreted to be due to fluid loss across the G fault to normally pressured sandstone units. Location of cross-section shown on Figure 4. (from H. Cander, M. Staines, R. Ames and K. Sawyer; Amoco Trinidad Oil Company)

that the asphalt is the result of devolatilization of the original oil charge during episodic secondary migration from the underlying and abnormally pressured shale seal below the 30B sand. The present-day pressure drop over the asphalt zone and pressure seal, a few tens of feet (meters), is approximately 2,700 psi (18.62 MPa).

Flamboyant Field

Flamboyant field is located 6 miles (9.6 km) east of Immortelle field, in 260 ft (80 m) of water (Figure 4). It was discovered in 1987 by the Amoco Trinidad WEQB #1 well which was re-entered and put on production in 1993. A second well, Flamboyant #2, was drilled in 1994. The two well Flamboyant field has produced over 150 billion ft³ (4.25 billion m³) of gas, with condensate, in two years.

All the gas reserves of the Flamboyant field are within the abnormally pressured section, in two intervals, in pressure compartments which are 1,311 and 1,498 psi (9.04 and 10.33 MPa) above normal pressure (Figures 18 and 19). The greater than 500 ft (152 m) gas column of the MP-50 sand is in a pressure enhanced trap. The informally named MP-50 sand and sandstone units immediately above and below it are within a pressure reversal from shallower and higher pressured shales and sandstones.

The top of abnormal pressure is dramatically shallower and in much younger rock in the Flamboyant field than it is in the South SEG/Immortelle area to the west across the regional G fault. In that area, the top of

abnormal pressure is below 15,000 ft (4,600 m) within the middle Pliocene—it "steps-up" to 6,655 ft (2,028 m), just above the top of the Pliocene, at Flamboyant. The deepest shale drilled to date in the Flamboyant field is at very high pressure and is probably an ineffective hydrocarbon seal.

Unlike previously described Trinidad fields, Flamboyant field had several periods of structural deformation. A sub-regional unconformity, named Textularia-Eggerella after the foraminifera assemblage, separates shallow marine upper Pliocene sediments from lower Pliocene sediments deposited in water depths >600 ft (180 m) (Figure 18). The field area was a topographic high at the onset of upper Pliocene sedimentation above the unconformity. A large increase in pore pressure occurs below the unconformity (Figures 18 and 19) where shale and siltstone form a pressure compartment 3,100 psi (21.4 MPa) above normal pressure.

Below this high pressure interval, pore pressure drops dramatically in the informally named MP series of sands over a zone between 12,200 ft (3718 m) to >13,600 ft (4145m) (Figure 19). Pressure within this interval decreases from 3,100 to 1,338 psi (21.4 to 9.22 MPa) above normal. Pressure increases within the zone of pressure reversal at about 1.75 psi/ft (39.6 kPa/m) through at least five sandstone pressure compartments. All the intervening shale beds within the pressure reversal show greater compaction than shales in the overlying and underlying high pressure zones. The pressure in the uppermost sandstone of this zone, the MP-30, was not measured and the true pore pressure is unknown. A reasonable range, though, is shown on

Figure 19. Drilling was halted on the second well of the field when pore pressure was estimated to be within 700 psi (4,826 kPa) of the overburden pressure gradient as there were low expectations of good hydrocarbon seals at this very high pore pressure and reservoir quality was decreasing.

The pressure reversal combined with normal faulting created a pressure-enhanced trap for the 500+ ft (152+ m) gas column of the MP-50 sand. The ME fault juxtaposed the MP-50 sand against higher-pressured shale of the overlying section at the crest of the structure (Figure 20). Due to the uncertainty of the pressure in the MP-30 sand, it is unclear whether the fault trap of the MP-50 is completely enhanced across the face of the fault throughout the trapped column of gas. The gas column in the underlying MP-60 sand is considerably smaller, probably due to the unfavorable pressure conditions across the ME fault (from high pressure on the trapping side to low pressure on the downthrown side [Figure 20]).

The evolution of abnormal pressure and hydrocarbon migration in Flamboyant field is uncertain. In order to create the pressure reversal, pressure-induced fluid flow must have occurred out of the sand-rich MP-40 to MP-80 section. The most likely avenue for high-pressure fluids out of the Flamboyant structure was westward across the regional G fault to normally pres-

sured or mildly overpressured, thick sandstones of the SEG/Immortelle area (Figure 18). Fluid flow could have occurred as pressure in the Flamboyant field built to higher levels than the G fault could sustain—either nearly continuously from the sand-rich MP-40 to MP-80 section, or episodically by hydraulic fracturing of normally pressured sandstone across the fault induced by high pressure at Flamboyant field. A dashed line on the pressure versus depth plot of Figure 19 shows the hydraulic fracture gradient of a normally pressured sandstone interpreted to exist west of the G fault. Based on the magnitude of abnormal pressure both above and below the zone of pressure reversal, it is evident that pressure could have built up in the MP-40 to MP-80 sands to greater than the hydraulic fracture gradient of normally pressured sandstone across the G fault.

Because of the uncertainty of the origin of the pressure reversal, the timing of gas migration into the field relative to the development of the pressure reversal is not known. If the release of pressure in the MP-40 through MP-80 zone began relatively early, then the zone of pressure reversal would have been a low pressure site for the accumulation of hydrocarbons migrating through the lower Pliocene section. If migration followed the pressure reversal, then it is unlikely that gas could have migrated vertically into the shallower

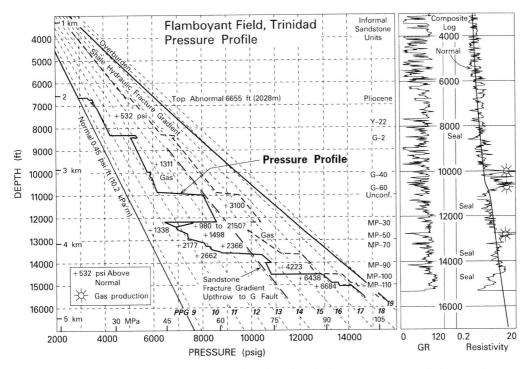

Figure 19. Pressure profile of the Flamboyant field showing the shallow occurrence of abnormal pressure near the top of the Pliocene at 6,655 feet (2,028 m), and a very gradual increase in pressure to a thick pressure compartment within the gas reservoirs of the G series of sands which are 1,311 psi (2,137 kPa) above normal pressure. A rapid increase in pressure is present below the upper to lower Pliocene unconformity at 11,000 feet (3,350 m). A pressure reversal is present within a zone of laterally continuous sandstones that apparently leaked fluid across the regional G fault to normally pressured sands to the west. Pressure at the base of the known section is within 700 psi (4,826 kPa) of overburden, and may be incapable of retaining significant hydrocarbons.

Figure 20. A detail of the geologic cross-section through the Flamboyant field, shown in Figure 18, illustrating the pressure differences across the trapping ME fault. The fault has about 800 feet (244 m) of throw. The pressures in equivalent sandstone units in the undrilled, downthrown side are assumed to be the same amount above normal as in the upthrown side. The 500+ ft (150 m) gas column of the MP-50 sand is a pressure enhanced trap juxtaposed to shale which is 1,320 psi (9.101 MPa) higher in pressure that the MP-50 sand. Little or no hydrocarbon charge is present in the upthrown block where high pressure is juxtaposed to low pressure.

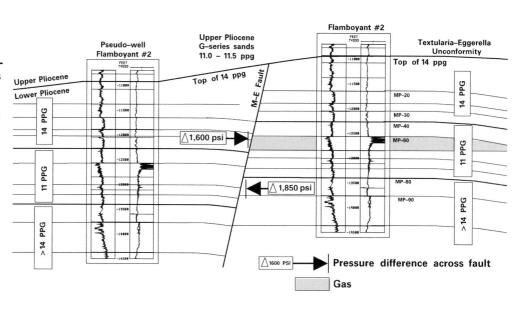

G-series sands from the MP-50 and MP-60 sands. It would have been difficult to transmit gas from the low pressure MP-40 zone through the overlying highly overpressured zone to accumulate in the overlying G-series of sands (Figures 18 and 19). In this case, the path of vertical gas migration must have diverged at some point well off the flank of the structure in order to charge the G-series of sands and the MP-50 sand in the field through lateral migration. Alternatively, hydrocarbons may have migrated into this zone when it was highly overpressured. If migration pre-dated the pressure reversal, then vertical migration into the MP-50 and MP-60 sands and the shallower G-series of sands would have been upward toward lower pressure potential, similar to other fields in the basin. Condensate compositions from the two accumulations are very similar, suggesting that the two gas accumulations must have commingled within or near the present structure. Whether hydrocarbon migration pre-dated or post-dated formation of the pressure reversal (in the MP-40 to MP-80 sand section), the Flamboyant field illustrates the relationship between abnormal pressure and hydrocarbon migration and entrapment. Further studies of pore pressure distribution, geochemistry, and the depositional and structural history of the basin will likely lead to greater understanding of the migration of oil and gas and indicate additional hydrocarbon prospects in the area.

ORIGIN OF ABNORMAL PRESSURE

Of the many proposed origins for abnormal pressures (Swarbrick and Osborne, 1998), most current

workers consider compaction disequilibrium, hydrocarbon generation within source rocks, and aquathermal expansion of pore fluids the most significant. These mechanisms for generation of abnormal pore pressures have likely been active in Trinidad but no one of them can account for all of the high pressures. Regardless of origin, high pore pressures require a seal having significant reduction in permeability necessary to trap or retard the flow of high pressure fluids. We suggest that clay rocks in the Trinidad region can form seals with near zero permeability. Furthermore, we conclude that the principal origin of abnormal pressure in Trinidad was the transfer of overburden stress to the pore system after pressure seals formed in clay rocks during subsidence. At depths of 10,000 to 15,000 ft (3,000 to 4,500 m) permeability in most clay rocks approaches zero forming effective pressure seals.

Hydrocarbon generation within the Upper Cretaceous source rocks may also have played a role in creating abnormal pressure conditions. But high pressured fluids originating from hydrocarbon generation probably only affected the overlying, relatively permeable sandstones and siltstones and did not contribute to abnormal pressures of the Tertiary shales. As will be presented later in this section, we consider the Tertiary shales within the abnormally pressured section to have had near zero permeability and, likely to have been too impermeable for high pressure fluids to have invaded from high pressure sandstone beds or during hydraulic fracturing events unless the relatively high threshold pressure gradient of the shale was reached. The Tertiary clay rocks are very lean in organics, but as argued by Hedberg (1974), the generation of minor volumes of gas within shale may contribute to abnormal pressure.

Aquathermal expansion of water and other fluids, as presented by Barker (1972) and Bradley (1975), was a likely cause of increasing pressure in the rock section that has been deeply buried following the formation of pressure seals. However, aquathermal fluid expansion alone can not account for all occurrences of high pressure in Trinidad. Abnormally high pressure is found at quite shallow depth in Trinidad within anticlines which have been uplifted over 10,000 ft (3,000 m). These uplifted rocks have presumably undergone a good deal of cooling and attendant fluid shrinkage, which should have greatly reduced the fluid pressure. The contribution of aquathermal expansion in the generation of abnormal pressure is controversial. This is because of the relatively small volume increase generated, the requirements of having a perfect seal, and that there be no change in volume of the pressure compartment (Mouchet and Mitchell, 1989, p. 36).

It is possible that tectonic stress also contributed to the generation of abnormal pressure within uplifted areas as suggested by Berry (1973) for some abnormally pressured zones adjacent to the San Andreas fault. However, some of the highest pressure gradients found in Trinidad occur in areas that have had little uplift and lie within areas subjected principally to tension (such as at Cassia and South SEG fields [Figure 4]) rather than compression. The alteration of smectite to illite as a source of high pressure fluids, as proposed by Powers (1967) and Burst (1969), does not seem to be applicable as the dominant clay minerals in the study area are illite and kaolinite with lesser amounts of mixed layer, illite-smectite, and chlorite (Harry, 1995).

It has been generally noted that for abnormal pressure to exist in nature for any reasonable length of time, a restriction of fluid movement must exist. A loss of only a few percent of the total fluid from a rock near the upper limit of pore pressure (lithostatic pressure) can reduce the pressure to normal hydrostatic (Barker, 1987). However, abnormal pore pressure is present in many sedimentary basins within rocks of all geologic eras. The concept of pressure seals as presented by Hunt (1990) has been used by some writers to account for enduring high pressures. Pressure seals are interpreted to be essentially impermeable zones and may be shale, salt, anhydrite, or extremely cemented and impermeable rocks (Bradley and Powley, 1994; Qin and Ortoleva, 1994). In the Trinidad area, the clay rocks are the pressure seals. There are no indications from well logs or cores of any extremely well-cemented, high density beds within the abnormally pressured section nor are there any deposits of salt or anhydrite.

Alternatively, pressure seals have been considered to be dynamic and to function only as zones of relatively low hydraulic conductivity which are unable to drain fluid at a sufficient rate to outpace the generation of abnormally pressured fluids and gases (Bredehoeft and Hanshaw, 1968; Deming, 1994). Where sedimentation rates are higher than the rates of fluid expulsion, a large part of the lithostatic load is transferred to the

pore system (Smith, 1970). This process is referred to as compaction disequilibrium. Pressure seals are considered transient due to all clay rocks possessing some very low-permeability. As a consequence, abnormal pore pressure should decline to normal hydrostatic pressure given sufficient geologic time following the cessation of the processes that produced it.

Since Miocene time the Eastern Venezuelan Basin has been an area of rapid deposition, and as a result, abnormally high pressures may have been due to compaction disequilibrium. It is the authors' contention that, given the longevity of abnormal pressures in old and stable basins, such as the Anadarko of Oklahoma (Al-Shaieb et al., 1994) and Delaware Basin of Texas and New Mexico (Luo et al., 1994), that abnormally high pressures sealed by shale beds have been relatively stable compared to other kinds of pressure seals.

The well log response of abnormally pressured Tertiary shales in the Eastern Venezuelan Basin indicates that they are very undercompacted for their depths. The relatively high porosity of the shales is a consequence of being abnormally pressured and undrained during burial. But the relatively high porosity of these rocks does not indicate their permeability, or the origin of the abnormal pressure (possibly through disequilibrium compaction).

In support of the disequilibrium compaction hypothesis, some workers have demonstrated that it is not feasible for rocks possessing the very lowest measured permeabilities to retain high abnormal pressure for long periods of geologic time (Deming, 1994; Bredehoeft et al., 1982; Bredehoeft and Hanshaw, 1968). However, Neuzil (1986) discussed the difficulties in measuring extremely low permeability in clay rocks and pointed out that pressure gradients higher than those found in nature were necessary to induce any measurable flow in tested samples. No flow was apparent for samples when more realistic potentiometric heads were applied. Zero permeability in clay rocks has been postulated to be due to the quasi-crystalline structure of water around clay minerals (Neuzil, 1986; Miller and Low, 1963). Miller and Low (1963) proposed that for high-porosity clay systems, a threshold pressure gradient exists below which no flow occurs due to the quasi-crystalline structure of water. In their experiments, thresholds equivalent to 30 psi/ft (679 kPa/m; 21 cm head for a 0.3 cm sample) were measured for clay samples with 44 to 54% porosity.

The authors suggest that the abnormally pressured shale seals of the Eastern Venezuelan Basin were formed by a combination of bound or crystalline water around clay minerals, decreased pore throat size through mechanical rearrangement, recrystallization of clay minerals, and limited cementation. These conditions may be sufficient for shale to retard fluid movement indefinitely within abnormally pressured sections where the average head (fluid potential) is only 1 psi/ft (22.6 kPa/m) of rock. Most shale beds within the abnormally pressured section in Trinidad have been observed to have pressure gradients of 4 psi/ft (90.5 kPa/m) or less from underlying to overlying sandstone. Based on Miller and Low's (1963) concept of a threshold gradient, we propose that the latter gradient may be within the threshold for these clay rocks and likely is stable for significant periods of geologic time.

MIGRATION OF OIL AND GAS

The migration of oil and gas within the offshore Trinidad basin was significantly affected by widespread, abnormally high pressures. Migration from the now deeply buried source rocks of the Upper Cretaceous Gautier and Naparima Hill Formations (Figure 6) is considered to have occurred in three phases. First there was primary migration, enhanced by the development of hydraulic fractures from highly pressured source rocks and initial secondary migration into overlying, and also highly pressured, Tertiary age clastics. Second, episodic vertical, secondary migration occurred within the abnormally pressured Tertiary clastic section through pressure-induced hydraulic fractures focused by dipping carrier beds. And third there was a final phase of migration out of the highly pressured section into the overlying pressure transition zone and normally pressured rocks as a result of Pleistocene structural uplift and complex normal faulting. Secondary migration within the pressure transition zone and normally pressured rocks occurred by flow across normal faults where sandstone beds were juxtaposed. Secondary migration across faults was enhanced where high pressure compartments were fault juxtaposed with lower pressure compartments—some of the best traps occur where structural traps formed in low pressure fault blocks. Where hydraulically possible, vertical migration occurred through normal fault zones (Gibson, 1994). Similar migration scenarios have been proposed for the Lake Maracaibo area (Talukdar et al., 1985) and, in a general sense, for the U.S. Gulf Coast (Hanor and Sassen, 1990).

The few wells that have penetrated the Upper Cretaceous source rocks in Trinidad show that it is currently overpressured and is overlain by overpressured Tertiary clastics. Wells from Soldado Marine field off the west coast of Trinidad (Figure 1) have penetrated several thousand feet of Upper Cretaceous source rocks. The topmost source rock, at about 11,000 ft (3,400 m), is overlain by 7,000 ft (2,100 m) of overpressured Tertiary clastics which were drilled with up to 17 PPG (0.88 psi/ft, 19.9 kPa/m) mud. The source rocks appear to also be abnormally pressured to similar levels.

The generation of liquid and gaseous hydrocarbons from organic material creates an increase in volume, and in a confined space must generate an increase in pore pressure (Meisner, 1978; Barker, 1987). This mechanism of abnormal pressure generation has likely caused natural hydraulic fracturing in Trinidad area source rocks and could have greatly facilitated prima-

ry migration. Pressure-induced hydraulic fracturing and migration likely extended into the overlying, highly overpressured Tertiary clastic section.

However, to the south, in the area near the outcrop of the Guyana Shield (Figure 1) the pressure in the Upper Cretaceous section is known to be normal (Figure 5). At some unknown time and geographic location, the initial secondary migration phase into the overlying and overpressured Tertiary rocks was likely to have ceased or diminished, giving way to lateral migration within the Upper Cretaceous rocks to the south charging the 3 trillion bbl (477 billion m³) Orinoco oil belt of Venezuela as suggested by Gallango and Parnaud (1995) and Talukdar et al. (1988).

During Miocene and Pliocene subsidence and infilling of the study area, the initial oil and gas charge likely had a long residence time within the abnormally pressured section. The initial migration occurred prior to deposition of many of the currently charged Pliocene reservoirs. A simple model of expulsion and secondary migration into the final reservoir does not apply in this area. Primary migration in the onshore area may have occurred during the Miocene (Rodrigues, 1985) and offshore, southeast of Trinidad, during the early Pliocene (D. Grass and P. Kaufman, Amoco, personal communication, 1995). Most of this oil and gas is of moderate to low thermal maturity. Furthermore, complex faulting within the fields, which occurred during the latest Pliocene to Pleistocene, predates the migration of oil and gas. This is demonstrated by significantly different oil chemistry in the same sandstone, in different fault blocks (Ames and Ross, 1985; Ross and Ames, 1988; Heppard et al., 1990).

During the extended period of oil and gas residence within the abnormally pressured section, episodic vertical migration probably occurred as a consequence of periodic pressure-induced hydraulic fracturing—first during subsidence and later during uplift and faulting. Oil and gas migrated during hydraulic fracturing events at the crests of existing structures or at the updip limit of sandstone carrier beds. Episodic vertical migration was likely halted within the transition zone between extreme and moderate abnormal pressures when hydraulic fracturing was no longer possible. Normal faults, at this time, were much less prevalent in the area and likely played only a minor role in vertical migration. Further subsidence and infilling led to increases in pore pressure in existing (temporary) oil and gas accumulations which eventually led to pressure induced hydraulic fracturing of the shale seals and renewed vertical migration.

During Pleistocene uplift and deformation the abnormally pressured section was uplifted thousands of feet leading to the final migration of oil and gas into the structural traps. During this period of deformation, the pore pressure gradient in the sandstone pressure compartments at the crests of the structures increased as a consequence of re-equalization of abnormal pressure along the tilted and now uplifted structures. The

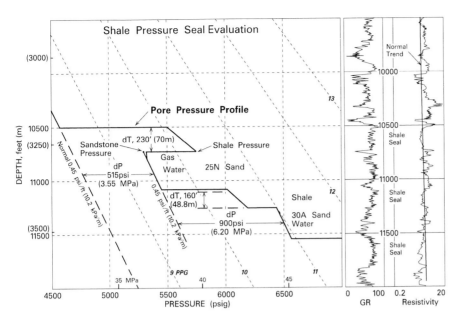

Figure 21. A typical pressure profile from the study area illustrating the method of describing the pressure conditions of shale seals in the basin. The pressure in the shale was calculated from well log responses of the slowest, or least resistive, shale and the pore pressure was assumed to change vertically within the shale unit at a gradient similar to lithostatic. The thickness, dT, of shale was interpreted based on log responses indicating abnormal pressure. The effective pressure difference across the shale, dP, is the difference in pressure from the overlying and the underlying sandstone pressure compartments in water, discounting the effects of hydrocarbon buoyancy, and taking into account the normal increase in pressure with depth due to the pressure gradient of water.

increased pressure gradient would eventually be greater than the hydraulic fracture gradient of the overlying shale and would lead to vertical migration through pressure-induced fracturing. The hydraulic fracture gradient of shale was also reduced in the field areas due to a reduction in the overburden stress by erosion at the surface. The complex normal faulting also led to numerous opportunities for cross-fault migration and, in some cases, migration via faults where the hydraulic properties of faults were sufficient to allow it to occur (Gibson, 1994). This cross-fault migration was enhanced or retarded by cross-fault pressure differences as previously discussed in the field examples.

The sudden drop in pressure, as the hydrocarbon charge migrated episodically out of the high pressure zone, is thought to have led to a general devolatilization of the oils causing separate migration of gas and oil charges into the fields (Heppard et al., 1990). However, Talukdar et al. (1990) suggest that evaporative fractionation—the selective removal in gaseous solution of some light hydrocarbons from oil by migrating gas through an oil accumulation—can best explain the composition of many of Trinidad's oils. Perhaps both processes have affected the compositions of the area's oils and gases.

EVALUATION OF PRESSURE SEALS

A study of the pressure seals in the offshore area southeast of Trinidad was undertaken to examine some general characteristics and possibly define some critical parameters applicable to exploration efforts in the region. For purposes of this study, shale pressure seals are defined as any clay rock unit that vertically separates porous and permeable sandstone beds with very different levels of abnormal pressure. The parameters

studied include: seal thickness; pressure difference across the seal (pressure differential); pore pressure of the shale; sandstone pore pressure; and the proximity of both shale and sandstone pore pressure to the hydraulic fracture gradient of shale and the overburden pressure gradient. A result of the study was that no proven hydrocarbon accumulations exist where an overlying shale seal has a pressure differential greater than 4 psi/ft (90.5 kPa per m).

To the authors' knowledge, no comprehensive examination of the clay rocks in the offshore Trinidad area has been undertaken. Within the abnormally pressured zones shale cuttings are often described as being well lithified, unusually large, and having blocky or splintery form. In other areas, mostly in the Pleistocene and upper Pliocene section in the eastern part of the study, abnormally pressured shale is sometimes poorly lithified and is described as "gumbo". The dominant clay minerals present are illite and kaolinite, with lesser amounts of chlorite and mixed-layer illite-smectite. In the Cassia field, where Harry (1995) studied the vertical distribution of clay minerals, there was no noticeable change in the relative amounts of these minerals with increasing depth, indicating that little diagenetic alteration has taken place.

The determination of some shale parameters, such as thickness and absolute pressure, is interpretive. Figure 21 illustrates a typical pore pressure profile from Trinidad and the data points used to describe shale pressure seals. The reservoir pressure used for the seal analysis discounted any buoyancy pressure due to a column of hydrocarbons. The pressure differential between pressure compartments was merely the difference in the pressure, above normal, of the compartments. Shale pore pressure was estimated using log methods which yielded consistent and reasonable results across the basin. The contact between a sandstone and the shale pressure seal was interpretive

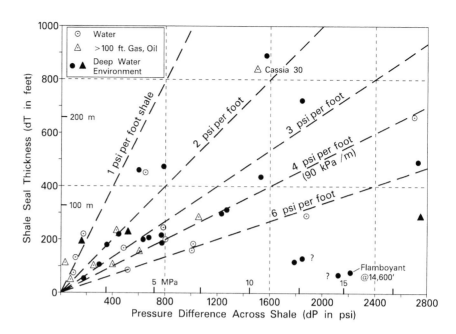

Figure 22. Cross-plot of shale seal thickness, dT, versus the pressure difference, dP, across shale units from within the overpressured section of offshore Trinidad. The plot shows that seal thickness is not a critical factor in seal efficiency. The environment of deposition is also not a factor. Some of the most effective seals are 100 feet (30 m) or less in thickness and separate pressure compartments with gradients of over 6 psi/ft (135 kPa/m). The plot shows that the upper limit of pressure difference across shale for trapping of hydrocarbons may be 4 psi/ft (90 kPa/m). Several points are marked as questionable (?) due to imprecise estimates of the reservoir pressure.

where the contacts were gradational. In such cases, the pressure contact was determined by a log response (gamma ray, resistivity, and travel time) which indicated an abnormally pressured shale (Figure 21). Some thick shale seals contained siltstone or thin sandstone beds. Where these non-shale lithologies were thicker than 50 ft (15 m) they were discounted and not considered to be part of the seal.

Table 1 presents the parameters for 42 pressure-sealing, clay rock units. Most of these shale units are about 200 ft (60 m) thick but range from 52 to 838 ft (16 to 255 m) in thickness. While it seemed reasonable that there should be a relationship between the thickness of a shale seal and the amount of pressure across it, there appeared to be little or no such relationship for these seals in Trinidad. Figure 22 is a plot of shale seal thickness and the pressure differential across them. Some of the largest pressure differences exist across the thinner seals. The most common pressure differences were between 2 and 4 psi/ft (45 and 90 kPa/m) but were as high as 27 psi/ft (610 kPa/m). Only 12% of the well documented pressure seals had pressure gradients higher than 6 psi/ft (136 kPa/m). Perhaps some flow occurred across the seals with the highest pressure differential until pressure gradients across the shale beds were reduced to 2 to 4 psi/ft (45 to 90 kPa/m).

During the pressure seal study it was observed that pore pressure increased sharply within shale that had been deposited in deep water (>300 ft, 90 m) in outer shelf, slope, and bathyal environments. However, examination of the seal efficiency of these shale beds (see Figure 22 and Table 1) revealed that no apparent relationship existed between seal efficiency and shales deposited in shallow marine versus deep-water environments. That is, shales deposited in deep-water environments were not more effective seals for abnormal pressure than shales deposited in other environments.

The ability of pressure sealing shales to trap hydrocarbons is of exploration significance. There does appear to be an upper limit to the pressure differential gradient—about 4 psi/ft (90 kPa/m)—at which hydrocarbons can be effectively sealed. The mechanism by which a hydrocarbon sealing shale would fail within an abnormally pressured section as a result of large pressure differences is not clear. A pressure differential greater than 4 psi/ft (90 kPa/m) may induce water flow after overcoming the shale's threshold gradient (Miller and Low, 1963) and perhaps overcome the capillary sealing properties for hydrocarbons of the shale at the crest of a trapping structure. This relationship is consistent with some analyses of Miocene seals in U.S. Gulf Coast conducted by the authors. In Jurassic and Cretaceous shales of offshore Nova Scotia and parts of the North Sea, however, much larger pressure differentials have been observed holding significant oil and gas columns.

None of the producing reservoirs studied approach the hydraulic fracture gradient of shale. However, a number of wells in the study area indicated very high pore pressure near this gradient at total depth. The hydraulic fracture gradient of shale is a generally recognized limit for pressure and hydrocarbon seals and was presented by Gaarenstroom et al. (1993) in their work on the North Sea. In Trinidad and other highly overpressured basins, the authors believe that at very high pore pressure—within 1,000 psi (6.9 MPa) of the hydraulic fracture gradient of shale—shale seals may have limited ability to seal hydrocarbons and fail episodically due to natural hydraulic fracturing. A thick zone of shale-dominated clastics, at very high pressure, is interpreted to underlie the oil and gas fields in Trinidad, forming a zone of rocks whose pressure and seal capacity for hydrocarbons is controlled by the hydraulic fracture gradient of shale.

Table 1. Seal efficiency data for shales in the Trinidad area:

Shale Bed	Depth (ft)	dT (ft)	dP (psi)	dP/dT (psi/ft)	Shale (psi)	Sandstone (psi)	Overburden (psi)	Reservoir Fluid Gas/Oil Column
Poui field, KB: 88 ft, water depth: 175 ft (within Z to Y fault block)								
17-2	7,930	242	785	3.240	4,325	* 4,322	7,415	70 ft oil
17-4	8,338	75	95	1.270	4,765	a 4,592	7,829	water
17-5	8,408	169	480	2.840	4,149	5,162	7,895	water
17-6	8,918	132	106	0.800	5,559	5,440	8,418	water
17-9	9,363	184	1,147	6.230	6,566	* 6,787	8,876	40 ft gas
18-0	9,530	110	28	0.254	6,188	* 7,136	8,883	350 ft G, 643 ft O
21-0	11,038	215	769	3.480	9,314	* 8,373	10,618	50 ft gas
Teak field, KB: 100 ft water depth: 190 ft (within F to F2 fault block)								
7AL	11,240	105	400	3.810	5,013	* 5,413	10,734	168 ft oil
9	12,200	155	700	4.520	6,344	* 6,545	11,773	210 ft G, 179 ft O
10	12,909	200	800	4.000	8,042	7,664	12,521	water
L	14,700	460	700	1.520	11,304	9,270	14,480	water
Teak field (upthrown to 19 Fault)								
L	12,715	490	2,727	5.560	9,778	a 10,404	12,295	water
Samaan field, KB: 95 ft, water depth: 185 ft (at structural crest)								
7A	8,555	550	653	1.180	3,807	4,460	8,119	water
7B	9,395	50	75	1.500	4,181	* 4,913	9,019	100 ft oil
9	9,695	101	248	2.450	4,983	* 5,296	9,317	225 ft G, 30 ft O
10	9,995	235	427	1.820	5,197	* 5,858	9,635	219 ft oil
10A	10,395	219	172	0.780	5,665	a 6,210	10,073	water
11U	10,780	284	1,053	3.710	6,155	* 7,436	10,478	428 ft oil
12	11,284	290	1,875	6.460	7,210	9,538	11,036	water
13	12,166	721	1,841	2.550	9,733	11,776	11,996	water
b13	12,166	b 130	1,841	14.160	9,733	11,776	11,996	water
19	13,635	297	1,236	4.160	12,531	13,673	13,608	water
Cassia field, KB: 102 ft, water depth 220 ft (West SEG #1 well)								
30	11,852	* 838	1,500	1.790	7,633	* 6,888	11,295	1,250 ft gas
31	13,142	659	2,692	4.085	11,039	a 10,160	12,590	? water
32	14,242	890	1,574	1.768	13,103	a 12,229	13,715	? water
33	14,670	312	1,272	4.080	13,878	a 13,694	14,142	? water
South SEG field, KB: 108 ft, water depth 230 ft (SEG #1 well)								
25N	10,727	230	515	2.239	5,750	* 5,293	10,116	>100 ft gas
30A	11,257	160	900	5.625	6,191	a 6,432	10,660	water
30B	11,800	200	529	2.645	8,248	a 7,205	11,245	water
31	12,616	390	2,745	7.038	9,676	a 10,318	12,149	? water
Flamboyant field, KB: 85 ft, water depth: 260 ft (Flamboyant #2 well)								
un-named	6,744	89	507	5.700	3,378	c 3,504	6,002	water
G-2	8,417	88	779	8.850	5,336	c 5,060	7,659	water
G-40	9,957	130	0	0.000	5,755	*5,753	9,230	360 ft gas
MP-10	10,986	120	1,789	14.910	7,987	8,005	10,327	water
MP-30	12,202	70	-2,119	-30.270	8,554	c 6,434	11,616	water
MP-40	12,421	180	357	1.983	5,777	6,889	11,825	water
MP-50	12,727	193	160	0.829	7,916	* 7,187	12,154	>500 ft gas
MP-60	13,140	205	679	3.310	6,701	* 8,052	12,628	50 ft gas
MP-70	13,313	52	189	3.630	8,001	8,319	12,807	water
MP-80	13,528	105	296	2.819	8,644	8,711	13,054	water
MP-90	14,012	432	1,525	3.530	10,804	a 10,490	13,564	water
un-named	14,600	80	2,215	27.690	12,775	a 12,970	14,220	? water
MP-100	14,911	220	446	2.027	13,286	a 13,556	14,568	? water
un-named	15,527	472	785	1.663	14,611	a 14,618	15,263	water

* dT and dP, thickness and pressure across shale, are defined in Figure 21. All depths are True Vertical Depth from Rig/Platform datum (KB). Shales are named after the underlying reservoir. Pressure difference in water-bearing portion of the reservoir only, buoyancy pressure ignored.

a Sandstone reservoir pressure estimated as the pressure from mud weight during flow and the shut-in drill pipe pressure, as one half PPG less than mud weight used to control fluid flow, or an estimate between mud weight prior to and after control of flow.

b Shale thickness corrected for non-sealing lithologies including sandstone, siltstone, and thick shale that does not show significant log indications of abnormal pressure.

c Reservoir pressure inferred from well log response (shale calculation) and inferred from pressure measurements in other sandstone units. Flamboyant's G-2 unit assumed in a pressure compartment with G-40 to G-60, and the MP-30 unit assumed in "leaked" zone like underlying MP-40 sand following a 1.75 psi/ft (39.6 kPa/m) interval gradient.

CONCLUSIONS

Significant oil and gas reserves are present within abnormally pressured sandstone reservoir compartments in Trinidad and Venezuela. 82.5 million bbl (13.1 million m^3) of oil and 1.79 trillion ft^3 (50.7 billion m^3) of gas have been produced from abnormally pressured Pliocene sandstones from six large fields within the study area offshore of the southeast coast of Trinidad. This production represents 11.4% and 49.8% of the total oil and gas production, respectively, from the area. The productive, abnormally pressured reservoirs are mildly to highly abnormally pressured and do not exceed a pressure gradient of 0.727 psi/ft (16.4 kPa/m). No proven, economic reserves have been found in the study area within very high pressure conditions that approach the ultimate seal limit of the hydraulic fracture gradient of shale.

Some of the longest oil and gas columns of the offshore area are within the abnormally pressured section. The longest gas column is 1,250 ft (380 m) in the Cassia field. The pressure gradient at the top of this column is 0.618 psi/ft (14.0 kPa/m) at 11,750 ft (3,580 m) in a pressure compartment 1,600 psi (11.0 MPa) above normal pressure. The common oil column of the 18 and 19 sands in Poui field is 643 ft (196 m) long. It occurs in a pressure compartment 2,650 psi (18.2 MPa) above normal at 8,200 ft (2,500 m). The pressure gradient at the structural crest is 0.720 psi/ft (16.3 kPa/m).

Shale seals within the abnormally pressured section commonly have pressure gradients across them (from sandstone to sandstone bed) of 1 to 4 psi/ft (22.6 to 90 kPa/m). Some gradients are as high as 14 to 27 psi/ft (317 to 611 kPa/m). All of the economically viable oil and gas within the abnormally pressured section have shale top seals with pressure gradients across them no higher than 4 psi/ft (90.5 kPa/m). This limiting gradient may be useful as a guideline for assessing seal risk within the basin.

The six studied fields in offshore Trinidad are complexly cut by normal faults. Within the abnormally pressured section the faults are pressure seals separating high-porosity sandstones which are at very different pressures. The pressure differences across faults may be as high as 1,856 psi (12.80 MPa). Large pressure differences across normal faults can aid trapping in the low pressure block, such as at Cassia and Flamboyant fields. Conversely, only small hydrocarbon columns can be expected in the high pressure block, such as at Poui field. Similar observations were made by Myers (1968) for the U. S. Gulf Coast.

Abnormal pressure is present within a large area of the Tertiary to Cretaceous sedimentary section of the Eastern Venezuelan Basin, including the Trinidad study area. The top of abnormal pressure crosses stratigraphic units from west to east, from within the Miocene to within the Pleistocene section, and appears controlled by the diachronous infilling of the foredeep basin, a result of tectonics associated with the movement of the Caribbean and South American Plates. Limited well control indicates that the Upper Cretaceous source rocks of the Naparima Hill and Gautier Formations in Trinidad (and equivalents in Venezuela) are highly abnormally pressured. Deep wells in Trinidad and Venezuela show that very high pore pressures, approaching the hydraulic fracture gradient of shale, are present throughout much of the basin overlying the Upper Cretaceous source rocks.

Within the offshore Trinidad area hydrocarbon migration from the now deeply buried Upper Cretaceous Gautier and Naparima Hill source rocks occurred in three phases:

1. primary migration, enhanced by natural hydraulic fractures, from highly pressured source rocks into overlying, and highly pressured, Tertiary age clastics;

2. long residence time of the hydrocarbon charge within the abnormally pressured Tertiary clastic section during basin infilling punctuated by episodic vertical migration, principally through hydraulic fracturing focused by dipping carrier beds; and

3. final pulses of migration out of the highly pressured section into the pressure transition zone and the normally pressured rocks during Pleistocene structural uplift and faulting.

Secondary migration within the pressure transition zone and normally pressured rocks occurred by cross fault juxtaposition of sandstones, and where hydraulically possible, migration may have occurred through fault zones.

ACKNOWLEDGMENTS   *The authors wish to thank Amoco Trinidad Oil Company and Amoco Corporation for permission to publish this study. Our thanks go to Ben Law, Charles Spencer, and Marty Albertin for their patience and constructive criticism. A great number of Amoco Trinidad workers have contributed to the general understanding of the geology of this area. Of these, the following individuals provided valuable assistance during the preparation of this paper: T. Cronin, J. Finneran, D. Grass, J. Krushin, K. Ortmann, C. Schroeder, R. Sels, F. Sobol, W. Skelton, M. Staines, and L. Wood. We also appreciated the work of K. Maxey, J. Grimes, K. Miller, M. Schendel, and P. Fullbright for the preparation of the figures.*

REFERENCES CITED

Abelwhite, K. and G.E. Higgins, 1968, A review of Trinidad, West Indies, oil development and the accumulations at Soldado, Brighton Marine, Grande Ravine, Barrackpore-Penal and Guayaguayare, *in* J.B. Saunders, ed., Transactions of the 4th Caribbean Geological Conference, p. 41–58.

Al-Shaieb, Z., J.O. Puckette, A.A. Abdalla, and P. B. Ely,

1994, Megacompartment complex in the Anadarko Basin: a completely sealed overpressured phenomenon, *in* P.J. Ortoleva, ed., Basin compartments and seals, AAPG Memoir 61: Tulsa, AAPG, p. 55–68.

Alison, G.B., and P.F. Farfan, 1989, Case history of the gas-condensate Cassia Field, offshore Trinidad: 21st Offshore Technology Conference, Houston, Texas, May 1–4, 1989, paper OTC 5897, p. 141–146.

Ames, R.L., and L.M. Ross, 1985, Petroleum geochemistry applied to oilfield development, offshore Trinidad, *in* Rodrigues, K., ed., Transactions of the First Geological Conference of the Geological Society of Trinidad and Tobago, Port-of-Spain, Trinidad, July 10–12, 1985, p. 227–236.

Arnold, R., and G.A. MacReady, 1956, Island-forming mud volcano in Trinidad, British West Indies: AAPG Bulletin, v. 40, p. 2748–2758.

Ave Lallemant, H.G., 1991, The Caribbean—South American plate boundary, Araya Peninsula, eastern Venezuela, *in* D.K. Larue and G. Draper, eds., Transactions of the 12th Caribbean Geological Conference, St. Croix, U.S.Virgin Islands, August 1989.

Bane, S.C., and R.R. Chanpong, 1980, Geology and development of the Teak oil field, Trinidad. West Indies, *in* Giant oil and gas fields of the decade, 1968–1978, AAPG Memoir 30, Tulsa, AAPG, p. 387–398

Barker, C., 1972, Aquathermal pressuring—role of temperature in development of abnormal pressure zones: AAPG Bulletin, v. 56, p. 2068–2071.

Barker, C., 1987, Generation of anomalous internal pressures in source rocks, in B. Doligez, ed., Migration of hydrocarbons in sedimentary basins, Collection 45, 2nd IFP Exploration Research Conference, Carcans, France, June 15–19, 1987, p. 223–235.

Barker, M.H.S., and K.H. Roberts, 1965, Excursion 6A South Trinidad: Transactions of the Fourth Caribbean Geological Conference, Trinidad, 1965, p. 438–441.

Barr, K.W., and J.B. Saunders, 1968, An outline of the geology of Trinidad: Transactions of the Fourth Caribbean Geological Conference, Trinidad, 1965, p. 1–10.

Berry, F.A.F., 1973, High fluid potentials in California Coast Ranges and their tectonic significance: AAPG Bulletin, v. 57, p. 1219–249.

Bradley, J.S., 1975, Abnormal formation pressure: AAPG Bulletin, v. 59, p. 957–973.

Bradley, J.S., and D.E. Powley, 1994, Pressure compartments in sedimentary basins: a review, *in* P.J. Ortoleva, ed., Basin compartments and seals, AAPG Memoir 61: Tulsa, AAPG, p. 3–26.

Bredehoeft, J.D., and B.B. Hanshaw, 1968, On the maintenance of anomalous fluid pressures: I. Thick sedimentary sequences: Geological Society of America Bulletin, v. 79, p. 1097–1106.

Bredehoeft, J.D., W. Back, and B.B. Hanshaw, 1982, Regional ground-water flow concepts in the United States: historical perspective, *in* T.N. Narasimhan,

ed., Recent trends in hydrogeology: Geological Society of America Special Paper 189, p. 297–316.

Burst, J.F., 1969, Diagenesis of Gulf Coast clayey sediments and its possible relation to petroleum migration: AAPG Bulletin, v. 53, p. 73–93

Carnevali, J.O., 1992, Monagas thrust-fold belt in the Eastern Venezuelan Basin: anatomy of a giant discovery of the 1980s: Proceeding of the 13th World Petroleum Congress, Buenos Aires, Argentina, October 20–25, 1992, p. 47–58.

Carr-Brown, B., 1971, The Holocene-Pleistocene contact in the offshore area east of Galeota Point, Trinidad West Indies: Transactions of the 6th Caribbean Conference, Isla Margarita, Venezuela, July 6–14, 1971, p. 381–397.

Davis, B, and T. Jones, 1994, Pore pressure prediction cuts exploration drilling risk: World Oil, September, 1994, p. 63–66.

de Landro, W.V.C., 1985, Petrophysical study of the Samaan field shaly sands, offshore Trinidad, in Rodrigues, K., ed., Transactions of the First Geological Conference of the Geological Society of Trinidad and Tobago, Port-of-Spain, Trinidad, July 10–12, 1985, p. 183–198.

Deming, D., 1994, Factors necessary to define a pressure seal: AAPG Bulletin, v. 78, p. 1005–1009.

Dewan, J.T., 1983, Essentials of modern open-hole log interpretation: Tulsa, Oklahoma, PennWell Books, 361 p.

Eaton, B.A., 1975, The equation for geopressure prediction from well logs: Society of Petroleum Engineers of AIME, paper SPE 5544, 11 p.

Eberhardt-Phillips, D., D-H Han, and M.D. Zoback, 1989, Empirical relationships among seismic velocity, effective pressure, porosity, and clay content in sandstone: Geophysics, v. 54, p. 82–89.

Erhlich, R.N., and S.F. Barrett, 1992, Petroleum geology of the Eastern Venezuela foreland basin, in R.W. MacQueen and D.A. Leckie, eds., Foreland Basins and Fold Belts, AAPG Memoir 55: Tulsa, p. 341–362.

Erhlich, R.N., P.F. Farfan, and P. Hallock, 1993, Biostratigraphic, depositional environments, and diagenesis of the Tamana Formation, Trinidad: a tectonic marker horizon: Sedimentology, v. 40, p. 743–768.

Eggertson, E.B., 1995, The Rio Claro boulder bed; evidence for an ancestral wrench fault zone across central Trinidad, West Indies: Transactions of the third Geological Conference of the Geological Society of Trinidad and Tobago, Port of Spain, Trinidad, June, 1995.

Farfan, P. and Y.K. Bally, 1990, An outline of the geology of Samaan field, Trinidad, West Indies, *in* K.A. Gillezeau, ed., Transactions of the Second Geological Conference of the Geological Society of Trinidad and Tobago, p. 139–156.

Foster, J.B., and H.E. Whalen, 1966, Estimation of formation pressures from electrical surveys—offshore Louisiana: Journal of Petroleum Technology, February 1966, p. 165–171.

Gaarenstroom, L., R.A.J. Tromp, C. de Jong, and A.M. Brandenburg, 1993, Overpressures in the Central North Sea: implications for trap integrity and drilling safety, *in* Parker, J.R., ed., Petroleum Geology of Northwest Europe: Proceedings of the 4th Conference, London, The Geological Society, p. 1305–1313.

Gallango, O., and F. Parnaud, 1995, Two-dimensional computer modeling of oil generation and migration in a transect of the Eastern Venezuela Basin, in A.J. Tankard, R. Sua'rez S., and H.J. Welsink, eds., Petroleum basins of South America, AAPG Memoir 62: Tulsa, AAPG, p. 727–740.

Gibson, R.G., 1994, Fault-zone seals in siliciclastic strata of the Columbus Basin, offshore Trinidad: AAPG Bulletin, v. 78, p. 1372–1385.

Hanor, J.S., and R. Sassen, 1990, Evidence for large-scale vertical and lateral migration of formation waters, dissolved salt, and crude oil in the Louisiana Gulf Coast, *in* Schumacher, D., and R.F., Perkins, eds., Gulf Coast oil and gases, their characteristics, origins, distribution, and exploration and production significance, Proceedings of the Ninth Annual Research Conference, Gulf Coast Section, Society of Economic Paleontologists and Mineralogists Foundation, p. 283–296.

Harry, B.E., 1995, The clay mineralogy of sediments in the West SEG-1 well, and implications for the diagenetic and sedimentary history of the southeastern Trinidad area: Transactions of the Third Geological Conference of the Geological Society of Trinidad and Tobago, Port-of-Spain, Trinidad, June, 1995.

Hedberg, H.D., 1974, Relation of methane generation to undercompacted shales, shale diapirs, and mud volcanos: AAPG Bulletin, v. 58, p. 661–673.

Heppard, P.D., R.L. Ames, and L.M. Ross, 1990, Migration of oils into Samaan Field, offshore Trinidad, West Indies, in K.A. Gillezeau, ed., Transactions of the Second Geological Conference of the Geological Society of Trinidad and Tobago, p. 157–168.

Higgins, G.E., and J.B. Saunders, 1967, Report on 1964 Chatham Mud Island, Erin Bay, Trinidad, West Indies: AAPG Bulletin, v. 51, p. 55–64.

Hottmann, C.E., and R.K. Johnson, 1965, Estimation of formation pressures from log-derived shale properties: Journal of Petroleum Technology, June 1965, p. 717–722.

Hubbert, M.K., and W.W. Rubey, 1959, Role of fluid pressure in mechanics of overthrust faulting: Bulletin of the Geological Society of America, v. 70, p. 115–166.

Hunt, J.M., 1990, Generation and migration of petroleum from abnormally pressured fluid compartments: AAPG Bulletin, v. 72, p. 1–12.

Jones, H.P., 1965, The geology of the Herrera sands in the Moruga West oilfield of south Trinidad, *in* J.B. Saunders, ed., Transactions of the Fourth Caribbean Conference, Trinidad, p. 91–100.

Kugler, H.G., 1965, Sedimentary volcanism, *in* J.B. Saunders, ed., Transactions of the Fourth Caribbean Conference, Trinidad, p. 11–13.

Lanan, B., S. Mohammed, and D. Therry, 1995, The Immortelle horizontal oil well program: using new technologies to optimize results: Transactions of the Third Geological Conference of the Geological Society of Trinidad and Tobago, Port-of-Spain, Trinidad, June 1995, (abstract).

Lantz, J.R., and N.X. Ali, 1990, Development of a mature giant offshore oil field, Teak Field, Trinidad, 22nd Annual Offshore Technology Conference, paper OTC 6267: Houston, 26 p.

Luo, M., M.R. Baker, and D.V. LeMone, 1994, Distribution and generation of the overpressure system, eastern Delaware basin, western Texas and southern New Mexico: AAPG Bulletin, v. 78, p. 1386–1405.

MacGregor, J.R., 1965, Quantitative determination of reservoir pressures from conductivity log: AAPG Bulletin, v. 49, p. 1502–1511.

McDougall, A.W., 1985, Geology of the East Soldado field: Transactions of the Fourth Latin American Conference: Port-of-Spain, Trinidad, July 7–15, 1979, p. 720–725.

Meissner, F.F., 1978, Patterns of source-rock maturity in nonmarine source-rocks of some typical western interior basins, in nonmarine Tertiary and Upper Cretaceous source rocks and the occurrence of oil and gas in west central U. S.: Rocky Mountain Association of Geologists, Continuing Education Lecture Series, p. 1–37.

Michelson, J.E., 1976, Miocene deltaic oil habitat, Trinidad: AAPG Bulletin, v. 60, p. 1502–1519.

Miller, R.J., and P.F. Low, 1963, Threshold gradient for water flow in clay systems: Soil Science Society of America Proceedings, v. 27, p. 605–609.

Miller, T.W., 1995, New insights on natural hydraulic fractures induced by abnormally high pore pressures: AAPG Bulletin, v. 79, p. 1005–1018.

Mouchet, J.P., and A. Mitchell, 1989, Abnormal pressures while drilling—manuels techniques 2: Boussens, France, Elf Aquitaine Edition, p. 264.

Myers, J.D., 1968, Differential pressures: a trapping mechanism in Gulf Coast oil and gas fields: Transaction, Gulf Coast Association of Geological Societies, v. XVIII, p. 56–80.

Neuzil, C.E., 1986, Groundwater flow is low-permeability environments: Water Resources Research, v. 22, p. 1163–1195.

Qin, C., and P.J. Ortoleva, 1994, Banded diagenetic pressure seals: types, mechanisms, and homogenized basin dynamics, *in* P.J. Ortoleva, ed., Basin compartments and seals, AAPG Memoir 61: Tulsa, AAPG, p. 3–26.

Parnaud, F., Y. Gou, J.C. Pascual, I. Truskowski, O. Gallango, H. Passalacqua, and F. Roure, 1995, Petroleum geology of the central part of the Eastern Venezuela basin, *in* A. J. Tankard, R. Sua'rez S., and H.J. Welsink, eds., Petroleum basins of South America, AAPG Memoir 62: Tulsa, AAPG, p. 741–756.

Pennebaker, E.S., 1968, An engineering interpretation of seismic data: Preprint, Society of Petroleum Engineers of AIME, paper number SPE 2165, p. 51–62.

Powers, M.C, 1967, Fluid-release mechanisms in compacting marine mudrocks and their importance in oil exploration: AAPG Bulletin, v. 51, p. 1240–1254.

Prieto Cedraro, R., 1987, Seismic stratigraphy and depositional systems of the Orinoco platform area, northeastern Venezuela: Austin, Texas, University of Texas, unpublished Ph.D. thesis, p. 144.

Prieto, R. and G. Valdes, 1992, El Furrial oil field: a new giant in an old basin, *in* Halbouty, M.T., ed., Giant oil and gas fields of the decade, 1978–1988, AAPG Memoir 54: Tulsa, AAPG, p. 155–161.

Radovsky, B., and J. Iqbal, 1985, Geology of the North Soldado field, Transactions of the Fourth Latin American Conference: Port-of-Spain, Trinidad, July 7–15, 1979, p. 759–769.

Reynolds, E.B., 1970, Predicting overpressured zones with seismic data: World Oil, v. 171, no. 5, p. 78–82.

Requejo, A.G., C.C. Wielchowsky, M.J. Klosterman and R. Sassen, 1994, Geochemical characterization and lithofacies and organic facies in Cretaceous organic—rich rocks from Trinidad, East Venezuela Basin: Organic Geochemistry, v. 22, p. 441–459.

Robertson, P. and K. Burke, 1989, Evolution of the southern Caribbean plate boundary, vicinity of Trinidad and Tobago: AAPG Bulletin, v. 73, p. 490–509.

Rodrigues, K., 1985, Thermal history modeling in petroleum exploration: examples from southern Trinidad, in Rodrigues, K., ed., Transactions of the First Geological Conference of the Geological Society of Trinidad and Tobago, Port-of-Spain, Trinidad, July 10–12, 1985, p. 217–226.

Rodrigues, K., 1988, Oil source bed recognition and crude oil correlation, Trinidad, West Indies: Advances in Organic Geochemistry 1987, Organic Geochemistry, v. 13, p. 365–371.

Rodrigues, K., 1995, Factors controlling API gravity variations among Trinidad crudes: Transactions of the Third Geological Conference of the Geological Society of Trinidad and Tobago and 14th Caribbean Geological Conference.

Rohr, G.M, 1990, Paleogeographic maps, Maturin Basin of eastern Venezuela and Trinidad, in K.A. Gillezeau, ed., Transactions of the Second Geological Conference of the Geological Society of Trinidad and Tobago, p. 88–105.

Ross, L.M., and R.L. Ames, 1988, Stratification of oil in Columbus Basin of Trinidad: Oil and Gas Journal, September 26, 1988, p. 72–76.

Ross, M.I. and Scotese, C.R., 1988, A hierarchical tectonic model of the Gulf of Mexico and Caribbean region: Tectonophysics, v. 155, p. 139–168

Russo, A.J., and O. Dumont, 1994, Reservoir characteristics using multiple formation pressure tester: El Furrial field, Venezuela: Society of Petroleum Engineers paper 26962, presented at 3rd Latin American and Caribbean Petroleum Engineering Conference, Buenos Aires, Argentina, April 27–29, 1994, p. 11.

Salvador, A., and R.M. Stainforth, 1965, Clues in Venezuela to the Geology of Trinidad, and vice versa: Transactions of the Fourth Caribbean Conference, Trinidad, 1965, p. 31–40.

Scott, D.R., and L.A. Thomsen, 1993, A global algorithm for pore pressure prediction: 8th Society of Petroleum Engineers Middle East Oil Show and Conference, Manama, Bahrain, April 3–6, Proceedings, v. 2, p. 645–654.

Sealy, E.C., and A. Ramlackhansingh, 1985, The geology of Trinidad-Tesoro's Palo Seco Field including South Erin and Central Los Bajos: Transactions of the Fourth Latin American Conference, Port-of-Spain, Trinidad, July 7–15, 1979, p. 796–802.

Smith, J.E., 1970, The dynamics of shale compaction and evolution of pore-fluid pressures: Mathematical Geology, v. 3, p. 239–263.

Swarbrick, R.E. and M.J. Osborne, 1998, Mechanisms that generate abnormal pressures: an overview, *in* Law, B.E., G.F. Ulmishek, and V.I. Slavin eds., Abnormal pressures in hydrocarbon environments: AAPG Memoir 70, p. 13–34.

Talukdar, S.C., O. Gallango, and M. Chin-A-Lien, 1986, Generation and migration of hydrocarbons in the Maracaibo Basin, Venezuela: an integrated basin study: Advances in Organic Geochemistry 1985, Organic Geochemistry, v. 10, p. 261–279.

Talukdar, S.C., O. Gallango, and A. Ruggiero, 1988, Generation and migration of oil in the Maturin Subbasin, Eastern Venezuela Basin: Advances in Organic Geochemistry 1987, Organic Geochemistry, v. 13, p. 537–547.

Talukdar, S.C., W.G. Dow, and K.M. Persad, 1990, Deep oil prospects in Trinidad: Houston Geological Society Bulletin, October 1990, p. 16–19.

Terzaghi, K. and R.B. Peck, 1948, Soil mechanics in engineering practice: New York, John Wiley & Sons, Inc., p. 556.

Tiratsoo, E.N., 1986, Oilfields of the world, 3rd edition: Houston, Texas, Gulf Publishing Company, p. 392.

Traugott, M.O., 1982, Rock mechanics, petrophysics, and stratigraphy of the Tuscaloosa trend: Journal of Petroleum Technology, February, 1982, p. 428–432.

Traugott, M.O., 1984, The application of geophysical and geomechanical concepts to drilling problems in the Tuscaloosa trend: University of Idaho, unpublished Masters thesis, p. 34.

Traugott, M.O., and Heppard, P.D., 1994, Prediction of pore pressure before and after drilling—taking the risk out of drilling overpressured prospects, *in* Law, B.E., G. Ulmishek, and V.I. Slavin, eds., AAPG Hedberg Conference, Abnormal Pressures in Hydrocarbon Environments: Golden, Colorado, June 8–10, 1994, (extended abstract).

Wharton, S., and Hudson, D., 1995, Report on a recent eruption of the mud volcano, Devil's Woodyard, south Trinidad: Transactions of the Third Geological

Conference of the Geological Society of Trinidad and Tobago, Port-of-Spain, Trinidad, June 1995.

Woodside, Philip R., 1981, Petroleum geology of Trinidad and Tobago: Oil and Gas Journal, Sept. 28, 1981, v. 79, no. 39, p. 364–389.

CONVERSIONS USED

0.3048 m = 1 ft

6.89476 kPa = 1 psi

22.6206 kPa/m = 1 psi/ft

19.25 PPG = 1 psi/ft

$0.15898 \text{ m}^3 = 1 \text{ bbl}$

$0.028317 \text{ m}^3 = 1 \text{ ft}^3$

Law, B.E., S.H.A. Shah, and M.A. Malik, 1998, Abnormally high formation pressures, Potwar Plateau, Pakistan, *in* Law, B.E., G.F. Ulmishek, and V.I. Slavin eds., Abnormal pressures in hydrocarbon environments: AAPG Memoir 70, p. 247–258.

Chapter 14

ABNORMALLY HIGH FORMATION PRESSURES, POTWAR PLATEAU, PAKISTAN

B.E. Law[1]
U.S. Geological Survey
Denver, Colorado, U.S.A.

S.H.A. Shah
M.A. Malik
Oil and Gas Development Corp. of Pakistan
Islamabad, Pakistan

Abstract

Abnormally high formation pressures in the Potwar Plateau of north-central Pakistan are major obstacles to oil and gas exploration. Severe drilling problems associated with high pressures have, in some cases, prevented adequate evaluation of reservoirs and significantly increased drilling costs. Previous investigations of abnormal pressure in the Potwar Plateau have only identified abnormal pressures in Neogene rocks. We have identified two distinct pressure regimes in this Himalayan foreland fold and thrust belt basin: one in Neogene rocks and another in pre-Neogene rocks. Pore pressures in Neogene rocks are as high as lithostatic and are interpreted to be due to tectonic compression and compaction disequilibrium associated with high rates of sedimentation. Pore pressure gradients in pre-Neogene rocks are generally less than those in Neogene rocks, commonly ranging from 0.5 to 0.7 psi/ft (11.3 to 15.8 kPa/m) and are most likely due to a combination of tectonic compression and hydrocarbon generation. The top of abnormally high pressure is highly variable and doesn't appear to be related to any specific lithologic seal. Consequently, attempts to predict the depth to the top of overpressure prior to drilling are precluded.

INTRODUCTION

For many years, abnormally high reservoir pressures have posed significant drilling and completion problems in oil and gas wells in Pakistan. The principal areas of abnormally high pressure in Pakistan are located in the onshore and offshore areas in the Makran Basin in southern Pakistan and the Potwar Plateau in north-central Pakistan (Figure 1, B and A, respectively). In these areas, abnormally high pressures occur in Neogene rocks and are as high as lithostatic. In the Potwar Plateau, abnormally high pressures also occur in pre-Neogene rocks.

This investigation is concerned with abnormal pressures in the Potwar Plateau. The Potwar Plateau, located in north-central Pakistan (Figures 1 and 2), is bounded by the Parachinar-Murree Fault to the north, the Jehlum fault to the east, the Salt Range to the south, and the Indus River to the west. It is the principal oil producing region in Pakistan. Table I lists oil and gas production, through June, 1994, from all the fields in the

Potwar Plateau study area. Oil production is from pre-Neogene carbonate and sandstone reservoir rocks. The petroleum potential of the Potwar Plateau and adjacent areas has been summarized by Khan et al. (1986).

Drilling costs on the Potwar Plateau are very high due, in part, to severe drilling and completion problems associated with the abnormally high formation pressures in Neogene rocks (Malick, 1979). Pressures in excess of lithostatic have been reported in these rocks (Malick, 1979; Sahay and Fertl, 1988; Kadri, 1991). The drilling problems associated with overpressuring have been difficult to avoid because of the inability to predict the occurrence of overpressuring.

The few published sources of information dealing with abnormal pressures in Pakistan are mainly concerned with the magnitude and consequences of overpressuring. These earlier investigations concluded that tectonic compression and compaction disequilibrium are the main causes of overpressuring in Neogene rocks. Unlike previous investigations of abnormal pressure in the Potwar Plateau, we have evaluated the roles

---

[1] *Present Affiliation: Consulting Petroleum Geologist, Lakewood, Colorado, U.S.A.*

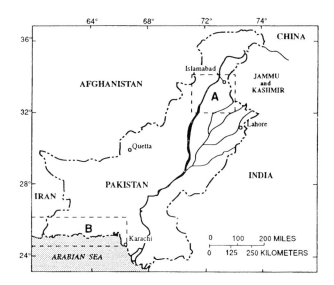

Figure 1. Map of Pakistan showing location of abnormally pressured rocks in the Potwar Plateau (A) and the Makran Basin (B) areas.

of temperature and burial history, clay mineral transformations, hydrocarbon source rock, sedimentation rates, and formation water quality in an attempt to determine the cause(s) of the abnormally high pressures. Because of the impact of overpressuring on oil and gas exploration on the Potwar Plateau, the purposes of this study were to: 1) determine the cause(s) of overpressuring and 2) attempt to provide a way of predicting where the top of overpressuring occurs.

STRUCTURE

The Potwar Plateau is a foreland basin and part of the Himalayan foreland fold- and thrust-belt in northern Pakistan (Figures 1 and 2). Structural deformation in the Potwar Plateau is a result of the ongoing collision between the Eurasian and Indian plates that began in early to middle Eocene time (Stoneley, 1974). Since collision began, about 1,240 mi (2,000 km) of convergence has occurred between India and Eurasia (Patriat and Achache, 1984). The plateau is underlain by a low-dipping thrust fault that has transported the

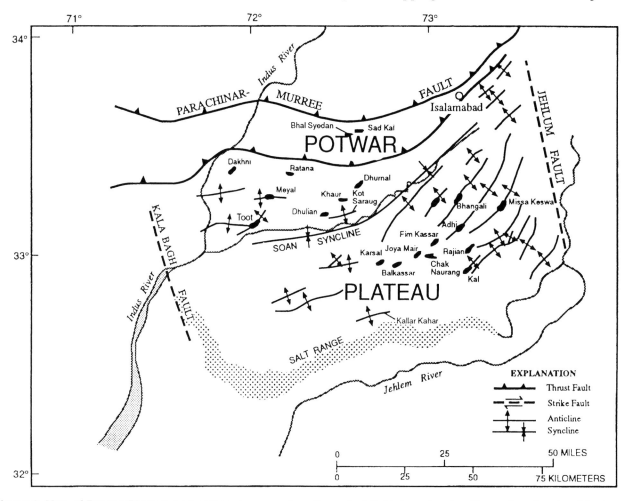

Figure 2. Map of Potwar Plateau and adjacent areas of north-central Pakistan showing the major structural features and oil and gas fields.

Table 1. Oil and gas fields in the Potwar Plateau and cumulative oil and gas production through June, 1994. Location of fields are shown on Figure 2.

Field	Oil Production x10⁶bbl [x10⁶m³]	Gas Production x10⁹ft³ [x10⁹m³]	Discovery Year
Khaur	4.19 [0.67]	–	1915
Dhulian	41.36 [6.58]	–	1936
Joya Mair	6.91 [1.10]	–	1944
Balkassar	32.70 [5.20]	–	1946
Meyal	36.04 [5.73]	228.00 [6.46]	1968
Toot	11.62 [1.85]	27.00 [0.76]	1968
Fim Kassar	5.88 [0.94]	2.00 [0.06]	1989
Dakhni	0.14 [0.02]	33.00 [0.94]	1983
Chak Naurang	2.55 [0.40]	–	1986
Bhal Syedan	0.07 [0.01]	–	1989
Missa Keswal	4.48 [0.71]	8.00 [0.23]	1991
Sadkal	1.21 [0.19]	7.00 [0.20]	1992
Dhurnal	44.16 [7.02]	110.00 [3.12]	1984
Bhangali	1.93 [0.31]	1.00 [0.03]	1989
Ratanna	0.32 [0.05]	–	1989
Adhi	4.95 [0.79]	–	1978
Karsal	0.21 [0.03]	–	1956
Rajian	–	–	1994
Kal	–	–	1995
TOTAL	198.73 [31.60]	416.00 [11.78]	

entire sedimentary section southward along a decollement in the Precambrian Salt Range Formation (Seeber and Armbruster, 1979; Lillie et al., 1987; Baker et al., 1988; Jaume and Lillie, 1988; and Pennock et al., 1989). At the southern margin of the plateau, the Precambrian Salt Range Formation and younger sedimentary rocks have been thrust southward over Neogene molasse rocks (Yeats et al., 1984). The timing of the thrusting has been determined to have begun between 2.1 and 1.6 Ma (Johnson et al., 1986).

Disharmonic folding of the sedimentary section relative to the underlying basement rocks has produced tight, salt-cored anticlines separated by broad synclines (Pennock et al., 1989). The structural axis of the basin trends in a east-northeast direction, in close proximity and nearly parallel to the Soan River (Figure 2). The north flank of the basin is defined by the Parachinar-Murree fault (Figure 2) and is characterized by high dips and intense faulting and folding, whereas the south flank of the basin is characterized by less intense structural deformation. The intervening area of the basin, in the vicinity of the Soan syncline, is relatively undeformed (Figures 2 and 3). Pre-collision structural history is poorly known. Angular unconformities at the base of Permian and Tertiary rocks (Figure 4) indicate two periods of regional tilting; the first occurring during Late Cambrian to Carboniferous time and the second during Late Cretaceous to early Paleocene time (Gee, 1980).

STRATIGRAPHY

Sedimentary rocks in the Potwar Plateau include siliciclastic, carbonate, and evaporite sequences of Precambrian through Pleistocene age (Figure 5) are as thick as 29,500 ft (9,000 m) (Fatmi et al., 1984). Precambrian rocks are composed predominantly of salt and gypsum with lesser amounts of claystone and shale. These rocks are locally intruded by igneous rocks (Martin, 1956). The thickness of Precambrian rocks is highly variable but, locally are known to be greater than 2,625 ft (800 m) thick (Fatmi et al., 1984). Paleozoic rocks are as thick as 4,035 ft (1,230 m) and are composed of conglomerate, sandstone, siltstone, shale, and carbonate sequences deposited in marine and nonmarine environments. Erosion at the base of the Lower Permian and base of Paleocene rocks accounts for most of the thickness variations in these sequences (Figure 4). Mesozoic rocks are as thick as 1,500 ft (460 m) and consist of sandstone, siltstone, shale, and carbonate sequences deposited in marine and nonmarine environments. Progressive eastward truncation of Mesozoic rocks at the base of the Tertiary section has resulted in the absence of Mesozoic rocks in the central and eastern parts of the Potwar Plateau (Figure 4).

The Tertiary section includes Paleocene through Pleistocene conglomerate, sandstone, siltstone, shale, carbonate, and coal that were deposited in marine and nonmarine environments. Paleogene rocks range in thickness from 790 to 985 ft (240–300 m) (Fatmi et al., 1984); the Neogene rocks are as thick as 21,000 ft (6,400 m) (Fatmi et al., 1984). Of particular importance to this study are the Neogene molasse rocks that include the Miocene Rawalpindi Group and Miocene to Pleistocene Siwalik Group (Figure 5). The stratigraphy and sedimentology of the Rawalpindi and Siwalik Groups

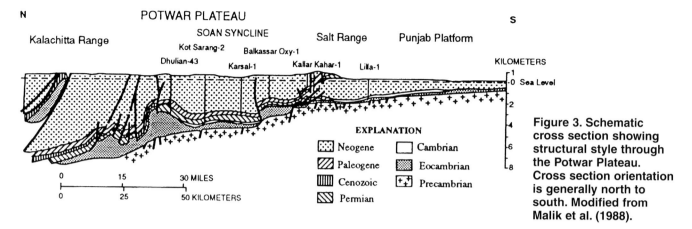

Figure 3. Schematic cross section showing structural style through the Potwar Plateau. Cross section orientation is generally north to south. Modified from Malik et al. (1988).

have been studied extensively (Pilbeam et al., 1977; Burbank and Raynolds, 1988; Johnson et al., 1986). They are composed of conglomerate, sandstone, siltstone, and shale deposited in fluvial environments in response to the southward-advancing Himalayan thrusting. Sediments derived from the thrust sheets, along the northern edge of the plateau, were deposited rapidly. Rates of sedimentation in the Siwalik Group have been determined to range from 50 to 125 in/1,000 yr (20–50 cm) (Raynolds and Johnson, 1985). Individual lithologic units within both groups exhibit variable thicknesses and are highly lenticular.

OVERPRESSURING

Abnormally high formation pressures in Neogene rocks in the Potwar Plateau have been reported by Malick (1979), Sahay and Fertl (1988), and Kadri (1991). Malick (1979) and Sahay and Fertl (1988) have speculated that the cause of overpressuring in Neogene rocks is due to tectonism associated with the collision of the Indian and Eurasian plates. Kadri (1991) suggested that in addition to tectonic compression, undercompaction may contribute to the development of high formation pressures. Our investigations show that, in addition to overpressuring in Neogene rocks, overpressuring also occurs in Paleogene and older rocks.

The criteria used to measure or identify overpressuring in this region are based primarily on drilling

mud weight and to a lesser extent, drillstem tests (DSTs). Shale density, drilling exponent, drilling speed, and flow line temperatures have also been used by the operators to detect abnormal pressures while drilling. Examples of pressure gradients using mud weight and DST data from some wells on the Potwar Plateau are shown on pressure vs. depth plots (Figures 6 through 10). There is also some indication that shale transit time may be of help in identification of overpressuring as shown on Figure 11. The various pore pressure indices used in the region indicate that the top of overpressuring occurs at shallow depths and continues to total depth. Malick (1979) reported the occurrence of overpressured rocks in the eastern part of the Potwar Plateau as shallow as 945 ft (290 m).

Based on mud weight and drilling exponent data, maximum pressure gradients are between 0.9 and 1.0 psi/ft (20.4 and 22.6 kPa/m). However, it is difficult to accurately evaluate pressures based on mud weight data because most operators in the Potwar Plateau region increase mud weight in anticipation of high pressures rather than adjust the mud system in response to pressure conditions. Therefore, rock units that are normally or only slightly above hydrostatic pressures are not detected. Drilling reports of lost circulation in Neogene rocks indicate that the occurrence and magnitude of pore pressures are highly variable and are probably related to lithologic variations. The practice by operators of increasing mud weights in anticipation of high pore pressures also precludes the

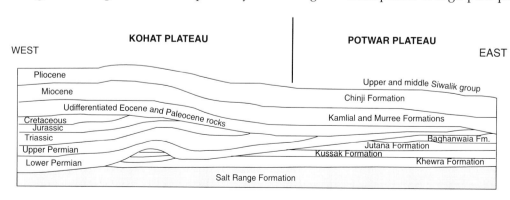

Figure 4. Diagrammatic east-west cross section through the Kohat and Potwar Plateaus showing regional relationships at the bases of Permian and Tertiary rocks. After Gee (1980).

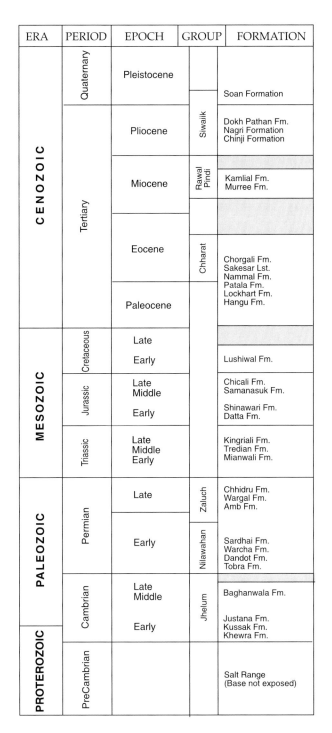

Figure 5. Generalized geologic column for the Potwar Plateau, Pakistan. Modified from Fatmi et al. (1984).

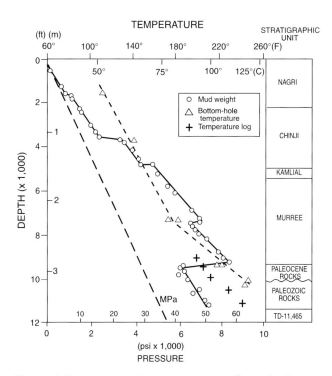

Figure 6. Pressure and temperature gradients in the Gulf Oil, Fim Kassar well. Approximate location of well shown on Figure 2.

identification of a pressure seal. It is quite probable, however, that a basin-wide lithologic pressure seal does not exist. The heterogeneous lithologic character of Neogene molasse rocks may have resulted in a compartmentalization of pressures having no recognizable, unique top of overpressuring.

Pressures are higher in the northern and eastern part of the Potwar Plateau than in the southern part. Kadri

(1991) reported that the highest pressures in the Potwar Plateau occur in wells drilled along the crests of anticlines in the eastern part of the plateau. Pressure gradients in the northern and eastern part of the Potwar Plateau are commonly as high as 0.9 psi/ft (20.4 kPa/m). Malick (1979) reported pressure gradients above lithostatic in the Adhi area (Figure 2) of northeastern Potwar. In contrast, pressure gradients in the southern part of the plateau are slightly lower, ranging from 0.7 to 0.8 psi/ft (15.8 to 18.1 kPa/m). The highest pressures also occur in structurally complex areas such as Dakhni, Adhi, Mianwala, Khaur, and Chak Naurang (Figure 2). However, some wells also drilled in structurally complex areas, such as the Qazian and Balkassar wells (Figures 8 and 9), exhibit relatively lower pressure gradients. Because nearly all wells drilled in the Potwar Plateau have been drilled on structures, it remains uncertain whether there is a relationship between structural deformation and overpressuring. However, available information indicates that there is a relationship among structure, structural intensity, and overpressuring. The fact that the Potwar Plateau is a seismically active area also must be taken into account when considering the cause(s) of overpressuring.

There are two distinct pressure regimes in the sedimentary rocks on the Potwar Plateau: one in Neogene rocks and one in pre-Neogene rocks. Pressures are generally higher in Neogene rocks than in pre-Neogene rocks. A few wells, however, such as the Occidental Khaur well (Figure 10), do not exhibit two distinct pressure regimes. The two pressure systems are fairly

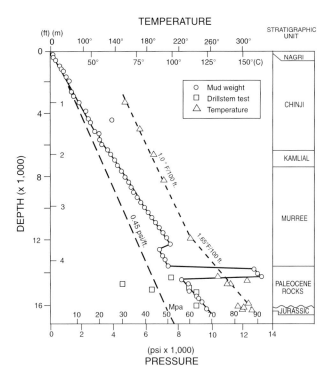

Figure 7. Pressure and temperature gradients in the Oil and Gas Development Corporation (OGDC), Dakhni No. I well. Approximate location of well shown on Figure 2.

well demonstrated by the pressure gradients shown in the wells in Figures 6 through 9. In these and other wells, the two pressure systems are distinguished by an abrupt decrease of pressure at, or near, the base of Neogene rocks.

Neogene Pressure System

Pore pressures in Neogene rocks are characterized by being very high. Pressure gradients calculated from mud weights and drilling exponents are as high as 0.9 to 1.0 psi/ft (20.4 to 22.6 kPa/m) (Figures 7 to 10). Only a few DSTs have been recorded in Neogene rocks and so it is difficult to corroborate these high pressures, however, DSTs recorded in the Murree Formation in the Qazian I-X well (Shown as Missa Keswal on Figure 8) indicate pressure gradients as high as 0.81 psi/ft (18.3 kPa/m).

Neogene rocks contain large amounts of relatively fresh water. The salinity of water recovered from DSTs in the Qazian I-X well range from 2,500 to 5,700 ppm. Furthermore, the low water salinity in these rocks is supported by neutral to reversed spontaneous potential (SP) curves as shown in the Qazian I-X well (Figure 12). Reversed SP curves have been observed in other overpressured rocks (Sahay and Fertl, 1988; Law et al., 1980) and have been interpreted to be due to the presence of fresh water.

Corrected thermal gradients through Neogene rocks are low, ranging from 1.0° to 1.3°F/100 ft (18.2° to 23.7°C/km), compared to average thermal gradients of

about 1.28°F/100 ft (23.4°C/km) in most basins of the World. Raza (1981) reported thermal gradients in Neogene rocks in the Potwar Plateau ranging from less than 1.0° to 1.5°F/100 ft (18.2° to 27.4°C/km). In the examples shown on Figures 6 through 9, moderately sloping thermal gradients extend upward from pre-Neogene rocks into Neogene rocks, where they intersect more steeply sloping low thermal gradients. We interpret the steeply sloping low thermal gradients to be due, in part, to convective heat transfer associated with vertically flowing water. Furthermore, we suggest that the high pressures in the lower part of Neogene rocks have created a pressure regime that forcibly expels water out of higher pressured rocks into overlying, lower pressured rocks. In the Balkassar well (Figure 9) the intersection of the temperature gradients is coincident with the Paleogene-Neogene boundary. The intersection in this well is also coincident with a lithologic boundary between the underlying carbonates, shales, and siliciclastic rocks and overlying siliciclastic rocks. In this particular example, thermal conductivity contrasts may account for some or all of the change in thermal gradient. In most wells, however, the change in thermal gradient is in the lower part of the Neogene section where there are no significant thermal conductivity contrasts. Based on time-temperature modeling (discussed later) present-day temperatures are maximum temperatures; paleotemperatures were lower than present-day temperatures.

Pre-Neogene Pressure System

Pressures in pre-Neogene rocks are lower than those in the overlying Neogene rocks. Pressure gradients,

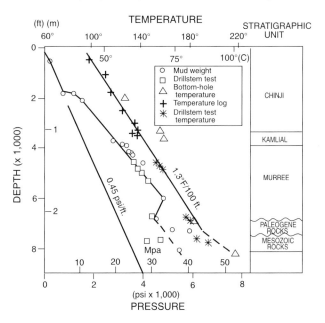

Figure 8. Pressure and temperature gradients in the Gulf Oil, Qazian I-X well. Approximate location of well shown as Missa Keswal on Figure 2.

based on mud weight and drilling exponent data, range from 0.5 to 0.9 psi/ft (11.3 to 20.4 kPa/m) and are commonly 0.6 to 0.7 psi/ft (13.6 to 15.8 kPa/m). Pressures recorded during DSTs range from 0.45 to 0.6 psi/ft (10.2 to 13.6 kPa/m). In contrast to Neogene rocks, formation water in pre-Neogene rocks is considerably more saline, ranging from 7,000 to 41,000 ppm. The higher salinity waters are reflected by the SP response of these rocks; the SP curve is most often normal, indicating the presence of saline formation water.

Thermal gradients through pre-Neogene rocks are high, ranging from 1.65° to 3.3°F/100 ft (30.1° to 60.2°C/km). Raza (1981) reported thermal gradients as high as 4.0°F/100 ft (73°C/km). We interpret the high thermal gradients to be due to relatively lower thermal conductivities than those in the overlying Neogene rocks. The thermal conductivity of siliciclastic rocks, the dominant lithology in the Neogene section, ranges from 2 to 5 watts per meter Kelvin (W/m°K) whereas, the thermal conductivity of shale and carbonate rocks, the dominant pre-Neogene lithology, ranges from 0.5 to 4.5 W/m°K (Gretener, 1981). The higher pressures in Neogene rocks may pose a barrier to vertically flowing fluids, precluding significant upward flow of fluids.

An important aspect when evaluating abnormal pore pressure in any basin is hydrocarbon generation potential. Paleogene rocks in the Potwar Plateau are more organic-rich and more thermally mature than the overlying Neogene rocks. Proprietary data show that

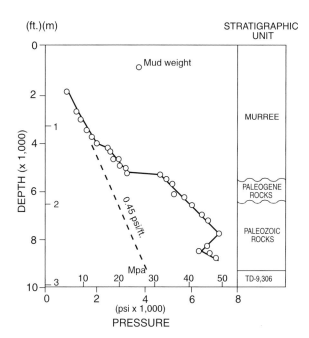

Figure 10. Pressure gradient in the Occidental, Khaur No. 1 well. Approximate location of well shown on Figure 2.

the total organic carbon (TOC) content of some of the pre-Neogene rocks range up to 1.5% and, in a few rock units, is as high as 30%. By comparison, Neogene rocks are organically lean (commonly less than 0.5% TOC) and less thermally mature. Mean random vitrinite reflectance ($\%R_o$) in these rocks is commonly greater than 0.6 $\%R_o$. The few vitrinite reflectance measurements made during this investigation in the Paleocene Lockhart Formation in the Fim Kassar well and in the Lower Cretaceous Lumshiwal Formation in the Dakhni-I well range from 0.94 to 1.52 $\%R_o$, respectively. Thus, most pre-Neogene rocks are well within the catagenic zone of hydrocarbon generation. This is further illustrated by burial and thermal history reconstructions discussed below, and on most mud logs by the top of sustained hydrocarbon shows. However, the high mud weights used to control overpressuring may have obscured hydrocarbon shows in the overlying Neogene rocks.

Time-Temperature Modeling

In order to further evaluate overpressuring in the Potwar Plateau, we reconstructed the burial and thermal history of the rock sequences penetrated in the Fim Kassar and Dakhni wells using BasinMod (Platte River Associates, 1997, v5.0, Denver, Colorado), a commercially available computer program. The program can calculate thermal maturity by either a time-temperature model (TTI) or by a kinetic model. The TTI model is based on the assumption that the maturity reaction doubles for every 10°C increase in temperature (Waples, 1980). Several important geologic variables

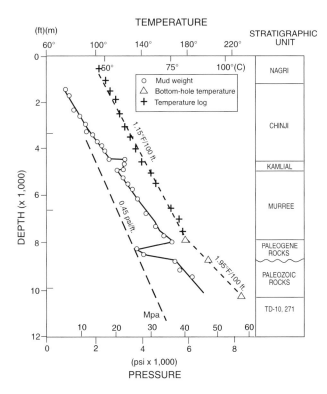

Figure 9. Pressure and temperature gradients in the Occidental, Balkassar No.1 well. Approximate location of well shown on Figure 2.

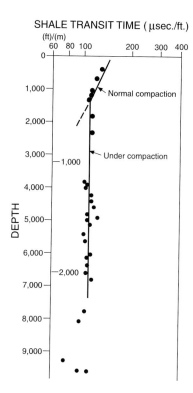

SHALE TRANSIT TIME (μsec./ft.)

Figure 11. Cross-plot of shale interval time vs. depth in the Gulf Oil, Fim Kassar well. Approximate location of well shown on Figure 2.

such as porosity, compaction, thermal conductivity, heat flow, and thermal gradient can be incorporated. The program assumes that heat transfer is accomplished by thermal conductivity; convective or advective heat transfer cannot be incorporated.

The accuracy of TTI modeling is dependent on accurate reconstruction of the burial, structural, and thermal history. In the Potwar Plateau, the timing and original thickness of rocks eroded at the base of Permian and Tertiary rocks is not well known. In addition, there is no information concerning pre-Paleogene, pre-collisional thermal events.

The pressure and temperature profiles for the Fim Kassar and Dakhni wells are shown on Figures 6 and 7. The burial history reconstructions of the two wells, shown on Figures 13 and 14, relied heavily on the stratigraphic work of Fatmi (1973) and Fatmi et al. (1984). The calculated thermal maturity profiles for the wells, utilizing present-day thermal gradients, are shown on Figures 15 and 16. With respect to measured levels of thermal maturity, the calculated values are nearly the same. In order to match the calculated and measured levels of thermal maturity, in both wells, it was necessary to reduce the present-day thermal gradients. The modeling of these wells indicate that present-day temperatures are as high as these rocks have ever been exposed to. In other words, paleotemperatures were lower than present-day temperatures. The modeling also shows that the top of the oil window occurs in the Miocene Murree Formation.

Probable Causes of Overpressuring

The proposed causes of overpressuring worldwide are many, and have been discussed by Fertl (1976), Gretener (1981), Gretener and Feng (1985), Chapman (1994), and Swarbrick and Osborne (1998-this volume). Some of the more commonly cited causes of overpressuring include compaction disequilibrium, hydrocarbon generation, aquathermal pressuring, tectonism, and mineral transformations. When considering the cause(s) of abnormal pressure, it is of paramount importance to do so within the context of the geologic history of the region under investigation. With this concern in mind, we discuss the causes of overpressuring with respect to a Neogene pressure regime and a pre-Neogene pressure regime consistent with their respective geologic histories. These two rock intervals have distinctly different depositional and structural histories that are important to understanding the causes of overpressuring.

There is no compelling evidence in support of any one dominant cause of overpressuring in the Potwar Plateau. However, some processes can be eliminated by a consideration of the geologic characteristics of these rocks. Overpressuring related to clay mineral transformations, originally proposed by Powers (1967) and Burst (1969), is based on the premise that the addition of water to the pore system from the transformation of smectitic shale to ordered, mixed-layered smec-

Gulf Oil Pakistan Lmt.
Qazian 1-x

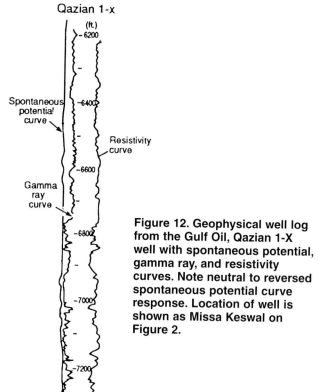

Figure 12. Geophysical well log from the Gulf Oil, Qazian 1-X well with spontaneous potential, gamma ray, and resistivity curves. Note neutral to reversed spontaneous potential curve response. Location of well is shown as Missa Keswal on Figure 2.

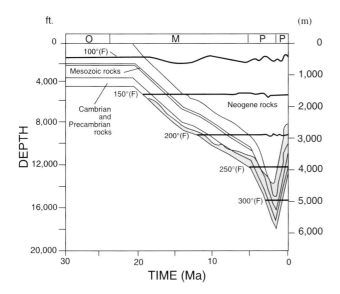

Figure 13. Burial curves for rocks penetrated in the Gulf Oil, Fim Kassar well. Shaded area is within the oil window. Location of well shown on Figure 2.

tite/illite shale will lead to overpressuring. This process is dependent on the degree to which the clay transformation has been completed and the existence of a large volume of clay (or shale) from which clay-derived water can be sourced. With respect to the Neogene rocks in the Potwar Plateau, the sand/shale ratio is very high (about 8:1) and the shales are thin and most often very sandy or silty.

X-ray diffraction analyses of whole rock rotary cuttings samples from the Occidental, Khaur well (Figure 17) by C.G. Whitney (U.S. Geological Survey, Denver, Colorado) shows that expandable clays are abundant only near the surface at < 2,280 ft (695 m) and there is no systematic variation in expandable clays with depth in the well. Furthermore, the top of overpressuring at about 4,200 ft (1,280 m), does not correspond to any significant change in clay mineralogy. Thus, these analyses coupled with the low volume of shale make it unlikely that clay mineral transformations have played a significant role in the origin of overpressuring.

Another process that is often cited as a cause of overpressuring is aquathermal pressuring. Aquathermal pressuring is the pressure effect produced by the thermal expansion of water in a relatively isolated, well-sealed volume of rock (Barker, 1972). Although there are no data available on the effectiveness of lithologic seals in the Potwar Plateau, the low thermal gradients through these rocks indicate that there may be a significant volume of water vertically flowing across lithologic boundaries, suggesting that the shale seals may not be as effective as required by the aquathermal pressuring process. Overpressuring caused by hydrocarbon generation can also be dismissed, primarily because the Neogene rocks are organically lean and are therefore, incapable of generating significant amounts of hydrocarbons. Furthermore, most of the Neogene

rocks are also thermally immature with respect to the oil window.

Previous workers in the Potwar Plateau have concluded that the most likely cause of overpressuring in the Neogene molasse rocks is tectonic stress (Malick, 1979; Sahay and Fertl, 1988) or a combination of tectonic stress and compaction disequilibrium (Kadri, 1991). The conclusion that tectonic compression has played an important role is based on the assumption that the involvement of Neogene rocks in the ongoing structural deformation associated with the collision of the Indian and Eurasian plates has produced abnormally high formation pressures. Presumably, compressive stress caused by plate collision would bring about abnormally high pressures. In this scenario, higher pressures should occur in proximity to more severely deformed areas. Indeed, high pressures are associated with intensely deformed areas as discussed previously and as noted by Kadri (1991). However, although it seems likely that compression may well have contributed to overpressuring, it is unlikely that the pervasive occurrence of overpressuring in the Potwar Plateau could be caused by tectonic compression alone.

Another commonly cited cause of abnormally high pore pressure is compaction disequilibrium. This process requires relatively rapid rates of sedimentation. In areas undergoing rapid sedimentation, the rate at which pore water is expelled from compacting sediment cannot keep pace with the rate of sedimentation; consequently the weight of the overburden is increasingly shared by the pore fluids, causing pore fluid pressures to rise above hydrostatic.

In the Neogene rocks of the Potwar Plateau the shales are generally thin (<50 ft) and contain large amounts of siltstone and sandstone. Raynolds and Johnson (1985) reported sedimentation rates ranging

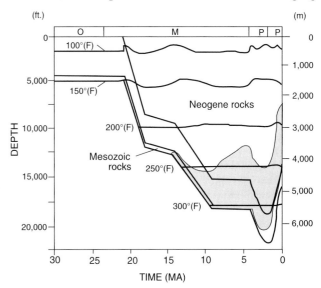

Figure 14. Burial curves for rocks penetrated in the OGDC, Dakhni No. I well. Shaded area is within the oil window. Location of well shown on Figure 2.

from 50 to l28 in./1,000 yr (20 to 50 cm) for the Pliocene and Pleistocene Siwalik Group on the Potwar Plateau. The underlying Miocene Kamlial and Murree Formations are lithologically similar to the Siwalik Group and were probably deposited at similar rates. Sedimentation rates for the molasse rocks in the Potwar Plateau are within the range of those reported in other overpressured rock sequences in the Gulf of Mexico, North Sea, and Haltenbanken in the Norwegian Sea (Mann and MacKenzie, 1990).

An analysis of shale travel time in the Fim Kassar well (Figure 11) provides some evidence for an undercompacted condition in these Neogene rocks. This method, developed by Hottman and Johnson (1965) in thick shale sequences in the U.S. Gulf Coast, is dependent on the recognition of departures from normal compaction trends as defined by interval transit times. The departure from a normal shale compaction trend is commonly interpreted to be a consequence of the retardation of the compaction process. In the Fim Kassar well, a well developed compaction trend occurs from the surface down to about 1,400 ft (430 m) (Figure 11). The departure of the compaction trend at this depth is interpreted here to reflect sediment undercompaction due to pore fluids that were unable to be expelled during sedimentation and burial because of the high rates of sedimentation. This analysis method should be used cautiously because of its sensitivity to borehole conditions. Apparent departures from normal compaction trends can be caused by hole irregularities. However, hole conditions in the Fim Kassar well, as indicated by the caliper log, do not indicate that the hole is irregularly shaped.

We conclude that the overpressuring in Neogene rocks is due to a combination of compaction disequilibrium and tectonic stress. Undercompaction is most likely the more fundamentally important cause of overpressuring. Tectonic stress appears to be of secondary importance and is most pronounced in areas of significant structural deformation.

In contrast to the overlying Neogene rocks, pre-Neogene rocks were deposited at much slower rates of sedimentation and are more thermally mature and organically-rich. These observations, in conjunction with the occurrence of sustained hydrocarbon shows on mud logs in Paleogene rocks, indicate that hydrocarbon generation has played a prominent role in the development of abnormally high pressure. This process is dependent on the addition of thermally generated hydrocarbons into the pore system and the subsequent increase in pore pressure above hydrostatic. The reader is referred to Meissner (1978), Hedberg (1979), Law and Dickinson (1985), and Spencer (1987) for more detailed information concerning hydrocarbon generation and overpressuring. The process also requires adequate seals to prevent the migration of hydrocarbons out of the pore system or a dynamic hydrocarbon generation process in which the rate of hydrocarbon generation exceeds the rate of hydrocarbon loss.

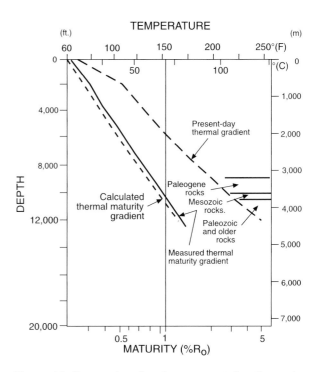

Figure 15. Cross plot showing present-day thermal gradient, measured thermal maturity gradient, and calculated thermal maturity gradient in the Gulf Oil, Fim Kassar well. Location of well shown on Figure 2.

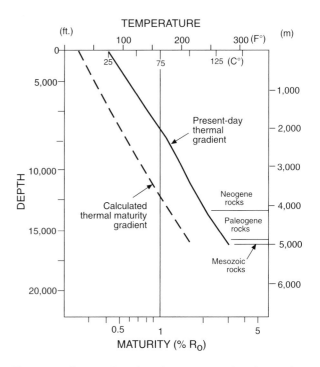

Figure 16. Cross plot showing present-day thermal gradient and calculated thermal maturity gradient in the OGDC, Dakhni No. 1 well. Location of well shown on Figure 2.

SUMMARY

Previous investigations of abnormal pressure in the Potwar Plateau of north-central Pakistan identified abnormal formation pressures only in Neogene rocks. The authors have identified two distinct pressure regimes in the Potwar Plateau—one in Neogene rocks and another in pre-Neogene rocks. Pore pressures in Neogene rocks are as high as lithostatic while pore pressures in pre-Neogene rocks are generally less than those in Neogene rocks, and commonly range from 0.5 to 0.7 psi/ft (11.3 to 15.8 kPa/m). Based on an evaluation of thermal maturity, temperature, clay mineral transformations, hydrocarbon source rock potential, sediment deposition rates, formation water quality, and structural aspects, we conclude that the abnormally high pressures in Neogene rocks are due to tectonic compression and undercompaction associated with high rates of sedimentation. Pore pressures in the pre-Neogene rocks are most likely due to a combination of hydrocarbon generation and tectonic compression. The authors were unable to identify a lithologic pressure seal for either Neogene or pre-Neogene rocks, in part because of the practice of increasing the drilling mud weight in anticipation of high pressures rather than responding to actual pressure changes during drilling.

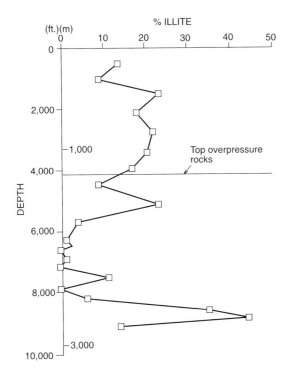

Figure 17. X-ray diffraction analyses showing percent illite vs. depth in drilling samples from the Occidental, Khaur No. I well. Location of well shown on Figure 2

REFERENCES CITED

Baker, D.M., Lillie, R.J., Yeats, R.S., Johnson, G.D., Yousuf, M., and Zamin, A.S.H., 1988, Development of the Himalayan frontal thrust zone: Salt Range, Pakistan: Geology, v. 16 p. 3–7.

Barker, C., 1972, Aquathermal pressuring - role of temperature in development of abnormal-pressure zones: AAPG Bulletin, v. 56, p. 2068–2071.

Burbank, D.W., and Raynolds, R.G.H., 1988, Stratigraphic keys to the timing of thrusting in terrestrial foreland basins: Applications to the northwestern Himalaya, *in* Paola, C. and Kleinspehn, K.L., eds., New Perspectives in Basin Analysis: New York, Springer-Verlag, p. 329–351.

Burst, J.F., 1969, Diagenesis of Gulf Coast clayey sediments and its possible relation to petroleum migration: AAPG Bulletin, v. 53, p. 673–683.

Chapman, R.E., 1994, Abnormal pore pressures: essential theory, possible causes, and sliding, *in* Fertl. R.E., Chapman, R.E., and Hotz, R.F., eds., Studies in abnormal pressures: Developments in Petroleum Science 38, Elsevier, p.51–91.

Fatmi, 1973, Lithostratigraphic units of the Kohat-Potwar Province, Indus Basin, Pakistan: Geological Survey of Pakistan Memoir no. 10, 80 p.

Fatmi, A.N., Akhtar, M., Alam, G.S., and Hussain, I., 1984, Guidebook to geology of Salt Range: Geological Survey of Pakistan, 14 p.

Fertl, W.H., 1976, Abnormal formation pressures, Elsevier, Amsterdam, 382 p.

Gee, E.R., t980, Map of the Salt Range: United Kingdom Directorate of Surveys for the Government of Pakistan, 6 sheets, 1:50,000 scale.

Gretener, P.E., 1981, Geothermics: Using temperature in hydrocarbon exploration: AAPG Education Course Note Series 17, 156 p.

Gretener, P., and Feng, Zeng-Mo, 1985, Three decades of geopressures - insights and enigmas: Bulletin Ver. Schweiz. Pet. Geol. Ing., 51, (120), p. 1–34.

Hedberg, H.D., 1979, Methane generation and petroleum migration: Oil and Gas Journal, v. 77, p. 186–192.

Hottman, C.E., and Johnson, R.K., 1965, Estimation of formation pressures from log-derived shale properties: Journal of Petroleum Technology, v. 17, p. 717–723.

Jaume, S.C., and Lillie, R.J., 1988, Mechanics of the Salt Range-Potwar Plateau, Pakistan: a fold-and-thrust belt overlain by evaporites: Tectonics, v. 7, p. 57–71.

Johnson, G.D., Raynolds, R.G., and Burbank, D.W., 1986, Late Cenozoic tectonics and sedimentation in the northwestern Himalaya foredeep: 1. Thrust ramping and associated deformation in the Potwar region, *in* Allen, P. and Homewood, P., eds., Foreland basins: International Association of Sedimentologists Special Publication 8, p. 273–291.

Kadri, I.B., 1991, Abnormal formation pressures in post-Eocene Formation, Potwar Basin, Pakistan: SPE/IADC Drilling Conference, 21920, p. 213–220.

Khan, M.A., Ahmed, R., Raza, H.A., and Kemal, A., 1986, Geology of petroleum in Kohat-Potwar depression, Pakistan: American Association of Petroleum Geologists Bulletin, v. 70, p. 396–414.

Law, B.E. and Dickinson, W.W., 1985, Conceptual model for origin of abnormal pressured gas accumulations in low-permeability reservoirs: American Association of Petroleum Geologists Bulletin, v. 69, p. 1295–1304.

Law, B.E., Spencer, C.W., and Bostick, N.H., 1980, Evaluation of organic matter, subsurface temperature, and pressure with regard to gas generation in low-permeability Upper Cretaceous and lower Tertiary strata in the Pacific Creek area, Sublette and Sweetwater Counties, Wyoming: Mountain Geologist, v. 17, no. 2, p. 23–35.

Lillie, R.J., Johnson, G.D., Yousuf, M., Zamin, A.S.H., and Yeats, R.S., 1987, Structural development within the Himalayan foreland fold- and thrust-belt of Pakistan, in Beaumont, C. and Tankard, A.J., eds., Sedimentary basins and basin-forming mechanisms: Canadian Society of Petroleum Geologists, Memoir 12.

Malick, A.M., 1979, Pressures plague Pakistan's Potwar: Petroleum Engineer, International, June, 1979, p. 26–36.

Malik, Z., Kemal, A., Malik, Azam M., and Bodenhausen, J.W.A., 1988, Petroleum potential and prospects in Pakistan, in Raza, H.A. and Sheikh, A.M., eds., Petroleum for the future, Islamabad.

Mann, D.M., and MacKenzie, A.S., 1990, Prediction of pore pressures in sedimentary basins: Marine and Petroleum Geology, v. 7, p. 55–65.

Martin, N.R., 1956, The petrology of the Khewra trap rock, Salt Range, West Pakistan: Geological Survey of Pakistan Records, v. 8 (pt. 1), p. 45–48.

Meissner, F.F., 1978, Patterns of source-rock maturity in nonmarine source-rocks of some typical Western Interior basins, in nonmarine Tertiary and Upper Cretaceous source rocks and the occurrence of oil and gas in west-central U.S.: Rocky Mountain Association of Geologists Continuing Education Lecture Series, p. 1–37.

Patriat, P. and Achache, J, 1984, Collision chronology and its implications for crustal shortening and the driving mechanisms of plates—India-Eurasia: Nature, v. 311, p. 615–621.

Pennock, E.S., Lillie, R.J., Zaman, A.S.H., and Yousaf, M., 1989, Structural interpretation of seismic reflection data from eastern Salt Range and Potwar Plateau, Pakistan: AAPG Bulletin, v. 73, p. 841–857.

Pilbeam, D.R., Barry, J., Meyer, G.E., Shah, S.M.I., Pickford, M.H.I., Bishop, W.W., Thomas, H., Jacobs, L.L., 1977, Geology and paleontology of Neogene strata of Pakistan: Nature, v. 270, p. 684–689.

Powers, M.C., 1967, Fluid-release mechanisms in compacting marine mudrocks and their importance in oil exploration: AAPG Bulletin, v. 51, p. 1240–1254.

Raynolds, R.G.H., and Johnson, G.D., 1985, Rates of Neogene depositional and deformational processes, north-west Himalayan foredeep margin, Pakistan, in N.J. Snelling, ed., The chronology of the geologic record: Geological Society of London Memoir 10, p. 297–311.

Raza, H.A., 1981, Geothermal gradients in Pakistan: Geological Bulletin of Punjab Univ., v. 16, p. 71–82.

Sahay, B., and Fertl, W.H., 1988, Origin and evaluation of formation pressures: Kluwer Academic Publishers, Dordecht, Boston, London, 292 p.

Seeber, L., and Armbruster, J., 1979, Seismicity of the Hazara Arc in northern Pakistan: Decollement vs. basement faulting, in Farah, A., and Dejong, K.A., eds., Geodynamics of Pakistan: Quetta, Geological Survey of Pakistan, p. 131–142.

Spencer, C.W., 1987, Hydrocarbon generation as a mechanism for overpressuring in Rocky Mountain region: AAPG Bulletin, v. 71, p. 368–388.

Stoneley, R., 1974, Evolution of the continental margins bounding a former southern Tethys, in Burk, C.A., and Drake, C.L., eds., The geology of continental margins: New York, Springer-Verlag, p. 889–903.

Swarbrick, R.E. and M.J. Osborne, 1998, Mechanisms that generate abnormal pressures: an overview, in Law, B.E., G.F. Ulmishek, and V.I. Slavin eds., Abnormal pressures in hydrocarbon environments: AAPG Memoir 70, p. 13–34.

Waples, D.W., 1980, Time and temperature in petroleum formation: Application of Lopatin's method to petroleum exploration: AAPG Bulletin, v. 64, p. 916–926.

Yeats, R.S., Khan, S.H., and Akhtar, M., 1984, Late Quaternary deformation of the Salt Range of Pakistan: Geological Society of America Bulletin, v. 95, p. 958–966.

INDEX